# Achieving sustainable cultivation of tree nuts

It is widely recognised that agriculture is a significant contributor to global warming and climate change. Agriculture needs to reduce its environmental impact and adapt to current climate change whilst still feeding a growing population, i.e. become more 'climate-smart'. Burleigh Dodds Science Publishing is playing its part in achieving this by bringing together key research on making the production of the world's most important crops and livestock products more sustainable.

Based on extensive research, our publications specifically target the challenge of climate-smart agriculture. In this way we are using 'smart publishing' to help achieve climate-smart agriculture.

Burleigh Dodds Science Publishing is an independent and innovative publisher delivering high quality customer-focused agricultural science content in both print and online formats for the academic and research communities. Our aim is to build a foundation of knowledge on which researchers can build to meet the challenge of climate-smart agriculture.

For more information about Burleigh Dodds Science Publishing simply call us on +44 (0) 1223 839365, email info@bdspublishing.com or alternatively please visit our website at www.bdspublishing.com.

**Related titles:**

Achieving sustainable cultivation of temperate zone tree fruits and berries
Volume 1: Physiology, genetics and cultivation
Print (ISBN 978-1-78676-208-5) Online (ISBN 978-1-78676-210-8; 978-1-78676-211-5)

Achieving sustainable cultivation of temperate zone tree fruits and berries
Volume 2: Case studies
Print (ISBN 978-1-78676-212-2) Online (ISBN 978-1-78676-214-6; 978-1-78676-215-3)

Agroforestry for sustainable agriculture
Print (ISBN 978-1-78676-220-7) Online (ISBN 978-1-78676-222-1; 978-1-78676-223-8)

Integrated management of diseases and insect pests of tree fruit
Print (ISBN 978-1-78676-256-6); Online (ISBN 978-1-78676-259-7, 978-1-78676-258-0)

Chapters are available individually from our online bookshop:
https://shop.bdspublishing.com

climate•SMART•publishing
IN AGRICULTURAL SCIENCE

BURLEIGH DODDS SERIES IN AGRICULTURAL SCIENCE
NUMBER 56

# Achieving sustainable cultivation of tree nuts

Edited by Professor Ümit Serdar,
Ondokuz Mayıs University, Turkey and Emeritus
Professor Dennis Fulbright, Michigan State University, USA

burleigh dodds
SCIENCE PUBLISHING

Published by Burleigh Dodds Science Publishing Limited
82 High Street, Sawston, Cambridge CB22 3HJ, UK
www.bdspublishing.com

Burleigh Dodds Science Publishing, 1518 Walnut Street, Suite 900, Philadelphia, PA 19102-3406, USA

First published 2019 by Burleigh Dodds Science Publishing Limited


Library of Congress Control Number: 2019941168

British Library Cataloguing in Publication Data
A catalogue record for this book is available from the British Library

ISBN 978-1-78676-224-5 (Print)
ISBN 978-1-78676-227-6 (PDF)
ISBN 978-1-78676-226-9 (ePub)
ISSN 2059-6936 (print)
ISSN 2059-6944 (online)

DOI 10.19103/AS.2018.0042

Typeset by Deanta Global Publishing Services, Chennai, India

Printed and bound by CPI Group (UK) Ltd, Croydon, CR0 4YY

# Contents

# Series list

| Title | Series number |
| --- | --- |
| **Achieving sustainable cultivation of maize - Vol 1**<br>From improved varieties to local applications<br>Edited by: Dr Dave Watson, CGIAR Maize Research Program Manager, CIMMYT, Mexico | 001 |
| **Achieving sustainable cultivation of maize - Vol 2**<br>Cultivation techniques, pest and disease control<br>Edited by: Dr Dave Watson, CGIAR Maize Research Program Manager, CIMMYT, Mexico | 002 |
| **Achieving sustainable cultivation of rice - Vol 1**<br>Breeding for higher yield and quality<br>Edited by: Prof. Takuji Sasaki, Tokyo University of Agriculture, Japan | 003 |
| **Achieving sustainable cultivation of rice - Vol 2**<br>Cultivation, pest and disease management<br>Edited by: Prof. Takuji Sasaki, Tokyo University of Agriculture, Japan | 004 |
| **Achieving sustainable cultivation of wheat - Vol 1**<br>Breeding, quality traits, pests and diseases<br>Edited by: Prof. Peter Langridge, The University of Adelaide, Australia | 005 |
| **Achieving sustainable cultivation of wheat - Vol 2**<br>Cultivation techniques<br>Edited by: Prof. Peter Langridge, The University of Adelaide, Australia | 006 |
| **Achieving sustainable cultivation of tomatoes**<br>Edited by: Dr Autar Mattoo, USDA-ARS, USA & Prof. Avtar Handa, Purdue University, USA | 007 |
| **Achieving sustainable production of milk - Vol 1**<br>Milk composition, genetics and breeding<br>Edited by: Dr Nico van Belzen, International Dairy Federation (IDF), Belgium | 008 |
| **Achieving sustainable production of milk - Vol 2**<br>Safety, quality and sustainability<br>Edited by: Dr Nico van Belzen, International Dairy Federation (IDF), Belgium | 009 |
| **Achieving sustainable production of milk - Vol 3**<br>Dairy herd management and welfare<br>Edited by: Prof. John Webster, University of Bristol, UK | 010 |
| **Ensuring safety and quality in the production of beef - Vol 1**<br>Safety<br>Edited by: Prof. Gary Acuff, Texas A&M University, USA & Prof. James Dickson, Iowa State University, USA | 011 |
| **Ensuring safety and quality in the production of beef - Vol 2**<br>Quality<br>Edited by: Prof. Michael Dikeman, Kansas State University, USA | 012 |
| **Achieving sustainable production of poultry meat - Vol 1**<br>Safety, quality and sustainability<br>Edited by: Prof. Steven C. Ricke, University of Arkansas, USA | 013 |
| **Achieving sustainable production of poultry meat - Vol 2**<br>Breeding and nutrition<br>Edited by: Prof. Todd Applegate, University of Georgia, USA | 014 |

# Acknowledgements

We wish to acknowledge the following for their help in reviewing particular chapters:

- Chapter 3: Prof Elsa Ramalhosa, Polytechnic Institute of Braganca, Portugal
- Chapter 5: Dr Nerea Becerra, Universitat Rovira i Virgili, Spain
- Chapter 6: Dr Carmen Cuadrado, INIA, Spain
- Chapter 10: Dr Richard Buchner, University of California Cooperative Extension – Tehama County, USA
- Chapter 11: Dr Bruce Lampinen, University of California-Davis, USA

# Introduction

Tree nuts such as almonds, walnuts and pistachio have long been grown for fresh consumption and as a food ingredient. It is estimated that 4 million metric tons of tree nuts are produced each year in over 40 countries. Demand is rising, in part because of research on the nutraceutical benefits of tree nuts. At the same time, these crops face particular challenges such as safety in the face of allergens and mycotoxin contamination. There is a need to increase production but in a more sustainable way in the face of climate change. This volume reviews the wealth of research addressing these challenges. The volume reviews current research on nutraceutical properties, as well as allergen and other safety issues relating to tree nuts. It also assesses advances in integrated disease and pest management to improve yields and sustainability. Finally, the volume summarises key advances in breeding and cultivation of the main tree nuts, from walnuts and almonds to chestnuts, pistachios and hazelnuts.

## Part 1 Improving health, safety and sustainability

The first part of the volume discusses the growing number of studies on the health benefits of tree nuts. The subject of Chapter 1 is advances in understanding the health benefits of walnuts. Walnut fruit is rich in vitamins, carbohydrates, minerals, protein and unsaturated fats with Omega 3, for example, linked to improving brain and nerve cell function. The fruit is also a rich source of phytochemicals, including antioxidant compounds such as melatonin, ellagic acid, vitamin E, carotenoids, polyphenols, walnut oligopeptides (WOPs) and phytosterols. These compounds have been shown to have protective health effects against heart disease, some types of cancer, type 2 diabetes as well as some neurological illnesses. Walnut kernels also contain a very high amount of serotonin, a neurotransmitter involved in regulating both mood and blood pressure. The chapter also explores the antimicrobial and antioxidant properties of walnut green husk (WGH).

Moving from walnuts to almonds, Chapter 2 focusses on advances in understanding the health benefits of almonds. Almond (*Prunus dulcis*) is widely grown in countries with a Mediterranean climate and is considered a healthy source of many nutrients. The chapter discusses the nutritional composition of almonds and, in particular, those nutrients with nutraceutical properties, from phenolics and fatty acids to phytosterols. The chapter discusses how this unique nutritional profile translates into health benefits related to almond consumption. The chapter reviews the range of phenolic compounds found in almonds, as well as lipid and fatty acid content in different cultivars.

Chapter 3 examines advances in understanding the health benefits of hazelnuts. Hazelnuts (*Corylus avellana* L.) are rich in fats, protein, phytonutrients and antioxidants such as Vitamin E. The health benefits of hazelnut consumption arise from the synergic action of unsaturated fat or bioactive compounds including plant sterols and dietary fiber. After introducing the key nutritional features of hazelnuts, the chapter describes advances in metabolomics and outlines the results of epidemiological and clinical studies on the health benefits of hazelnuts.

As Chapter 4 sows, the production and consumption of chestnuts has shown a steady increase in recent years, driven by both increased consumer awareness about food composition and the health benefits of a nut-rich diet. The chapter provides an overview of the uses and health benefits of chestnuts (*Castanea* spp.). These nuts are rich in carbohydrates, minerals, vitamins, protein and fibre, and low in fat content, as well as being important sources of phenolics and flavonoids. The chapter also explores the effects of factors such as cultivar, post-harvest handling and storage as well as cooking on nutritional and nutraceutical properties. It discusses the potential value and uses co-products (bur, shell, leaves, curing wastewater, etc.) left behind by processing.

Chapter 5 looks at advances in understanding the health benefits of pistachio nuts. Pistachio is particularly rich in vegetable protein, fiber and bioactive compounds such as lutein-zeaxanthin and beta-carotene, all of which have been linked to potential benefits to health. The chapter summarises what we know about the nutritional composition of pistachio nuts and reviews the range of human, animal and *in vitro* research exploring the health benefits of pistachio nut consumption.

Moving on from health benefits to consider the challenge caused by allergies, Chapter 6 explains advances in detecting and identifying nut allergens. Together with peanut, tree nuts belong to a small group of allergens responsible for the great majority of food allergies. The chapter begins by providing an overview of tree nut allergies and their prevalence. It then discusses allergenic proteins and thresholds, as well as the required sensitivity of detection methods. Finally, the chapter reviews both general issues in developing effective detection methods as well the range of techniques available for particular tree nuts.

The penultimate chapter in Part 1, Chapter 7, addresses the subject of integrated disease management in tree nut cultivation. Integrated disease management techniques to minimize environmental impact is key to the sustainable cultivation of tree nuts. The chapter explores the principles and practice of integrated disease management, taking as an example one of the most threatening diseases affecting chestnuts worldwide: ink disease caused by *Phytophthora* spp. The chapter introduces the concept of area-wide integrated ink disease management, including the use of techniques such as monitoring and GIS mapping of the disease. *Phytophthora* is easily dispersed via water and infected soil. The chapter therefore covers tree and soil treatments used to prevent as well as treat the disease, and reviews advances in breeding varieties resistant to the disease (breeding of interspecific hybrids has led to resistant rootstocks). The chapter concludes with a discussion on the importance of quality and certification of material for propagation.

The final chapter of the first part of the volume, Chapter 8, draws connections between tree nut cultivation, biodiversity and conservation. There is growing concern about the impact of agriculture on the environment as well as the consequences of this damage for sustainable production in the face of increased abiotic and biotic threats. It is now well accepted that a holistic approach is needed to ensure management techniques that ensure both conservation of the natural world and stable production levels. In nut crop production the complex interconnections among physical and biological structures, agricultural practices, functions and services strongly influence, and may be influenced by, factors such as orchard design, ground cover management, and pest management. By analysing the ecosystem services and dis-services that may affect, and can in turn be affected by, agricultural practices, the chapter shows how to develop more sustainable cultivation methods.

## Part 2 Improving individual types of tree nuts

The second part of the volume addresses advances in breeding ad cultivation of particular nuts. Chapter 9 looks at breeding of walnut. Walnut breeding programs are facing the challenge of increasing production to meet growing demand but in a more sustainable way which adapts to but does not exacerbate climate change. The chapter reviews the development of new cultivars with genetic resistance to biotic and abiotic stresses, as well as the exploration and conservation of genetic diversity in walnut. Chapter 9 is complemented by Chapter 10 which reviews advances in the cultivation of walnuts. The chapter reviews the expansion of walnut production, examining key challenges such as the control of blight (*Xanthomonas arboricola pv. Juglandis*). The chapter reviews optimal methods for propagation, managing water, nutrients and planting density as well as recent advances in molecular genetics, including marker assisted selection.

Chapter 11 shifts the focus on to almond. The chapter offers an overview of the most recent advances in almond scion and rootstock genotypes as well as developments in orchard management, including irrigation and fertilization techniques. The focus is on understanding almond tree eco-physiology and yield potential which provides the foundation for more sustainable cultivation methods less reliant on environmentally-damaging agrochemical inputs.

Chapter 12 moves on to consider advances in the breeding of chestnuts. Over the two last centuries, diseases and pests have been introduced from Asia to chestnut plantations in both northern America and southern Europe, jeopardizing indigenous populations and some local chestnut varieties. Recent developments in tree genome mapping and sequencing have opened up new opportunities for breeders. The chapter describes developments in breeding varieties with resistance to diseases threatening chestnut orchards and includes the example of a French program for breeding ink-resistant rootstocks. The chapter then examines breeding for resistance to gall wasps, breeding for fruit quality, and breeding improved 'peel-ability' of chestnuts. Complementing Chapter 12, Chapter 13 looks at advances in the cultivation of chestnuts. The chapter includes sections on ecological conditions affecting the growth of chestnuts, chestnut management as coppice forest or orchards, propagation methods (including grafting, layering, cutting and micropropagation), and orchard establishment and management (including training, pruning, fertilization and irrigation). The chapter also discusses the harvesting and storage of chestnuts, as well as pest management in chestnut orchards.

Chapter 14 looks at advances in the breeding of pistachio. *Pistacia vera* L. (pistachio) has the only commercially important edible nuts in the genus, with wild species used for rootstock. Breeding in pistachio is complicated by a long juvenile period and the plant's dioecious character. The chapter describes the genetic resources available for pistachio breeding, and the importance of cytogenetics to this field. The chapter then examines the use of molecular markers, genome and transcriptome sequencing, the role of micropropagation and gives examples of new cultivars with improved properties. Building on this, Chapter 15 addresses advances in the cultivation of pistachio. The chapter includes a section on pistachio genetics and climatic requirements, which requires a period of chilling as well as a hot, dry climate for fruiting. The chapter discusses flower and fruit characteristics of pistachio and what this means for orchard design. The chapter also looks at propagation techniques (both by seed and budding), followed by a review of key research on establishing, planting and managing orchards. The chapter concludes with sections on managing pests and diseases and how to improve crop yield.

The subject of Chapter 16 is the development of hazelnuts as a crop. European hazelnut (*Corylus avellana*) is a major species of interest for food use and one of the few economically valuable commercial tree nut crops within the Betulaceae family. The chapter reviews the development of European hazelnut as industrial nut crop, highlighting its environmental requirement for good yields. The chapter discusses propagation and orchard management techniques, focusing on recent innovations in research including the exploitation of hazelnut biomass, and the application of precision farming strategies in newer large-scale hazelnut orchards. The chapter also reviews integrated pest management techniques for hazelnuts.

Complementing the previous chapter on hazelnut production, the volume's final chapter (Chapter 17) addresses advances in the breeding of hazelnuts. Cultivated hazelnut varieties differ from each other in such characteristics as nut size, shape, kernel ratio, shell thickness, etc. Key target characteristics for breeding not only include high yield but also good morphological and vegetative traits as well as resistance to pests and diseases. The chapter discusses advances in the breeding of hazelnuts. The chapter reviews floral biology, pollination and compatibility in hazelnut. The chapter examines the development of hazelnut varieties, focusing on key hazelnut breeding programmes in Turkey, Europe and the US (where the focus has been for breeding for resistance to Easter filbert blight).

# Part 1

## Improving health, safety and sustainability

# Chapter 1

## The nutritional and nutraceutical value of walnut

*Turan Karadeniz, Bolu Abant Izzet Baysal University, Turkey; Faik Ekmel Tekintaş, Adnan Menderes University, Turkey; and Seyit Mehmet Şen, Kastamonu University, Turkey*



### 1 Introduction

Walnut is one of the most popular and widely consumed tree nuts. It can be consumed in both raw, roasted or salted forms. Walnut is also widely used, in both shredded and oil form, as a flavour enhancer in biscuits, cakes and desserts as well as yoghurt, pizza and ice cream. As an example, in Middle Eastern Muslim countries, walnut is added to almond, date and grape to prepare a cake known as maamoul for Ramadan (Şen, 2011).

Walnut is rich in monounsaturated fatty acids. Walnut is a good source of Omega 3 (an essential fatty acid) and a source of arachidonic acid. Walnut is the source of many phytochemicals (Şen, 2013). These antioxidant compounds include melatonin, ellagic acid, vitamin E, carotenoids and polyphenols. These compounds have potential protective health effects against ageing, cancers, inflammations and neurological illnesses (Şen, 2013) (Table 1). Since the morphological structure of walnut resembles the human brain, it has traditionally been regarded as improving brain function and mood (Fig. 1). This chapter reviews the nutritional composition of walnuts and the evidence for their health benefits.

### 2 Walnut as a source of vitamins, protein, fat and carbohydrate

In terms of nutritional value, walnut is superior to pistachio, almond, hazelnut, pine nut and peanut, containing twice the level of antioxidants and vitamins

http://dx.doi.org/10.19103/AS.2018.0042.01

**Figure 1** Anatolian walnut (*Juglans regia*).

(Vinson and Cai, 2012). Walnut consumption is therefore seen as an important component in healthy nutrition (Preedy et al., 2011). Walnut is a particularly good source of vitamin E. Vitamin E in walnut fat has two forms: alpha-tocopherol and gamma-tocopherol. A 100 g of walnut has 21 mg gamma-tocopherol (vitamin E), providing 140% of the daily requirement. Vitamin E is a strong fat-soluble antioxidant and is needed to protect the mucus and skin cell membranes against the harmful effects of free radicals (Martin and Preedy, 2015; Şen, 2013). Walnut also contains important vitamins such as riboflavin, niacin, thiamine, pantothenic acid, vitamin B6 and folate/B9 (Şen, 2013). As walnut has a preventive effect on inflammation, it decreases the risk of high blood pressure; it also helps to prevent arterial illnesses such as heart disease and stroke as well as chest diseases, colon and prostate cancers (Marangoni et al., 2007; Amaral et al., 2003).

Walnut contains significant amounts of single and multi-unsaturated fats. Whilst proteins and carbohydrates provide 4 calories/g, fats provide 9 calories/g. In consuming hard shell fruits, consumers therefore need to be careful about exceeding their recommended calorie intake (Tapsell et al., 2009; Şen, 2013; Crews et al., 2005; Martin and Preedy, 2015). Generally, there are 50 g (47.14) of multi-unsaturated fatty acids in 100 g of walnut. 40 g (38.09 g) of this acid content is Omega 6 (linoleic acid) whilst 10 g (9.08 g) is Omega 3 (linolenic acid) (Şen, 2013; Ma et al., 2010). Although calorie content is high, this is counterbalanced by the presence of healthier unsaturated which can help in weight control (Vinson and Cai, 2012; Şen, 2013). Research has shown that the Omega 3 content in walnuts can offset the harmful effects of high saturated fat diets (Roman et al., 2011; Martin and Preedy, 2015).

**Table 1** The chemical structure of walnut

| Energy/constituents | Nutritional value/100 g |
|---|---|
| Energy | 2738 kJ (654 kcal) |
| Carbohydrates | 13.71 |
| Starch | 0.06 |
| Sugars | 2.61 |
| Lactose | 0 |
| Dietary fibre | 6.7 |
| Total fat | 65.21 |
| Saturated fat | 6.126 |
| Monounsaturated fat | 8.933 |
| Polyunsaturated fat | 47.174 |
| Protein | 15.23 |
| **Vitamins** | **Amount (%)** |
| Vitamin A equiv. | 1 μg (0) |
| Beta-carotene | 12 μg (0) |
| Lutein zeaxanthin | 9 μg (0) |
| Vitamin A | 20 IU |
| Thiamine (B1) | 0.341 mg (30) |
| Riboflavin (B2) | 0.15 mg (13) |
| Niacin (B3) | 1.125 mg (8) |
| Pantothenic acid (B5) | 0.570 mg (11) |
| Vitamin B6 | 0.537 mg (41) |
| Folate (B9) | 98 μg (25) |
| Vitamin B12 | 0 μg (0) |
| Vitamin C | 1.3 mg (2) |
| Vitamin D | 0 μg (0) |
| Vitamin E | 0.7 mg (5) |
| Vitamin K | 2.7 μg (3) |
| **Trace metals** | **Amount (%)** |
| Calcium | 98 mg (10) |
| Iron | 2.91 mg (22) |
| Magnesium | 158 mg (45) |
| Manganese | 3.414 mg (163) |
| Phosphorus | 346 mg (49) |
| Potassium | 441 mg (9) |
| Sodium | 2 mg (0) |
| Zinc | 3.09 mg (33) |
| **Other constituents** | **Amount (%)** |
| Water | 4.07 |
| Alcohol (ethanol) | 0 |
| Caffeine | 0 |

Percentages are roughly approximated using US recommendations for adults.
*Source*: USDA Nutrient Database.

## 3 Health-promoting properties of walnuts

As has been noted, walnuts are rich in antioxidants (Reiter et al., 2005; Canales et al., 2007; Fukuda et al., 2003; Negi et al., 2011). As a result, they have been shown to have positive effects on health (Tapsell et al., 2009; Ueshima et al., 2007; Ros et al., 2004). Research suggests that consuming walnut daily protects the body against heart disease, some types of cancer, type 2 diabetes and other health problems. Research has shown that walnut reduces the risk of heart attack since it improves blood flow (Şen, 2013; Fukuda et al., 2004; Gillen et al.,2005; Ma et al., 2010; Anderson et al., 2001). Walnut reduces cholesterol and the oxidative stresses caused by free radicals (Anderson et al., 2001; Morgan et al., 2002). Use of walnut oil has also been associated with improved skin health (Şen, 2013). Walnut oil alleviates skin conditions such as sunburn, skin ulcers and blisters (Anon., 2018b).

Research has established a link between Omega 3, brain and nerve cell function (Martin and Preedy, 2015). Adequate amounts of Omega 3 have been shown to assist in the transmission of electrical signals in the nervous system (Preedy et al., 2011). Studies have shown a correlation between an increase in depression rates and a decrease in Omega 3 fat intake (Mercola, 2004; Şen, 2013; Şen and Karadeniz, 2015). Other studies have linked hyperactivity, learning difficulties and behavioural problems in children with a low intake of Omega-3 in their diet (Richardson and Puri, 2002; Bélanger et al., 2009; Kirby and Derbyshire, 2018).

In a 2004 study, it was found that walnut extract was capable of inhibiting the fibrous structure of the amyloid beta protein, which is the origin of the amyloid plaques in the brains of Alzheimer's patients (Anon., 2018a). Another study has suggested that walnut may reduce or delay the risk of Alzheimer's disease by retaining amyloid-β protein in solution and preventing the deterioration of acetylcholine (Anon., 2018b). Walnut has also been linked to the prevention of some nervous disorders such as Parkinson's disease (Anon., 2018c).

The biological activity of aqueous extracts from walnut green husk (WGH) was evaluated which exhibited both antimicrobial and antioxidant properties. Therefore, walnut husk can be used as an alternative, cheap source of antioxidant and antimicrobial agents (Fernández-Agulló et al., 2013).

Walnut has also exclusive medicinal and biological properties. Recent studies show that walnut has small molecule peptides which are known as walnut oligopeptides (WOPs). The WOPs exhibit anti-fatigue properties which were examined in mice. The mice were orally administered with different doses of WOPs. After a month of treatment, the anti-fatigue activity was monitored by employing weight-loaded swimming test and determining particular biochemicals such as energy metabolism enzymes and glycogen storage, mitochondrial function and antioxidative capacity. It was noticed that the mice treated with WOPs exhibited long/extended swimming time, showed enhanced glycogen storage in liver and gastrocnemius muscle and diminished level of lactate dehydrogenase (LDH), blood lactic acid (BLA), creatine kinase (CK) and blood urea nitrogen (BUN). Furthermore, WOPs markedly repressed fatigue-induced oxidative stress by rising the activity of glutathione peroxidase (GPX), superoxide dismutase (SOD) and depressing the content of cell membrane injury marker malondialdehyde (MDA). Strikingly, WOPs augmented the action of $Na^+$-$K^+$-ATPase, pyruvate kinase (PK) and succinate dehydrogenase (SDH), and boosted the mRNA level of mitochondrial biogenesis factors and mitochondrial DNA content in skeletal muscles of mice. Thus these findings clearly indicate that WOPs have positive anti-fatigue effects, which may be associated with their specific impacts on enhancing glycogen storage, suppressing oxidative stress, fixing energy metabolism and boosting mitochondrial function in skeletal muscles and alleviating cell and muscular injury (Liu et al., 2018).

Serotonin is a famous neurotransmitter derived via the serotonin pathway from tryptophan (Guillén-Casla et al., 2012). This is found in the serotonergic neurons in the central nervous system and mainly produced in the enterochromaffin cells of the gastrointestinal tract (Watanabe et al., 2010). This neurotransmitter has a very important role in the regulation of moods, feelings of happiness, sleep, anxiety, appetite and blood pressure (Watts et al., 2012). Suppressed levels or diminishing of its synthesis can promote several diseases like obesity, depression and schizophrenia (Spadaro et al., 2015). Thanks to plants, serotonin is not only produced in humans but also in different tissues of plants such as leaves, flowers, fruits, seeds and roots (Kang and Back, 2006; Kang et al., 2007). Recently, serotonin content was determined in different nuts such as pine nut, walnut, macadamia nut and pecan. It was found that in walnut, serotonin content (155 ± 57 μg/g) was at the highest level among all other nuts (Yılmaz et al., 2019). Therefore, consumption of walnut on a daily basis can promote human well-being efficiently. Albeit, walnut has some allergenic protein, however, a few small molecules like gallic acid and/or ellagic acid have immunomodulatory functions. Recently, the compound effects on mice T cells (immune cells) have been tested for their growth and activation. It was found that the compounds effectively decreased the T-cell activation/growth and the secretion of interleukin-4, interleukin-10 and interferon-γ. These findings implied that walnuts also contained compounds that can alleviate allergens' effects. Fortification of hydrolysable tannins to walnut-containing foods might mitigate allergic reaction by walnut consumption (Hamada et al., 2019).

Phytosterols have health-promoting effects such as alleviating diabetes mellitus, obesity, hyperlipidaemia and hypertension (Marinangeli and Jones, 2010; Misawa et al., 2012; Mohamed, 2014). Luckily recent reports revealed that walnut kernel is rich in phytosterols. There exist three phytosterols, namely, $\Delta^{5,23}$-stigmastadienol, cycloeucalenol and 28-methylobtusifoliol, reported for the first time in black walnut. Further, these findings illustrated that walnut is rich in compounds which are also bioactivities (Vu et al., 2019).

WGH is a pericarp of walnut fruit which is also rich in polyphenols, naphthoquinones (Zhou et al., 2015a), diarylheptanoid and other active components (Huo et al., 2017) and flavones and terpenoids (Qiu et al., 2017). Further, WGH also has anti-inflammatory, anti-cancer (Zhou et al., 2015b), antitumor (Hu, 2016), antibacterial (Zhang et al., 2017) and anti-ageing effects (Lu et al., 2011). Obesity is recognized as a serious problem of modern lifestyle (Eid et al., 2017; Yanovski, 2018) due to the consumption of a high-fat diet (HFD). The high-fat diet can disturb the ratio of Bacteroidetes and Firmicutes, leading to the onset of obesity and other metabolic disorders (Choi et al., 2017). Recently, the WGH effect on the reduction of body weight and improvement of total cholesterol and total triglycerides has been reported. Further, fat accumulation, hepatic steatosis and adipose tissue hypertrophy of HFD were diminished by WGH. Thus the positive effect of WGH on the alleviation of fatness and fat disorders could be correlated with the modulation of the gut microbiota (Wang et al., 2019).

## 4 Conclusion

Walnut is a very healthy fruit for humans and contains nutrients like vitamins, minerals, proteins, unsaturated fats and phytosterols. It has been suggested that walnut should be consumed from childhood onwards to increase heart and brain health, with a suggested intake of 20–30 g/day. Further, WGH has alleviating effect on obesity and well-being. Health promoting contents can be improved in walnut by different breeding methods. Methods such as introduction, hybridization, selection and mutation breeding can enhance nutritional level as well as size and yield of the walnuts. In Turkey, the Horticulture Department, Agriculture Faculty of Sütçü Imam University, Kahramanmaraş, Turkey, and in the United States, the Walnut Breeding Program of The University of California-Davis is actively involved in breeding high quality walnut cultivars in order to improve human health.

## 5 References

Amaral, J. S., Casal, S., Pereira, J. A., Seabra, R. M. and Oliveira, B. P. P. 2003. Determination of sterol and fatty acid compositions, oxidative stability, and nutritional value of six walnut (*Juglans regia* L.) cultivars grown in Portugal. *Journal of Agricultural and Food Chemistry* 51(26), 7698–702. doi:10.1021/jf030451d.

Anderson, K. J., Teuber, S. S., Gobeille, A., Cremin, P., Waterhouse, A. L. and Steinberg, F. M. 2001. Walnut polyphenolics inhibit in vitro human plasma and LDL oxidation. *The Journal of Nutrition* 131(11), 2837–42. doi:10.1093/jn/131.11.2837.

Anon. 2018a. Available at: https://www.health.ny.gov (accessed on 10 January 2019).

Anon. 2018b. Available at: https://www.drugs.com (accessed on 10 January 2019).

Anon. 2018c. Available at: https://heartlandnutsnmore.com. Accessed on 10 January, 2019.

Bélanger, S. A., Vanasse, M., Spahis, S., Sylvestre, M. P., Lippé, S., L'heureux, F., Ghadirian, P., Vanasse, C. M. and Levy, E. 2009. Omerga-3 fatty acid treatment of children with attention-deficit hyperactivity disorder: a randomized, double-blind, placebo-controlled study. *Paediatrics and Child Health* 14(2), 89–98. doi:10.1093/pch/14.2.89.

Canales, A., Benedi, J., Nus, M., Librelotto, J., Sánchez-Montero, J. M. and Sánchez-Muniz, F. J. 2007. Effect of walnut-enriched restructured meat in the antioxidant status of overweight/obese senior subjects with at least one extra CHD-risk factor. *Journal of the American College of Nutrition* 26(3), 225–32. doi:10.1080/07315724.2007.107 19605.

Choi, S., Hwang, Y. J., Shin, M. J. and Yi, H. 2017. Difference in the gut microbiome between ovariectomy-induced obesity and diet-induced obesity. *Journal of Microbiology and Biotechnology* 27(12), 2228–36. doi:10.4014/jmb.1710.10001.

Crews, C., Hough, P., Godward, J., Brereton, P., Lees, M., Guiet, S. and Winkelmann, W. 2005. Study of the main constituents of some authentic walnut oils. *Journal of Agricultural and Food Chemistry* 53(12), 4853–60. doi:10.1021/jf0478354.

Eid, H. M., Wright, M. L., Anil Kumar, N. V., Qawasmeh, A., Hassan, S. T. S., Mocan, A., Nabavi, S. M., Rastrelli, L., Atanasov, A. G. and Haddad, P. S. 2017. Significance of microbiota in obesity and metabolic diseases and the modulatory potential by medicinal plant and food ingredients. *Frontiers in Pharmacology* 8, 387. doi:10.3389/fphar.2017.00387.

Fernández-Agulló, A., Pereira, E., Freire, M. S., Valentão, P., Andrade, P. B., González-Álvarez, J. and Pereira, J. A. 2013. Influence of solvent on the antioxidant and antimicrobial properties of walnut (*Juglans regia* L.) green husk extracts. *Industrial Crops and Products* 42, 126–32. doi:10.1016/j.indcrop.2012.05.021.

Fukuda, T., Ito, H. and Yoshida, T. 2003. Antioxidative polyphenols from walnuts (*Juglans regia* L.). *Phytochemistry* 63(7), 795–801. doi:10.1016/S0031-9422(03)00333-9.

Fukuda, T., Ito, H. and Yoshida, T. 2004. Effect of the walnut polyphenol fraction on oxidative stress in type 2 diabetes mice. *BioFactors* 21(1–4), 251–3. doi:10.1002/biof.552210148.

Gillen, L. J., Tapsell, L. C., Patch, C. S., Owen, A. and Batterham, M. 2005. Structured dietary advice incorporating walnuts achieves optimal fat and energy balance in patients with type 2 diabetes mellitus. *Journal of the American Dietetic Association* 105(7), 1087–96. doi:10.1016/j.jada.2005.04.007.

Guillén-Casla, V., Rosales-Conrado, N., León-González, M. E., Pérez-Arribas, L. V. and Polo-Díez, L. M. 2012. Determination of serotonin and its precursors in chocolate samples by capillary liquid chromatography with mass spectrometry detection. *Journal of Chromatography A* 1232, 158–65. doi:10.1016/j.chroma.2011.11.037.

Hamada, Y., Haramiishi, R., Ojima, Y., Amakura, Y., Yoshimura, M., Sawamoto, A., Okuyama, S., Furukawa, Y. and Nakajima, M. 2019. Hydrolysable tannins, gallic acid, and ellagic acid in walnut reduced 3-(4,5-Dimethylthiazol-2-yl)-2,5-Dipheny ltetrazolium bromide (MTT) reduction in T-Cells cultured from the spleen of mice. *PharmaNutrition* 7, 100140. doi:10.1016/j.phanu.2018.100140.

Hu, Z. 2016. Effects of juglandis immaturum exocarpium polysaccharide on proliferation, apoptosis and PI3K/Akt signaling pathway of human colon carcinoma cell line HCT-116. *Chinese Journal of Experimental Traditional Medical Formulae* 18, 136–9.

Huo, J., Sun, G., Dong, W. and Wang, W. 2017. Dynamic variation of components in exocarp of *Juglans mandshurica* with browning based on UPLC-Q-TOF/MS. *China Journal of Chinese Materia Medica* 16, 3112–20.

Kang, S. and Back, K. 2006. Enriched production of N-hydroxycinnamic acid amides and biogenic amines in pepper (*Capsicum annuum*) flowers. *Scientia Horticulturae* 108(3), 337–41. doi:10.1016/j.scienta.2006.01.037.

Kang, S., Kang, K., Lee, K. and Back, K. 2007. Characterization of tryptamine 5-hydroxylase and serotonin synthesis in rice plants. *Plant Cell Reports* 26(11), 2009–15. doi:10.1007/s00299-007-0405-9.

Kirby, A. and Derbyshire, E. 2018. Omega-3/6 fatty acids and learning in children and young people: a review of randomized controlled trials published in the last five years. *Journal of Nutrition and Food Sciences* 8(2), 1–10.

Liu, R., Wu, L., Du, Q., Ren, J. W., Chen, Q. H., Li, D., Mao, R. X., Liu, X. R. and Li, Y. 2018. Small molecule oligopeptides isolated from walnut (*Juglans regia* L.) and their anti-fatigue effects in mice. *Molecules* 24(1). doi:10.3390/molecules24010045.

Lu, H., She, G., Guo, Z. and She, D. 2011. Study on capacities of free radical effects by walnut peel in vitro. *West China Journal of Pharmaceutical Sciences* 1, 32–3.

Ma, Y., Njike, V. Y., Millet, J., Dutta, S., Doughty, K., Treu, J. A. and Katz, D. L. 2010. Effects of walnut consumption on endothelial function in type 2 diabetic subjects: a and omizedcontrolled crosso vertrial. *Diabetes Care* 33, 227–32.

Marangoni, F., Colombo, C., Martiello, A., Poli, A., Paoletti, R. and Galli, C. 2007. Levels of the n–3 fatty acid eicosapentaenoic acid in addition to those of alpha linolenic acid are significantly raised in blood lipids by the intake of four walnuts a day in humans. *Nutrition, Metabolism and Cardiovascular Diseases* 17(6), 457–61. doi:10.1016/j.numecd.2006.02.004.

Marinangeli, C. P. and Jones, P. J. 2010. Plant sterols, marine-derived omega-3 fatty acids and other functional ingredients: a new frontier for treating hyperlipidemia. *Nutrition and Metabolism* 7, 76. doi:10.1186/1743-7075-7-76.

Martin, C. R. and Preedy, V. 2015. *Diet and Nutrition in Dementia and Cognitive Decline*. Elsevier Academic Press, Waltham, MA.

Mercola, J. 2004. *Dr. Mercola's Total Health Cookbook & Program* (1st edn.). Fifth Printing, Schaumburg, IL.

Misawa, E., Tanaka, M., Nomaguchi, K., Nabeshima, K., Yamada, M., Toida, T. and Iwatsuki, K. 2012. Oral ingestion of *Aloe vera* phytosterols alters hepatic gene expression profiles and ameliorates obesity-associated metabolic disorders in zucker diabetic fatty rats. *Journal of Agricultural and Food Chemistry* 60(11), 2799–806. doi:10.1021/jf204465j.

Mohamed, S. 2014. Functional foods against metabolic syndrome (obesity, diabetes, hypertension and dyslipidemia) and cardiovasular disease. *Trends in Food Science and Technology* 35(2), 114–28. doi:10.1016/j.tifs.2013.11.001.

Morgan, J. M., Horton, K., Reese, D., Carey, C., Walker, K. and Capuzzi, D. M. 2002. Effects of walnut consumption as part of a low-fat, low-cholesterol diet on serum cardiovascular risk factors. *International Journal for Vitamin and Nutrition Research. Internationale Zeitschrift für Vitamin- und Ernährungsforschung. Journal International de Vitaminologie et de Nutrition* 72(5), 341–7. doi:10.1024/0300-9831.72.5.341.

Negi, A. S., Luqman, S., Srivastava, S., Krishna, V., Gupta, N. and Darokar, M. P. 2011. Antiproliferative and antioxidant activities of *Juglans regia* fruit extracts. *Pharmaceutical Biology* 49(6), 669–73. doi:10.3109/13880209.2010.537666.

Preedy, V. R., Watson, R. R. and Patel, V. B. 2011. *Nuts and Seeds in Health and Disease Prevetion*. Elsevier Academic Press, San Diego, CA.

Qiu, J. Y., Wang, W. M., Li, J., Zhao, M., Wang, J. L. and Zhang, S. J. 2017. Chemical constituents in green walnut husks of *Juglans regia*. *Chinese Traditional and Herbal Drugs* 48, 2385–9.

Reiter, R. J., Manchester, L. C. and Tan, D. X. 2005. Melatonin in walnuts: influence on levels of melatonin and total antioxidant capacity of blood. *Nutrition* 21(9), 920–4. doi:10.1016/j.nut.2005.02.005.

Richardson, A. J. and Puri, B. K. 2002. A randomized double-blind, placebo-controlled study of the effects of supplementation with highly unsaturated fatty acids on ADHD-related symptoms in children with specific learning difficulties. *Progress in Neuro-Psychopharmacology and Biological Psychiatry* 26(2), 233–9. doi:10.1016/S0278-5846(01)00254-8.

Roman, P., Iglesias, P., Leal, M., Martinez, E., Mellado, L., Garcia de la Corte, F., Ortega, M., Romero, N.,Corchado, Y., Miro, L., et al. 2011. Efectos De Una Dieta Mediterranea, Suplementada Con Aciete De Oliva O Con Frutos Secos, Sobre La Presıon Arterial Ambulatoria En Pacientes Hipertensos De Alto Riesgo Cardiovascular. Conference: *IV Simposio CIBER-OBN At: Malaga (Spain)*.

Ros, E., Nunez, I., Perez-Heras, A., Serra, M., Gilabert, R., Casals, E. and Deulofeu, R. 2004. A walnut diet improves endothelial function in hypercholesterolemic subjects: a randomized crossover trial. *Circulation* 109(13), 1609–14. doi:10.1161/01.CIR.0000124477.91474.FF.

Şen, S. M. 2011. *Walnut, Cultivation, Nutritional Value, Folklore* (4th edn.) (in Turkish). ICC Publication, Ankara, Turkey, p. 220.

Şen, S. M. 2013. *Walnut Diet, Eating Walnut Living Healthy* (in Turkish). Alper Publishing, Ankara, Turkey, p. 216S.

Şen, S. M. and Karadeniz, T. 2015. The Nutrıtıonal value of walnut. *Journal of Hygienic Engineering and Design* 11, 68–71.

Spadaro, P. A., Naug, H. L., Du Toit, E. F., Donner, D. and Colson, N. J. 2015. A refined high carbohydrate diet is associated with changes in the serotonin pathway and visceral obesity. *Genetics Research* 97, e23. doi:10.1017/S0016672315000233.

Tapsell, L., Batterham, M., Tan, S. Y. and Warensjö, E. 2009. The effect of a calorie controlled diet containing walnuts on substrate oxidation during 8-hours in a room calorimeter. *Journal of the American College of Nutrition* 28(5), 611–7. doi:10.1080/07315724.2009.10719793.

Ueshima, H., Stamler, J., Elliott, P., Chan, Q., Brown, I. J., Carnethon, M. R., Daviglus, M. L., He, K., Moag-Stahlberg, A., Rodriguez, B. L., et al. 2007. Food Omega-3 fatty acid intake of individuals (total, linolenic acid, long-chain) and their blood pressure: INTERMAP study. *Hypertension* 50(2), 313–9. doi:10.1161/HYPERTENSIONAHA.107.090720.

Vinson, J. A. and Cai, Y. 2012. Nuts, especially walnuts, have both antioxidant quantity and efficacy and exhibit significant potential health benefits. *Food and Function* 3(2), 134–40. doi:10.1039/c2fo10152a.

Vu, D. C., Lei, Z., Sumner, L. W., Coggeshall, M. V. and Lin, C.-H. 2019. Identification and quantification of phytosterols in black walnut kernels. *Journal of Food Composition and Analysis* 75, 61–9. doi:10.1016/j.jfca.2018.09.016.

Wang, X., Chen, D., Li, Y., Zhao, S., Chen, C. and Ning, D. 2019. Alleviating effects of walnut green husk extract on disorders of lipid levels and gut bacteria flora in high fat diet-induced obesity rats. *Journal of Functional Foods* 52, 576–86. doi:10.1016/j.jff.2018.11.022.

Watanabe, H., Akasaka, D., Ogasawara, H., Sato, K., Miyake, M., Saito, K., Takahashi, Y., Kanaya, T., Takakura, I., Hondo, T., et al. 2010. Peripheral serotonin enhances lipid metabolism by accelerating bile acid turnover. *Endocrinology* 151(10), 4776–86. doi:10.1210/en.2009-1349.

Watts, S. W., Morrison, S. F., Davis, R. P. and Barman, S. M. 2012. Serotonin and blood pressure regulation. *Pharmacological Reviews* 64(2), 359–88. doi:10.1124/pr.111.004697.

Yanovski, J. A. 2018. Obesity: trends in underweight and obesity – scale of the problem. *Nature Reviews Endocrinology* 14(1), 5–6. doi:10.1038/nrendo.2017.157.

Yılmaz, C., Taş, N. G., Kocadağlı, T. and Gökmen, V. 2019. Determination of serotonin in nuts and nut containing products by liquid chromatography tandem mass spectrometry. *Food Chemistry* 272, 347–53. doi:10.1016/j.foodchem.2018.08.064.

Zhang, L., Qi, S.-C., Iwanaga, K., Uemura, K., Zhang, L.-X., Kudo, S., Hayashi, J.-i., Furuya, K. and Norinaga, K. 2017. An approach for on-line analysis of multi-component volatiles from coal pyrolysis with Li+-attachment ionization mass spectrometry. *Fuel Processing Technology* 158, 141–5. doi:10.1016/j.fuproc.2016.12.001.

Zhou, Y., Yang, B., Jiang, Y., Liu, Z., Liu, Y., Wang, X. and Kuang, H. 2015a. Studies on cytotoxic activity against HepG-2 cells of naphthoquinones from green walnut husks of *Juglans mandshurica* Maxim. *Molecules* 20(9), 15572–88. doi:10.3390/molecules200915572.

Zhou, Y., Yang, B., Liu, Z., Jiang, Y., Liu, Y., Fu, L., Wang, X. and Kuang, H. 2015b. Cytotoxicity of triterpenes from green walnut husks of *Juglans mandshurica* Maxim in HepG-2 cancer cells. *Molecules* 20(10), 19252–62. doi:10.3390/molecules201019252.

# Chapter 2

## Advances in understanding the nutritional profile and health benefits of almonds

*Özlem Tokuşoğlu, Celal Bayar University, Turkey*



### 1 Introduction

Almond is one of the oldest shell nut crops [*Prunus dulcis* (Mill.); syn. *P. amygdalus Batsch*]. The almond tree is thought to have originated in the mountainous regions of Central Asia (Grasselly, 1976). It is believed that the fruit was spread throughout the Mediterranean by the Phoenicians, Greeks, and Romans. Almonds were introduced to North America by Spanish colonists in the 16th century and spread from there to Southern hemisphere countries with Mediterranean climates including Australia, Chile, Argentina, and South Africa. World almond production is now concentrated particularly in California in the United States, followed by countries such as Spain, Iran, Morocco, and Turkey (Lopez-Ortiz et al., 2008; Tokuşoğlu, 2018).

Almonds consist of a hull (pericarp), shell (endocarp), and seed contained in a skin (Fig. 1). Almonds are a good source of health-promoting bioactive micronutrient compounds such as phytochemical phenolics and tocopherols and are rich in plant protein, unsaturated fatty acids (FAs), plant sterols, and dietary fibre (Kris-Etherton et al., 2001; Tokuşoğlu et al., 2005; Tokuşoğlu, 2011, 2018; Tokuşoğlu and Hall, 2011). This chapter discusses the nutritional composition of almonds and, in particular, those nutrients with nutraceutical properties, from phenolics and FA to phytosterols. It then shows how this unique nutritional profile translates into health benefits related to almond consumption (Fig. 2).

http://dx.doi.org/10.19103/AS.2018.0042.03

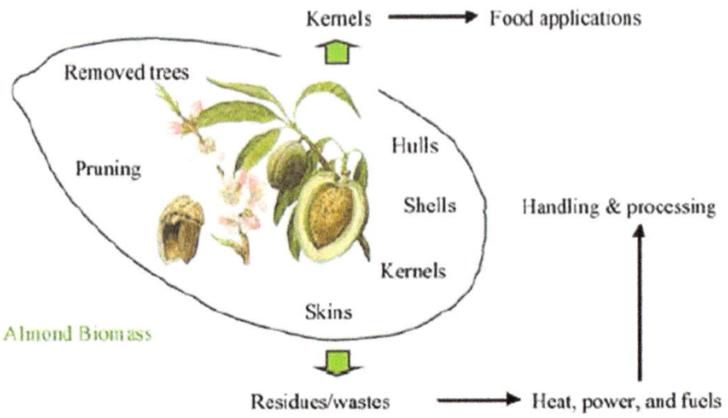


**Figure 1** Constituents of almond and their uses. Source: adapted from Chen et al. (2010).

## 2 Phenolics

Almonds have been found to have a higher content of antioxidants than many other nuts (Table 1) (Mishra et al., 2010). The phenolics in almond are a mixture of flavonoids, phenolic acids, and tannins that contribute to their antioxidant capacity (Milbury et al., 2006; Chen and Blumberg, 2008; Garrido et al., 2008).

Almonds contain a variety of flavonoids including flavanols (catechin and epicatechin), flavonols (kaempferol, isorhamnetin, and quercetin), flavanones (naringenin and eriodictyol), anthocyanins (cyanidin and delphinidin), and proanthocyanidins (of both B- and A-types), as well as phenolic acids (caffeic acid, chlorogenic acid, ferulic acid, *p*-coumaric acid, *p*-hydroxybenzoic acid, protocatechuic acid, and vanillic acid), and some alcohols and benzoic aldehydes (eugenol, *p*-hydroxybenzaldehyde, and pro- tocatechuic aldeide) (Amarowicz et al., 2005; Monagas et al., 2007; Chen and Blumberg, 2008; Garrido et al., 2008). The most abundant group of almond skin phenolics are flavanols (~50%), followed by flavonol glycosides (~25%), and phenolic acids (~20%). The remaining phenolic compounds (flavonol aglycones, flavanone glycosides, and aglycones) represent ~5% of the total phenolic compounds (Monagas et al., 2007; Garrido et al., 2008). Mass spectrometric analysis has identified new compounds like catechin dihexoside, caffeoyl-6'-secologanoside, and (epi)catechin-ethyl trimer (Qureshi et al., 2016).

The skin, while representing only approximately 4% of the total weight, contains 70–100% of the total phenolics present in the nut (Milbury et al., 2006). The total content of almond skin phenolics varies from about 160 to 800 μg/g, but this range is affected both by variety and processing (roasting and blanching). The mean total concentration of phenolic compounds has been found to be significantly higher in the skins of Spanish almonds (~410 μg/g) compared to American almonds (~270 μg/g). The content of phenolic compounds in American almonds has a high variability (Milbury et al., 2006;

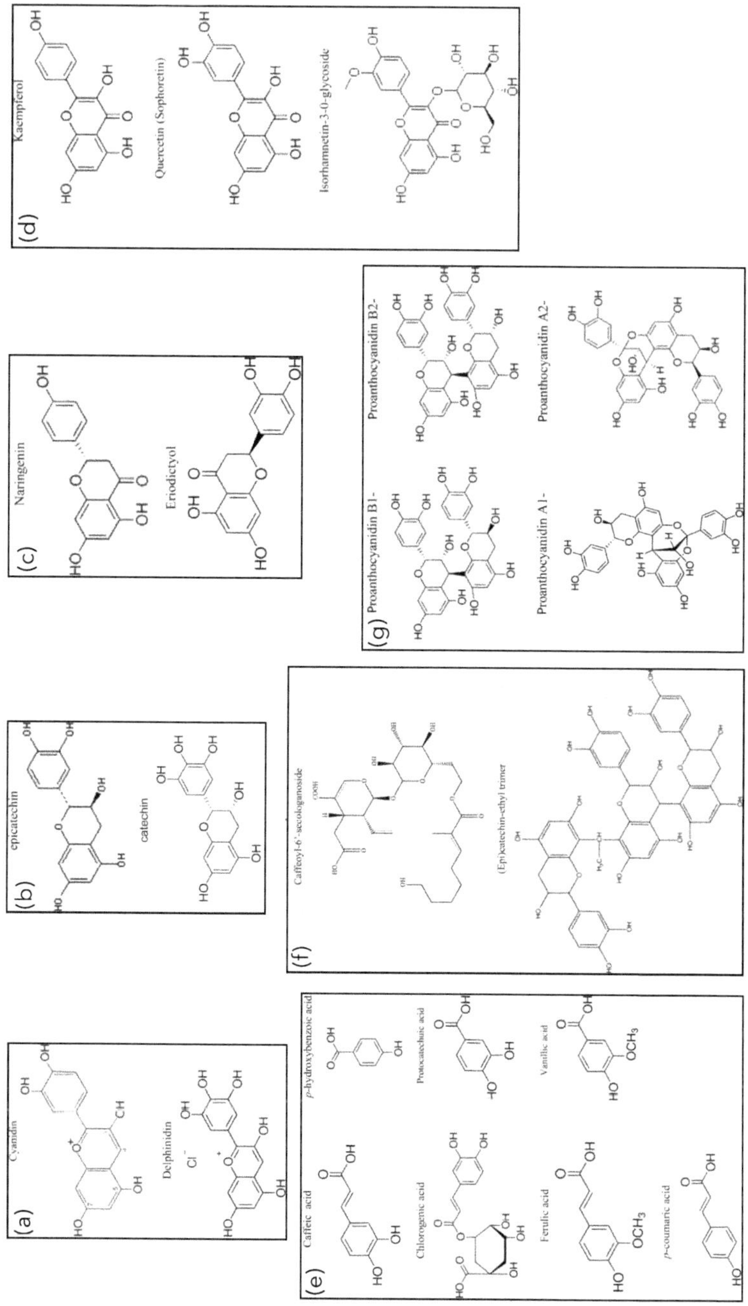


**Figure 2** Bioactive compounds in almond. (a) Major anthocyanidins in almonds. (b) Major flavonols in almonds. (c) Major flavanones in almonds. (d) Major flavonol glycosides in almonds. (e) Major phenolic acids in almonds. (f) Newly determined compounds in almonds. (g) Procyanidins in almonds.

**Table 1** Total phenolics of almond and pistachios*

| Total phenolics (mg of GAE/100 g fresh weight) | Almond with skin | Almond without skin | Pistachio with skin |
|---|---|---|---|
| Mean value | 313.6 ± 1.28 | 57.44 ± 3.1 | 922.6 ± 2.8 |
| Range values | 148.8 ± 1.76/497.3 ± 2.03 | 52.7 ± 0.5/58.4 ± 0.9 | 548.7 ± 3.3/1542.3 ± 2.6 |

* Data are the mean ± SD for three independent experiments; ten group of almonds, ten group of pistachio nuts were used and each group was 2 kg.
ND = Not detected; $P < 0.05$; n = 3.

Garrido et al., 2008). A study of almonds from Datça in Turkey by Tokuşoğlu (2011) used gallic acid equivalents (GAEs) to measure phenolic content (as mg of GAE/100 g fresh weight). Total phenolics (TPs) of almonds with skin (white seed and yellow orange/beige skin) ranged as 148.8 ± 1.76/497.3 ± 2.03 (avg. 313.6 ± 1.28) mg of GAE/100 g fresh weight whereas TP contents of almonds without skin (white seed) ranged as 52.7 ± 0.5/58.4 ± 0.9 (avg. 57.44 ± 1.1) mg of GAE/100 g fresh weight ($P$<0.05) (Table 2).

Table 3 summarises the content (µg/g) of phenolic compounds of almond skins from Spain and the United States after blanching and drying. In a study by Garrido et al. (2008), the phenolic composition of almond skins obtained from different processes (blanching, drying, and roasting) was evaluated. The mean content of total polyphenols was higher (>two-fold) in the roasted samples (28.4 mg/g) than in the blanched samples (13.3 mg/g). The drying of the blanched samples produced an increase (+34%) in the content of phenolic compounds of blanched samples. Moreover, drying and roasting processes also

**Table 2** Total lipid levels, peroxide values, tocopherol and sterol levels of almond and pistachios*

| | Almond | Pistachio |
|---|---|---|
| Total oil (g/100 g) | 42.3 ± 1.9 | 54.6 ± 1.1 |
| Peroxide value (meq $O_2$/kg oil) | 0.21 ± 0.01 | 0.24 ± 0.05 |
| *Tocopherol (mg/100 g oil)* | | |
| $\alpha$ (Alfa)-Toc | 36.8 ± 4.7 | 1.64 ± 0.8 |
| $\gamma$ (Gamma)-Toc | 8.4 ± 0.9 | 37.7 ± 3.4 |
| $\delta$ (Delta)-Toc | ND | 2.02 ± 1.3 |
| *Sterol (mg/g oil)* | | |
| $\beta$ (Beta)-Sterol | 2155.2 ± 11.3 | 4706.5 ± 73.2 |
| Stigmasterol | 53.6 ± 0.8 | 703.4 ± 10.5 |
| Campasterol | 56.2 ± 7.0 | 247.0 ± 9.2 |

* Data are the mean ± SD for three independent experiments; ten group of almonds, ten group of pistachio nuts were used and each group was 2 kg.
ND = Not detected; $P < 0.05$; n = 3.

**Table 3** Content (μg/g) of phenolic compounds in almond skins from Spain and the United States

| Compound | Spain[a] | United States[b] |
|---|---|---|
| *Hydroxybenzoic acids and aldehydes* | | |
| *p*-hydroxybenzoic acid | 6.9 | 3.1 |
| Vanillic acid | 14.5 | 6.0 |
| Protocatechuic acid | 32.0 | 9.7 |
| Protocatechuic aldehyde | 20.1 | 9.9 |
| *Hydroxycinnamic acids* | | |
| *trans*-*p*-coumaric | - | 0.7 |
| Chlorogenic acid | 10.6 | 12.1 |
| *Flavan-3-ols* | | |
| (+)-catechin | 90.1 | 37.5 |
| (−)-epicatechin | 36.6 | 18.7 |
| B-type procyanidins | 77.7 | 36.2 |
| A-type procyanidins | 28.9 | 11.7 |
| *Flavonol glycosides* | | |
| Kaempferol-3-O-rutinoside | 12.8 | 29.2 |
| Kaempferol-3-O-glucoside | - | 5.0 |
| Kaempferol-3-O-galactoside | - | 0.1 |
| Isorhamnetin-3-O-rutinoside and glucoside | 43.2 | 79.1 |
| Isorhamnetin-3-O-galactoside | - | 5.7 |
| Quercetin-3-O-glucoside | 2.4 | 1.6 |
| Quercetin-3-O-galactoside | - | 7.9 |
| Quercetin-3-O-rutinoside | - | 2.6 |
| *Flavanone glycosides* | | |
| Naringenin-7-O-glucoside | 22.1 | 9.8 |
| Eriodictyol-7-O-glucoside | 1.6 | 1.1 |
| *Flavonol aglycones* | | |
| Kaempferol | 1.7 | 2.0 |
| Quercetin | 1.8 | 1.7 |
| Isorhamnetin | 4.9 | 4.0 |
| *Dihydroflavonol aglycones* | | |
| Dihydroquercetin | - | 9.0 |
| Dihydrokaempferol | - | 1.1 |
| *Flavanone aglycones* | | |
| Naringenin | 2.8 | 4.2 |
| Eriodictyol | 2.4 | 2.7 |
| *Total* | 413.0 | 274.8 |

Data reported by: [a,b] Monagas et al. (2007), [a,b] Garrido et al. (2008), [b] Milbury et al. (2006).

increase antioxidant activity by affecting the concentration of some phenolic compounds (Piga et al., 2003). The oxygen radical absorbance capacity (ORAC) values of almond skins (ORAC values measure antioxidant content) are within 0.331–3.0 mmol Trolox/g range, suggesting the potential nutraceutical use of almond skins (Friedrich et al., 2000; Monagas et al., 2007; Chen and Blumberg, 2008; Garrido et al., 2008).

## 3 Lipids and fatty acids

The total oil content of almonds ranges from 41.2% to 44.2% (avg. 42.3 ± 1.9) with peroxide values (PV – which measure oxidative stability) of almonds ranging from 0.13 to 0.25 mEq $O_2$/kg oil (avg. 0.21 ± 0.01) (Fourie and Basson, 1989; Tokuşoğlu, 2011). Almond is a rich source of fat and FA (Table 4). A study by Ros (2009) reported the overall fat content of almond as 50.6%, saturated fatty acid (SFA) content was reported as 3.9%, monounsaturated fatty acid (MUFA) content was reported as 32.2%, and polyunsaturated fatty acid (PUFA) content was reported as 12.2%. High concentrations of MUFA and PUFA have been found in other studies (Venkatachalam and Sathe, 2006). The high level of unsaturated FA improves the nutritional value of almonds (Table 5).

**Table 4** Fatty acid (FA) levels of almond and pistachios (as % of total oil)*

| | | Almond | Pistachio |
|---|---|---|---|
| Miristic | 14:0 | 0.05 ± 0.01 | 0.08 ± 0.01 |
| Palmitic | 16:0 | 6.92 ± 0.3 | 7.67 ± 0.2 |
| Palmitoleic | 16:1 | 0.60 ± 0.05 | 0.69 ± 0.1 |
| Stearic | 18:0 | 1.31 ± 0.2 | 0.85 ± 0.1 |
| Oleic | 18:1 | 69.86 ± 0.62 | 60.55 ± 0.43 |
| Linoleic | 18:2 | 21.76 ± 0.12 | 32.13 ± 0.7 |
| Linolenic | 18:3 | 0.17 ± 0.06 | 0.48 ± 0.03 |
| Arachidic | 20:0 | 0.15 ± 0.01 | 0.6 ± 0.02 |
| Eicosenoic | 20:1 | ND | 0.58 ± 0.05 |
| Behenic | 22:0 | 0.05 ± 0.0 | 0.35 ± 0.02 |
| Erucic | 22:1 | ND | 0.54 ± 0.03 |
| *Saturated* | SFA | 8.48 ± 0.52 | 9.55 ± 0.35 |
| *M.Unsatu* | MUFA | 70.46 ± 0.67 | 62.36 ± 0.61 |
| *P.Unsatu.* | PUFA | 21.93 ± 0.18 | 32.61 ± 0.73 |
| | Unsat./satur. ratio | 10.9 | 9.94 |

*Data are the mean ± SD for three independent experiments; ten group of almonds, ten group of pistachio nuts were used and each group was 2 kg.
ND = Not detected; $P < 0.05$; n = 3.

**Table 5** Mean content (%) of major fatty acid of almond cultivars from different origins

| | Fatty acids | | | | |
|---|---|---|---|---|---|
| Origin | Palmitic C 16:0 | Palmitoleic C 16:1 | Stearic C 18:0 | Oleic C 18:1 | Linoleic C 18:2 |
| American cultivars[a] | 5.76 | 0.369 | 1.72 | 62.75 | 20.42 |
| Italian cultivars[a] | 5.03 | 0.401 | 1.77 | 64.53 | 11.87 |
| Spanish cultivars[b] | 7.42 | 0.632 | 2.03 | 68.46 | 20.45 |
| Tunisian cultivars[a] | 7.50 | 0.368 | 1.70 | 73.10 | 19.50 |
| Turkish cultivars[c] | 7.50 | 0.724 | 1.83 | 70.84 | 18.49 |

Data reported by: [a,b] García-López et al. (1996), [b] Miraliakbari and Shahidi (2007), [c] Askin et al. (2007).

Triacylglycerols (TGs) represent the major lipid class in tree nut oils (>95%). Among nuts, almonds have one of the lowest oil yields (on average ~50%), but its oil contains the highest TG content (~98 g/100 g oil) (Aparicio and Aparicio-Ruiz, 2000; Ulberth and Buchgraber, 2000; Miraliakbari and Shahidi, 2007). Table 6 shows the TG content (%) of almonds of different geographic origins. There are five major TGs: OLL, OLO, PLO, OOO, and POO (P = palmitic, O = oleic, S = stearic, L = linoleic), representing ~80% of total TG content. There are four minor TGs: LLL, PLL, PLP, and SOO, representing ~20% of total TG content. POP and LnOO have also been found in small amounts (<1% and <3%, respectively) (Cherif et al., 2004). The quantitative differences in the TGs composition can be useful for distinguishing almonds of different cultivars (Martin-Carratala et al., 1999; Cherif et al., 2004).

The composition (%) of the major fatty acids (FAs) of some almond cultivars is reported in Table 7. The major fatty acid in almonds is oleic acid (C 18:1),

**Table 6** Mean content (%) of major triacylglycerols of almonds of different geographic origins

| Triacylglycerols | Italy[a] | Spain[b] | France[c] | Tunisia[d] | United States[e] | Australia[f] |
|---|---|---|---|---|---|---|
| LLL | 1.21 ± 0.27 | 1.87 ± 0.37 | 1.78 ± 0.02 | 1.96 ± 0.02 | 2.51 ± 0.77 | 2.24 ± 0.03 |
| OLL | 6.59 ± 0.33 | 10.95 ± 1.96 | 9.46 ± 0.05 | 11.34 ± 0.09 | 11.67 ± 3.70 | 12.68 ± 0.08 |
| PLL | 1.08 ± 0.10 | 1.76 ± 0.29 | 1.53 ± 0.02 | 1.73 ± 0.02 | 1.83 ± 0.65 | 2.08 ± 0.02 |
| OLO | 17.72 ± 1.28 | 22.65 ± 2.61 | 22.99 ± 0.13 | 25.68 ± 0.16 | 24.14 ± 3.18 | 26.25 ± 0.16 |
| PLO | 5.37 ± 0.37 | 8.31 ± 1.13 | 7.27 ± 0.06 | 9.54 ± 0.09 | 8.28 ± 1.92 | 10.69 ± 0.05 |
| PLP | 0.35 ± 0.02 | 0.45 ± 0.05 | 0.40 ± 0.01 | 0.44 ± 0.01 | 0.39 ± 0.03 | 0.49 ± 0.01 |
| OOO | 52.06 ± 2.79 | 39.27 ± 5.64 | 41.68 ± 0.15 | 34.27 ± 0.17 | 37.71 ± 9.49 | 30.29 ± 0.21 |
| POO | 12.14 ± 0.72 | 11.88 ± 0.32 | 11.35 ± 0.11 | 12.61 ± 0.14 | 10.94 ± 0.48 | 12.28 ± 0.07 |
| SOO | 3.47 ± 0.61 | 2.82 ± 0.50 | 3.53 ± 0.04 | 2.43 ± 0.05 | 2.54 ± 0.48 | 2.99 ± 0.05 |

Data reported by: Martin-Carratala et al. (1999).
Cultivars examined: [a] Genco, Tuono, Cristomorto; [b] Malaguena, Peraleja, Atocha, Del Cid, Desmayo Largueta, Ramillete, Marcona; [c] Ferragnes; [d] Achaak; [e] Texas, Nonpareil, Titan, Wawona; [f] Chellaston.

**Table 7** Bioactive substances of selected nuts per 100 g of dry weight

| Nut item (*specie*) | Anthocyanins | Flavonoids | Stilbenes | Total phenols | Phytoestrogens | Tocopherols | Chlorophylls | Xanthophylls |
|---|---|---|---|---|---|---|---|---|
| Almonds (*Prunus dulcis*) | (2.46)[a] | (28.7) 0.018-93.5[b] | ND | (90.0) 12.4-239[c] | (40.0) 0.112-120[d] | (23.6) 17-27.44[e] | ND | ND |
| Hazelnuts (*Corylus avellana* L.) | (6.7)[a] | (37.9) 0.027-113.7[f] | ND | (210.9) 8.0-425[g] | (0.093) 0.080-0.107[h] | (30.6) 15.36-45.5[i] | ND | ND |
| Peanuts (*Arachis hypogea* L.) | ND | (94.9) 0.094-189.8[j] | (0.096) 0.002-0.2[k] | (357.9) 8.0-645.9[l] | (110.0) 0.173-220[m] | (9.7) 4.8-16[n] | ND | ND |
| Pistachios (*Pistacia vera* L.) | (26.2) 24.3-28.1[o] | (26.7) 0.033-143.3[p] | (0.24) 0.009-0.71[q] | (335.0) 15.8-867[r] | (36.14) 0.062-108[s] | (24.6) 8.4-53.0[t] | (14.3)[u] | (3.8)[u] |
| Walnuts (*Juglans regia* L.) | ND | (333.4) 0.031-744.8[v] | ND | (814.5) 21.0-1625[w] | (35.93) 0.139-107.5[x] | (24.3) 14.9-31.72[y] | ND | ND |

Note: Data are expressed as means (in parentheses) and range; ND: Not detected.
[*] Data reported by Ballistreri et al. (2011).
Data reported by: [a] Harnly et al. (2006), [b,c] Milbury et al. (2006), [b,d,f,h,p,s,v,x] Thompson et al. (2006), [b,d,f,h,j,m,p,s,v,x] Kuhnle et al. (2008), [b,c,f,g,j,l,p,r,v,w] Yang et al. (2009), [c,e,g,i,l,n,r,t,w,y] Kornsteiner et al. (2006), [c,e,g,i,r,t,w,y] Miraliakbari and Shahidi (2008), [c,r,v] Arcan and Yemenicioglu (2009), [d,e,i,m,n,s,t,x,y] Kocygit et al. (2006), [e,g,i,l,n,r,t,w,y] Arranz et al. (2008), [k,q] Tokuşoğlu et al. (2005), [k,q] Baur and Sinclair (2006), [k] Hurst et al. (2008), [o,u] Bellomo and Fallico (2008), [o,p,q,r,t] Ballistreri et al. (2009), [p] Seeram et al. (2006), [p,q,r,t] Gentile et al. (2007), [q] Grippi et al. (2008).

representing ~60–70% of total FA content, followed by linoleic acid (C 18:2) at ~10–20%, and palmitic acid at 5–8% (García-López et al., 1996; Askin et al., 2007; Miraliakbari and Shahidi, 2007). Small amounts (below 1%) of myristic acid (C 14:0), stearic acid (C18:0), and linolenic acid (C 18:3) have also been detected (Martin-Carratala et al., 1999; Kris-Etherton et al., 1999; Zacheo et al., 2000). Other SFAs in almonds include miristic acid (C14:0), arachidic acid (C20:0), behenic acid (C22:0), while other MUFAs include palmitoleic acid (C16:1), eicosenoic acid (C20:1), and erucic acid (C22:1) (Tokuşoğlu, 2011). Askin et al. (2007) have observed that the content of the major FAs is influenced by kernel weight. The amount of palmitic, stearic, and oleic acid is positively correlated with kernel weight, contrary to linoleic acid content. In addition, shell thickness is negatively correlated with palmitic and stearic acid contents, but positively correlated with oleic acid content. Among polar lipids, phosphatidylserine, phosphatidylinositol, and phosphatidylcholine are present at 0.13, 0.09, and 0.14 g/100 g of almond oil, respectively. Sphingolipids are also present (0.37 g/100 g of oil; Miraliakbari and Shahidi, 2007, 2008).

## 4 Phytosterols and tocopherols

Almond is considered a good source of sterols (Tokuşoğlu, 2011). Almond oil contains between 2.2 and 2.8 g/kg of phytosterols, mainly β-sitosterol (~2.0 g/kg), with trace amounts of stigmasterol, Δ5-avenasterol, and campesterol (Table 8) (Quilez et al., 2003; Maguire et al., 2004; Miraliakbari and Shahidi, 2007). Philips et al. (2005) reported that almond contains 143.4 mg/100 g β-sitosterol,

**Table 8** Fatty acid composition of selected nuts and selected oils as percentage of total fat by weight[a]

| | Fatty acid | | | | | | | | |
|---|---|---|---|---|---|---|---|---|---|
| Total fat | 10:0 | 12:0 | 14:0 | 16:0 | 18:0 | 18:1 | 18:2 | 18:3 | % of total fat by wt |
| *Nuts* | | | | | | | | | |
| Almonds | 52.2 | 0.0 | 0.0 | 0.6 | 6.6 | 1.9 | 63.7 | 20.1 | 0.7 |
| Hazelnuts | 62.6 | 0.0 | 0.0 | 0.2 | 5.0 | 2.0 | 77.7 | 9.3 | 0.2 |
| Peanuts | 49.2 | 0.0 | 0.0 | 0.1 | 10.5 | 2.2 | 48.1 | 31.6 | 0.0 |
| Pistachios[b] | 50.0 | 0.0 | 0.0 | 0.0 | 9.9 | 2.1 | 69.6 | 15.4 | 0.5 |
| Walnuts | 56.6 | 0.0 | 0.0 | 0.0 | 3.7 | 2.5 | 21.0 | 59.2 | 5.8 |
| Mean | 54.1 | - | - | 0.2 | 7.1 | 2.1 | 56.0 | 31.7 | 1.4 |
| *Oils* | | | | | | | | | |
| Olive oil | 100 | 0.0 | 0.0 | 0.0 | 11.0 | 2.2 | 72.5 | 7.9 | 0.6 |
| Canola oil | 100 | 0.0 | 0.0 | 0.0 | 4.0 | 1.8 | 56.1 | 20.3 | 9.3 |
| Safflower oil | 100 | 0.0 | 0.0 | 0.0 | 4.8 | 1.3 | 75.3 | 14.2 | 0.0 |

Data reported: [a] Kris-Etherton et al. (1999), [b] Arena et al. (2007).

4.9 mg/100 g campesterol, 5 mg/100 g stigma-sterol, and 19.7 mg/100 g Δ5-avenasterol. A study by Qureshi et al. (2016) isolated stigmasitosterol-3-*O*-b-D-glucoside and a-D-glucopyranosyl-(1/2)-b-D-fructofuranoside for the first time from almond skin.

Almonds stand out for having a particularly high $\alpha$-tocopherol content compared to other nuts, ranging from about 8.0 to 25.0 mg/100 g of oil (Kornsteiner et al., 2006; Miraliakbari and Shahidi, 2007; Lopez-Ortiz et al., 2008). Almond oil contains smaller amounts of $\beta$ + $\gamma$-tocopherol, ranging from 0.1 to 3-0 mg/100 g of oil, and $\delta$-tocopherol present in traces. Almonds are dietary sources of $\alpha$-, $\beta$-, and $\gamma$-tocopherol and can contribute to a balanced intake of vitamin E. This explains why almonds have been included in the recommendations of *The Dietary Guidelines for Americans* (USDA, 2005) in the context of enhancing the intake of this vitamin.

## 5 Health benefits of almonds

The nutritional composition of almonds gives them a broad range of nutraceutical properties (Tokuşoğlu, 2011, 2018; Tokuşoğlu and Hall, 2011). Sterols, for example, have been reported to lower blood cholesterol levels and have anticancer and immune-boosting properties (Award et al., 2001; Plat and Mensink, 2001). Phytosterols have been shown to diminish blood cholesterol as well as to reduce the risk of certain types of cancer and enhance immune function (Ling and Jones, 1995; Awad and Fink, 2000; Bouic, 2001; Moreau et al., 2002; Ostlund, 2004). Phytosterol compounds, especially beta-sitosterol, campesterol, and stigmasterol, can reduce serum LDL cholesterol and atherosclerotic risk (de Jong et al., 2003). Epidemiological studies have also shown that phenolic intake is significantly and inversely associated with coronary heart disease (CHD) mortality (Hertog et al., 1993, 1995). The functionality of phenolics has been expressed through their action as an inhibitor or an activator for a large variety of enzyme systems and as metal chelators and scavenger of free oxygen radicals (Shahidi and Naczk, 1995; Russo et al., 2000; Garbisa et al., 2001). Oxygen-free radicals are involved in many pathological conditions such as cancer and chronic inflammation (Briviba and Sies, 1994).

These nutraceutical properties explain the long-established links between nut consumption and improved health, including the Nurses Health Study, which found that women who consumed nuts five or more times a week were seen to have a 35% lower risk of total coronary heart disease (CHD) as compared with women who rarely ate nuts (Hu et al., 1998). There is increasing evidence that diets including nuts have cardio-protective effects (Kris-Etherton et al., 2001). A number of studies have shown an inverse correlation between nut consumption and cardiovascular disease (CVD) risk

(Fraser et al., 1992; Fraser, 1999; Hu et al., 1998; Ellswort et al., 2001; Albert et al., 2002). Regular consumption of nuts is also associated with beneficial plasma lipid profiles, enhancement of immune function, and decreased risk of cancer, inflammation, stroke, type-2 diabetes, and diverse other chronic diseases (Grunwald, 1975; Phillips et al., 1999; Moreau et al., 2002). The beneficial effects of healthy diets including tree nuts are attributed to high levels of MUFAs or PUFAs, known for their favourable effects on blood lipids, and low levels of SFAs.

There have been a range of studies that link almonds in particular to improved cardiovascular health (Chen et al., 2015; Lee et al., 2017; Jung et al., 2018; Kalita et al., 2018). Almond consumption has been shown to reduce low-density lipoprotein cholesterol (LDL-C), which is a known risk factor in coronary heart disease (Rusinger et al., 2015; Musa-Veloso et al., 2016; Berryman et al., 2015). Consumption has also been associated with beneficial levels of high-density lipoprotein cholesterol (HDL-C) (Nishi et al., 2014; Berryman et al., 2017). Almond consumption has also been associated with reducing the risk of diabetes, by reducing levels of fasting insulin and fasting glucose (Chen et al., 2015; Gulati et al., 2017; Beatrice and Shivaji, 2015; Sweazea et al., 2014). Almonds have also been shown to help reduce weight gain due to factors such as increasing satiety (from high levels of fat and fibre content) and optimising gut function (due to high fibre content) (Dhillon et al., 2017, 2018; Holsher et al., 2018; Liu et al., 2016, 2017, 2018; Burns et al., 2016; Grundy et al., 2015; Hull et al., 2015). A good source of information on the broader health benefits of almonds is the Almond Board of California (http://www.almonds.com/consumers/health-and-nutrition/references).

## 6 References

Albert, C. M., Gaziano, J. M., Willet, W. C. and Manson, J. E. 2002. Nut consumption and decreased risk of sudden cardiac death in the Physicians' Health Study. *Archives of Internal Medicine* 162(12), 1382–7. doi:10.1001/archinte.162.12.1382.

Allen, K. G. D., Klevay, L. M. and Springer, H. L. 1977. The zinc and copper content of seeds and nuts. *Nutrition Reports International* 16, 227–30.

Amarowicz, R., Troszynska, A. and Shahidi, F. 2005. Antioxidant activity almond seed extract and its fraction. *Journal of Food Lipids* 12(4), 344–58. doi:10.1111/j.1745-4522.2005.00029.x.

Anderson, J. W., Smith, B. M. and Gustafson, N. J. 1994. Health benefits and practical aspects of high-fiber diets. *The American Journal of Clinical Nutrition* 59(Suppl.), 1242S–7S. doi:10.1093/ajcn/59.5.1242S.

Aparicio, R. and Aparicio-Ruiz, R. 2000. Authentication of vegetable oils by chromatographic techniques. *Journal of Chromatography. A* 881(1–2), 93–104. doi:10.1016/S0021-9673(00)00355-1.

Aparicio, R. and Luna, G. 2002. Characterization of monovarietal virgin olive oils. *European Journal of Lipid Science and Technology* 104, 614–27.

Arcan, I. and Yemenicioglu, A. 2009. Antioxidant activity and phenolic content of fresh and dry nuts with or without the seed coat. *Journal of Food Composition and Analysis* 22(3), 184–8. doi:10.1016/j.jfca.2008.10.016.

Arena, E., Campisi, S., Fallico, B. and Maccarone, E. 2007. Distribution of fatty acids and phytosterols as a criterion to discriminate geographic origin of pistachio seeds. *Food Chemistry* 104(1), 403–8. doi:10.1016/j.foodchem.2006.09.029.

Arranz, S., Cert, R., Perez-Jimenez, J., Cert, A. and Saura-Calixto, F. 2008. Comparison between free radical scavenging capacity and oxidative stability of nut oils. *Food Chemistry* 110(4), 985–90. doi:10.1016/j.foodchem.2008.03.021.

Askin, M., Balta, M., Tekintas, F., Kazankaya, A. and Balta, F. 2007. Fatty acid composition affected by kernel weight in almond [*Prunus dulcis* (Mill.) D. A. Webb.] genetic resources. *Journal of Food Composition and Analysis* 20, 7–12.

Awad, A. B. and Fink, C. S. 2000. Phytosterols as anticancer dietary components: evidence and mechanism of action. *The Journal of Nutrition* 130(9), 2127–30. doi:10.1093/jn/130.9.2127.

Awad, A. B., Williams, H. and Fink, C. S. 2001. Phytosterols reduce *in vitro* metastatic ability of MDA-MB-231 human breast cancer cells. *Nutrition and Cancer* 40(2), 157–64. doi:10.1207/S15327914NC402_12.

Ayorinde, F. O., Eribo, B. E., Balan, K. V., Johnson Jr., J. H. and Wan, L. W. 1999. Determination of major triacylglycerol components of polyunsaturated specialty oils using matrix-assisted laser desorption/ionization time-of-flight mass spectrometry. *Rapid Communications in Mass Spectrometry* 13(10), 937–42. doi:10.1002/(SICI)1097-0231(19990530)13:10<937::AID-RCM590>3.0.CO;2-T.

Ballistreri, G., Arena, E. and Fallico, B. 2009. Influence of ripeness and drying process on the polyphenols and tocopherols of *Pistacia vera* L. *Molecules* 14(11), 4358–69. doi:10.3390/molecules14114358.

Ballistreri, G., Arena, E. and Fallico, B. 2011. Lipid composition of Italian pistachios. *Acta Horticulturae* 912, 455–66.

Baur, J. A. and Sinclair, D. A. 2006. Therapeutic potential of resveratrol: the *in vivo* evidence. *Nature Reviews Drug Discovery* 5(6), 493–506. doi:10.1038/nrd2060.

Beatrice, A. and Shivaji, G. 2015. Effect of almond supplementation on the anthropometric measurements and blood pressure levels of men with metabolic syndrome. *Indian Journal of Nutrition and Dietetics* 52(2), 184–91.

Belitz, H. D., Grosch, W. and Schieberle, P. 2004. *Food Chemistry* (3rd edn.). Springer Verlag, Berlin Heidelberg.

Bellomo, M. and Fallico, B. 2008. Anthocyanins, chlorophylls and xanthophylls in pistachio nuts. *Journal of Food Composition and Analysis* 20, 352–9.

Berryman, C. E., West, S. G., Fleming, J. A., Bordi, P. L. and Kris-Etherton, P. M. 2015. Effects of daily almond consumption on cardiometabolic risk and abdominal adiposity in healthy adults with elevated LDL-cholesterol: a randomized controlled trial. *Journal of the American Heart Association* 4(1), e000993. doi:10.1161/JAHA.114.000993.

Berryman, C. E., Fleming, J. A. and Kris-Etherton, P. M. 2017. Inclusion of almonds in a cholesterol-lowering diet improves plasma HDL subspecies in normal weight individuals with elevated LDL cholesterol. *The Journal of Nutrition* 147(8), 1517–23. doi:10.3945/jn.116.245126.

Bolling, B. W., Dolnikowski, G., Blumberg, J. B. and Oliver Chen, C. Y. 2009. Quantification of almond skin polyphenols by liquid chromatography-mass spectrometry. *Journal of Food Science* 74, C326–32.

Bouic, P. J. 2001. The role of phytosterols and phytosterolins in immune modulation: a review of the past 10 years. *Current Opinion in Clinical Nutrition and Metabolic Care* 4(6), 471–5. doi:10.1097/00075197-200111000-00001.

Bowers, J. L., Tyulmenkov, V. V., Jernigan, S. C. and Klinge, C. M. 2000. Resveratrol acts as a mixed agonist/antagonist for estrogen receptors α and β. *Endocrinology* 141(10), 3657–67. doi:10.1210/endo.141.10.7721.

Bramley, P. M., Elmadfa, I., Kafatos, A., Kelly, F. J., Manios, Y., Roxboroug, H. E., Schuch, W., Sheehy, P. and Wagner, K. 2000. Vitamin E. *Journal of the Science of Food and Agriculture* 80(7), 913–38. doi:10.1002/(SICI)1097-0010(20000515)80:7<913::AID-JSFA600>3.0.CO;2-3.

Briviba, K. and Sies, H. 1994. Nonenzymatic antioxidant defense system. In: Frei, B. (Ed.), *Natural Antioxidants in Human Health and Disease*. Academic Press, New York.

Burns, A. M., Zitt, M. A., Rowe, C. C., Langkamp-Henken, B., Mai, V., Nieves, C., Ukhanova, M., Christman, M. C. and Dahl, W. J. 2016. Diet quality improves for parents and children when almonds are incorporated into their daily diet: a randomized, crossover study. *Nutrition Research* 36(1), 80–9. doi:10.1016/j.nutres.2015.11.004.

Chen, C. Y. and Blumberg, J. B. 2008. *In vitro* activity of almond skin polyphenols for scavenging free radicals and inducing quinone reductase. *Journal of Agricultural and Food Chemistry* 56(12), 4427–34. doi:10.1021/jf800061z.

Chen, C. Y., Milbury, P. E., Lapsley, K. and Blumberg, J. B. 2005. Flavonoids from almond skins are bioavailable and act synergistically with vitamins C and E to enhance hamster and human LDL resistance to oxidation. *The Journal of Nutrition* 135(6), 1366–73. doi:10.1093/jn/135.6.1366.

Chen, P., Cheng, Y., Deng, S., Lin, X., Huang, G. and Ruan, R. 2010. Utilization of almond residues. *International Journal of Agricultural and Biological Engineering* 3(4), 1–18.

Chen, C. Y., Holbrook, M., Duess, M. A., Dohadwala, M. M., Hamburg, N. M., Asztalos, B. F., Milbury, P. E., Blumberg, J. B. and Vita, J. A. 2015. Effect of almond consumption on vascular function in patients with coronary artery disease: a randomized, controlled, cross-over trial. *Nutrition Journal* 14, 61. doi:10.1186/s12937-015-0049-5.

Cherif, A., Sebei, K., Boukhchina, S., Kallel, H., Belkacemi, K. and Arul, J. 2004. Kernel fatty acid and triacylglycerol composition for three almond cultivars during maturation. *Journal of the American Oil Chemists' Society* 81, 901–5.

Chung, K. T., Wong, T. Y., Wei, C. I. Y., Huang, Y. W. Y. and Lin, Y. Y. 1998. Tannins and human health: a review. *Critical Reviews in Food Science and Nutrition* 38(6), 421–64. doi:10.1080/10408699891274273.

Dhillon, J., Tan, S. Y. and Mattes, R. D. 2017. Effects of almond consumption on the post-lunch dip in overweight and obese adults. *The British Journal of Nutrition* 117(3), 395–402. doi:10.1017/S0007114516004463.

Dhillon, J., Thorwald, N., De La Cruz, E., Vu, E., Asghar, S. A., Kuse, Q., Diaz Rios, L. K. and Ortiz, R. M. 2018. Gluco-regulatory and cardiometabolic profiles of almond v. cracker snacking in young adults. *Nutrients* 10(8), 1–8.

Ellswort, J. L., Kushi, L. H. and Folsom, A. R. 2001. Frequent nut intake and risk of death from coronary heart disease and all causes in postmenopausal women: the Iowa Women's Health Study. *Nutrition, Metabolism and Cardiovascular Diseases* 11, 362–71.

Elmadfa, I. and Wagner, K. H. 2003. Non-nutritive bioactive food constituents of plants: tocopherols (vitamin E). *International Journal for Vitamin and Nutrition Research.*

*Internationale Zeitschrift Fur Vitamin- und Ernahrungsforschung. Journal International de Vitaminologie et de Nutrition* 73(2), 89–94. doi:10.1024/0300-9831.73.2.89.
Fallico, B., Ballistreri, G., Arena, E. and Tokuşoğlu, Ö. 2011. Nut bioactives: phytochemicals and lipid-based components of almond, hazelnut, peanut, pistachio, walnut. (Chapter 9 - Part II. Chemistry and mechanisms of beneficial bioactives in fruits & cereals). In: Tokuşoğlu, Ö. and Hall, C. (Eds), *Fruit and Cereal Bioactives: Sources, Chemistry & Applications*. CRC Press, Taylor & Francis Group, Boca Raton, FL, pp. 185–212.
Fraser, G. E. 1999. Nut consumption, lipids and risk of a coronary event. *Clinical Cardiology* 22(7 Suppl.), III11–5.
Fraser, G. E., Sabatè, J., Beeson, W. L. and Strahan, T. M. 1992. A possible protective effect of nut consumption on risk of coronary heart disease. The Adventist Heart Study. *Archives of Internal Medicine* 152(7), 1416–24. doi:10.1001/archinte.152.7.1416.
Friedrich, W., Eberhardt, A. and Galesa, R. 2000. Investigation of proanthocyanidins by HPLC with electrospray ionization mass spectrometry. *European Food Research and Technology* 211, 54–64.
Galvano, F., Pietri, A., Fallico, B., Bertuzzi, T., Scirè, S., Galvano, M. and Maggiore, R. 1996. Activated carbons: *in vitro* affinity for aflatoxin $B_1$ and relation of absorption ability to physicochemical parameters. *Journal of Food Protection* 59(5), 545–50. doi:10.4315/0362-028X-59.5.545.
Garbisa, S., Sartor, L., Biggin, S., Salvato, B., Benelli, R. and Albini, A. 2001. Tumor gelatinasse and invasion inibite by the green tea flavanol epigallocatechin-3-gallate. *Cancer* 91(4), 822–32. doi:10.1002/1097-0142(20010215)91:4<822::AID-CNCR1070>3.0.CO;2-G.
García-López, C., Grané-Teruel, N., Berenguer-Navarro, V., García-García, J. E. and Martín-Carratalá, M. L. 1996. Major fatty acid composition of 19 almond cultivars of different origins. A chemometric approach. *Journal of Agricultural and Food Chemistry* 44(7), 1751–5. doi:10.1021/jf950505m.
Garrido, I., Monagas, M., Gomez-Cordoves, C. and Bartolome, B. 2008. Polyphenols and antioxidant properties of almond skins: influence of industrial processing. *Journal of Food Science* 73(2), C106–15. doi:10.1111/j.1750-3841.2007.00637.x.
Gentile, C., Tesoriere, L., Butera, D., Fazzari, M., Monastero, M., Allegra, M. and Livrea, M. A. 2007. Antioxidant activity of Sicilian pistachio (*Pistacia vera* L. var. Bronte) nut extract and its bioactive components. *Journal of Agricultural and Food Chemistry* 55(3), 643–8. doi:10.1021/jf062533i.
Grasselly, C. 1976. Origine Et Evolution De L'amandier Cultive. *Options Méditerranéennes* 32, 45–9.
Grippi, F., Crosta, L., Aiello, G., Tolomeo, M., Oliveri, F., Gebbia, N. and Curione, A. 2008. Determination of stilbenes in Sicilian pistachio by high-performance liquid chromatographic diode array (HPLC-DAD/FLD) and evaluation of eventually mycotoxin contamination. *Food Chemistry* 107(1), 483–8. doi:10.1016/j.foodchem.2007.07.079.
Grundy, M. M., Grassby, T., Mandalari, G., Waldron, K. W., Butterworth, P. J., Berry, S. E. and Ellis, P. R. 2015. Effect of mastication on lipid bioaccessibility of almonds and its implications for digestion kinetics, metabolizable energy and postprandial lipemia. *The American Journal of Clinical Nutrition* 101(1), 25–33. doi:10.3945/ajcn.114.088328.

Grunwald, C. 1975. Plant sterols. *Annual Review of Plant Physiology* 26(1), 209–36. doi:10.1146/annurev.pp.26.060175.001233.
Gulati, S., Misra, A. and Pandey, R. M. 2017. Effect of almond supplementation on glycemia and cardiovascular risk factors in Asian Indians in North India with type 2 diabetes mellitus: a 24-week study. *Metabolic Syndrome and Related Disorders* 15(2), 98–105. doi:10.1089/met.2016.0066.
Guo, D. A., Venkatramesh, M. and Nes, W. D. 1995. Developmental regulation of sterol biosynthesis in *Zea mays*. *Lipids* 30(3), 203–19. doi:10.1007/BF02537823.
Hagerman, A. E., Riedl, K. M., Jones, G. A., Sovik, K. N., Ritchard, N. T. and Hartzfeld, P. W. 1998. High molecular weight plant polyphenolics (tannins) as biological antioxidants. *Journal of Agricultural and Food Chemistry* 46, 1887–92.
Halvorsen, B. L., Carlsen, M. H., Phillips, K. M., Bohn, S. K., Holte, K., Jacobs, D. R. and Blomhoff, R. 2006. Content of redox-active compounds (i.e., antioxidants) in foods consumed in the United States. *The American Journal of Clinical Nutrition* 84(1), 95–135. doi:10.1093/ajcn/84.1.95.
Harnly, J. M., Doherty, R. F., Beecher, G. R., Holden, J. M., Haytowitz, D. B., Bhagwat, S. and Gebhardt, S. 2006. Flavonoid content of U.S. fruits, vegetables, and nuts. *Journal of Agricultural and Food Chemistry* 54(26), 9966–77. doi:10.1021/jf061478a.
Hertog, M. G. L., Feskens, E. J. M., Hollman, P. C. H., Katan, M. B. and Kromhout, D. 1993. Dietary antioxidant flavonoids and cancer risk in the Zutphen Elderly Study. *Lancet* 342(8878), 1007–11. doi:10.1016/0140-6736(93)92876-U.
Hertog, M. G. L., Kromhout, D., Aravanis, C., Blackburn, H., Buzina, R., Fidanza, F., Giampaoli, S., Jansen, A., Menotti, A. and Nedeljkovic, S. 1995. Flavonoid intake and long-term risk of coronary heart disease and cancer in the Seven Countries Study. *Archives of Internal Medicine* 155(4), 381–6.
Holcapek, M., Jandera, P., Zderadicka, P. and Hruba, L. 2003. Characterization of triacylglycerol and diacylglycerol composition of plant oils using high-performance liquid chromatography-atmospheric pressure chemical ionization mass spectrometry. *Journal of Chromatography. A* 1010(2), 195–215. doi:10.1016/S0021-9673(03)01030-6.
Holscher, H. D., Taylor, A. M., Swanson, K. S., Novotny, J. A. and Baer, D. J. 2018. Almond consumption and processing affects the composition of the gastrointestinal microbiota of healthy adult men and women: a randomized controlled trial. *Nutrients* 10(2), 126–32. doi:10.3390/nu10020126.
Hu, F. B., Stampfer, M. J., Manson, J. E., Rimm, E. B., Colditz, G. A., Rosner, B. A., Speizer, F. E., Hennekens, C. H. and Willet, W. C. 1998. Frequent nut consumption and risk of coronary heart disease in women: prospective cohort study. *British Medical Journal* 317(7169), 1341–5. doi:10.1136/bmj.317.7169.1341.
Hull, S., Re, R., Chambers, L., Echaniz, A. and Wickham, M. S. 2015. A mid-morning snack of almongs generates satiety in healthy women. *European Journal of Nutrition* 54(5), 803–10. doi:10.1007/s00394-014-0759-z.
Hurst, W. J., Glinski, J. A., Miller, K. B., Apgar, J., Davey, M. H. and Stuart, D. A. 2008. Survey of the transresveratrol and trans-piceid content of cocoa-containing and chocolate products. *Journal of Agricultural and Food Chemistry* 56, 8374–8.
Jung, H., Chen, C. Y. O., Blumberg, J. B. and Kwak, H. 2018. The effects of almonds on vitamin E status and cardiovascular risk factors in Korean adults. *European Journal of Nutrition* 57(6), 2069–79. doi:10.1007/s00394-017-1480-5.

Kalita, S., Khandewal, S., Madan, J., Pandya, H., Sesikeran, B. and Krishnaswamy, K. 2018. Almonds and cardiovascular health: a review. *Nutrients* 10(4), 468–80. doi:10.3390/nu10040468.

Karamac, M., Kosinska, A., Rybarczyk, A. and Amarowicz, R. 2007. Extraction and chromatographic separation of tannin fractions from tannin-rich plant material. *Polish Journal of Food and Nutrition Sciences* 57, 471–4.

Karchesy, J. J. and Hemingway, R. W. 1986. Condensed tannins: (4b-8;2b-O-7)-linked procyanidins in *Arachis hypogaea* L. *Journal of Agricultural and Food Chemistry* 34, 966–70.

Klevay, L. M. 1993. Copper in nuts may lower heart disease risk. *Archives of Internal Medicine* 153(3), 401–2. doi:10.1001/archinte.153.3.401.

Kocygit, A., Koylu, A. A. and Keles, H. 2006. Effects of pistachio nuts consumption on plasma lipid profile and oxidative status in healthy volunteers. *Nutrition, Metabolism and Cardiovascular Diseases* 16(3), 202–9. doi:10.1016/j.numecd.2005.08.004.

Kornsteiner, M., Wagner, K. H. and Elmadfa, I. 2006. Tocopherols and total phenolics in 10 different nut types. *Food Chemistry* 98(2), 381–7. doi:10.1016/j.foodchem.2005.07.033.

Kris-Etherton, P. M., Yu-Poth, S., Sabate, J., Ratcliffe, H. E., Zhao, G. and Etherton, T. D. 1999. Nuts and their bioactive constituents: effects on serum lipids and other factors that affect disease risk. *The American Journal of Clinical Nutrition* 70(3 Suppl.), 504S–11S. doi:10.1093/ajcn/70.3.504s.

Kuhnle, G. G. C., Dell'Aquila, C., Aspinall, S. M., Runswick, S. A., Mulligan, A. A. and Bingham, S. A. 2008. Phytoestrogen content of beverages, nuts, seeds, and oils. *Journal of Agricultural and Food Chemistry* 56(16), 7311–5. doi:10.1021/jf801534g.

Kuhnle, G. G. C., Dell'Aquila, C., Runswick, S. A. and Bingham, S. A. 2009. Variability of phytoestrogen content in foods from different sources. *Food Chemistry* 113(4), 1184–7. doi:10.1016/j.foodchem.2008.08.004.

Kurowski, E. M. and Carroll, K. K. 1992. Effect of high levels of selected dietary essential amino acids on hypercholesterolemia and down-regulation of hepatic LDL-receptors in rabbits. *Biochimica et Biophysica Acta* 1126(2), 185–91. doi:10.1016/0005-2760(92)90289-8.

Lee, Y., Berryman, C. E., West, S. G., Chen, C. O., Blumberg, J. B., Lapsley, K. G., Preston, A. G., Fleming, J. A. and Kris-Etherton, P. M. 2017. Effects of dark chocolate and almonds on cardiovascular risk factors in overweight and obese individuals. *Journal of the American Heart Association* 6, 1–12.

Ling, W. H. and Jones, P. J. 1995. Dietary phytosterols: a review of metabolism, benefits and side effects. *Life Sciences* 57(3), 195–206. doi:10.1016/0024-3205(95)00263-6.

Liu, Z., Wang, W., Huang, G., Zhang, W. and Ni, L. 2016. *In vitro* and *in vivo* evaluation of the prebiotic effect of raw and roasted almonds (*Prunus amygdalus*). *Journal of the Science of Food and Agriculture* 96(5), 1836–43. doi:10.1002/jsfa.7604.

Liu, Y., Hwang, H. J., Ryu, H., Lee, Y. S., Kim, H. S. and Park, H. 2017. The effects of daily intake of almonds on the body composition and blood lipid profile of healthy adults. *Nutrition Research and Practice* 11(6), 479–86. doi:10.4162/nrp.2017.11.6.479.

Liu, Y., Hwang, H. J., Kim, H. S. and Park, H. 2018. Time and intervention effects of daily almond intake on changes in lipid profile and body composition of healthy adults. *Journal of Medicinal Food* 21(4), 340–7. doi:10.1089/jmf.2017.3976.

Lopez-Ortiz, C. M., Prats-Moya, S., Sanahuja, A. B., Maestre-Perez, S. E., Grané-Teruel, N. and Martin-Carratala, M. L. 2008. Comparative study of tocopherol homologue content in four almond oil cultivars during two consecutive years. *Journal of Food Composition and Analysis* 21(2), 144–51. doi:10.1016/j.jfca.2007.09.004.

Martin-Carratala, M. L., Llorens-Jordá, C., Berenguer-Navarro, V. and Grané-Teruel, N. 1999. Comparative study on the triacylglycerol composition of almond kernel oil. A new basis for cultivar chemometric characterization. *Journal of Agricultural and Food Chemistry* 47(9), 3688–92. doi:10.1021/jf981220n.

Milbury, P. E., Chen, C. Y., Dolnikowski, G. G. and Blumberg, J. B. 2006. Determination of flavonoids and phenolics and their distribution in almonds. *Journal of Agricultural and Food Chemistry* 54(14), 5027–33. doi:10.1021/jf0603937.

Miller III, E. R., Pastor-Barriuso, R., Dalal, D., Riemersma, R. A., Appel, L. J. and Gullar, E. 2005. Meta-analysis: high-dosage vitamin E supplementation may increase all-cause mortality. *Annals of Internal Medicine* 142(1), 37–46. doi:10.7326/0003-4819-142-1-200501040-00110.

Miraliakbari, H. and Shahidi, F. 2007. Lipid class compositions, tocopherols and sterols of tree nut oils extracted with different solvents. *Journal of Food Lipids* 15(1), 81–96. doi:10.1111/j.1745-4522.2007.00104.x.

Miraliakbari, H. and Shahidi, F. 2008. Antioxidant activity of minor components of tree nut oils. *Food Chemistry* 111(2), 421–7. doi:10.1016/j.foodchem.2008.04.008.

Mishra, N., Dubey, A., Mishra, R. and Barik, N. 2010. Study on antioxidant activity of common dry fruits. *Food and Chemical Toxicology* 48(12), 3316–20. doi:10.1016/j.fct.2010.08.029.

Monagas, M., Garrido, I., Lebron-Aguilar, R., Bartolome, B. and Gomez-Cordoves, C. 2007. Almond (*Prunus dulcis* (Mill.) D. A. Webb) skins as a potential source of bioactive polyphenols. *Journal of Agricultural and Food Chemistry* 55, 8498–507.

Monagas, M., Garrido, I., Lebron-Aguilar, R., Gomez-Cordoves, M. C., Rybarczyk, A., Amarowicz, R. and Bartolome, B. 2009. Comparative flavan-3-ol profile and antioxidant capacity of roasted peanut, hazelnut, and almond skins. *Journal of Agricultural and Food Chemistry* 57(22), 10590–9. doi:10.1021/jf901391a.

Moreua, R. A., Whitaker, B. D. And Hicks, K. B. 2002. Phytosterols, phytostanols and their conjugates in foods. *Progress in Lipid Research* 41(6), 457–500. doi:10.1016/S0163-7827(02)00006-1.

Moure, A., Pazos, M., Medina, I., Dominguez, H. and Parajo, J. C. 2007. Antioxidant activity of extracts produced by solvent extraction of almond shells acid hydrolysates. *Food Chemistry* 101(1), 193–201. doi:10.1016/j.foodchem.2006.01.017.

Musa-Veloso, K., Paulionis, L., Poon, T. and Lee, H. Y. 2016. The effects of almond consumption on fasting blood lipid levels: a systematic review and meta-analysis of randomised controlled trials. *Journal of Nutritional Science* 5(34), e34. doi:10.1017/jns.2016.19.

Nishi, S., Kendall, C. W., Yoon, A. M., Bazinet, R. P., Bashyam, B., Lapsley, K. G., Augustin, L. S., Sievenpiper, J. L. and Jenkins, D. J. 2014. Effect of almond consumption on the serum fatty acid profile: a dose-response study. *The British Journal of Nutrition* 112(7), 1137–46. doi:10.1017/S0007114514001640.

Ostlund Jr., R. E. 2004. Phytosterols and cholesterol metabolism. *Current Opinion in Lipidology* 15(1), 37–41. doi:10.1097/00041433-200402000-00008.

Phillips, K. M., Tarragò-Trani, M. T. and Stewart, K. K. 1999. Phytosterol content of experimental diets differing in fatty acid composition. *Food Chemistry* 64(3), 415–22. doi:10.1016/S0308-8146(98)00090-9.

Phillips, K. M., Ruggio, D. M. and Ashraf-Khorassani, M. 2005. Phytosterol composition of nuts and seeds commonly consumed in the United States. *Journal of Agricultural and Food Chemistry* 53(24), 9436–45. doi:10.1021/jf051505h.

Piga, A., Del Caro, A. and Corda, G. 2003. From plums to prunes: influence of drying parameters on polyphenols and antioxidant activity. *Journal of Agricultural and Food Chemistry* 51(12), 3675–81. doi:10.1021/jf021207+.

Pinelo, M., Rubilar, M., Sineiro, J. and Nunez, M. J. 2004. Extraction of antioxidant phenolics from almond hulls (*Prunus amygdalus*) and pine sawdust (*Pinus pinaster*). *Food Chemistry* 85(2), 267–73. doi:10.1016/j.foodchem.2003.06.020.

Plat, J. and Mensink, R. P. 2001. Effects of plant sterols and stanols on lipid metabolism and cardiovascular risk. *Nutrition, Metabolism, and Cardiovascular Diseases* 11(1), 31–40.

Quilez, J., Garcia-Lorda, P. and Salas-Salvado, J. 2003. Potential uses and benefits of phytosterols in diet: present situation and future directions. *Clinical Nutrition* 22(4), 343–51. doi:10.1016/S0261-5614(03)00060-8.

Qureshi, M. N., Numonov, S., Abudurexiti, A. and Aisa, H. A. 2016. Phytochemical investigations and evaluation of antidiabetic potential of *Prunus dulcis* nuts. *LWT - Food Science and Technology* 66, 311–7. doi:10.1016/j.lwt.2015.08.076.

Ranalli, A., de Mattia, G., Ferrante, M. L. and Giansante, L. 1997. Incidence of olive cultivation area on the analytical characteristics of the oil. Note 1. *La Rivista Italiana delle Sostanze Grasse* 74, 501–8.

Rapagna, S. and Latif, A. 1997. Steam gasification of almond shells in a fluidised bed reactor: the influence of temperature and particle size on product yield and distribution. *Biomass and Bioenergy* 12(4), 281–8. doi:10.1016/S0961-9534(96)00079-7.

Ros, E. 2009. Nuts and novel biomarkers of cardiovascular disease. *The American Journal of Clinical Nutrition* 89(5), 1649S–56S. doi:10.3945/ajcn.2009.26736R.

Ruggeri, S., Cappelloni, M., Gabelli, L. and Carnevale, E. 1998. Chemical composition and nutritive value of nuts grown in Italy. *Italian Journal of Food Science* 10, 243–52.

Ruisinger, J. F., Gigson, C. A., Backes, J. M., Smith, B. K., Sullivan, D. K., Moriarty, P. M. and Kris-Etherton, P. 2015. Statins and almonds to lower lipoproteins (the STALL Study). *Journal of Clinical Lipidology* 9(1), 58–64. doi:10.1016/j.jacl.2014.10.001.

Russo, A., Acquaviva, R., Campisi, A., Sorrenti, V., Di Giacomo, C., Virgata, G., Barcellona, M. L. and Vanella, A. 2000. Bioflavonoids as antiradicale, antioxidants and DNA cleavage protectors. *Cell Biology and Toxicology* 16, 91–8.

Seeram, N. P., Zhang, Y., Henning, S. M., Lee, R., Niu, Y., Lin, G. and Heber, D. 2006. Pistachio skin phenolics are destroyed by bleaching resulting in reduced antioxidative capacities. *Journal of Agricultural and Food Chemistry* 54(19), 7036–40. doi:10.1021/jf0614948.

Selhub, J., Jacques, P. F., Bostom, A. G., D'Agostino, R. B., Wilson, P. W., Belanger, A. J., O'Leary, D. H., Wolf, P. A., Schaefer, E. J. and Rosenberg, I. H. 1995. Association between plasma homocysteine concentrations and extracranial carotid-artery stenosis. *The New England Journal of Medicine* 332(5), 286–91. doi:10.1056/NEJM199502023320502.

Senter, S. D., Horvat, R. J. and Forbus, W. R. 1983. Comparative GLC-MS analysis of phenolic acids of selected tree nuts. *Journal of Food Science* 48(3), 798–9, 798–9. doi:10.1111/j.1365-2621.1983.tb14902.x.

Shahidi, F. and Naczk, M. 1995. *Food Phenolics. Sources, Chemistry, Effects, Applications*. Technomic Publishing, Lancaster, PA.

Sosulski, F., Krygier, K. and Hogge, L. 1982. Free, esterified, and insoluble-bound phenolic acids. 3. Composition of phenolic acids in cereal and potato flours. *Journal of Agricultural and Food Chemistry* 30(2), 337–40. doi:10.1021/jf00110a030.

Statista. 2018. Almond production worldwide in 2017/2018, by country. Available at: https://www.statista.com/statistics/632865/almond-production-worldwide-by-country/.

Sweazea, K., Jonston, C., Ricklefs, K. and Petersen, K. 2014. Almond supplementation significantly reduces C-reactive protein in subjects with type 2 diabetes. *Journal of Functional Foods* 10, 252–9.

Sze-Tao, K. W. C. and Sathe, S. K. 2000. Walnut (*Juglans regia* L.): proximate composition, protein solubility, protein amino acid composition and protein *in vitro* digestability. *Journal of the Science of Food and Agriculture* 80, 1393–401.

Thompson, L. U., Boucher, B. A., Liu, Z., Cotterchio, M. and Kreiger, N. 2006. Phytoestrogen content of foods consumed in Canada, including isoflavones, lignans and coumestan. *Nutrition and Cancer* 54(2), 184–201. doi:10.1207/s15327914nc5402_5.

Tokuşoğlu, Ö. 2011. The oxidative stability, lipid profiles and total phenolics of almond (*Prunus amygdalus* L.) and pistachio (*Pistachia vera* L.). *Acta Horticulturae* 912, 781–9. In: Ak, B. E., Wirthensohn, M. and Gradziel, T. (Eds), *Proceedings of the Vth IS on Pistachio and Almonds*. International Society for Horticultural Science (ISHS).

Tokuşoğlu, Ö. 2018. Food chemistry and minor food constituents. *Course Notes*. Dokuz Eylul University Technology Development Zone, Depark Technopark, Spillnnova LLC, Izmir,Turkey.

Tokuşoğlu, Ö. and Hall, C. 2011. *Fruit and Cereal Bioactives: Sources, Chemistry & Applications*. CRC Press, Taylor & Francis Group, Boca Raton, FL, 459pp. ISBN: 9781439806654; ISBN-10: 1439806659.

Tokuşoğlu, O., Unal, M. K. and Yemis, F. 2005. Determination of the phytoalexin resveratrol (3,5,4′-trihydroxystilbene) in peanuts and pistachios by high-performance liquid chromatographic diode array (HPLC-DAD) and gas chromatography-mass spectrometry (GC-MS). *Journal of Agricultural and Food Chemistry* 53(12), 5003–9. doi:10.1021/jf050496+.

Tomas-Barberan, F. A. and Espin, J. C. 2001. Phenolic compounds and related enzymes as determinants of quality in fruits and vegetables. *Journal of the Science of Food and Agriculture* 81(9), 853–76. doi:10.1002/jsfa.885.

Ulberth, F. and Buchgraber, M. 2000. Authenticity of fats and oils. *European Journal of Lipid Science and Technology* 102, 687–94.

USDA (United States Department of Agriculture). 2005. *Dietary Guidelines for Americans*. U.S. Department of Health and Human Services. Available at: http://www.Healthierus.gov/dietaryguidelines.

Venkatachalam, M. and Sathe, S. K. 2006. Chemical composition of selected edible nut seeds. *Journal of Agricultural and Food Chemistry* 54(13), 4705–14. doi:10.1021/jf0606959.

Wang, T., Hicks, K. B. and Moreau, R. 2002. Antioxidant activity of phytosterols, oryzanol and other phytosterols conjugates. *Journal of the American Oil Chemists' Society* 79(12), 1201–6. doi:10.1007/s11746-002-0628-x.

Wijeratne, S. S. K., Amarowicz, R. and Shahidi, F. 2006. Antioxidant activity of almonds and their by-products in food model systems. *Journal of the American Oil Chemists' Society* 83(3), 223–30. doi:10.1007/s11746-006-1197-8.

Willet, W. C. 2001. *Eat, Drink, and Be Healthy–the Harvard Medical School Guide to Healthy Eating*. Simon & Schuster, New York.

Yang, J., Liu, R. H. and Halim, L. 2009. Antioxidant and antiproliferative activities of common edible nut seeds. *LWT - Food Science and Technology* 42(1), 1–8. doi:10.1016/j.lwt.2008.07.007.

Zacheo, G., Cappello, M. S., Gallo, A., Santino, A. and Cappello, A. R. 2000. Changes associated with post-harvest ageing of almond seeds. *LWT - Food Science and Technology* 33(6), 415–23. doi:10.1006/fstl.2000.0679.

## *Additional references to integrate into reference section:*

AOCS. 1999. Official methods of analysis of the Association of Official Analytical Chemists. Method N. 92307. AOCS, Washington DC.

de Jong, A., Plat, J. and Mensink, R. P. 2003. Metabolic effects of plant sterols and stanols (Review). *The Journal of Nutritional Biochemistry* 14(7), 362–9.

Fourie, P. C. and Basson, D. S. 1989. Predicting rancidity in stored nuts and beans by means of chemical analysis. *Lebensmittel-Wissenschaft und -Technologie* 22, 251–3.

IDF. 1991. International Dairy Federation, Standard Method No. 74A, Brussels, Belgium.

Kris-Etherton, P. M., Zhao, G., Binkoski, A. E., Coval, S. M. and Etherton, T. D. 2001. The effects of nuts on coronary heart disease risk. *Nutrition Reviews* 59(4), 103–11. doi:10.1111/j.1753-4887.2001.tb06996.x.

Maguire, L. S., O'Sullivan, S. M., Galvin, K., O'Connor, T. P. and O'Brien, N. M. 2004. Fatty acid profile, tocopherol, squalene and phytosterol content of walnuts, almonds, peanuts, hazelnuts and the macadamia nut. *International Journal of Food Sciences and Nutrition* 55(3), 171–8. doi:10.1080/09637480410001725175.

Spiller, G. A., Jenkins, D. J., Cragen, L. N., Gates, J. E., Bosello, O., Berra, K., Rudd, C., Stevenson, M. and Superko, R. 1992. Effect of a diet high in monounsaturated fat from almonds on plasma cholesterol and lipoproteins. *Journal of the American College of Nutrition* 11(2), 126–30.

# Chapter 3

## Advances in understanding health benefits of hazelnuts

*Loretta Bacchetta, Silvia Procacci and Barbara Benassi, ENEA-Italian National Agency for New Technologies, Energy and Sustainable Economic Development, Italy*



## 1 Introduction

Hazelnuts (*Corylus avellana* L.), like other nuts, are a high energy food rich in fats, protein, fibre content, phytonutrients and antioxidants such as vitamin E (Table 1). The health benefits of hazelnut consumption arise from the synergistic action of unsaturated fats or bioactive compounds including plant sterols and dietary fibre.

### 1.1 Fatty acids

The lipid portion is the main component of the hazelnut kernel, and represents more than 60% of dry weight (Rustan and Drevon, 2005). By analysing 75 European cultivars for 3 years Bacchetta et al. (2013) found significant differences in oil content, ranging from 42% to 64%. Among primary fatty acids, oleic acid was the predominant one in all the extracted hazelnut oils, ranging from 73% to 85% and 49 accessions were characterized by C18:1 content >80%. Including palmitoleic, the monounsaturated fatty acids (MUFA) made up the largest portion (mean value of 80.85%), followed by polyunsaturated fatty acids (PUFA = linoleic + linolenic) of about 10.70%. It is well known that fatty acids, both individually and as part of complex lipids, play a number of key roles in human metabolism – as major metabolic fuel (storage and transport of energy), as essential components of all membranes and as gene regulators (Table 1). In particular, unsaturated fatty acids, which are produced

http://dx.doi.org/10.19103/AS.2018.0042.04

**Table 1** Hazelnut nutrient content of raw kernel per 100 g[a]

| *Fatty acids and phytosterols* | | *Proteins and vitamins* | | *Minerals* | |
|---|---|---|---|---|---|
| Energy (kcal) | 628 | Protein (g) | 14.95 | Zinc (mg) | 2.45 |
| Total fat (g) | 60.75 | Lysine (g) | 0.42 | Selenium (μg) | 2.4 |
| Saturated fat (g) | 4.464 | Arginine (g) | 2.211 | Potassium (g) | 634.4[b] |
| MUFA (g) | 45.6 | α-tocopherol (mg) | 15.03 | Iron (mg) | 3.3 |
| PUFA (g) | 7.92 | β-tocopherol (mg) | 0.33 | Calcium | 395.4[b] |
| Phytosterol (mg) | ~122 | Folate (μg) | 113 | Phosphorus (mg) | 322 |
| Stigmasterol | 1 | Vit B6 (mg) | 0.563 | Magnesium (mg) | 166.5[b] |
| Campesterol | 7 | Vit A (μg) | 30 | Copper (mg) | 1.3 |
| B-sitosterol | 102 | Thiamine (mg) | 0.64 | *Carbohydrates, fibre, carotenoids* | |
| δ 5-avenasterol | 2.6 | *Polyphenols* | | Carbohydrate (g) | 16.7 |
| B-sitostanol | 3.9 | Total polyphenols | 687 | Total fibre (g) | 9.7 |
| Campestanol | 3 | Flavonoids (mg) | 113.7 ± 30.2 | Carotenoids (μg) | 106 |
| Others | 2.5 | Proanthocyanidins (mg) | 491 | Lutein + zeaxanthin (μg) | 92 |

[a] Source: United States Department of Agriculture Nutrient Database for Standard Reference.
[b] Data were obtained from Bacchetta et al. (2015).

only by plants and phytoplankton, are essential to all higher organisms, being precursors of powerful locally acting metabolites, that is the eicosanoids (Rustan and Drevon, 2005). These signalling molecules are called leukotrienes, prostaglandins, thromboxanes, prostacyclins, lipoxins and hydroperoxy fatty acids. Eicosanoids are important for several cellular functions such as platelet agreeability (ability to clump and fuse), chemotaxis (movement of blood cells) and cell growth (Das et al., 2003). Furthermore, an increased amount of PUFA may change the physical characteristics of the membranes with positive impact on microcirculation and properties of low-density lipoproteins (LDL), which lower the melting point of core cholesteryl esters (Drevon et al., 1995). There is convincing evidence that the fatty acids profile of hazelnut has favourable effects on blood lipid profile (Sabaté and Wien, 2010; Orem et al., 2013). Park and Carr (2013) demonstrated that PUFAs decreased the expression of cholesterol transporter NPC1Ll in Caco-2 cells (human colon adenocarcinoma cell line) and HepG2 cells (human liver carcinoma cell line). The expression of several genes involved in lipid metabolism can be mediated by PUFAs via nuclear receptors including the nuclear receptors (PPAR), liver X receptor and hepatocyte nuclear factor (HNF)-4α (Kim et al., 2017). These aspects are particularly interesting, because saturated fatty acids (SFA = palmitic + stearic) in hazelnuts are minor components, accounting for only 8.43% of the total fatty acids (Parcerisa et al., 1995; Erdogan and Aygun, 2005; Bacchetta et al., 2013). The unique fatty acid profile, characterized by a higher content of MUFA and PUFA versus SFA, are the main contributing factor to the beneficial health effects of frequent hazelnut consumption (Sabaté and Wien, 2010). Nevertheless, unsaturated fatty acids of hazelnut may play a role in glucose control and appetite suppression (Kim et al., 2017) (Table 2). From a quality point of view, Bacchetta et al. (2015) discussed the effects of environmental condition on fatty acid profiles in hazelnuts. Significant differences in fatty acids content of seven hazelnut cultivars cultivated in six countries were found. The findings suggested harvest year, genotype and geographical location of cultivations playing a significant role in fatty acid profiling expression. These results agree with previous work (Parcerisa et al., 1993, 1995) which reported the type of soils and cultural practices as the main factors affecting the nutritional properties of hazelnuts with an important role in stability and global quality of the crop. On the other hand, Kodad et al. (2011) showed no significant differences in the almond fatty acid profiles among different geographic origins whereas in other species such as olive, the amount of unsaturated fatty acids increased in cool areas (Lotti et al., 1985).

### *1.2 Phytosterols*

Nuts are cholesterol-free, but their fatty fraction contains sizeable amounts of chemically related non-cholesterol sterols, belonging to a heterogeneous

**Table 2** Influence of hazelnut functional composition on metabolic events

| Hazelnut constituents | Functional effects | References |
|---|---|---|
| High in monounsaturated and polyunsaturated fatty acids versus low saturated | Lower plasma concentration of total and LDL cholesterol<br>Favourable impact on fatty acid composition of erythrocyte phospholipids<br>Activation of glycolytic and lipogenic genes including PPAR<br>Role in glucose control and appetite suppression | Rustan and Devron (2005), Mercanligil et al. (2007), Sabaté and Wien (2010), Yücesan et al. (2010), Tey et al. (2011a,b), Orem et al. (2013), Perna et al. (2016), Deon et al. (2018), Tey et al. (2017), Kim et al. (2017) |
| Phytosterols | Interfering with cholesterol absorption | Hartmann (1998), Segura et al. (2006) |
| Vegetable protein (amino acids): low lysine/arginine ratio | Reduction in serum LDL levels; arginine is a precursor of nitric oxide<br>74% decreased risk of CVD events<br>Lowering plasma uric acid | Huynh and Chin-Dusting (2006), Brufau et al. (2008), Mirmiran et al. (2017), Santi et al. (2017) |
| Tocopherols | Increase in LDL oxidative resistance<br>Modulating inflammation | Salas-Salvadó et al. (2006), Yücesan et al. (2010), Orem et al. (2013) |
| Folic acid | Decrease in homocysteine levels | Welch and Loscalzo (1998), Segura et al. (2006) |
| Polyphenols and phytochemicals | Antioxidant potential<br>Positive effects on T2DM Type 2 diabetes mellitus<br>Pro-apoptotic effect and cytotoxic effect in human cervical cancer<br>Anti-diabetic effect by altering gut microbiota | Segura et al. (2006), Escurriol et al. (2009), Sabaté and Wien (2010), Askari et al. (2013), Zamora-Ros et al. (2013), Esposito et al. (2017), Kim et al. (2017), Gargari et al. (2018) |
| Magnesium | Prevents ventricle arrhythmia<br>Modulating inflammation | Salas-Salvadó et al. (2006), Sabaté and Wien (2010) |
| Copper | Reduce blood cholesterol and pressure | Sabaté and Wien (2010) |
| Dietary fibre | Inverse association between dietary fibre intake and CVD risk<br>Anti-diabetic effect by altering gut microbiota<br>Lower serum LDL | Salas-Salvadó et al. (2006), Satija and Hu (2012), Kim et al. (2017) |

group of compounds known as plant sterols or phytosterols (Segura et al., 2006). Bolling et al. (2011) reported hazelnuts contained 122 mg/100 g of phytosterols, with the highest content of B-sitosterol and a number of minor sterols, sterol esters and stanols (Table 1). They are non-nutritive components of all plants that play an important structural role in membranes, where they serve to stabilize phospholipid bilayers just as cholesterol does in animal cell membranes (Hartmann, 1998). Phytosterols interfere with cholesterol absorption and thus help lower blood cholesterol when present in sufficient amounts in the intestinal lumen. The mechanism of action of phytosterols has been linked to their hydrophobicity, which is higher than cholesterol because of a bulkier hydrocarbon molecule and entails a higher affinity for micelles than has cholesterol. Consequently, cholesterol is displaced from micelles and the amount available for absorption is limited (Calpe-Berdiel et al., 2009; Kim et al., 2017). In a placebo-controlled, crossover feeding trial, phytosterol intakes at moderate (458 mg/day) and high (2059 mg/day) doses significantly increased total faecal cholesterol excretion and biliary cholesterol excretion and decreased intestinal cholesterol absorption (Racette et al., 2010).

## 1.3 Proteins

Hazelnut seeds also contain a moderate quantity of proteins (15%) which mainly belong to the globulin group. Some authors (Serra Bonvehí and Ventura-Coll, 1993) emphasized increased protein content in Spanish watered cultivar, due to the beneficial effects of the cultural practice on seeds' protein composition. However, protein content of hazelnut kernels ranged from 4.8% to 19.7% on a dry weight, depending on the cultivars, this was assessed in 92 cultivars of hazelnut evaluated for 3 years (Bacchetta et al., 2015). The major protein components are albumin, globulin, glutelin and prolamin with a low amount of lysine and methionine and good content of L-arginine. As this amino acid is the precursor of the endogenous vasodilator, nitric oxide (NO) (Huynh and Chin-Dusting, 2006), nut intake might help improve vascular reactivity (Table 1); decrease in bioavailability of NO is one of the factors associated with endothelial dysfunction (Pober et al., 2009). Further, this high level of L-arginine yields a preferentially low arginine to lysine amino acid ratio (range 0.01–0.57) that can contribute to reducing blood LDL concentrations (Jenkins et al., 1989). Orem et al. (2013) reported increased levels (1.0–1.9 g/day) of L-arginine in the plasma profile of hypercholesterolaemic subjects after a hazelnut-enriched diet. L-arginine supplementation (2–2.5 g/day) improved the lipidaemia-induced endothelial dysfunction in healthy volunteers and in subjects with hypercholesterolaemia associated with endothelial dysfunction. Furthermore Santi et al. (2017) showed that hazelnut intake can lower serum urate levels (6.42 ± 12.6 Δ%) in healthy adults. After the hazelnut-enriched diet, uric acid and creatininaemia mean values significantly decreased. The quantity and the quality of hazelnut pro-anthocyanidins could be involved in this process. As reported by Sonlimar and Sarmalina (2010), grape seeds intake in gout-animals model were able to exert inhibitory effects on transcription factors like nuclear factor kappa b (NF-KB) or activator protein-1 (AP-1), with consequent beneficial effects in preventing high uric acid level and high blood pressure.

## 1.4 Vitamins

Among antioxidant vitamins (A, C and E), vitamin E, involved in the prevention of oxidation processes, is the most abundant. Among the three major tocopherols identified (α-, β-, γ-tocopherols), α-tocopherol was the most dominant isomer in hazelnuts representing approximately 97.5% of the total tocopherols identified (Sivakumar et al., 2005; Sivakumar and Bacchetta, 2006). Moreover α-tocopherol is the most active form in human and animal tissues (Kornsteiner et al., 2006; Alasalvar et al., 2009) due to the substitution pattern of the methyl groups on the chromanol ring making the hydrogen of the C-6 hyroxy group especially active that is facilitating the transfer of the hydrogen to a peroxyl radical (Sivakumar et al., 2005). The natural isomer (R,R,R) disappears more quickly by the liver after supplementation than enantiomer (S,S,S) (Ajjawi and Shintani, 2004). This implies that the stereoisomers of α-tocopherol are discriminate from each other and it suggests that there is an interaction with another chiral entity, such as a macromolecule. It is impossible to obtain equally high bioactive α-tocopherol through chemical synthesis. A previous study carried out in ENEA Casaccia in 2005 (Sivakumar et al., 2005) showed variability in the content of α-tocopherol in nuts of 12 traditional and local hazelnut cultivars. These differences were supposed to be used in a simple way in hazelnut authentication (Sivakumar and Bacchetta, 2006). The average content of α-tocopherol in 75 European hazelnut cultivars was 195.98 ppm ranging from 102.34 to 390.97 ppm. About 30% of the analysed accessions showed the α-tocopherol content >200 ppm (Bacchetta et al., 2013). As for γ-tocopherol, it represented the second major tocopherol isomer found in hazelnuts accounting for 2.7% of the total tocopherol content. The average content of δ-tocopherol was 0.36 mg/kg oil. Potential health benefits of tocopherol intake are numerous (Shahidi and Costa de Camargo, 2016); potential health benefits of tocopherols include prevention of heart disease, inflammation and certain types of cancer; furthermore, the synergism of α-tocopherol and polyphenols is well documented (Altunkaya et al., 2016). A recent study attributed the slow in vitro oxidation rate of human LDL cholesterol (LDL-c) as being dependent on the depletion of lipid-soluble antioxidants such as tocopherol, therefore suggesting its critical role in the resistance of LDL-c to oxidation (Ayoub et al., 2016). The role of α-tocopherol in decreasing the susceptibility of LDL-c to oxidation and in reducing myocardial infarction has also been demonstrated in human trials (Orem et al., 2013; Loffredo et al., 2015; Ayoub et al., 2016). Furthermore, vitamin E preserves bioavailability of nitric oxide in animal studies (Keaney, 2000) and prevents the haemolysis of erythrocytes, and thus protects the body against anaemia (Alphan et al., 1997). The kernels are also rich in vitamin B6 with a concentration ranging between 0.55 and 0.88 mg/100 g (Richardson, 1997). Only a few nuts contain vitamin A, and hazelnuts are among them. Their loss is dependent on kernel processing (blanching or roasting), due to the temperature effects on vitamin degradation. Nuts contain sizeable amounts of folate (113 µg; Segura et al., 2006), a B-vitamin necessary for normal cellular function that plays an important role in detoxifying homocysteine, a sulphur-containing amino acid with athero thrombotic properties that accumulates in plasma when folate status is subnormal (Welch and Loscalzo, 1998).

## 1.5 Polyphenols

Among the bioactives and health-promoting components, hazelnuts contain phytochemicals with different antioxidants properties. Data from the Folin-Ciocalteu assay, a non-specific method to characterize the content of phenols, indicate that hazelnuts have a significant phenolic content ranging from 291 to 835 mg of gallic acid equivalents/100 g (Alasalvar and Bolling, 2015), as a consequence of the different antioxidant and nutritional value of cultivars. Bacchetta et al. (2015) identified and quantitatively determined 18 phenolic compounds in 74 hazelnut cultivars from 13 European collections. Among polyphenols 10 phenolics (quercetin pentoside, quercetin 3-O-rhamnoside, catechin, epicatechin, myricetin-3-rhamnoside, procyanidin dimer-1, procyanidin dimer-2, procyanidin trimer-1, -2 and -3) were identified to belong to a wider group of flavonoids; 7 belonging to the group of phenolic acids (gallic, protocatechuic, chlorogenic, *p*-coumaric, ferulic, ellagic and caffeic acid), and among dihydrochalcones the phloridzin. Proanthocyanidins are the most abundantly found class of polyphenols in hazelnuts, and include both B-type and A-type proanthocyanidins. Epidemiological studies have shown an inverse association between proanthocyanidin intake and incident Type 2 diabetes mellitus (T2DM) (Zamora-Ros et al., 2013). Furthermore, Gargari et al. (2018) examined the prebiotic effects of regular hazelnut intake (approximately 0.43 g of hazelnuts with skin per kg of body weight) on the intestinal microbial ecosystem of 15 children and adolescents with hyperlipidaemia for 8 weeks (Table 2). They observed that hazelnut intervention induces limited changes in faecal microbiota composition, but can significantly modulate the faecal levels of acetate. Finally, correlation analyses indicated modifications of the abundance of specific bacterial taxa, such as the families *Lachnospiraceae* and *Ruminococcaceae* and the genera *Akkermansia*, *Bacteroides*, *Roseburia* and *Faecalibacterium*. A favourable prebiotic effect of proanthocyanidin derived from grape seeds (Yamakoshi et al., 2001) and cocoa (Tzounis et al., 2011) was demonstrated in healthy adults. It is worth emphasizing that hazelnuts generally are consumed dried or toasted in order to enhance their organoleptic properties. However, Pelvan et al. (2012) showed that roasting leads to significant loss in the antioxidant status and phenolic profiles because of the removal of the pellicle, which is a rich source of phenolic. On the other hand,

Schmitzer et al. (2011) reported that roasting had a negative effect on individual phenolics but not on the total phenolic content and TAC (determined by DPPH) of hazelnuts. They found that the roasting increased gallic acid and decreased protocatechulic acid, phloretin-2-O-glucoside, catechin and epicatechin content in whole hazelnuts of six genotypes harvested in Maribor, Slovenia. Polyphenols, a primary role in conservation and stability of kernel, contribute to protect LDL from oxidative damage in a synergic work with vitamin E when hazelnuts are consumed (Table 2). Recent findings (Valdes et al., 2015) showed phenolic compounds as potential modulators of the microbiota; Gargari et al. (2018) investigated the potential role of hazelnuts as an intestinal microbial ecosystem modulator, suggesting a dietary intervention with hazelnuts as part of a controlled diet for hyperlipidaemia in children and adolescents.

## 1.6 Minerals

Hazelnut is also a good source of minerals like iron, magnesium, calcium, zinc and copper, and contains good levels of potassium as well as low sodium (Özdemir et al., 2001). Bacchetta et al. (2015) analysed the mineral nutrient content in 92 accessions for 3 years, which showed that the potassium content varied 1.9 times, ranging from 440.5 to 828.4 mg/100 g dry weight, the calcium content varied 2.3 times, ranging from 238.8 to 552.0 mg/100 g dry weight, while the magnesium content varied 1.7 times, ranging from 121.3 to 211.0 mg/100 g dry weight. These data agreed with those found by Silva et al. (2005), Özdemir and Akinci (2004) and Açkurt et al. (1999). To understand more fully the contribution of the individual mineral nutrient to the variation observed, a principal component analysis on seven cultivars (*Barcelona*, *Gironell*, *Merveille de Bollwiller*, *Negret*, *Pauetet*, *Tonda di Giffoni* and *Tonda Gentile Langhe*) analysed in six different collection sites (France, Southern and Northern Italy, Portugal, Slovenia and Spain) during 2 years of cultivation was elaborated. Samples were clearly separated between harvesting years, among minerals, potassium and magnesium content were the most variable and susceptible to the effect of environment. The sodium content of hazelnut is almost undetectable, thus high intake of calcium, magnesium and potassium, together with a low sodium intake, is associated with protection against bone demineralization, arterial hypertension, insulin resistance and overall cardiovascular risk (Cordain et al., 2005). Furthermore, magnesium and copper are associated with improved insulin resistance, reduced arterial hypertension and prevention of ventricular arrhythmias (Table 1).

## 1.7 Carbohydrates and total fibre

In the kernel 4–5% are sugars, consisting mainly of sucrose (which is 80–90% of total sugars), followed by stachyose and raffinose with 6% and 3%, respectively.

Glucose, fructose and inositol are present in low quantities (50–400 μg/g p.s.) as soluble sugars and as polysaccharides often associated with fibre. Along with the polysaccharides in hazelnut, there are about 10 g of dietary fibre per 100 g of hazelnut kernel. About 80% of the fibre is insoluble with a large variability among cultivars. Through a variety of mechanisms, dietary fibre negatively affects the intestinal carbohydrate absorption increasing faecal bile acid excretion (Table 1). Therefore, consumption of some types of soluble fibre can enhance satiety, which is associated with a lower BMI, and reduce blood cholesterol and the post-prandial glucose response (Salas-Salvadó et al., 2006). These positive effects of fibre may partly contribute to the cholesterol-lowering effects of hazelnut. Fibre in synergy with polyphenols may also have an anti-diabetic effect by altering gut microbiota (like prebiotic effect) (Kim et al., 2017).

## 2 Advances in metabolomics

Cappelli et al. (2018) reported a metabolomic characterization of hazelnut hydro alcoholic extract (1:10 w/v of raw kernel/ethanol 60%), by using an LC-HRMS system, operating in both positive and negative ion modes. A total number of 79 primary (nine acids, four amino acids, nine sugars, seven vitamins) and secondary (three alkaloids, forty-two phenylpropanoids, five taxanes) metabolites were tentatively identified, quantified and validated. To our knowledge, it represents the largest metabolomics resource currently available for hazelnut kernel studies; in literature, only a targeted metabolomics LC-ESI-QqQ-MS application for hazelnut geographical origin discrimination based on 20 non-polar key metabolites was developed by Klockmann et al. (2017).

Phenylpropanoids represented the most abundant metabolic class (49.4%) of the detected molecules. Korkina (2006) discussed their potential applications as anti-cancer, anti-virus, anti-inflammatory, wound healing and antibacterial agents as a consequence of their strong antioxidant and free radical scavenging properties. Among phenylpropanoids, flavonoids were the most abundant subclass ($C_{21}H_{22}O_{12}$/$C_{23}H_{22}O_{12}$), showing the highest content (fold change (FC), equal to 94); $C_{27}H_{30}O_{16}$, $C_{16}H_{14}O_7$ and hexose ester of syringic acid, with FCs equal to, respectively, 42, 20 and 13. The second largest categories were acids and sugars (10.6% each). In their experimental conditions, the authors found a very high accumulation of oxoglutarate (FC over the internal standard: 68), bis-D-fructose 2′,1:2,1′-dianhydride (FC: 26) and vitamin B2 (FC: 1), which were the primary metabolites showing the highest levels in the acid, sugar and vitamin groups, respectively. With respect to the secondary metabolism, three alkaloids were identified ($C_{26}H_{28}N_2O_5$, the most abundant; $C_{26}H_{28}N_2O_6$ and $C_{29}H_{25}NO_9$). Using ionization parameters, traces of four amino acids (L-threo-3-phenylserine, hydroxyl-phenylalanine, phenylalanine and tyrosine) and of five taxanes (baccatin III, 7-epi-10-deacetyltaxol, 10-deacetylbaccatin III, 10-deacetyl-7-xylosyltaxol and taxinine M) were detected. The last ones are interesting molecules involved in the production of anti-cancer drugs (Kingston and Newman, 2007).

## 3 Epidemiological and clinical studies on healthy and unhealthy volunteers

Epidemiological studies have reported beneficial effects of highest and lowest frequency of nut intake on cardiovascular diseases (CVDs); however, scientific evidences reporting specifically benefits of hazelnut consumption are mainly available as hazelnut feeding trials, aimed to assess the effects on cardiovascular risk factors and to understand the underlying mechanisms that explained the observational data. Most clinical studies, reported in Tables 3–5, have been short term and have compared diets supplemented with hazelnuts with control diets in healthy and unhealthy subjects. There are a few studies on hazelnut intake in patients with obesity, the metabolic syndrome or type-2 diabetes investigating insulin sensitivity or glycaemic control besides the lipid profile. No studies were found to influence hazelnut intake on athlete performance. More recent clinical trials have dealt with oxidative markers and immune response as well as DNA damage, inflammation status and effect on dysbiosis. Some studies have focused on the relationship of hazelnut-enriched diet and body weight changes.

### *3.1 Improvements on blood profile*

Perna et al. (2016) presented a systematic review and meta-analysis to combine the results from several trials and to estimate the pooled effect of hazelnut on blood lipids. Nine studies (reported in Table 3) representing 425 participants were included in the analysis. The intervention diet lasted 28–84 days with a supplement of hazelnuts ranging from 29 to 86 g/day. Out of nine studies, a significant decrease ($P < 0.05$) in total cholesterol (minimum −0.19 and maximum −0.47 mmol/L) and LDL in four studies (minimum −0.22 and maximum 0.36 mmol/L) (Durak et al., 1999; Yücesan et al., 2010; Tey et al., 2011a; Damavandi et al., 2013) were observed. Four studies (Durak et al., 1999; Mercanligil et al., 2007; Tey et al., 2011b; and Damavandi et al., 2013) found a significant increment in high-density lipoprotein (HDL) and two works (Yücesan et al., 2010 and Orem et al., 2013) showed a significant increment in ApoA (0.06 and 0.15 g/L respectively). Tey et al. (2015) compared the effects of consuming 30 g/day of hazelnuts for 4 weeks on risk factors for CVD and acceptance in Māori and European healthy participants. The main outputs were the absence of statistically significant differences in plasma TC, LDL-C, HDL-C and TC/HDL-C ratio between Māori and European (all $P \geq 0.145$) with no statistically significant differences in these variables within the groups from baseline to the end of the intervention (all $P \geq 0.114$). Santi et al. (2017) confirmed

**Table 3** Summary of clinical trials examining the effect of hazelnut consumption on plasma lipid profile of healthy and unhealthy subjects

| Subjects | Study design | Diet intervention | Biological end point | References |
|---|---|---|---|---|
| Healthy volunteers: 18 males and 12 females aged 18–19 years | No randomized clinical trial | 69 g/days for 30 days | −0.20 ($P < 0.005$) in TC<br>−0.36 ($P < 0.0005$) of LDL-C<br>+0.09 ($P < 0.05$) in HDL-C | Durak et al. (1999) |
| 21 hypercholesterolemic volunteers (18 men and 3 women) | Double control sandwich model intervention study with a single group and three isoenergetic diet periods | • Control diet I (4 weeks)<br>• Hazelnut-enriched diet (4 weeks with hazelnut contributing 18–20% of the total daily energy intake)<br>• Control diet period II (4 weeks) | Significant improvement of total cholesterol (27.8%), triacylglycerol (27.3%), LDL cholesterol (26.17%) and high-density lipoprotein cholesterol (6.07%) compared with the control diet. Significant change in ApoA (+0.15 with $P < 0.05$) | Orem et al. (2013) |
| 15 hypercholesterolemic men aged 48 ± 8 years | 2-period (P1 and P2) study design with a total of 8 weeks | • P1: subjects consumed a control diet (low-fat, low-cholesterol and high-carbohydrate)<br>• P2: the control diet was supplemented with MUFA-rich hazelnut (40 g/day), which provided 11.6% of total energy content | Favourable plasma lipid profiles and reduced risk of coronary heart disease (CHD)<br>Decreased ($P<0.05$) concentrations of LDL cholesterol, triacylglycerol, Apo B by 29.5%, 31.8% and 9.2% respectively<br>Increased HDL cholesterol concentrations by 12.6%. Total/HDL cholesterol and LDL/HDL cholesterol ratios favourably decreased ($P<0.05$) | Mercanlıgil et al. (2007) |
| 21 healthy individuals (13 women and 8 men) with a mean age 28 ± 5 years | Single intervention study design | Hazelnut-enriched diet (1 g/kg/day) during a 4-week period | Lowering of LDL oxidation<br>−0.36 ($P < 0.001$) in TC<br>−0.21 ($P = 0.008$) of LDL-C<br>+0.06 ($P = 0.086$) in HDL-C<br>−0.07 ($P = 0.005$) in Apo B<br>0.06 g/L ($P = 0.005$) in ApoA | Yücesan et al. (2010) |
| 46 individuals (27 women and 19 men) with a mean age 49.9 ± 9.4 years | Randomized crossover | Trial duration: 84 days<br>Hazelnut diet: 30 g/day | −0.19 ($P < 0.001$) in TC<br>−0.22 ($P < 0.001$) of LDL-C<br>+0.03 ($P = 0.023$) in HDL-C<br>−0.04 ($P = 0.002$) in Apo B | Tey et al. (2011a) |

*(Continued)*

**Table 3** (*Continued*)

| Subjects | Study design | Diet intervention | Biological end point | References |
|---|---|---|---|---|
| 48 individuals with 55.68 ± 7.74<br>HZN: 7 males, 16 females<br>Control: 8 males; 17 females | Randomized parallel study | Trial duration: 56 days<br>Hazelnut diet: 29 g/day | −0.12 ($P > 0.05$) in TC<br>+0.03 ($P > 0.05$) of LDL-C<br>−0.06 ($P > 0.05$) in HDL-C | Damavandi et al. (2013) |
| 66 subjects | 8-week randomized, single-blind, controlled, three-arm, parallel group study | 1. Three groups receiving:Hazelnuts (15–30 g/day, depending on patient weight) with skin (HZN + S);<br>2. Without skin (HZN-S);<br>3. Dietary advices for hyperlipidaemia only (controls) | Significantly reduced the concentrations of LDL-C and increased HDL-C/LDL-C ratio. HZNs also had a favourable impact on fatty acids composition of erythrocyte phospholipids, as demonstrated by time × treatment interaction, with a significant increase of monounsaturated fatty acids (MUFAs) ($P = 0.008$) and MUFAs/saturated fatty acids (SFAs) ratio ($P = 0.002$) with respect to the control group | Deon et al. (2018) |
| 107 individuals with mean age 42.5 ± 12.4 years<br>HZN 30 g: 13 males, 20 females<br>HZN 60 g: 17 males, 20 females<br>Control: 16 males, 21 females | Randomized controlled parallel | Trial duration: 84 days<br>Hazelnut diet: 30 or 60 g/day | 30 g/day — 60 g/day<br>−0.14 ($P > 0.05$) — −0.13 ($P > 0.05$) of TC<br>−0.14 ($P > 0.05$) — −0.09 ($P > 0.05$) of LDL-C<br>+0.04 ($P > 0.05$) — +0.00 ($P > 0.05$) in HDL-C<br>−0.03 ($P > 0.05$) — −0.02 ($P > 0.05$) in Apo B | Tey et al. (2011a,b) |
| 24 volunteers (10 females and 14 males) | Crossover scheme study with a double sandwich control | • Three study periods during 14 weeks:T0 baseline<br>• T1 6 weeks of 40 g hazelnut daily administration<br>• T2 6-week washout period | LDL, GGT, uric acid and serum creatinine mean values significantly decreased after hazelnut intake and an increment was observed for ALT | Santi et al. (2017) |

BMI, Body Mass Index; HDL-C, High-Density Lipoprotein Cholesterol; LDL-C, Low-Density Lipoprotein Cholesterol; GGT, Gamma-Glutamyl Transferase Lipoprotein; TC, Total Cholesterol; Apo B, Apolipoprotein B; Apo A, Apolipoprotein A; ALT, Alanine Transaminase.

**Table 4** Summary of clinical trials examining the effect of hazelnut consumption on metabolic functions

| Metabolic functions | Subjects | Study design | Diet intervention | Biological end point | References |
|---|---|---|---|---|---|
| Oxidative stress DNA damage | 50 subjects with metabolic syndrome | Randomized, controlled, parallel/4 weeks | 30 g/day mix nuts (15 g walnuts, 7.5 g almonds and 7.5 g hazelnuts) vs. control diet without nuts | Decrease of DNA damage ($P < 0.001$) No changes in antioxidant capacity, oxidized LDL, conjugated diene, 8-iso-prostanes and endothelial functions | López-Uriarte et al. (2010) |
| | 66 hyperlipidaemic children and adolescents (mean age 11.6 ± 2.6 years) | 8-week, randomized, single-blind, controlled trial, with three parallel treatment arms | Group 1 received hazelnuts with skin (HZN + S); group 2 hazelnuts without skin (HZN-S), at equivalent doses (15–30 g/day, based on body weight); and group 3 (controls) received only dietary recommendations (no nuts) | A short-term hazelnut intervention improves cell DNA protection and resistance against oxidative stress in hyperlipidaemic paediatric patients | Guaraldi et al. (2018) |
| Gut microbiota modification | 15 children and adolescents with primary hyperlipidaemia | 8-week randomized, single-blind, controlled, three-arm, parallel group study | 8-week intake of HZN + S approximately 0.43 g/kg body weight up to a maximum of 30 g | Hazelnut intake induced limited changes in bacterial abundances but modulated SCFA levels in faeces. The modification of several taxa of the faecal microbiota correlates with changes in the host's hyperlipidaemia biomarkers | Gargari et al. (2018) |
| Glycaemic control | Subjects with T2DM | 12 randomized controlled trials/3-week follow-up period | Diet supplemented with three nuts between almonds, pistachios, walnuts, pecans, hazelnuts, peanuts, cashews, macadamias vs. isocaloric diet without nuts | Decrease HbA1c ($P = 0.0003$) Decrease fasting glucose No changes in fasting insulin, HOMA-IR after consumption of tree nuts at a median dose of 56 g/day | Viguiliouk et al. (2014) |
| | 36 healthy adults aged between 13 and 65 years | Randomized, controlled, four-arm, crossover design over a period of 50 days in total | Control bread with no hazelnuts and hazelnut-enriched breads (30 g hazelnut flour defatted per 120 g bread), for a week, followed by a 6-day washout period | Blood glucose response was measured over 120 min was significantly lower for the nut breads compared to the control bread (all $P < 0.001$) | Devi et al. (2016) |

SCFA, Short Chain Fatty Acid; T2DM, Type 2 Diabetes Mellitus; HOMA-IR, Homeostasis Model Assessment of Insulin Resistance.

**Table 5** Summary of epidemiological and clinical trials examining the effect of hazelnut consumption on body weight and BMI

| Subjects | Study design | Diet intervention | Biological endpoint | References |
|---|---|---|---|---|
| 4242 subjects | Randomized controlled/3 years | Mediterranean diets enriched with 30 g/day of nuts (15 g/day walnuts, 7.5 g/day hazelnuts and 7.5 g/day almonds) vs. Mediterranean diets enriched with 50 mL/day of extra virgin olive oil vs. low-fat diet | Decrease of body weight (decrease ~5 kg in the two lowest quintiles of energy density; decrease ~4 kg the highest quintile of energy density; decrease ~2 kg in the two middle quintiles of energy density) | Razquin et al. (2017) |
| 101 overweight subjects | Randomized controlled/18 months | Moderate fat low-energy diet (fat: 35% of energy; predominantly PUFAs; nut-enriched diet) vs. low-fat, low-energy diet (fat: 20% of energy) supplemented with peanuts, almonds, cashews, hazelnuts, macadamias, pecans, pistachios, walnuts | Decrease of 4.1 kg body weight<br>Decrease of 1.6 kg/m$^2$ BMI<br>Decrease of 6.9 cm waist circumference in moderate fat group<br>Increase of 2.9 kg body weight<br>Increase of 1.4 kg/m$^2$ BMI<br>Increase of 2.6 cm waist circumference in low-fat group | McManus et al. (2001) |
| 30 healthy subjects | Pre-post supplemental, dietary advice | 1 g/kg body weight/day of hazelnuts supplemented with a habitual diet vs. baseline | Increase of 0.5 kg body weight after hazelnut supplementation | Durak et al. (1999) |
| 46 individuals with mean age 37.5 ± 14 years<br>HZN: 15 males, 17 females<br>Control: 12 males, 17 females | Randomized controlled parallel | Trial duration: 84 days<br>Hazelnut diet: 42 g/day | Increase of BMI and body weight | Tey et al. (2011b) |
| 48 individuals with 55.68 ± 7.74<br>HZN: 7 males, 16 females<br>Control: 8 males, 17 females | Randomized parallel study | Trial duration: 56 days<br>Hazelnut diet: 29 g/day | Significant decrease ($P < 0.05$) of BMI and body weight | Damavandi et al. (2013) |

BMI, Body Mass Index; PUFAs, Polyunsaturated Fatty Acids.

a positive effect of hazelnut-enriched diet (40 g/days) on the plasma lipid profile of 24 healthy volunteers following a crossover scheme built as a double sandwich control for a total of 14 weeks. Significant differences from baseline values were found in total cholesterol, LDL fraction and LDL/HDL ratio ($P < 0.01$, 0.03 and 0.02, respectively) after hazelnut intake. No significant differences were observed in HDL, nor in triacylglycerol contents. The Mann–Whitney U test showed Δ% LDL between baseline LDL (lower or higher than 130 mg/dL), BMI (higher or lower than 25 kg/m$^2$), age (older or younger than 55 years) and gender subgroups. A higher decrease was observed in females versus males (−22.61 vs. −8.37) and in the group with higher versus lower baseline LDL values (−33.14 vs. −20.02), but both differences were not statistically significant. In addition, no difference was observed between age and BMI subgroups. Since the aim of the research was to assess the influence of hazelnut intake on plasma key parameters commonly used to ascertain the individual's well-being status, a multiple reversible effect on blood parameters, a part of lipid content, was observed. Whereas no significant changes were observed in azotaemia, serum iron, glycaemia, aspartate aminotransferase (AST), total bilirubin, haemoglobin, white blood cells, red blood cells, platelet count and total plasma proteins, hazelnut intake appeared to decrease uric acid concentration from 4.66 mg/dL (before the enriched diet - T0) to 4.31 mg/dL (after hazelnut consumption - T1) ($P < 0.025$). Significantly as well, creatinine decreased from 0.94 mg/dL (T0) to 0.93 mg/dL (T1). Within the normal ranges, a significant increment was observed for alanine aminotransferase (ALT) during hazelnut-enriched diet (from 30.9 to 35.22 U/L, $P < 0.01$), while gamma-glutamyl transferase (GGT) significantly decreased from 38.04 mg/dL at T0 to 35.27 mg/dL at T1 ($P < 0.001$). As reported by Orem et al. (2013), this study confirmed the reversibility of hazelnut-enriched diet benefits. With the aim to investigate the effect of a dietary intervention with hazelnuts (15–30 g/day, depending on patient weight) in children and adolescents with primary hyperlipidaemia, Deon et al. (2018) carried out an 8-week randomized, single-blind, controlled, three-arm, parallel group study on 66 subjects. Two-way ANOVA analysis showed a significant effect of time on serum LDL-c, HDL-C/LDL-c ratio and non-HDL-c ($P \leq 0.001$), but not of treatment and time × treatment interaction. In particular, hazelnuts significantly reduced the concentrations of LDL-c and increased HDL-c/LDL-c ratio with a favourable impact on FAs composition of erythrocyte phospholipids in children with primary hyperlipidaemia.

## 3.2 Immune adjuvant effects

Among traditional uses of nuts, Casas-Agustench et al. (2011b) reported improvement of immunity; however, only a few studies are available on hazelnut as immune adjuvant factors. Some years later Ren et al. (2016) published a scientific report on the immunomodulatory activity of hydrolysed peptides from *Corylus heterophylla* Fisch dregs, the by-products of hazelnut oil expression. Mice were fed daily with hydrolysed peptide fractions at doses of 200, 400 and 800 mg/kg body weight for 10, 20 and 30 days. The results showed an improvement of most immune indexes (such as organ index, spleen lymphocyte proliferation, macrophage activity, secretory immunoglobulin A content), which supported potential immunomodulatory properties of the by-product for promising implications for the development of new functional foods. Recently, Cappelli et al. (2018) discussed the effect of raw hazelnut extracts on human innate immune response. One of the main features of the innate immune response players is the capacity to face bacterial infections. Among the bacterial pathogens, *Staphylococcus aureus* (*S. aureus*) is one of the principal causes of both community-acquired and nosocomial infections, responsible for severe life-threatening diseases associated with significant morbidity and mortality. To depict the intracellular macrophage milieu in different experimental conditions, the authors selected a group of 29 genes involved in pro- and anti-inflammatory response (towards infection and/or oxidized lipids) and intracellular iron metabolism. They investigated, via quantitative real-time PCR (RT-PCR), the influence of peripheral blood mononuclear cells' initial pro-inflammatory status on monocyte-derived macrophage response after hazelnut extract treatment by studying both the inflammation and iron metabolism genes regulation and, the cell bactericidal response to *S. aureus*. Based on the intracellular bacteria colony-forming units reduction and on the treatment doses used, the donors were grouped into high-dose and low-dose responders. Expression profile for inflammatory and iron metabolism genes, both for peripheral blood mononuclear cells and macrophages, highlighted that low-dose responders came from a more pronounced pro-inflammatory milieu as compared with high-dose responders. Blood test results revealed that low-dose responders had higher LDL-c and triglycerides and lower HDL-c. At present, it is commonly accepted and extensively documented that the pro-inflammatory role of high-fat diet, obesity and atherosclerosis have a detrimental role on innate immune response, by altering its components' reactivity (Lumeng et al., 2007). On the other hand, the authors favoured the hypothesis that a pro-inflamed macrophage, with a high content of ROS, low pHi and metabolically active $Fe^{2+}$, when treated with hazelnut extract will tolerate a much smaller dose as compared to a much lower inflamed macrophage. In this scenario, the individual initial inflammatory status and the lipid metabolism are critical factors to be analysed in order to optimize the phytocomplex effect.

### *3.3 Antioxidant activity and decreased oxidative stress*

Studies investigating the relationship between hazelnut consumption and oxidative stress in humans are limited. López-Uriarte et al. (2010) evaluated the role of mixed nuts, including hazelnuts, in adult patients with metabolic syndrome, and found a significant reduction of DNA damage. It is well known that oxidative stress results in damage of different cellular components by excessive generation of pro-oxidant species, as well as deficiency of antioxidant defence mechanisms. Lipids are susceptible targets of oxidation, and plasma oxidized LDLs (ox-LDLs) play a major role in atherosclerotic plaque formation. Several studies in adults demonstrated significant correlations between high circulating levels of ox-LDL and prevalence of CVDs, diabetes and metabolic syndrome (Holvoet et al., 2001). Recently, 66 hyperlipidaemic children and adolescents (mean age 11.6 ± 2.6 years) with diagnosis of primary hyperlipaemia were enrolled to participate in a trial to evaluate the effect of hazelnuts intake on plasma ox-LDL concentrations and oxidative levels of DNA damage in lymphocytes; the intervention study was an 8-week, randomized, single-blind, controlled trial, with three parallel treatment arms (Guaraldi et al., 2018). Among the main results (Table 4) the regular intake of hazelnuts was associated with the reduction (time effect) of the levels of DNA strand breaks by 18.8% ($P < 0.003$) whole hazelnuts intake. A significant decrease (time × treatment interaction, $P < 0.0006$) was also observed for the levels of enzyme formamidopyrimidine glycosylase (FPG)-sensitive sites (endogenous DNA damage). On the other hand, the authors could not demonstrate an effect of whole hazelnuts on ox-LDLs, whose levels remained unchanged in treated and untreated patients. No effect was also observed in the ratio of ox-LDL/HDL. Recently, in a controlled, single-blind, crossover study, 14 healthy individuals were randomized to consume 100 g of either antioxidant ice cream containing dark cocoa powder and hazelnut and green tea extracts or milk chocolate ice cream (control ice cream). Participants were studied at baseline and 2 h after ingesting ice cream. Serum polyphenols ($P < 0.001$), nitric oxide ($P < 0.001$), antioxidant status (ferric-reducing ability of plasma-FRAP) ($P < 0.005$), endothelium function (flow-mediated dilation) ($P < 0.001$) and reactive hyperaemia index ($P < 0.05$) increased significantly while oxidative stress decreased (reactive oxygen metabolites (d-ROMs) $P < 0.001$ and hydrogen peroxide $P < 0.001$). The study demonstrated that natural ice cream rich in polyphenols, including those of hazelnuts, acutely improved vascular function and physical performance in healthy individuals through a reduction in oxidative stress (Sanguigni et al., 2017).

### 3.4 Gut microbiota modification

Hyperlipidaemia starts at a paediatric age and represents a potential factor for CVD risk. Modulation of the intestinal microbial ecosystem can be considered additional strategy to ameliorate lipid profiles. In fact, scientific evidences indicate that the gut microbiota mediates pathophysiological mechanisms that alter lipid metabolism and other related metabolic traits (Ghazalpour et al., 2016; Brahe et al., 2016). Numerous metabolites produced by intestinal bacteria, such as trimethylamine/trimethylamine-N-oxide, secondary bile acids, catecholamines and short chain fatty acids (SCFAs), act as hormone-like factors that impact numerous aspects of host physiology (Boulange et al., 2016). Particularly, SCFAs, including acetate, butyrate, propionate and valerate, are involved in the regulation of energy metabolism, insulin sensitivity and immune responses (Brown and Hazen, 2015). Therefore, hyperlipidaemic subjects could receive both specific and overall benefits by consuming dietary patterns/food products that can affect lipid metabolism through modulation of the intestinal microbiota. Gargari et al. (2018) characterized the intestinal microbial ecosystem in children and adolescents with primary hyperlipidaemia (15 volunteers, 7 and 17 years old; hazelnut intake 0.43 g/kg body weight up to a maximum of 30 g for 8 weeks); the aim was to investigate the potential modulatory effects of regular hazelnut intake on microbiota composition and SCFA levels. Among the main results (Table 4) hazelnut intake induced limited changes in bacterial abundances but modulated SCFA levels in faeces. After hazelnut consumption eight SCFAs and lactate were found. Moreover, acetate increased significantly over the intervention ($P < 0.05$ according to paired Wilcoxon test) and an increasing trend ($P = 0.079$) was also observed for propionate. Overall, these results indicate that inclusion of hazelnuts in the diet for an 8-week intervention period may induce limited changes in the faecal microbiota composition but significantly modulate faecal levels of the predominant intestinal SCFA acetate. Furthermore, the correlation analyses showed a relationship between the variations observed in the abundance of bacterial taxa and changes in the lipid profile of hyperlipidaemic subjects. An inverse correlation was observed between the changes in total cholesterol and the variation of several *Ruminococcaceae*, particularly *Faecalibacterium prausnitzii*. The changes in HDL-C inversely correlated with fluctuations in the abundance of several *Clostridiales* and *Collinsella* OTUs; on the contrary, a positive correlation for HDL-C was observed with variations in a smaller number of *Clostridiales* and mostly *Lachnospiraceae*. The properties of hazelnut skin extract administrated during the 8 weeks to Golden Syrian hamsters fed a high-fat diet showed a decrease of the lithocholic/deoxycholic bile acid faecal ratio, which is a risk factor for colon cancer (Caimari et al., 2015).

### 3.5 Glucose lowering effect

According to recent estimates, there were 285 million people worldwide with diabetes in 2010, with that number projected to increase to 439 million cases by 2030. Type 2 diabetes, which accounts for 90–95% of diabetic cases, is an important cardiovascular risk factor (Damavandi et al., 2013). Type 2 diabetes induces an abnormal lipid profile called atherogenic dyslipidaemia, which is characterized by elevated triglycerides (TGs), reduced HDL-c and small, dense LDL particles. Few nut feeding trials have been performed in type 2 diabetes with contradictory results regarding glycaemic control (Tapsell et al., 2009; Li et al., 2011; Casas-Agustench et al., 2011a). A meta-analysis of 12 randomized controlled trials with a ≥ 3-week follow-up period in subjects with type 2 diabetes compared a diet supplemented with tree nuts (almonds, Brazil nuts, cashews, hazelnuts, macadamia nuts, pecans, pine nuts, pistachios and walnuts) and an isocaloric diet without tree nuts. The results (Table 4) showed that consumption of tree nuts at a median dose of 56 g/day improved glycaemic control, showing significantly decreased glycosylated haemoglobin levels (HbA1c; mean difference 0.07%, $P < 0.0003$) and fasting glucose ($P < 0.03$). Very few outcomes are available on the evaluation of the effect of hazelnuts-enriched diet on glucose control. Alphan et al. (1997) discussed the nutritional composition of hazelnuts intake and its effects on glucose and lipid metabolism, concluding the goal of diet therapy in patients with diabetes was to improve the control of hyperglycaemia but also to reduce the risk of coronary heart disease by optimizing plasma lipid levels; in this regard, the incorporation of hazelnuts in the diet represents a cornerstone of therapy for patients with non-insulin-dependent diabetes mellitus. Mercanlígil et al. (2007) reported no changes in fasting levels of glucose after a diet enriched with 40 g/day in 15 hyper-cholesterolaemic subjects. The same results were discussed by Santi et al. (2017) in healthy volunteers. Replacing 10% of the total daily calorie intake with hazelnuts (mean amount of hazelnuts consumed 29 g/day) for 8 weeks in 50 type 2 diabetic patients (16 men and 34 women, mean age 55.68 ± 7.74 years), Damavandi et al. (2013) confirmed significant increment in HDL-c and greater reduction in triglyceride concentration (Table 3). However, neither between-group changes nor within-group changes were significant for fasting blood sugar. The authors concluded the effects of hazelnuts consumption showed a dose-dependent pattern which could partially explain the differences in outcomes among the different scientific works. Recently a crossover study with 36 participants was performed to analyse blood glucose response after a consumption of bread supplemented with 30 g of hazelnuts for 8 days (Devi et al., 2016). The area under the blood glucose curve was significantly lower for volunteers who consumed the nut breads compared to the control (all $P < 0.001$), no significant differences in satiety (all $P \geq 0.135$) or gastrointestinal symptoms (all $P \geq 0.102$) were found between the bread consumption.

### 3.6 Effect on body weight

The common perception that fatty foods provide excess energy and thus promote obesity has had a negative effect on the image of nuts. The question of whether increasing the intake of nuts and therefore calories could lead to unwanted weight gain and related health problems is a critical one. However, as thoroughly reviewed (Table 5) there is considerable scientific evidence indicating that there are no adverse effects of frequent nut consumption on energy balance or body weight. Perna et al. (2016) presented a systematic review and meta-analysis to combine the results from nine trials and to estimate the pooled effect of hazelnut on body weight through a Bayesian meta-analysis. Even in the presence of high and prolonged caloric supplementation, the statistical elaboration showed an absence of body weight change (evaluated on means of BMI data).

All studies showed the absence of difference and one showed a significant decrease ($P < 0.05$) of $-0.5$ kg/m$^2$ whereas Tey et al. (2011b) reported an increase of BMI and body weight. Also Santi et al. (2017) showed no changes in body weight among the 24 healthy volunteers.

## 4 Molecular studies: nutrigenomics and nutri-epigenomics

Food cannot be considered as a mere source of nutrients: recent clinical and experimental findings have highlighted additional functions exerted by food that positively impact on human health. Specifically, nutrients have been characterized as antioxidant molecules and transcriptional and epigenetic modulators (Crescenti et al., 2013; Pan et al., 2013; Remely et al., 2015a,b; Declerck et al., 2016; Abdul et al., 2017). These molecular pathways can actively cross-talk in tissues to drive beneficial effects on physiological processes and disease prevention (Angelini et al., 2017). The study of nutrients and their effects on human health through interaction with genes, and consequent modification of gene expression, has been termed 'nutrigenomics' (Dang et al., 2000; Sales et al., 2014). A subset of such interaction implies epigenetic changes are triggered by food, and has been defined as 'nutri-epigenomics' (Tammen et al., 2013). Some emerging nutrigenomic and nutri-epigenomic findings, including preliminary results from our group, have been produced in response to hazelnuts supplementation. They will be discussed in the next paragraph and summarized in Table 6.

### *4.1 Nutrigenomics and hazelnuts*

Nutrigenomics is the area of nutrition that applies molecular tools to elucidate the plethora of responses obtained through a specific diet applied between individuals or population groups. In particular, nutrigenomics attempts to understand how the bioactive components of food may affect (increase or suppress) the expression of genes and impact on human health. An example of such gene-nutrient interaction is the ability to bind to transcription factors and thus either stimulate or interfere with gene promoter regulation. Compounds

**Table 6** Nutrigenomic and nutriepigenomic data in response to hazelnut supplementation

| Hazelnut source | Hazelnut administration | Experimental model | Molecular endpoint | References |
|---|---|---|---|---|
| *Nutrigenomics* | | | | |
| *Corylus* avellana *L.* cv *Tonda Gentile delle Langhe* | SD (4 weeks) + washout (3 weeks) + SD plus 40 g HZN/day (40 days) | 24 healthy volunteers: blood collection for RNA extraction and qRT-PCR analysis | ↑BNIP3, GPX2, GSR, HSPA1A, TTN TXNRD1<br>↓CCL5, KRT1, MBL2, PRDX6, SOD1 after HZN supplementation | Di Renzo et al. (2014) |
| *Corylus* avellana *L.* cv *Tonda Gentile delle Langhe* | Fast food meal (F) or fast food meal plus 40 g HZN (F-HZN) | 22 healthy volunteers: blood collection for RNA extraction and RT-PCR analysis | Gene variation in F-HZN vs. F: 17.48% Hip genes 33.98% HOSp genes | Di Renzo et al. (2017) |
| *Corylus* avellana *L.* cv *Tonda Gentile Romana* (liquid extract) | HZN administered on day 7 of the *in vitro* culturing and/or infection with *S. aureus* | Monocyte-derived-macrophages collected from whole blood of healthy donors: RNA extraction and qRT-PCR | ↑CD39, ADORA2B, LOX1, HOX1, CCR1, TREM1, NRF2, HAMP<br>↓IL-6, IL-10, TNF-alpha, PAI1 in the high-dose responders after HZN administration | Cappelli et al. (2018) |
| *Nutri-epigenomics* | | | | |
| Not specified | Diet with vegetables and legumes (C) plus folate and 25 g hazelnut oil capsule (8 weeks) (C+S) | 40 female subjects: methylation analysis performed by bisulphite conversion and pyrosequencing in genomic DNA from blood | ↓Methylation pattern of ADRB3 gene in (C+S) subjects vs. (C) ones | Lima et al. (2017) |
| *Corylus* avellana *L.* cv *Tonda Gentile Romana* (liquid extract) | 24 and 72 h HZN administration to cells in vitro (concentration range: 0.004–4 mg/mL) | HepG2 human liver carcinoma cells: RNA extraction/qRT-PCR and bisulphite conversion/pyrosequencing from genomic DNA | ↑LDLR expression and ↓methylation status of LDLR promoter gene in the HZN-treated cells compared to solvent-treated ones | Benassi (2018) submitted |

Hip, Human Inflammasome Pathway; HOSp, Oxidative stress; HZN, Hazelnut; SD, Standard diet; qRT-PCR, Real-time RT-PCR.

such as resveratrol, as well as vitamins A, D and fatty acids, have been shown to directly interact and activate nuclear receptors and induce gene transcription, and to positively affect pathological states such as inflammation and cancer (Barnes, 2008; Fialho et al., 2008; Virmani et al., 2013; Schwartz, 2014; Rodríguez-Cruz and Serna, 2017).

In terms of hazelnut supplementation and nutrigenomics, the group of Di Renzo et al. (2014) performed a first clinical trial in a cohort of 24 participants (18–50 years, with a BMI ≥ 19 $kg/m^2$) undergoing a supplementation with 40 g/day of hazelnuts (*Corylus avellana* L.) over a 40-day period. They demonstrated a significant lowering of the LDL oxidized proteins, in association with the consumption of hazelnuts, accompanied by a significant expression change of genes regulating redox state in peripheral blood mononuclear cells (PBMCs). Different lines of evidence indicate that molecular signs in PBMCs, such as transcript profiles and epigenetic changes, can be used as a reliable 'surrogate' marker of pathological state in tissues (Masliah et al., 2013; San-Cristobal et al., 2016; Noé et al., 2016). The variation in blood cells upon hazelnut supplementation included stimulation of glutathione peroxidase 2 (GPX2), glutathione reductase (GSR) and thioredoxin reductase 1 (TXNRD1), and inhibition of peroxiredoxine 6 (PRDX6). All hazelnut-affected genes play unique functions in fighting redox imbalance and act synergistically to face the oxidative stress underlying many pathologies, such as CVDs and atherosclerosis, thus suggesting a strong antioxidant ability of the hazelnut consumption and a potential protective role against such diseases (Di Renzo et al., 2014). These findings were further corroborated by the same group in a recent clinical study, where the anti-atherogenic effect of 40 g raw hazelnut intake was assessed in 22 healthy human volunteers, by evaluating the post-prandial plasma level of ox-LDL after a commercial fast food meal (Di Renzo et al, 2017). The hazelnut-enriched dietary treatment triggered a reduction of the ox-LDL particles produced after the high-fat meal consumption. In terms of LDL oxidation, both studies from Di Renzo et al. are particularly relevant, as the formation of ox-LDL is a risk factor in atherosclerosis (Rafieian-Kopaei et al., 2014). The oxidation of LDL converts the native LDL into pathogenic, immunogenic and atherogenic particles; they recall macrophages, with the consequent activation of monocyte–macrophage system and formation of foam cells, contributing to the development of atherosclerosis and CVD in association with vascular inflammation, lipid deposition and smooth muscle cell differentiation (Maiolino et al., 2013). The clinical trial also implied the expression profiling in PBMCs of 103 genes related to antioxidant defences and inflammasome pathway, after the high-fat meal with or without 40 g of hazelnut supplementation (Di Renzo et al., 2017). Still, the hazelnut supplementation was effective in modulating the expression of genes positively involved in the antioxidant and anti-inflammatory response, such as catalase (CAT), copper chaperone for superoxide dismutase (CCL5), g-glutamylcysteine ligase (GCLC), glutathione peroxidase 1 (GPX1), superoxide dismutase 1 soluble (SOD1), nitric oxide synthase 2 inducible (NOS2), cysteine-aspartic acid protease 5 (CASP5) and thioredoxin (TXN). As overall, hazelnut consumption might help in reducing post-prandial risk factors of atherosclerosis, including ox-LDL, inflammation and oxidative stress-related genes, and as future perspective, provide functional molecules contrasting those pathological states that occur upon sustained and impaired oxidative processes.

In a completely different biological scenario, Cappelli et al. (2018) demonstrated that a long-term maceration liquid extract obtained from hazelnut (*Corylus avellana* L., cultivar *Tonda Gentile Romana*) is able to stimulate the in vitro response of macrophages against *Staphylococcus aureus*, through induction of the anti-inflammatory and iron metabolism genes, as previously described. Human macrophages obtained from healthy donors were pre-treated with the hazelnut liquid extract, and then infected with the *Staphylococcus*. According to the different bactericidal response, donors were divided into high-dose and low-dose responders. Their gene expression profiling underlined that low-dose responders came from a more pronounced pro-inflammatory milieu if compared to high-dose responders, and displayed higher blood LDL-c and triglycerides and lower HDL-c levels. Authors conclude that hazelnut can exert an immune adjuvant effect, thus supporting those studies showing the immune modulating properties of tree nuts, but it strongly depends on the intracellular inflammatory status of the donor.

## 4.2 Nutri-epigenomics and hazelnuts

Epigenetics refers to those heritable modifications in gene expression that occur independently of changes in DNA sequence. It is often described as a change in phenotype without a corresponding change in genotype, and identified as the molecular interface between the environment, including life style and nutrition, and the genome (Sun et al., 2013; van Soom et al., 2014; Herceg, 2016). Epigenetic mechanisms are well characterized and classified in three major classes: (1) DNA methylation, (2) histone post-translational modifications and (3) non-coding RNA-mediated gene silencing, mainly operated by microRNAs (miRNAs). These hierarchical covalent modifications interact with each other: both DNA methylation and histone modifications regulate the expression of miRNAs and long non-coding RNAs (lncRNAs) as much as specific miRNAs control the expression of key epigenetic regulators, including DNA methyltransferases (DNMTs) and histone deacetylases (Iorio et al., 2010; Su et al., 2011; Osella et al., 2014).

1 DNA methylation consists in the covalent addition of a methyl group (-CH3) to the 5′-position of cytosine residues at cytosine-guanine dinucleotide (CpG) islands, catalysed by a family of DNMTs. CpG dinucleotides are often clustered in the promoter region of human genes, called CpG islands; following promoter methylation, gene expression is epigenetically controlled either by downregulating gene expression or blocking the binding of transcription factor(s) to DNA (Crider et al., 2012).
2 Histones are highly conserved nuclear proteins, principally devoted to package DNA into the specialized nucleosomes compartments (Marinõ-Ramírez et al., 2005). They exert a structural function within the nucleus, but also play different functional roles in the transition between active and inactive chromatin states that directly affect gene expression. Amino termini of histones can undergo post-translation modifications (including acetylation, methylation, phosphorylation, ubiquitination and sumoylation) (Kouzarides, 2007; Ruthenburg et al., 2007), which are responsible for controlling chromatin structure and, therefore, downstream gene expression. For instance, acetylation eliminates positive charges from histones (lysine residues) and decreases the interaction between histones and the negatively charged DNA. De-condensed chromatin leads to a relaxed DNA form, and transcriptional factors can better access promoter regions (Jenuwein and Allis, 2001).
3 miRNAs are a class of short non-coding RNAs (ncRNAs) made of about 18–25 nucleotides. They are transcribed by RNA polymerase II into primary miRNAs which are then processed by ribonuclease III Drosha and DGCR8 into precursor miRNAs. The latter translocate into the cytoplasm though Exportin-5, where ribonuclease III Dicer further modifies miRNAs, releasing mature products (Cai et al., 2004). miRNAs exert their epigenetic function by post-transcriptionally interfering with gene expression. They recognize and bind to the 5′-untranslated region of target mRNAs, and drive either transcript degradation or translation inhibition (Dong et al., 2013; Mohr and Mott, 2015; Gorski et al., 2017).

Nutrition can affect the epigenetic machinery by directly modifying DNA methylation, histone modifications and miRNA expression, thereby influencing gene expression. Epigenetics controls the physiological development and tissue functions over lifetime, and underlies ageing as well as many pathological conditions, including cancer, CVDs and neurodegeneration (Brunet and Berger 2014; Kanwal et al., 2015). By tuning epigenetics, the diet can directly contribute to physiological pathways and act in disease prevention (Gerhauser, 2013; Remely et al., 2015b; Kalea et al., 2018). The epigenome is susceptible to changes and can be shaped by nutritional states, especially in the prenatal period through transgenerational mechanisms and in early postnatal life when critical developmental processes are taking place. Although more stable, the epigenetic marks in adulthood are also dynamic and modifiable by environmental factors, including diet.

Many vitamins have been studied in terms of epigenetic modulation (Camarena and Wang, 2016; Mentch and Locasale, 2016), as well as different polyphenols, such as curcumin, quercetin, genistein, epigallocatechin gallate and resveratrol (Choi et al., 2012; Abbas et al., 2013; Pan et al., 2013; Tenore et al., 2015; Declerck et al., 2016; Giglio et al., 2018). Since nuts (including hazelnuts) have favourable nutrient profiles, rich in fibre and vitamins, low in saturated fats, high in beneficial unsaturated fats and very high in antioxidants, including polyphenols (Ros, 2009), they are likely modulators of the epigenome.

Only very recently has the association between hazelnut intake and epigenome modulation been addressed.

In this context, the study from Lima et al. (2017) evaluated the effect of a diet containing folate and hazelnut oil capsule on lipid profile and oxidative stress as well as on the methylation levels of the beta-3 adrenergic receptor (ADRB3) gene, in overweight and obese women. Results showed a reduction in the methylation levels of the ADRB3 gene by dietary factors, driven in particular by monounsaturated and polyunsaturated fats contained in hazelnuts, associated with increased HDL-C values and decreased oxidative stress. The ADRB3 regulates lipolysis in the brown and white adipose tissue of humans, thus providing free fatty acids for thermogenesis. Additionally, there is indirect evidence that ADRB3 may participate in the regulation of body weight in humans, suggesting that epigenetic changes in this locus may be involved in the development of obesity and its associated metabolic complications.

Also our group is currently studying the epigenetic changes potentially driven by hazelnut (*Corylus avellana* L., cultivar *Tonda Gentile Romana*) consumption, in an attempt to elucidate the molecular mechanism(s) responsible for the health benefits driven by hazelnut in terms of improved lipid profile and CVD prevention (Orem et al., 2013; Deon et al., 2018; Di Renzo et al., 2014, 2017; Perna et al., 2016; Santi et al., 2017). The LDL is a major blood cholesterol carrier, and, in liver, the low-density lipoprotein receptor (LDLR) triggers the endocytosis of the LDL particles from plasma, thus regulating the blood cholesterol levels (Goldstein and Brown, 2009). We are assessing the ability of a hazelnut extract to regulate the expression of the LDLR in the HepG2 human liver carcinoma cells, and verifying whether any epigenetic pathway is involved in such regulation. Our preliminary results have identified a deregulated DNA methylation pattern in response to hazelnut administration, and different dedicated experiments are ongoing to specifically map the DNA regions tuned by epigenetic modulation.

## 5 Conclusion and future trends

Hazelnuts, like other nuts, are a high energetic food rich in MUFA, PUFA and several essential proteins and minerals; they are a valuable source of fibre, phytonutrients and antioxidants, such as vitamin E, including a rich array of phytochemicals with functional activities. Concentrations and profiles of phytochemicals are related to genetic intraspecific variability and may be affected by (1) cultivation techniques; (2) processing; and (3) storage. Although many hazelnut bioactive components remain to be fully identified and characterized, the use of advanced techniques, such as metabolome analysis, could improve our knowledge of secondary metabolite properties and their functional role.

Despite the still limited number of scientific data, hazelnut intake has been associated with an array of bioactivities impacting on human health, including hypo-cholesterolaemic, antioxidant, anti-inflammatory, anti-proliferative actions, as well as the potential ability of preventing the onset of many pathogenic processes. In some cases, such as glucose control, the data are still contradictory and additional studies should be undertaken at the molecular level, along with a detailed investigation of different amounts, study duration, background and diet. Interestingly, research has recently shown that hazelnut (*Corylus avellana* L.) is also able to produce Taxol, thus highlighting its chemotherapeutic properties; the cytotoxic effects of hazelnut cell extract were even stronger than that of pure Taxol (standard), suggesting hazelnut cell extract as a novel and suitable probe for treating human cancer (Bemani et al., 2015).

The new scenario underlying the nutraceutical properties of hazelnuts implies the involvement of gene expression (nutrigenomics) and epigenetic regulation (nutria-epigenetics) by nutrients. This new level of action discloses new functions exerted by food components in the prevention of ageing and chronic diseases, that need to be further clarified to better explain the broad range of beneficial functions exerted by nuts on human health.

## 6 References

Abbas, A., Patterson 3rd, W. and Georgel, P. T. (2013), The epigenetic potentials of dietary polyphenols in prostate cancer management, *Biochem. Cell Biol.*, 91(6), 361–8.

Abdul, Q. A., Yu, B. P., Chung, H. Y., Jung, H. A. and Choi, J. S. (2017), Epigenetic modifications of gene expression by lifestyle and environment, *Arch. Pharm. Res.*, 40(11), 1219–37.

Açkurt, F., Özdemir, M., Biringen, G. and Löker, M. (1999), Effects of geographical origin and variety on vitamin and mineral composition of hazelnut (*Corylus avellana* L.) varieties cultivated in Turkey, *Food Chem.*, 65, 309–13.

Ajjawi, I. and Shintani, D. (2004), Engineered plants with elevated vitamin E: A nutraceutical success story, *Trends Biotechnol.*, 22(3), 104–7.

Alasalvar, C. and Bolling, B. W. (2015), Review of nut phytochemicals, fat-soluble bioactives, antioxidant components and health effects, *Br. J. Nutr.*, 113, S68–78.
Alasalvar, C., Amaral, J. S. Satir, G. and Shahidi, F. (2009), Lipid characteristics and minerals of native Turkish hazelnut essential varieties (*Corylus avellana* L.), *Food Chem.*, 113(4), 919–25.
Alphan, E., Ackurt, F. and Yamilaz, T. (1997), Nutritional composition of hazelnut and its effects on glucose and lipid metabolism, in Proceedings of the Fourth International Symposium on Hazelnut, *Acta Hort.*, 445, 305–10.
Altunkaya, A., Gökmen, V. and Skibsted, L. H. (2016), pH dependent antioxidant activity of lettuce (*L. sativa*) and synergism with added phenolic antioxidants, *Food Chem.*, 190, 25–32.
Angelini, F., Pagano, F., Bordin, A., Milan, M., Chimenti, I., Peruzzi, M., et al. (2017), The impact of environmental factors in influencing epigenetics related to oxidative states in the cardiovascular system, *Oxid. Med. Cell. Longev.*, 2017, 2712751.
Askari, G., Yazdekhasti, N., Mohammadifard, N., Sarrafzadegan, N., Bahonar, A., Badiei, M., et al. (2013), The relationship between nut consumption and lipid profile among the Iranian adult population, Isfahan Healthy Heart Program, *Eur. J. Clin. Nutr.*, 385–9.
Ayoub, M., de Camargo, A. C. and Shahidi, F. (2016), Antioxidants and bioactivities of free, esterified and insoluble-bound phenolics from berry seed meals, *Food Chem.*, 197, 221–32.
Bacchetta, L., Aramini, M., Zini, A., Di Giammatteo, V., Spera, D., Drogoudi, P., Rovira M., et al. (2013), Fatty acids and alpha-tocopherol composition in hazelnut (*Corylus avellana* L.): A chemometric approach to emphasize the quality of European germplasm, *Euphytica*, 191, 57–73.
Bacchetta, L., Rovira, M., Tronci, C., Aramini, M., Drogoudi, P., Silva, A. P., Solar, A., et al. (2015), A multidisciplinary approach to enhance the conservation and use of hazelnut genetic resources, *Genet. Resour. Crop Evol.*, 62, 649–63.
Barnes, S. (2008), Nutritional genomics, polyphenols, diets, and their impact on dietetics, *J. Am. Diet Assoc.*, 108(11), 1888–95.
Bemani, E., Ghanati, F., Rezaei, A. and Jamshidi, M. (2015), Effect of phenylalanine on Taxol production and antioxidant activity of extracts of suspension-cultured hazel (*Corylus avellana* L.) cells, *J. Nat. Med.*, 67(3), 446–51.
Bolling, B. W., Oliver Chen, C. Y., McKay, D., et al. (2011), Tree nut phytochemicals: Composition, antioxidant capacity, bioactivity, impact factors: A systematic review of almonds, Brazils, cashews, hazelnuts, macadamias, pecans, pine nuts, pistachios and walnuts, *Nutr. Res. Rev.*, 24, 244–75.
Boulange, C. L., Neves, A. L., Chilloux, J., et al. (2016), Impact of the gut microbiota on inflammation, obesity, and metabolic disease, *Genome Med.*, 8, 42.
Brahe, L. K., Astrup, A. and Larsen, L. H. (2016), Can we prevent obesity-related metabolic diseases by dietary modulation of the gut microbiota?, *Adv. Nutr.*, 7, 90–101.
Brown, J. M. and Hazen, S. L. (2015), The gut microbial endocrine organ: Bacterially derived signals driving cardiometabolic diseases, *Annu. Rev. Med.*, 66, 343–59.
Brufau, G., Boatella, J. and Rafecas, M. (2008), Nuts: Source of energy and macronutrients, *Br. J. Nutr.*, 99(2), 447–8.
Brunet, A. and Berger, S. L. (2014), Epigenetics of aging and aging-related disease, *J. Gerontol. A Biol. Sci. Med Sci.*, 69 Suppl 1, S17–20.

Cai, X., Hagedorn, C. H. and Cullen, B. R. (2004), Human microRNAs are processed from capped, polyadenylated transcripts that can also function as mRNAs, *RNA*, 10, 1957–66.

Caimari, A., Puiggròs, F., Suárez, M., Crescenti, A., Laos, S., Ruiz, J. A., et al. (2015), The intake of a hazelnut skin extract improves the plasma lipid profile and reduces the lithocholic/deoxycholic bile acid faecal ratio, a risk factor for colon cancer, in hamsters fed a high-fat diet, *Food Chem.*, 167, 138–44.

Calpe-Berdiel, L., Escola-Gil, J. C. and Blanco-Vaca, F. (2009), New insights into the molecular actions of plant sterols and stanols in cholesterol metabolism, *Atherosclerosis*, 203, 18–31.

Camarena, V. and Wang, G. (2016), The epigenetic role of vitamin C in health and disease, *Cell Mol. Life Sci.*, 73(8), 1645–58. Review.

Cappelli, G., Giovannini, D., Basso, A., Demurtas, O. C., Diretto, G., Santi, C., et al. (2018), A *Corylus avellana* L. extract enhances human macrophage bactericidal response against *Staphylococcus aureus* by increasing the expression of anti-inflammatory and iron metabolism genes, *J. Funct. Foods*, 45, 499–511.

Casas-Agustench, P., López-Uriarte, P., Bulló, M., Ros, E., Cabré-Vila, J. J. and Salas-Salvadó, J. (2011a), Effects of one serving of mixed nuts on serum lipids, insulin resistance and inflammatory markers in patients with the metabolic syndrome, *Nutr. Metab. Cardiovasc. Dis.*, 21, 126–35.

Casas-Agustench, P., Salas-Huetos, A. and Salas-Salvado, J. (2011b), Mediterranean nuts: Origins, ancient medicinal benefits and symbolism, *Public Health Nutr.*, 14, 2296–301.

Choi, D. Y., Lee, Y. J., Hong, J. T. and Lee, H. J. (2012), Antioxidant properties of natural polyphenols and their therapeutic potentials for Alzheimer's disease, *Brain Res. Bull.*, 87(2–3), 144–53.

Cordain, L., Eaton, S. B., Sebastian, A., Mann N. Lindeberg, S., Watkins, B. A., et al. (2005), Origins and evolution of the Western diet: Health implications for the 21st century, *Am. J. Clin. Nutr.*, 81, 341–54.

Crescenti, A., Solà, R., Valls, R. M., Caimari, A., Del Bas, J. M., Anguera, A., et al. (2013), Cocoa consumption alters the global DNA methylation of peripheral leukocytes in humans with cardiovascular disease risk factors: A randomized controlled trial, *PLoS One*, 8(6), e65744.

Crider, K. S., Yang, T. P., Berry, R. J. and Bailey, L. B. (2012), Folate and DNA methylation: A review of molecular mechanisms and the evidence for folate's role, *Adv. Nutr.*, 3, 21–38.

Damavandi, R. D., Eghtesadi, S., Shidfar, F., Heydari, I. and Foroushani A. R. (2013), Effects of hazelnuts consumption on fasting blood sugar and lipoproteins in patients with type 2 diabetes, *J. Res. Med. Sci.*, 18, 314–21.

Dang, T. S., Walker, M., Ford, D. and Valentine, R. A. (2000), Nutrigenomics: The role of nutrients in gene expression, *Periodontol.*, 64(1), 154–60.

Das, U. N., Ramos, E. J. and Meguid, M. M. (2003), Metabolic alterations during inflammation and its modulation by central actions of omega-3 fatty acids, *Curr. Opin. Clin. Nutr. Metab. Care*, 6(4), 413–9.

Declerck, K., Szarc vel Szic, K., Palagani, A., Heyninck, K., Haegeman, G., Morand, C., et al. (2016), Epigenetic control of cardiovascular health by nutritional polyphenols involves multiple chromatin-modifying writer-reader-eraser proteins, *Curr. Top. Med. Chem.*, 16(7), 788–806.

Deon, V., Del Bo', C., Guaraldi, F., Abello, F., Belviso, S., Porrini, M., et al. (2018), Effect of hazelnut on serum lipid profile and fatty acid composition of erythrocyte phospholipids in children and adolescents with primary hyperlipidemia: A randomized controlled trial, *Clin. Nutr.*, 37(4), 1193–201.

Devi, A., Chisholm, A., Gray, A., Tey, S. L., Williamson-Poutama, D., Cameron, S. L., et al. (2016), Nut-enriched bread is an effective and acceptable vehicle to improve regular nut consumption, *Eur. J. Nutr.*, 55(7), 2281–93.

Di Renzo, L., Carraro, A., Minella, D., Botta, R., Contessa, C., Sartor, C., et al. (2014), Nutrient analysis critical control point (NACCP): Hazelnut as a prototype of nutrigenomic study, *Food Nutr. Sci.*, 5, 79–88.

Di Renzo, L., Merra, G., Botta, R., Gualtieri, P., Manzo, A., Perrone, M. A., et al. (2017), Post-prandial effects of hazelnut-enriched high fat meal on LDL oxidative status, oxidative and inflammatory gene expression of healthy subjects: A randomized trial, *Eur. Rev. Med. Pharmacol. Sci.*, 21(7), 1610–26.

Dong, H., Lei, J., Ding, L., Wen, Y., Ju, H. and Zhang, X. (2013), MicroRNA: Function, detection, and bioanalysis, *Chem Rev.*, 113(8), 6207–33.

Drevon, C. A., Nenseter, M. S., Brude, I. R., et al. (1995), Omega-3 fatty acids – nutritional aspects, *Can. J. Cardiol.*, 11 Suppl G, 47G-54G.

Durak, I., Köksal, I., Kaçmaz, M., Büyükkoçak, S., Cimen, B. M. and Oztürk, H. S. (1999), Hazelnut supplementation enhances plasma antioxidant potential and lowers plasma cholesterol levels, *Clin. Chim. Acta*, 284, 113–15.

Erdogan, V. and Aygun, A. (2005), Fatty acid composition and physical properties of Turkish tree Hazelnuts, *Chem. Nat. Comp.*, 41(4), 378–81.

Escurriol, V., Cofan, M., Serra, M., Bullo, M., Basora, J., Salas-Salvado, J., et al. (2009), Serum sterol responses to increasing plant sterol intake from natural foods in the Mediterranean diet, *Eur. J. Nutr.*, 48, 373–82.

Esposito, T., Sansone, F., Franceschelli, S., Del Gaudio, P., Picerno, P., Aquino, R. P., et al. (2017), Hazelnut (*Corylus avellana* L.) shells extract: Phenolic composition, antioxidant effect and cytotoxic activity on human cancer cell lines, *Int. J. Mol. Sci.*, 18(2), 392.

Fialho, E., Moreno, F. S. and Ong, T. P. P. (2008), Nutrition in the post-genomics: Fundamentals and applications of omics tools, *J. Nutr.*, 21(6), 757–66.

Gargari, G., Deon, V., Taverniti, V., Gardana, C., Denina, M., Riso, P., et al. (2018), Evidence of dysbiosis in the intestinal microbial ecosystem of children and adolescents with primary hyperlipidemia and the potential role of regular hazelnut intake, *FEMS Microbiol. Ecol.*, 94(5).

Gerhauser, C. (2013), Cancer chemoprevention and nutriepigenetics: State of the art and future challenges, *Top. Curr. Chem.*, 329, 73–132.

Ghazalpour, A., Cespedes, I., Bennett, B. J., et al. (2016), Expanding role of gut microbiota in lipid metabolism, *Curr. Opin. Lipidol.*, 27, 141–7.

Giglio, R. V., Patti, A. M., Cicero, A. F. G., Lippi, G., Rizzo, M., Toth, P. P., et al. (2018), Polyphenols: Potential use in the prevention and treatment of cardiovascular diseases, *Curr. Pharm. Des.*, 24(2), 239–58.

Goldstein, J. L. and Brown, M. S. (2009), The LDL receptor, *Arter. Thromb. Vasc. Biol.*, 29(4), 431–8.

Gorski, S. A., Vogel, J. and Doudna, J. A. (2017), RNA-based recognition and targeting: Sowing the seeds of specificity, *Nat. Rev. Mol. Cell Biol.*, 18(4), 215–28.

Guaraldi, F., Deon, V., Del Bo', C., Vendrame, S., Porrini, M., Riso, P., et al. (2018), Effect of short term hazelnut consumption on DNA damage and oxidized-LDL in children and adolescents with primary hyperlipidemia: A randomised controlled trial, *J. Nutr. Biochem.*, 57, 206–21.

Hartmann, M. A. (1998), Plant sterols and the membrane environment, *Trends Plant. Sci.*, 3, 170–5.

Herceg, Z. (2016), Epigenetic mechanisms as an interface between the environment and genome, *Adv. Exp. Med. Biol.*, 903, 3–15.

Holvoet, P., Mertens, A., Verhamme, P., Bogaerts, K., Beyens, G., Verhaeghe, R., et al. (2001), Circulating oxidized LDL is a useful marker for identifying patients with coronary artery disease, *Arter. Thromb. Vasc. Biol.*, 21, 844–8.

Huynh, N. N. and Chin-Dusting, J. (2006), Amino acids, arginase and nitric oxide in vascular health, *Clin. Exp. Pharmacol. Physiol.*, 33, 1–8.

Iorio, M. V., Piovan, C. and Croce, C. M. (2010), Interplay between microRNAs and the epigenetic machinery: An intricate network, *Biochim. Biophys. Acta.*, 1799(10–12), 694–701.

Jenkins, D. J., Wolever, T. M., Spiller, G., Buckley, G., Lam, Y., Jenkins, A. L., et al. (1989), Hypocholesterolemic effect of vegetable protein in a hypocaloric diet, *Atherosclerosis*, 78, 99–107.

Jenuwein, T. and Allis, C. D. (2001), Translating the histone code, *Science*, 293, 1074–80.

Kalea, A. Z., Drosatos, K. and Buxton, J. L. (2018), Nutriepigenetics and cardiovascular disease, *Curr. Opin. Clin. Nutr. Metab. Care*, 21(4), 252–9.

Kanwal, R., Gupta, K. and Gupta, S. (2015), Cancer epigenetics: An introduction, *Methods Mol. Biol.*, 1238, 3–25.

Keaney, J. F. (2000), Atherosclerosis: From lesion formation to plaque activation and endothelial dysfunction, *Mol. Asp. Med.*, 21, 99–166.

Kim, Y., Keogh, J. B. and Clifton, P. M. (2017), Benefits of nut consumption on insulin resistance and cardiovascular risk factors: Multiple potential mechanisms of actions, *Nutrients*, 9(11), E1271.

Kingston, D. G. and Newman, D. J. (2007), Taxoids: Cancer-fighting compounds from nature, *Curr. Opin. Drug Discov. Devel.*, 10, 130–44.

Klockmann, S., Reiner, E., Cain, N. and Fischer, M. (2017), Geographical origin determination of hazelnuts (*Corylus avellana*) by LC-QqQ-MS/MS-based targeted metabolomics application, *J. Agric. Food Chem.*, 65(7), 1456–65.

Kodad, O. G., Estopan, T., Molino, J. F., Mamouni, A., Messaoudi, Z., Lahlou, M., et al. (2011), Plasticity and stability in the major fatty acid content of almond kernels grown under two Mediterranean climates, *J. Hort. Sci. Biotechnol.* 85, 381–6.

Korkina, G. L. (2006), Phenilpropanoids as naturally occurring antioxidants: From plant defence to human health, *Cell Mol. Biol. (Noisy-le-grand)*, 53(1), 15–25.

Kornsteiner, M., Wagner, K. H. and Elmadfa, I. (2006), Tocopherols and total phenolics in 10 different nuts types, *Food Chem.*, 98, 381–7.

Kouzarides, T. (2007), Chromatin modifications and their function, *Cell*, 128(4), 693–705. Review.

Li, S. C., Liu, Y. H., Liu, J. F., Chang, W. H., Chen, C. M. and Chen, C. Y. (2011), Almond consumption improved glycemic control and lipid profiles in patients with type 2 diabetes mellitus, *Metabolism*, 60, 474–9.

Lima, R. P. A., do Nascimento, R. A. F., Luna, R. C. P., Persuhn, D. C., da Silva, A. S., da Conceição Rodrigues Gonçalves, M., et al. (2017), Effect of a diet containing folate and hazelnut oil capsule on the methylation level of theADRB3gene, lipid profile and oxidative stress in overweight or obese women, *Clin. Epigen.*, 9, 110.

Loffredo, L., Perri, L., di Castelnuovo, A., Iacoviello, L., de Gaetano, G. and Violi, F. (2015), Supplementation with vitamin E alone is associated with reduced myocardial infarction: A meta-analysis, *Nutr. Metab. Cardiovasc. Dis.*, 25, 354–63.

López-Uriarte, P., Nogués, R., Saez, G., Bulló, M., Romeu, M., Masana, L., et al. (2010), Effect of nut consumption on oxidative stress and the endothelial function in metabolic syndrome, *Clin. Nutr.*, 29, 373–80.

Lotti, G., Izzo, R. and Riu, P. (1985), Effects of climate on acid and sterol composition of olive oil, *Riv. Soc. Ital. Sci.*, 11, 126.

Lumeng, C. N., Bodzin, J. L. and Saltiel, A. R. (2007), Obesity induces a phenotypic switch in adipose tissue macrophage polarization, *J. Clin. Invest.*, 117, 175–84.

Maiolino, G., Rossitto, G., Caielli, P., Bisogni, V., Rossi, G. P. and Calò, L. A. (2013), The role of oxidized low-density lipoproteins in atherosclerosis: The myths and the facts, *Mediators Inflamm.*, 2013, 714653.

Marinõ-Ramírez, L., Kann, M. G., Shoemaker, B. A. and Landsman, D. (2005), Histone structure and nucleosome stability, *Exp. Rev. Prot.*, 2, 719–29.

Masliah, E., Dumaop, W., Galasko, D. and Desplats, P., (2013), Distinctive patterns of DNA methylation associated with Parkinson disease: Identification of concordant epigenetic changes in brain and peripheral blood leukocytes, *Epigenetics*, 8(10), 1030–8.

McManus, K., Antinoro, L. and Sacks, F. (2001), A randomized controlled trial of a moderate-fat, low-energy diet compared with a low fat, low-energy diet for weight loss in overweight adults, *Int. J. Obes. Relat. Metab. Disord.*, 25, 1503–11.

Mentch, S. J. and Locasale, J. W. (2016), One-carbon metabolism and epigenetics: Understanding the specificity, *Ann. N. Y. Acad. Sci.*, 1363, 91–8.

Mercanligil, S. M., Arslan, P., Alasalvar, C., Okut, E., Akgül, E., Pinar, A., et al. (2007), Effects of hazelnut-enriched diet on plasma cholesterol and lipoprotein profiles in hypercholesterolemic adult men, *Eur. J. Clin. Nutr.*, 61(2), 212–20.

Mirmiran, P., Bahadoran, Z., Ghasemi, A. and Azizi, F. (2017), Contribution of dietary amino acids composition to incidence of cardiovascular outcomes: A prospective population-based study, *Nutr. Metab. Cardiovasc. Dis.*, 27(7), 633–41.

Mohr, A. M. and Mott, J. L. (2015), Overview of microRNA biology, *Semin. Liver Dis.*, 35(1), 3–11.

Noé, G., Bellesoeur, A., Thomas-Schoemann, A., Rangarajan, S., Naji, F., Puszkiel, A., et al. (2016), Clinical and kinomic analysis identifies peripheral blood mononuclear cells as a potential pharmacodynamic biomarker in metastatic renal cell carcinoma patients treated with sunitinib, *Oncotarget*, 7(41), 67507–20.

Orem, A., Yucesan, F. B., Orem, C., Akcan, B., Kural, B. V., Alasalvar, C. and Shahidi, F. (2013), Hazelnut-enriched diet improves cardiovascular risk biomarkers beyond a lipid-lowering effect in hypercholesterolemic subjects, *J. Clin. Lipid.*, 7(2), 123–31.

Osella, M., Riba, A., Testori, A., Corà, D. and Caselle M. (2014), Interplay of microRNA and epigenetic regulation in the human regulatory network, *Front. Genet.*, 6(5), 345.

Özdemir, F. and Akinci, I. (2004), Physical and nutritional properties of four major commercial Turkish hazelnut varieties, *J. Food Eng.*, 63, 341–7.

Özdemir, M., Filiz Açkurt, F., Kaplan, M., Yíldíz, M., Löker, M., Gürcan, T., et al. (2001), Evaluation of new Turkish hybrid hazelnut (*Corylus avellana* L.) varieties: Fatty acid

composition, α-tocopherol content, mineral composition and stability, *Food Chem.*, 73(4), 411-15.
Pan, M. H., Lai, C. S., Wu, J. C. and Ho, C. T. (2013), Epigenetic and disease targets by polyphenols, *Curr. Pharm. Des.*, 19(34), 6156-85.
Parcerisa, J., Boatella, J., Codony, R., Farran, A., Garcia, J. and Lopez, A. (1993), Influence of variety and geographical origin on the lipid fraction of hazelnuts (*Corylus avellana* L.) from Spain: I. Fatty acid composition, *Food Chem.*, 43, 411-14.
Parcerisa, J. M., Rafecas, A. I., Castellote, R., Codony, A., Ferràn, J,. Garcia, C., et al. (1995), Influence of variety and geographical origin on the lipid fraction of hazelnuts (*Corylus avellana* L.) from Spain: III Oil stability, tocopherol content and some mineral contents (Mn, Fe, Cu), *Food Chem.*, 53, 71-7.
Park, Y. and Carr, T. P. (2013), Unsaturated fatty acids and phytosterols regulate cholesterol transporter genes in Caco-2 and HepG 2 cell lines, *Nutr. Res.*, 33, 154-61.
Pelvan, E., Alasalvar, C. and Uzman, S. (2012), Effects of roasting on the antioxidant status and phenolic profiles of commercial Turkish hazelnut varieties (*Corylus avellana* L.), *J. Agric. Food Chem.*, 60, 1218-23.
Perna, S., Giacosa, A., Bonitta, G., Bologna, C., Isu, A., Guido, D., et al. (2016), Effects of hazelnut consumption on blood lipids and body weight: A systematic review and Bayesian meta-analysis, *Nutrients*, 8(12), 747.
Pober, J. S., Min, W. and Bradley, J. R. (2009), Mechanisms of endothelial dysfunction, injury and death, *Annu. Rev. Pathol. Mech. Dis.*, 4, 71-95.
Racette, S. B., Lin, X., Lefevre, M., Spearie, C. A., Most, M. M., Ma, L. and Ostlund, R. E., (2010), Dose effects of dietary phytosterols on cholesterol metabolism: A controlled feeding study, *Am. J. Clin. Nutr.*, 91, 32-8.
Rafieian-Kopaei, M., Setorki, M., Doudi, M., Baradaran, A. and Nasri H. (2014), Atherosclerosis: Process, indicators, risk factors and new hopes, *Int. J Prev. Med.*, 5(8), 927-46.
Razquin, C., Sanchez-Tainta, A., Salas-Salvado, J., Buil-Cosiales, P., Corella, D., Fito, M., Ros, E., Estruch, R., Aros, F., Gomez-Gracia, E., et al. (2017), Dietary energy density and body weight changes after 3 years in the PREDIMED study, *Int. J. Food Sci. Nutr.*, 68, 865-72.
Remely, M., Lovrecic, L., de la Garza, A. L., Migliore, L., Peterlin, B., Milagro, F. I., et al. (2015a), Therapeutic perspectives of epigenetically active nutrients, *Br. J. Pharmacol.*, 172(11), 2756-68.
Remely, M., Stefanska, B., Lovrecic, L., Magnet, U. and Haslberger, A. G. (2015b), Nutriepigenomics: The role of nutrition in epigenetic control of human diseases, *Curr. Opin. Clin. Nutr. Metab. Care*, 18(4), 328-33.
Ren, D., Wang, M., Shen, M., Liu, C., Liu, W., Min, W. and Liu, J. (2016), *In vivo* assessment of immunomodulatory activity of hydrolysed peptides from Corylus heterophylla, *Fisch. J. Sci. Food Agric.*, 96(10), 3508-14.
Richardson, D. G. (1997), The health benefits of eating hazelnuts: Implications for blood lipid profiles, coronary heart disease, and cancer risks, *Acta Hortic.*, 415, 295-7.
Rodríguez-Cruz, M. and Serna, D. S. (2017), Nutrigenomics of ω-3 fatty acids: Regulators of the master transcription factors, *Nutrition*, 41, 90-6.
Ros, E. (2009), Nuts and novel biomarkers of cardiovascular disease, *Am. J Clin. Nutr.* 89, 1649S-56S.
Rustan, A. C. and Devron, C. A. (2005), *Fatty Acids: Structures and Properties, Encyclopedia of Life Sciences & 2005*, John Wiley & Sons, Ltd. www.els.net

Ruthenburg, A. J., Allis, C. D. and Wysocka, J. (2007), Methylation of lysine 4 on histone H3: Intricacy of writing and reading a single epigenetic mark, *Mol. Cell*, 25(1), 15–30.

Sabaté, J. and Wien, M. (2010), Nuts, blood lipids and cardiovascular disease, *Asia Pac. J. Clin. Nutr.*, 19(1), 131–6.

Salas-Salvadó, J., Bulló, M., Pérez-Heras, A. and Ros, E. (2006), Dietary fibre, nuts and cardiovascular disease, *Br. J. Nutr.*, 96, S45–51.

Sales, N. M., Pelegrini, P. B. and Goersch, M. C. (2014), Nutrigenomics: Definitions and advances of this new science, *J. Nutr. Metab.*, 2014, 202759.

San-Cristobal, R., Navas-Carretero, S., Milagro, F. I., Riezu-Boj, J. I., Guruceaga, E., Celis-Morales, C., et al. (2016), Gene methylation parallelisms between peripheral blood cells and oral mucosa samples in relation to overweight, *J. Physiol. Biochem.*, 73(3), 465–74.

Sanguigni, V., Manco, M., Sorge, R., Gnessi, L. and Francomano, D. (2017), Natural antioxidant ice cream acutely reduces oxidative stress and improves vascular function and physical performance in healthy individuals, *Nutrition*, 33, 225–33.

Santi, C., Giorni, A., Terenzi, C. T., Altavista, P. and Bacchetta, L. (2017), Daily hazelnut intake exerts multiple reversible effects on plasma profile of healthy subjects, *Food Nutr. Sci.*, 8, 633–46.

Satija, A. and Hu, F. B. (2012), Cardiovascular benefits of dietary fiber, *Curr. Atheroscler. Rep.*, 14(6), 505–14.

Schmitzer, V., Slatnar, A., Veberic, R., et al. (2011), Roasting affects phenolic composition and antioxidative activity of hazelnuts (*Corylus avellana* L.), *J. Food Sci.*, 76, S14–19.

Schwartz, B. (2014), New criteria for supplementation of selected micronutrients in the era of nutrigenetics and nutrigenomics, *Int. J. Food Sci. Nutr.*, 65(5), 529–38.

Segura, R., Javierre, C., Lizarraga, M. A. and Ros, E (2006), Other relevant components of nuts: Phytosterols, folate and minerals, *Br. J. Nutr.*, 96(Suppl 2), S36–44.

Serra Bonvehí, J. and Ventura-Coll, F. (1993), Oil content, stability and fatty acid composition of the main varieties of Catalonian hazelnuts (*Corylus avellana* L.), *Food Chem.*, 48, 237–41.

Shahidi, F. and Costa de Camargo, A. (2016), Tocopherols and tocotrienols in common and emerging dietary sources: Occurrence, applications, and health benefits, *Int. J. Mol. Sci.*, 17, 1745.

Silva, A. P., Gonçalves, B., Santos, F., Rosa, E. and Santos, A. (2005), Effect of cultivar and year on the quality of hazelnut fruits (*Corylus avellana* l.), *Acta Hort. (ISHS)*, 686, 469-76.

Sivakumar, G. and Bacchetta, L. (2006), Alpha-Tocopherol from Italian hazelnut germoplasm, *Chem. Nat. Compd.*, 42(1), 266–8.

Sivakumar, G., Bacchetta, L., Gatti, R. and Zappa, G., (2005), HPLC screening of natural vitamin E from Mediterranean plant biofactories a basic tool for pilot-scale bioreactors production of a-tocopherol, *J. Plant Physiol.*, 162, 1280-3.

Sonlimar, M. and Sarmalina, S. (2010), Effect of grape (*Vitis vinifera* L.) seed on reducing serum uric acid level in gout-animals model, *Majalah Kesehatan Pharmamedika*, 2, 106–9.

Su, Z., Xia, J. and Zhao, Z. (2011), Functional complementation between transcriptional methylation regulation and posttranscriptional microRNA regulation in the human genome, *BMC Genom.*, 12(Suppl 5), S15.

Sun, C., Burgner, D. P., Ponsonby, A. L., Saffery, R., Huang, R. C., Vuillermin, P. J., et al. (2013), Effects of early-life environment and epigenetics on cardiovascular disease risk in children: Highlighting the role of twin studies, *Pediatr. Res.*, 73(4 Pt 2), 523–30.

Tammen, S. A., Friso, S. and Choi, S. W. (2013), Epigenetics: The link between nature and nurture, *Mol. Aspects Med.*, 34, 753–64.

Tapsell, L. C., Batterham, M. J., Teuss, G., Tan, S. Y., Dalton, S., Quick, C. J., et al. (2009), Long-term effects of increased dietary polyunsaturated fat from walnuts on metabolic parameters in type II diabetes, *Eur. J. Clin. Nutr.*, 63, 1008–15.

Tenore, G. C., Daglia, M., Ciampaglia, R. and Novellino, E. (2015), Exploring the nutraceutical potential of polyphenols from black, green and white tea infusions – an overview, *Curr. Pharm. Biotechnol.*, 16(3), 265–71.

Tey, S. L., Brown, R., Gray, R., Chisholm, A. and Delahunty, C. (2011a), Nuts improve diet quality compared to other energy-dense snacks while maintaining bodyweight, *J. Nutr. Metab.*, 2011, 357350.

Tey, S. L., Brown, R. C., Chisholm, A. W., Delahunty, C. M., Gray, A. R. and Williams, S. M. (2011b), Effects of different forms of hazelnuts on blood lipids and α-tocopherol concentrations in mildly hypercholesterolemic individuals, *Eur. J. Clin. Nutr.*, 65(1), 117–24.

Tey, S. L., Robinson, T., Davis, H., Gray, A. R., Chisholm, A. W. and Brown, R. C. (2015), The effect of regular hazelnut consumption on cardiovascular risk factors and acceptance in Maori and European, *J. Nutr.*, 2, 39–53.

Tey, S. L., Robinson, T., Gray, A. R., Chisholm, A. W., Brown, R. C. (2017), Do dry roasting, lightly salting nuts affect their cardioprotective properties and acceptability?, *Eur J. Nutr.*, 56(3), 1025–36.

Tzounis, X., Rodriguez-Mateos, A., Vulevic, J., Gibson, G. R., Kwik-Uribe, C. and Spencer, J. P. E. (2011), Prebiotic evaluation of cocoa-derived flavanols in healthy humans by using a randomized, controlled, double-blind, crossover intervention study, *Am. J. Clin. Nutr.*, 93, 62.

Valdes, L., Cuervo, A., Salazar, N., Ruas-Madiedo, P., Gueimonde, M. and Gonzalez, S. (2015), The relationship between phenolic compounds from diet and microbiota: Impact on human health, *Food Funct.*, 6, 2424–39.

van Soom, A., Peelman, L., Holt, W. V. and Fazeli, A. (2014), An introduction to epigenetics as the link between genotype and environment: A personal view, *Reprod. Domest. Anim.*, 49 Suppl 3, 2–10.

Viguiliouk, E., Kendall, C. W. C., Blanco Mejia, S., Cozma, A. I., Ha, V., Mirrahimi, A., et al. (2014), Effect of tree nuts on glycemic control in diabetes: A systematic review and meta-analysis of randomized controlled dietary trials, *PLoS One*, 9(7), e103376.

Virmani, A., Pinto, L., Binienda, Z. and Ali, S. (2013), Food, nutrigenomics, and neurodegeneration–neuroprotection by what you eat!, *Mol. Neurobiol.*, 48(2), 353–62.

Welch, G. N. and Loscalzo, J. (1998), Homocysteine and atherothrombosis, *N. Engl. J. Med.*, 338, 1042–50.

Yamakoshi, J., Tokutake, S., Kikuchi, M., Kubota, Y., Konishi, H. and Mitsuoka, T. (2001), Effect of proanthocyanidin-rich extract from grape seeds on human fecal flora and fecal odor, *Microb. Ecol. Health Dis.*, 13, 25–31.

Yücesan, F. B., Orem, A., Kural, B. V., Orem, C. and Turan, I. (2010), Hazelnut consumption decreases the susceptibility of LDL to oxidation, plasma oxidized LDL level and

increases the ratio of large/small LDL in normolipidemic healthy subjects, *Anatol. J. Cardiol.*, 10(1), 28–35.
Zamora-Ros, R., Forouhi, N. G., Sharp, S. J., Gonzalez, C. A., Buijsse, B., Guevara, M., van der Schouw, Y. T., Amiano, P., Boeing, H., Bredsdorff, L., et al. (2013), The association between dietary flavonoid and lignan intakes and incident type 2 diabetes in European populations: The epic-interact study, *Diabetes Care*, 36, 3961–70.

# Chapter 4

## Uses and health benefits of chestnuts

*Eduardo Augusto dos Santos Rosa, Maria Cristina Seixas Martins Morais, Ivo Vaz Oliveira, Berta Maria de Carvalho Gonçalves Macedo and Ana Paula Calvão Moreira da Silva, Centre for the Research and Technology of Agro-Environmental and Biological Sciences (CITAB), University of Trás-os-Montes e Alto Douro (UTAD), Portugal*



## 1 Introduction

According to the WHO (http://www.who.int/dietphysicalactivity/fruit/en/), insufficient intake of fruit and vegetables on a global scale is estimated to cause approximately 14% of gastrointestinal cancer deaths, 11% of ischaemic heart disease deaths and 9% of stroke deaths; worldwide, 3.4 million deaths can be attributed to the low consumption of fruit and 1.8 million to diets low in vegetables (FAO, 2017). This information, combined with the increased literacy of consumers regarding the relationship between food and health, has gradually made consumers more aware of the relevance of diets on health and their respective impact on the environment and sustainability of the planet. Following extensive research supporting the benefits of the daily consumption of fruits and vegetables based on their supply of a range of nutritional compounds (vitamins, minerals, dietary fibre and later extended to several secondary plant metabolites), during the 1990s several countries initiated campaigns and recommendations to promote fruit and vegetable consumption. This was reinforced in 2003 by a joint action by the FAO and WHO

http://dx.doi.org/10.19103/AS.2018.0042.05

that aimed to raise awareness and to boost fruit and vegetable production and consumption (400 g or five 80 g portions a day) to improve people's health and farmers' income. However, the inclusion of nuts in these references was minimal, relegating these fruits to a minor level. In the following sections it will be shown that in recent years there have been a number of research findings involving chestnuts, which support their importance and complementary role in our diet. There is readily available information focused on basic nutritional data along with more recent findings, which reveals the presence of bioactive compounds and their enzymatic metabolites, emphasizing the relevance of including chestnuts in the consumption pool of fruits and vegetables, as well as their respective health benefits. The following sections will also show that the higher consumption of chestnuts is a sustainable practice as this crop can be part of a more sustainable agricultural production system. Indeed, chestnut production, apart from being well adapted to soils of lower quality, requires less general resource input, such as water, labour and pest and disease management, when compared with other fruit crops; overall, it is better adapted to climate changes (Rackham, 1994; pers. comm.).

## 2 Recent research achievements

The production and consumption of chestnuts has steadily increased over the last few years, driven by an increased consumer awareness regarding food composition and the health benefits of a nut-rich diet. This has been reflected in the production area of *Castanea sativa* doubling over the last 20 years (1996–2016, the latest FAO data), from 299 911 to 602 718 ha, while the total recorded production showed a greater than threefold increase, from 678 446 t in 1996 to 2 261 589 t in 2016 (FAOstat, 2018) (Fig. 1). Similarly, from 2004 to 2012, there

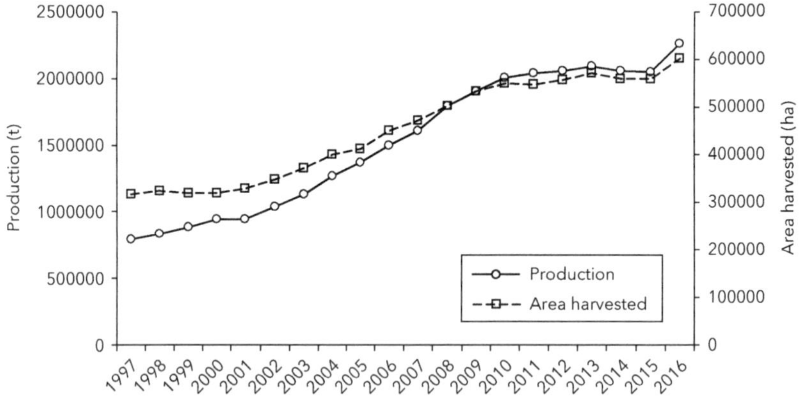


**Figure 1** Evolution of production and harvested area of *C. sativa* from 1997 to 2016 in the world. Source: adapted from FAOstat (2018).

was an average increase in the global consumption of chestnuts of around 59% (Table 1) (Bento et al., 2016). Some *Castanea* species can strictly be used for their nuts; however, there is a wide range of species with other multipurpose uses that are worth noting (Table 2).

Research involving chestnuts is quite limited, but over the last two decades it has received increased attention based on the development of new equipment and consequent methodologies being focused on *C. sativa*. For *C. mollissima* and *C. crenata*, information is still very scarce despite the recent increase in the number of studies; for these two species and for *C. dentata* further studies are necessary given their potential crop expansion, consumption and uses.

Recent work involving *C. sativa* has focused on composition. Data indicate a considerable variation of moisture (41.9-64.4%), with the major component of dry nuts being carbohydrates, ranging from 42% up to 94.2% (Table 3) and starch being the most abundant carbohydrate. Other compounds, such as fat (0.7-5.37%) and protein (2.2-12.4%) also present a significant variation of their content. These variations are linked to cumulative differences between cultivars, climate and growing conditions and harvest timing.

Total phenolics are the major group of secondary plant metabolites in chestnuts. Over the last decade, the quantification of total phenolics and flavonoids in *C. sativa* has been evaluated in tree wood and bark, but also in leaves, flowers and pollen. The nut has been the focus of many studies, either as a whole or in separate fractions, such as the inner or outer skin.

Phenolics are present in a higher amount in chestnut wood (40.8 g GAE/g d.w.), bark (23.9-56.1 g GAE/g d.w.), leaves (413 mg GAE/g d.w.) and nut seed coat or testa (710 mg GAE/g d.w.), with catkins also containing a considerable amount of these compounds (251.6-327.2 mg/g) (Table 4).

**Table 1** Chestnut consumption in 2004 and 2012 (t) and variation (%) between 2004 and 2012

| Countries | 2004 (t) | 2012 (t) | Variation (%) |
|---|---|---|---|
| China | 904962 | 1629788 | +80.1 |
| South Korea | 57251 | 61317 | +7.1 |
| Bolivia | 52758 | 57000 | +8.0 |
| Turkey | 41798 | 52603 | +25.9 |
| Italy | 24071 | 55995 | +132.6 |
| Japan | 48625 | 31159 | −35.9 |
| Greece | 19367 | 27737 | +43.2 |
| Portugal | 26082 | 8408 | −67.8 |
| France | 24763 | 9300 | −62.4 |
| Spain | 3282 | 8436 | +157.0 |
| *World* | *1256762* | *2000061* | *+59.1* |

*Source*: adapted from Bento et al. (2016).

**Table 2** Origin of *Castanea* species and major uses

| Origin | Species | Common name | Orchards | Predominant use |
|---|---|---|---|---|
| Europe | *C. sativa* Mill. | European or sweet chestnut | Europe, Asia Minor, North Africa | Nut, timber |
| Asia | *C. crenata* Sieb. et Zucc | Japanese chestnut | Japan, Korea | Nut |
| | *C. mollissima* Blume | Chinese chestnut | China | Nut |
| | *C. seguinii* Dode | - | China | Firewood |
| | *C. davidii* Dode | - | China | Firewood |
| | *C. henryi* (Skan) Rehd. et E. H. Wils | Willow leaf or pearl chestnut | China | Timber |
| America | *C. dentata* (Marsh.) Borkh | American chestnut | North America | Timber |
| | *C. pumila* Mill. var. *pumila* | Allegheny chinkapin | Southeast USA | Nut |
| | *C. pumila* Mill. var. *ozarkensis* (Ashe) Tucker | Ozark chinkapin | USA | Timber |
| | *C. floridana* (Sarg.) Ashe | Florida chinkapin | Southeast USA | Ornamental |
| | *C. ashei* (Sudw.) Ashe | Ashe chinkapin | Southeast USA | Ornamental |
| | *C. alnifolia* Nutt | Creeping chinkapin | Southern USA | - |
| | *C. paucispina* Ashe | - | Southern USA | - |

*Source*: adapted from Mellano et al. (2012).

The presence of amino acids and vitamins has also been studied in *C. sativa* nuts (Table 5). Recent work has shown the variation in the amino acid content (from 2.3 mg/100 g to 8.7 g/100 g d.w.), while considerable variation of vitamins C and E has also been found, which depends on several factors, particularly cultivars and growth conditions. The quantitative and qualitative mineral profiles of chestnuts are affected by the same factors, which has received further clarification over the last few years (Table 6), with K identified as the major mineral. The mineral composition of leaves and catkins is known, with Ca being the major mineral present in both plant organs, which also contains interesting amounts of other important minerals (Portela and Louzada, 2012). Another topic of study has been the effect of cooking on several quality parameters of chestnuts, namely the chemical composition and antioxidant activity, with several, not always comparable, effects (see details in Section 5). Considering the number of compounds with possible positive biological activities, many authors have focused their research on evaluating the bioactive properties of this nut and related plant tissues and organs. Over the last two decades, several studies have shown the possible positive effects of chestnuts, including antioxidant activities and antimicrobial properties, with promising results as neuroprotective or anticancer agents. Considering only *C. sativa*

**Table 3** Proximate analyses previously reported for chestnut fruits

| Species | Moisture (g/100 g f.w.) | Total fat (g/100 g d.w.) | Crude protein (g/100 g d.w.) | Total ash (g/100 g d.w.) | Total starch (g/100 g d.w.) | Carbohydrates (g/100 g d.w.) | Crude energy (Kcal/100 g d.w.) | Reference |
|---|---|---|---|---|---|---|---|---|
| *C. sativa* | 52-52.6 | 1.81-2.07 | 5.10-5.35 | 2.33-2.40 | 52.6-53.6 | - | - | Mota et al. (2018) |
| | 10.1 | 4.81 | 4.73 | 2.30 | - | 79.4 | - | Wani et al. (2017) |
| | 46.5-59.5 | 0.87-2.61 | 6.15-12.4 | 2.09-4.39 | 40.99-53.16 | 58.1-69.8 | - | Mert and Ertürk (2017) |
| | 56.7-62.64 | 2.06-2.58 | 3.52-6.39 | 2.43-2.68 | - | 88.9-91.5 | - | Poljak et al. (2016) |
| | 50.8-58.4 | 2.7-4.6 | 5.5-7.2 | 2.2-2.9 | - | 78.4-93.3 | - | Özel (2015) |
| | 49 | 2.3 | 5 | 2.3 | - | 92 | 410 | Barreira et al. (2012a) |
| | 53-54 | 0.7-0.8 | 2.4-3 | 0.8-1.1 | - | 42-43 | 184-189 | Barreira et al. (2012b) |
| | 52.8 | 2.95 | 5.0 | 2.23 | - | 60.3 | - | Dinis et al. (2012a,b) |
| | 49 | 2.0 | 4.0 | 2.0 | - | 92 | 402 | Fernandes et al. (2011) |
| | 50-60 | 1.88-4.71 | 4.51-9.89 | 1.69-2.56 | - | | | Gonçalves et al. (2010) |
| | 41.5 | | 3.68 | - | 54.98 | 79.7 | | Glushkova et al. (2010) |
| | 51.2-54.7 | 0.90-2.47 | 5.35-8.17 | - | - | 81.1-94.2 | - | Öztürk et al. (2010) |
| | 51.9-54.6 | 0.78-0.84 | 2.29-3.13 | 0.68-0.90 | - | 41.6-44.1 | 182.6-193.2 | Barreira et al. (2009) |
| | 41.9 | 2.1 | 2.2 | - | - | 52.4 | 237 | Bellini et al. (2008) |
| | 46-53 | 1.73-3.10 | 4.87-7.37 | 1.53-2.2 | 38.6-47.9 | - | - | Borges et al. (2008) |
| | 45.78-58.5 | 1.89-3.68 | 4.20-8.12 | - | 45.37-66.18 | 45.4-66.18 | - | Cristofori et al. (2007) |
| | 46.13-51.27 | 1.56-1.72 | 3.89-5.13 | 1.87-2.34 | 64.15-64.86 | - | - | Vasconcelos et al. (2007) |
| | 54 | 3 | 5.8 | 2.3 | - | - | - | Pereira-Lorenzo et al. (2006) |
| | 48.4-59.3 | 1.3-3 | 6.02-8.58 | 1.8-3 | 56.7-81.7 | - | - | Míguelez et al. (2004) |

(*Continued*)

**Table 3** (*Continued*)

| Species | Moisture (g/100 g f.w.) | Total fat (g/100 g d.w.) | Crude protein (g/100 g d.w.) | Total ash (g/100 g d.w.) | Total starch (g/100 g d.w.) | Carbohydrates (g/100 g d.w.) | Crude energy (Kcal/100 g d.w.) | Reference |
|---|---|---|---|---|---|---|---|---|
| | 50.7–53.6 | 3.69–5.37 | 3.98–5.48 | 2.30–2.51 | 42.2–66.5 | - | - | Sacchetti and Pinnavaia (2004) |
| | 52.7–64.4 | 0.9–1.7 | 5.4–12.0 | 2.2–2.9 | 60.7 | - | - | Sequeira et al. (2004) |
| *C. mollissima* | 46.43–49.75 | - | 7.54–9.74 | 1.44–1.81 | - | - | - | Yang et al. (2018) |
| | - | 1.87–2.98 | 4.42–8.27 | - | 48.20–71.08 | - | 370–405 | Hou et al. (2016) |
| | - | 2.27 | 8.27 | - | 71.08 | - | - | Li et al. (2016) |
| | 34.2–63.6 | 4.3–10.2 | 6.1–12.2 | 0.8–2.2 | 45.7–62.8 | 51.2–81.6 | - | Yang et al. (2015) |
| | 31.9 | - | 3.88 | - | 40.8 | 65.49 | - | Glushkova et al. (2010) |
| | 43.95 | 1.11[a] | 4.20[a] | 1.67[a] | - | 49.07[a] | - | McCarthy and Meredith (1988) |
| | 57.6 | 0.9 | 4.4 | 1.1 | | 34.6 | | Payne et al. (1983) |
| *C. crenata* | 57.7–60.5[b] | 0.3–0.8[b] | 2.0–2.7[b] | 1.0–1.6[b] | - | - | - | Jeong et al. (2012) |
| | 36.8–44.15 | - | 4.42–4.88 | - | 36.09–56.32 | 68.56–88.43 | - | Glushkova et al. (2010) |
| | 39.1–67.3[b] | 0.1–1.7[b] | 2.2–5.3[b] | 0.8–2.6[b] | - | 28.7–53.8[b] | - | Seo et al. (2009) |
| *C. dentata* | 43.7 | 1.32[a] | 4.83[a] | 1.58[a] | - | 48.57[a] | - | McCarthy and Meredith (1988) |
| | 33.7–47.0 | 8.4–10.2 | 9.1–10.2 | 1.1–1.9 | - | 32.4–42.3 | - | Payne et al. (1983) |

[a] Expressed as g/100 g d.w.
[b] Expressed as percentage.

**Table 4** Content of phenolics (mg GAE/g d.w.) and flavonoids (mg GAE/g d.w.) previously reported for fruits and other chestnut tissues

| Species | Tissue/organ | Total phenolics | Total flavonoids | Reference |
|---|---|---|---|---|
| *C. sativa* | Wood | 40.8[a] | - | Sanz et al. (2010) |
| | Bark | 23.9-56.1[a] | - | Comandini et al. (2014) |
| | Bark | 24.01[b] | - | Chiarini et al. (2013) |
| | Leaves | 412.96 | 82.87 | Barreira et al. (2010) |
| | Leaves | 1.4 | 0.33 | Živković et al. (2009) |
| | Leaves | 103[c] | 54.5[c] | Barreira et al. (2008) |
| | Flowers | 251.62-327.17[c] | 27.93-31.60[d] | Sapkota et al. (2010) |
| | Flowers | 298[c] | 160[c] | Barreira et al. (2008) |
| | Pollen | 64.02-103.80 | - | Avşar et al. (2016) |
| | Peeled seed | 0.59 | 0.007 | Živković et al. (2009b) |
| | Outer brown peel | 1.2 | 0.65 | Živković et al. (2009b) |
| | Outer skin | 510[c] | 503[c] | Barreira et al. (2008) |
| | Inner skin | 475[c] | 330[c] | Barreira et al. (2008) |
| | Skin | 709.96 | 901.59 | Barreira et al. (2010) |
| | Fruits | 1.60[b] | - | Wani et al. (2017) |
| | Fruits | 1580-3673[e] | 0.02[e] | Alasalvar and Bolling (2015) |
| | Fruits | 5-32.82 | - | Otles and Selek (2012) |
| | Fruits | 16.2 | - | Gonçalves et al. (2010) |
| | Fruits | 0.42 | 0.17 | Živković et al. (2009b) |
| | Fruits | 3.73[c] | 2.30[c] | Barreira et al. (2008) |
| *C. mollissima* | Fruits | 43.77e | - | Liu et al. (2017) |
| | Fruits | 2.24[c] | 2.62[c] | Li et al. (2016) |
| | Fruits | 20.4-27.7 × $10^3$[f] | 1.13-2.16 × $10^3$[f] | Yang et al. (2015) |
| | Burs | 3.31-7.12[e] | - | Zhao et al. (2011) |
| | Skin | 136-147[c] | 39-45[c] | Sato et al. (2010) |
| *C. crenata* | Leaves | 256.98[c] | - | Woo et al. (2017) |
| | Shell | 136.12-353.92[e] | 367.43-459.09[e] | Seo et al. (2016) |
| | Shell | 264.10-558.12[c] (extract) | 47.41-166.28[c] (extract) | Ham et al. (2015) |
| | Skin | 198-209[c] | 69-76[e] | Sato et al. (2010) |

[a] Expressed as g GAE/g wood.
[b] Expressed as g GAE/100 g d.w.
[c] Expressed as mg GAE/g d.w.
[d] Expressed as mg QE/g f.w.
[e] Expressed as mg GAE/100 g d.w.
[f] Expressed as mg/kg d.w.

**Table 5** Content of amino acids and vitamins previously reported for chestnut fruits

| Species | Free amino acids | Vitamin E (as tocols) | Vitamin C (as ascorbic and dehydroascorbic acid) | Reference |
|---|---|---|---|---|
| *C. sativa* | - | 16.42-16.53[a] | - | Delgado et al. (2016) |
| | - | 739-854[b] | - | Carocho et al. (2014a) |
| | - | 43[c] | - | Kalogeropoulos et al. (2013) |
| | - | 4347-5388[d] | - | Barreira et al. (2012a) |
| | - | 1.09[a] | - | Barreira et al. (2012b) |
| | - | 1039-1213[b] | - | Carocho et al. (2012) |
| | 374.55[a] | - | - | Gonçalves et al. (2012) |
| | - | - | 400-693[e] | Barros et al. (2011) |
| | - | 960.7-1201.5[b] | - | Fernandes et al. (2011) |
| | - | 4.33-23.95[f] | 8.76-12.87[c] | Vasconcelos et al. (2010) |
| | - | 4188-5719[d] | - | Barreira et al. (2009) |
| | 67.45[c] | - | - | Vasconcelos et al. (2009) |
| | - | 72[g] | - | Bellini et al. (2008) |
| | 5.5-8.7[h] | - | - | Borges et al. (2008) |
| | - | - | 0.028-0.128[h] | Dimitri et al. (2008) |
| | - | - | 30.8-36.3[c] | Pena-Méndez et al. (2008) |
| | - | - | 45.2-164.3[e]. | Ribeiro et al. (2007) |
| | 261.97-420.84[c] | - | - | Vasconcelos et al. (2007) |
| *C. mollissima* | 7.67[i] | - | $0.59\text{-}1.34 \times 10^{3}$[e] | Li et al. (2016) |

[a] Expressed as mg/100 g d.w.
[b] Expressed as µg/100 g d.w.
[c] Expressed as mg/100 g f.w.
[d] Expressed as ng/g f.w.
[e] Expressed as mg/kg d.w.
[f] Expressed as µg/g f.w.
[g] Expressed as mg/kg f.w.
[h] Expressed as g/100 g d.w.
[i] Expressed as mg/g d.w.

**Table 6** Content of minerals previously reported for fruits and other chestnut tissues

| Species | Ca | Mg | Na | K | P | Zn | B | Fe | Cu | Mn | Reference |
|---|---|---|---|---|---|---|---|---|---|---|---|
| *C. sativa* fruits | 2040-2937[a] | 1364-1852[a] | 245-526[a] | 10633-15871[a] | 1741-1879[a] | 54-85[a] | - | - | - | - | Özel (2015) |
| | 0.55[b] | 0.54[b] | - | 9.36[b] | 1.13[b] | 8.91[a] | 7.07[a] | 25.91[a] | 8.22[a] | 36.61[a] | Gonçalves et al. (2012) |
| | 13.2[c] | 27.6[c] | 1.7[c] | 351.1[c] | - | 0.7[c] | - | 0.8[c] | 0.3[c] | 0.8[c] | Bellini et al. (2008) |
| | 40.8-50.6[d] | 63.3-93.3[d] | - | 473-974[d] | 104-148[d] | - | - | 5.33-10.87[d] | 1.30-2.53[d] | 3.07-8.00[d] | Borges et al. (2008) |
| | 0.032-0.067[e] | - | - | - | 0.10-0.14[e] | - | - | - | - | - | Dimitri et al. (2008) |
| | 43-230[d] | 70-160[d] | 6-41[d] | 761-1271[d] | 107-191[d] | 1.8-9.1[d] | - | 0.4-5.7[d] | 0.6-3.8[d] | 0.7-5.5[d] | Ertürk et al. (2006) |
| | 0.026-0.072[e] | 0.049-0.100[e] | 0.003-0.017[e] | 0.789-1.130[e] | 0.068-0.305[e] | 10.0-19.0[f] | - | 13.5-23.8[f] | 5.5-10.3[f] | 17.0-124.6[f] | Pereira-Lorenzo et al. (2006) |
| | 37.1[d] | 67.2[d] | 2.7[d] | 1202.7[d] | 181.6[d] | 1.1[d] | - | 3.3[d] | 0.7[d] | - | Sequeira et al. (2004) |
| | 23[c] | 32[c] | 3[c] | 378[c] | 50[c] | 0.44[c] | - | 0.71[c] | 0.208[c] | 0.563[c] | McCarthy and Meredith (1988) |
| *C. sativa* catkins | 16.6[b] | 3.52[b] | - | 8.1[b] | 1.70[b] | 66[a] | 9.0[a] | 28[a] | 10.0[a] | 1041[a] | Portela and Louzada (2012) |
| *C. sativa* leaves | 9.3[b] | 2.00[b] | - | 7.0[b] | 1.54[b] | 25[a] | 8.8[a] | 53[a] | 8.0[a] | 880[a] | Portela and Louzada (2012) |
| *C. mollissima* | 287-1038[a] | - | - | - | 17-43[a] | 15-36[a] | - | 6-23[a] | 7-15[a] | - | Yang et al. (2015) |
| | 86.22[e] | 96.44[e] | 4.49[e] | 79033[e] | 114.33[e] | 5.07[e] | - | 2.71[e] | 2.37[e] | 5.17[e] | Glushkova et al. (2010) |
| | 18[c] | 84[c] | 3[c] | 447[c] | 96[c] | 0.87[c] | - | 1.41[c] | 0.363[c] | 1.601[c] | McCarthy and Meredith (1988) |
| *C. crenata* | 13.6-20.3[e] | 28.7-32.7[e] | 5.2-6.9[e] | 257.3-380.7[e] | 49.7-68.1[e] | 0.1-0.5[e] | - | 1.0-3.7[e] | - | 1.2-4.4[e] | Jeong et al. (2012) |
| | 41.14-86.08[e] | 88.84-96.36[e] | 3.39-429[e] | 893.67-1034.67[e] | 135.00-173.67[e] | 2.08-10.36[e] | - | 2.04-3.34[e] | 0.592.28[e] | 3.674.55[e] | Glushkova et al. (2010) |
| | 24[c] | 79[c] | 3[c] | 504[c] | 96[c] | 1.16[c] | - | 1.52[c] | 0.390[c] | 2.164[c] | McCarthy and Meredith (1988) |

[a] Expressed as mg/kg d.w.
[b] Expressed as g/kg d.w.
[c] Expressed as mg/100 g f.w.
[d] Expressed as mg/100 g d.w.
[e] Expressed as g/100 g d.w.
[f] Expressed as ppm.

(Table 7), almost all plant tissues have been studied with regard to health-promoting properties. Flower and pollen, tree bark and all nut parts (bur, seed coat or testa, shell and kernel) show diverse effects, with the most common being an interesting antioxidant activity against several oxidative radicals recorded for nuts. Besides those studies, some of the most relevant recent work shows the neuro- and cardioprotective effects of bark, the antitumoral and anti-inflammatory effect of flower extracts, the anti-angiogenic properties of shell or the antidiabetic effect of burs and seed coat or testa.

For *C. mollissima*, recent work has also focused on the composition of these nuts (Table 3). These recent investigations are complementary to a few previous studies, showing moisture content between 34.2% and 63.6%, while carbohydrates can range from 34.6% to 81.6%. Minor components, such as fat and protein, range from 0.9% to 10.2% and from 4.2% to 12.2%, respectively. As for *C. sativa*, some researchers have quantified the phenolic and flavonoid fraction present in *C. mollissima* (Table 4). Both phenolics and flavonoids can be found in higher amounts in the seed coat or testa (as much as 147 mg GAE/g f.w. and 45 mg GAE/g f.w., respectively), while burs and nuts have a lower amount, respectively, of phenolics and flavonoids. As for *C. sativa*, the mineral profiles of *C. mollissima* are highly variable, but K is the major mineral, although some authors refer to Ca as being at the highest levels (Table 6). For *C. crenata* and *C. dentata*, the existing information is scarce, with no recent work available for *C. dentata*. Although, these studies show a moisture content ranging from 39.1% to 67.3% for *C. crenata*, for *C. dentata*, the moisture range is lower (33.7–47.0%). As for other chestnuts, carbohydrates are the major components (28.7–88.4% in *C. crenata* and ranging from 32.4% to 48.6% in *C. dentata*). The fat content is higher for *C. dentata* (1.32–10.2%) than in *C. crenata* (0.1–1.7%), a pattern also found for protein (2.0–5.3% and 4.83–10.2% in *C. crenata* and *C. dentata*, respectively). For these two species, further studies are lacking and information is scarce. Total phenolics and flavonoids have been quantified in *C. crenata*, and show a higher amount of those compounds (up to 209 mg GAE/g f.w.) in skin, for phenolics and 459.1 mg GAE/100 g d.w. in shell, for flavonoids, while the mineral content of *C. crenata* has been recently described, indicating a predominance of K and P. The mineral composition of leaves and catkins of *C. dentata* is known, with an interesting amount of important minerals, and K being the major mineral present in both plant organs (Cho and Jo, 2003).

## 3 General uses of chestnuts

Sweet chestnuts used to be called the 'bread-tree', providing staple food with nutritious and health-giving properties for people in the Mediterranean and Europe. Chestnuts are widely recognized for their nutritive value (Wani et al., 2017) and distinctive flavour and taste (Ertan et al., 2015). These nuts are rich

**Table 7** Health effects previously reported for extracts of fruits and other chestnut tissue/organ

| Tissue/organ | Negative | Beneficial | References |
|---|---|---|---|
| Pollen | Allergy | | Sánchez-Monge et al. (2006) |
| Pollen | | Antimicrobial and antioxidant activities | Avşar et al. (2016) |
| Pollen | | Anti-inflammatory | Avşar et al. (2016) |
| Bark | | Preventive neuroprotective effects | Brizi et al. (2016) |
| Bark | | Anticancer agent | Lenzi et al. (2017) |
| Bark | | Antioxidant activity | Chiarini et al. (2013) |
| Bark | | Cardioprotective effect | Chiarini et al. (2013) |
| Bark | | Antimicrobial activity | Chiarini et al. (2013), Živković et al. (2009a) |
| Flowers | | Antitumor effects | Carocho et al. (2014b) |
| Flowers | | Antimicrobial activity | Carocho et al. (2014b) |
| Flowers | | Antioxidant activity | Carocho et al. (2014c,d), Barreira et al. (2008), Jukić et al. (2013), Sapkota et al. (2010) |
| Flowers | | Anti-inflammatory | Jukić et al. (2013) |
| Leaves | | Antibacterial activity | Quave et al. (2016), Basile et al. (2000) |
| Leaves | | Antioxidant activity | Almeida et al. (2008), Barreira et al. (2008), Vella et al. (2018) |
| Leaves | | Allelopathic activity | Basile et al. (2000) |
| Leaves | | Antimicrobial activity | Živković et al. (2009a) |
| Fruits | | Antioxidant activity | Barreira et al. (2008, 2010, 2011), Akbulut et al. (2017), Antonio et al. (2011), Carocho et al. (2012, 2014a), Dimitri et al. (2008), Dinis et al. (2012a,b), Galiñanes et al. (2015), Kalogeropoulos et al. (2013), Kirbaslar et al. (2012), Mujić et al. (2009), Otles and Selek (2012), Wani et al. (2017), Zeković et al. (2009), Živković et al. (2010) |
| Fruits | | Antimicrobial activity | Kirbaslar et al. (2012), Živković et al. (2010) |
| Burs | | Antidiabetic | Grdović et al. (2012) |
| Shell | | Anti-angiogenic | Sorice et al. (2016) |
| | | Anti-inflammatory | Sorice et al. (2016) |
| Skin | | Antidiabetic | Tsujita et al. (2008) |

in carbohydrates, minerals, vitamins, protein and fibre, and low in fat content (see Tables 3–6), which make them highly suitable for human nutrition and well-being (Liu et al., 2017; Mert and Ertürk, 2017).

Most chestnuts are sold fresh in autumn in the Mediterranean region (Delgado et al., 2016). However, the high tannin content in the nut is responsible for its astringent bitterness and lower palatability (Wani et al., 2017), making it highly suitable for processing and an array of novel uses. Chestnuts can be home processed or can undergo industrial processes in order to improve its organoleptic properties (aroma, flavour, texture) and digestibility (Vasconcelos et al., 2010). In the market, chestnuts are available in different industrial forms such as boiled, roasted, frozen at −40°C, dried or crystallized as the typical 'marrons glacés' (Vasconcelos et al., 2010). They can also be found as 'cured chestnuts', which involves immersing the nuts in cold water (hydrocooling) or hot water (thermohydrotherapy) to prolong the shelf life, a process that can result in changes to the chemical composition (Neri et al., 2010). Chestnuts are also used as an ingredient in many different bread and cake recipes (Dall'Asta et al., 2013), soups or other edible foods (Vossen, 2000) and gluten-free beers (Chenlo et al., 2009), a diversity of uses that are expanding. For example, in the Canary Islands, chestnuts are one of the ingredients used in the typical sweet 'morcillas' (Ríos-Mesa et al., 2011). In addition, novel products have been developed, such as yoghurt with candied chestnut (Sakin-Yilmazer et al., 2014). Chestnut flowers are also an important source of nectar and pollen for honey (Vasconcelos et al., 2010). Chestnut honey possesses a distinctive aroma and is very rich in K, Ca and Mn (Küçük et al., 2007) and has high antibacterial and antioxidant activities enabling its use for dressing chronic wounds, burns or skin ulcers (Vasconcelos et al., 2010). Moreover, chestnut flower extracts possess antimelanogenic effects, making them highly suitable for use in food, cosmetic and pharmaceutical applications (Sapkota et al., 2010).

## 4 Specific nutritional features

Folk tradition and the recognized nutritional value of chestnuts have led to them being widely consumed, especially in Europe, America and Asia (Vasconcelos et al., 2010). In some regions of such continents, chestnuts are still used as a potato substitute due to their high starch content (Vossen, 2000; Vasconcelos et al., 2007; Borges et al., 2008; Korel and Balaban, 2008; Silva et al., 2016). Although differences can be found among species/cultivars (Míguelez et al., 2004; Sequeira et al., 2004; Ertürk et al., 2006; Cristofori et al., 2007; Borges et al., 2008), origin (Míguelez et al., 2004; Ertürk et al., 2006; Cristofori et al., 2007; Yang et al., 2015), year of harvest (Sequeira et al., 2004), genetic factors (Borges et al., 2008) and/or fruit processing (Hou et al., 2016; Li et al., 2016), starch is the most abundant carbohydrate constituent of chestnuts, accounting for almost 50% of the total composition (Table 3). Amylose and amylopectin are the main types of starch in these nuts (Vasconcelos et al., 2010) and are a critical part of the functional and physico-chemical properties of starch (Muzzaffar et al., 2016). When compared with other staple starch foods such as potatoes, the starch content is approximately double or triple in chestnuts (Sato et al., 2017), making them a very popular and appreciated nut with potential health benefits. Moreover, the chestnut starch content exhibits lower gelatinization (56.1°C) and peak viscosity (79.5°C) temperatures, which compare favourably with corn starch, especially in applications where lower processing temperatures are needed (Cruz et al., 2013). Associated with a high starch content, chestnuts are also rich in sugars (Borges et al., 2008; Korel and Balaban, 2008; Vasconcelos et al., 2010), making this fruit an energetically valuable nut. Among the simple sugars chestnuts contain a high content of sucrose, which can be up to one-third of the total sugars (Vasconcelos et al., 2010) and is advantageous when compared with potatoes, wheat and walnuts (Künsch et al., 1999), for example. These complex (starch) and simple (sugar) carbohydrates are responsible for the sweetness (Pereira-Lorenzo et al., 2006; Yang et al., 2015) and palatability of chestnuts (Yang et al., 2015), which make them suitable for human and animal diets. On the contrary, the ash content of chestnuts is considered low, accounting for 0.8–4.4 g/100 g d.w. in various species (Table 3).

Water is also an important component in chestnuts and acts as an indicator of the shelf life of fresh nuts (Yang et al., 2015). In general, fresh chestnuts contain about 50% moisture (Table 3), which makes them highly suitable for industrial processing (Míguelez et al., 2004) but has a negative impact on their storage ability (Borges et al., 2008). Comparing the most cultivated *Castanea* species, McCarthy and Meredith (1988) found that *C. sativa* exhibits the highest moisture content, followed by *C. mollissima* and *C. dentata*. However, Glushkova et al. (2010) recorded the highest moisture content (50.4–55.4%) for two interspecific hybrid cultivars (*C. crenata* × *C. sativa*), and the lowest value (31.9%) for *C. mollissima*, while intermediate values (36.8–44.1%) were found for *C. crenata* and *C. sativa* cultivars. According to Pereira-Lorenzo et al. (2006) and Neri et al. (2010), the moisture content of chestnuts tends to change with the chestnut origin, which is probably due to the different growing and climate conditions between regions.

Chestnuts are also characterized by a low fat content (Table 3) but of relatively high quality, ranging from 0.7 to 10 g/100 g d.w., depending on the species/cultivar (Míguelez et al., 2004; Mert and Ertürk, 2017). Compared with other nuts, for example, hazelnut, macadamia, pecan or almond, chestnuts exhibit the lowest fat content (Alasalvar and Bolling, 2015). However, the fat-soluble bioactive compounds, such as tocols and phytosterols, are present in chestnuts in higher amounts, when compared with fat-rich nuts. Despite the low fat content, chestnuts contain a high amount of essential fatty acids

(Künsch et al., 1999; Borges et al., 2007; Poljak et al., 2016), either saturated or unsaturated, which has been linked to several processes involved in health and chronic diseases (Benatti et al., 2004). Among them, linoleic and linolenic acids are the most important unsaturated fatty acids (Künsch et al., 1999; España et al., 2011). The beneficial effects of fatty acids on human health are well known, as anticancer (Vasconcelos et al., 2010) and neuroprotective (Künsch et al., 1999) compounds, and they are also associated with protection against cardiovascular diseases (Braga et al., 2015).

Chestnuts are considered to be an excellent source of dietary fibre (Mert and Ertürk, 2017), which does not appear to be influenced by cultivar or origin (Borges et al., 2008). According to Pereira-Lorenzo et al. (2006), chestnuts contain a similar amount of fibre to the majority of nuts, but lower than hazelnuts. The presence of fibre in chestnuts make them more attractive to consumers since it is associated with favourable health effects, for example, in the intestine by influencing metabolic activities (Gonçalves et al., 2010). Dietary fibre also promotes, apart from other properties, cardiovascular health (Sathe et al., 2009), cholesterol reduction, regulation of insulin and metabolism of blood lipids (Prosky, 2000).

The protein content of chestnuts is considered to be moderate (Poljak et al., 2016), with some differences among species/cultivars (Mert and Ertürk, 2017), which is associated with the nature of the soil the chestnuts are grown in (Míguelez et al., 2004; Poljak et al., 2016). Considering the recommended daily intake (RDI) of protein, the protein content of these nuts represents approximately 9.2% of the RDI for females and 7.6% of the RDI for males (Vasconcelos et al., 2010), which highlights the relevance of these nuts in the human diet. Although the protein content in chestnuts is in the range of 2.0–8.6 g/100 g d.w. (Table 3) and is comparatively lower than other tree nuts such as peanuts (25.8 g/100 g d.w.), almonds (21.3 g/100 g d.w.) and hazelnuts (15.0 g/100 g d.w.) (Ros, 2010); it is considered to be high quality due to the presence of important amino acids (Vasconcelos et al., 2010). Aspartic acid and glutamic acid were detected as the main protein-derived amino acids (Vasconcelos et al., 2010) present in chestnuts. Borges et al. (2008) also identified arginine as one of the most predominant amino acids in different chestnut cultivars from various Portuguese regions. Other amino acids, such as γ-aminobutyric acid (GABA), occur naturally in chestnuts at a concentration of 188 nmole/g d.w. (Oh et al., 2003), making them extremely relevant for the human diet. This non-proteinogenic amino acid is considered to be effective in lowering blood pressure and heart rate (Oh et al., 2003). GABA has also been regarded as an important neurotransmitter inhibitor in the mammalian central nervous system (Bettler et al., 2004), which is relevant for the maintenance of cerebral activity (Gonçalves et al., 2012). In addition, GABA can also be helpful in some anxiety and neurological disorders such as Huntington's chorea, and

Parkinson's and Alzheimer's diseases (Gajcy et al., 2010). The protein content, as well as the amino acid profile of chestnuts, indicates the high nutritional quality of these nuts (Yang et al., 2018) with positive health effects on the brain and energy metabolism (Lau and Tymianski, 2010).

In addition to investigating the composition, recent approaches to chestnut analysis have revealed the content of total phenolics and flavonoids, not only in nuts but also in other plant tissues, including tree bark or burs (Table 4). As for many other nuts, like almonds (Milbury et al., 2006), these bioactive compounds appear to be present preferentially in the seed coat or testa, rather than in the kernel itself. This should be taken into consideration, not only from a consumer's point of view, regarding the intake of those health-promoting compounds, but also from a research-driven assessment, as compounds of interest are found in low added-value by-products. Some of the phenolics of interest in chestnuts are gallic and ellagic acids, with the content in *C. sativa* cultivars ranging between 3.46 and 9.07 g and 2.71 and 9.64 g/100 g d.w., respectively (Vasconcelos et al., 2007), with strong evidence regarding the positive effects on cardiovascular function, and anticarcinogenic and antiplasmodial activity (Okuda, 2005). Chestnut catkins, also a by-product, are nowadays regarded as an interesting source of compounds. Indeed, catkins of *C. sativa* contain soluble sugars, with special reference to fructose, but also glucose and sucrose, organic acids (quinic, oxalic, malic and shikimic acids) and about 27 phenolic compounds, with the predominance of trigalloyl-HHDP-glucoside, followed by pentagalloyl glucoside (Carocho et al., 2014c). According to Jukić et al. (2013), dehydrodigallic acid dimethyl ester is the major phenolic in chestnut catkins, which has a considerable amount of gallic acid derivatives. Flowers of *C. crenata* also contain sugars, with a predominance of sucrose and maltose, and organic acids (malic, citric, oxalic and succinic acids). Amino acids are also found in interesting amounts in flowers of *C. crenata*, with a higher content of aspartic acid and proline, and these plant organs are also rich in vitamin C (Cho and Jo, 2003).

Chestnuts are also rich sources of vitamins, particularly vitamins C and E, mainly tocols (Table 5), which are responsible for their beneficial health effects with respect to antioxidant and antimicrobial properties (Table 7). Vitamin E has also been associated with reducing the risk of cancer and cardiovascular diseases (Vasconcelos et al., 2010). In addition, such nuts are characterized by considerable amounts of other vitamins such as vitamin A, which has multiple actions in the human body, and B vitamins including thiamine, riboflavin, niacin and folate (Vasconcelos et al., 2010). These B vitamins are essential for brain health and act against neurological diseases. Chinese chestnuts contain a higher amount of vitamin A, riboflavin, pantothenic acid and pyridoxine, while European chestnuts are high in vitamin C and thiamine, and Japanese chestnuts are rich in niacin. Portuguese cultivars of *C. sativa* exhibit an interesting amount of vitamin C and, although their content varies significantly among samples (Vasconcelos et al., 2009; Barros et al., 2011), a clear relationship between the content of vitamin C and the antioxidant activity (Barros et al., 2011) has been recorded. There are a limited number of studies focusing on the presence of vitamins in *C. crenata*, *C. dentata* and *C. mollissima*, while for *C. sativa*, some work has been performed over the last few years to elucidate the extent of the presence of vitamins, namely C and E, in nuts of this species.

Chestnuts are also a good source of minerals (Table 6), that when deficient in the diet, may have a severe impact on human health (Belitz and Grosch, 1999; FAO/WHO, 2004). Ca, Mg, Na, K, P, Zn, B, Fe, Cu and Mn have been found in nuts of *C. sativa*, mainly in recently performed studies, while for *C. crenata* and *C. mollissima* information is still very limited. Although the content of macronutrients, particularly K (the most abundant one), P, Mg and Ca found in Portuguese *C. sativa* cultivars (Borges et al., 2008) are comparable with those of other cultivars (Pereira-Lorenzo et al., 2006), and even for Chinese ones (McCarthy and Meredith, 1988), values for micronutrients (Fe, Zn, Cu or Mn) appear to change substantially among cultivars (Borges et al., 2008).

## 5 Effects of cooking on quality

Chestnuts can be consumed raw but are usually processed. Consumer preference is either boiling or roasting (Silva et al., 2001), but other alternatives exist, such as frying or curing. As for any other food commodity, cooking may change the quantitative and qualitative composition of the chestnut, which will ultimately result in variations in actual nutrient intake by consumers. Considering these facts, several authors have focused their research on the effect of different cooking methods on chestnut composition (Table 8). This issue has been thoroughly reviewed by Zhu (2016) and will be briefly addressed here, considering proximate composition and also bioactive compounds. For one of the major components of chestnuts, that is carbohydrates, cooking considerably alters its quantity as well as other parameters. For *C. sativa*, insoluble, soluble and total fibre increases with roasting, but boiling only increases soluble and total fibre (Gonçalves et al., 2010). In *C. mollissima*, boiling reduces the amount of reducing sugars, sucrose, glucose and fructose (Li et al., 2016), a fact attributed to the solubilization of those carbohydrates. Starch is also markedly affected by cooking and decreases upon roasting and boiling as recorded in *C. mollissima* (Li et al., 2016), whereas the increase found for some cultivars of *C. sativa* (Silva et al., 2016) led to changes in the structure of starch. Indeed, roasting and boiling have proven to lead to the fusion of individual granules, and gelatinization occurring with the formation of elongated clusters (Silva et al., 2016). These changes can be responsible for the modification of starch properties, ultimately influencing sensory, rheological

**Table 8** Effects of cooking on compositional and bioactive characteristics of chestnut

| Species | Type of processing | Highlights | Reference |
|---|---|---|---|
| *C. sativa* | Boiling and roasting | Roasting increased protein and fibre, as well as gallic acid and total phenolics and lowered fat contents. Boiling increased fat, gallic and ellagic acids contents. Cooking increased citric acid contents, reducing malic acid. | Gonçalves et al. (2010) |
| | Roasting | Depending on cultivar, variations in fat, starch and sugars, as well as individual fatty acids. | Künsch et al. (2001) |
| | Boiling and roasting | Depending on cultivar and type of processing, variations of total and individual amino acids and minerals. | Gonçalves et al. (2012) |
| | Boiling, roasting and curing | Decrease in total phenolics and antioxidant activity with cooking. Depending on type of processing, variations in specific polyphenol compounds, minerals and trace elements. | Nazzaro et al. (2011) |
| | Pan and microwave roasting | Increase in carbohydrate content and antioxidant activity. | Wani et al. (2017) |
| | Boiling and roasting | Depending on cultivar and type of processing, variations in vitamin C, antioxidant activity, gallic and ellagic acid content. | Barros et al. (2011) |
| *C. mollissima* | Boiling, roasting and frying | Reduction in fat, protein, sugars, amino acids, total polyphenols, total flavonoids and organic acid with cooking. | Li et al. (2016) |
| | Frying | Decrease in total phenolics and flavonoids, ascorbic acid and reduction of antioxidant activity. | Hou et al. (2016) |

and chemical characteristics, but also digestibility (Noda et al., 2008). The protein content appears to be more resistant to cooking than other chestnut components. In fact, previous work shows that roasting or boiling of the *C. sativa* or *C. mollissima* nut only slightly alters the protein content (Gonçalves et al., 2010, 2012; Li et al., 2016; Wani et al., 2017). Although without an apparent effect on protein content, changes in the amount of total amino acids and their composition occur with cooking (Gonçalves et al., 2012; Li et al., 2016), possibly due the degradation of peptides and proteins and to Maillard reactions (Gonçalves et al., 2012). One of the amino acids that is most affected is L-aspartic acid (Li et al., 2016), which is one of the most abundant in chestnuts

(Vasconcelos et al., 2010), although several others are also affected (Gonçalves et al., 2012; Li et al., 2016). Considering that amino acids are biologically active, with several functions in the body, variations of their content alter the possible health-related effects of these compounds. Although having a low fat content, chestnuts present fat-soluble bioactive compounds in higher amounts than fat-rich nuts, as well as essential fatty acids, which are linked to several processes involved in health and chronic diseases. The fat fraction of chestnuts decreases upon boiling, frying or roasting (Gonçalves et al., 2010; Li et al., 2016). However, it appears that boiling leads to fewer changes, probably due to the lower temperature of the process (Das et al., 2013). Despite limited information, the fatty acid profile also appears to be slightly modified after roasting (Künsch et al., 2001), an effect that can change the nutritional value of chestnuts and is a topic that deserves further investigation.

Besides the effect on composition, cooking may also result in changes to the content of bioactive compounds, such as phenolics or vitamins. Some work has highlighted the negative effect of roasting, boiling or frying on total phenolic content (Barros et al., 2011; Nazzaro et al., 2011; Hou et al., 2016; Li et al., 2016), with a parallel reduction on the antioxidant activity. This decrease may be due to the fact that temperature leads to the oxidation of phenolics. However, an increase of the total phenolic content and antioxidant activity has also been found (Gonçalves et al., 2010; Wani et al., 2017), which may be linked to the release of tightly bound phenolics that, in this way, can be extracted and quantified (Liu et al., 2017). The vitamin content of chestnuts is also altered by cooking procedures. Although studies are lacking on this specific topic, a considerable decrease in the amount of vitamin C after cooking has been recorded (Barros et al., 2011; Li et al., 2016), and is probably linked to the conversion of ascorbic acid to dehydroascorbic acid, which was further degraded to 2,3-diketogulonic acid. Data regarding other important vitamins present in chestnuts is limited and should be considered an important major research line, but changes are to be expected, although dependent on time and conditions, as recorded for vitamin E (Delgado et al., 2016). Considering the health-promoting capacity of the vitamins present in chestnuts, all changes, either increasing or decreasing these compounds are of great importance to producers and consumers. Knowledge of their behaviour under certain cooking conditions, besides being a major research theme, can also be an added-value factor for this nut.

Other chestnut components are also affected by cooking, namely organic acids and minerals. The organic acid content decreases considerably after boiling, roasting or frying (Ribeiro et al., 2007; Li et al., 2016), but interestingly, Gonçalves et al. (2010) showed that individual organic acids are affected differently: cooking increased the content of citric, gallic and ellagic acids, while reducing the malic acid content. These variations of the content and organic acid profile after cooking must be considered from two points of view: from an organoleptic perspective and for improving health, as they are linked to the sensory characteristics of nuts (Vaughan and Geissler, 1997), while presenting antioxidant activity with a protective role against various diseases (Silva et al., 2004). Finally, the mineral content is known to decrease after cooking, with the exception of P, Fe, Cu and Mn (Gonçalves et al., 2010), although other studies show that Ca may not be lost (Künsch et al., 2001). The work of Nazzaro et al. (2011) shows a more complex pattern of cooking effects on chestnut minerals, with increases or decreases recorded for different minerals.

## 6 Post-harvest handling and storage

Due to their high moisture and sugar content (Attanasio et al., 2004; Jermini et al., 2006; Gounga et al., 2008, 2017), chestnuts are characterized by a short shelf life (Gounga et al., 2008). They are extremely susceptible to damage during storage (Ertan et al., 2015), which is reflected in important economic losses. Moreover, the peeled chestnut tends to be more perishable when compared with the intact nut (Gounga et al., 2008).

Weight loss is one of the major problems in chestnut preservation (Antonio et al., 2011; Ertan et al., 2015) as it promotes the condensation and growth of fungi (Antonio et al., 2011) with a negative influence on its availability and certain nutritional and physical properties of the nuts. Mencarelli (2001) reported weight losses of around 1% in fresh chestnuts after 1-day storage at 20°C and 70% relative humidity (RH). In studies where the recently harvested Portuguese varieties (Judia, Longal and Martaínha) were evaluated, a weight loss of up to 26% between the end of October and January at room temperature (13°C, 65% RH) was observed for the Judia variety; however, when kept under a modified atmosphere (1.7°C, 88% RH, 7% $O_2$ and 14% $CO_2$) the weight loss was only 6.6% up to the end of April, and even lower (1.3%) in the Martaínha variety (VALCAST project data, 2008). Data from the same project with the Lada variety, showed a weight loss variation from 35.9% at room temperature to as low as 1.4% under a modified atmosphere (1.7°C, 88% RH, 7% $O_2$ and 21% $CO_2$). Thus, apparently an increase in the $CO_2$ concentration of up to 21% can reduce weight loss, which has a tremendous impact on the yield of the processing industry, as it needs to process chestnuts from harvest up to a few months later. Similar findings were reported by Gounga et al. (2008) in *C. mollissima* peeled nuts, where a moisture loss of about 25% was observed following storage for 6 days at 30°C and 80% RH. Optimizing storage conditions is thus essential to reduce the incidence of nut decay and loss of quality attributes. According to Jermini et al. (2006), the most suitable conditions for preserving chestnuts ranged from 0°C and 2°C with a high RH of around 90%. The same conclusion was reached by Ertan et al. (2015) who found that buried nuts of *C. sativa* in the orchard had deleterious effects on their quality when compared with storage under low temperatures and high RH. These authors also concluded that temperature is the key factor in the preservation of chestnut quality. However, these methodologies are not always effective in suppressing fungal spoilage and need to be supplemented. One of the most common methods in the past consisted of the application of fungicides (Lee et al., 2016b) but the potential harm to health and environmental risk led to it being banned (Carocho et al., 2013). Thus, it is crucial to use novel technologies that do not cause environmental harm or negatively affect health.

Among the alternative methodologies, the use of electron beam irradiation has gained interest as it is considered to be a safe, cheap and environmentally friendly technology that is able to minimize the detrimental problems that can occur during the chestnut post-harvest process (Carocho et al., 2013). Another promising technology for the preservation of chestnuts consists of using a natural coating (Gounga et al., 2008), which acts as a barrier to water loss and gas exchange (Lin and Zhao, 2007), hence minimizing nut deterioration and, at the same time, acts as a carrier of health-promoting compounds. Gounga et al. (2008) investigated the application of this technology on fresh-roasted chestnuts and concluded that the use of a whey protein isolate-pullulan (WPI–Pul) coating combined with freeze-drying and low temperature storage was responsible for reducing microbial growth in nuts, thus extending the shelf life. According to Gounga et al. (2007), the use of a WPI–Pul coating enhanced the sensory attributes of fresh and dried chestnuts. Additionally, the use of chitosan coatings on Chinese chestnuts reduced respiratory rates with positive effects on shelf life (Tian et al., 2009). More recently, Gounga et al. (2017) tested chocolate as a new chestnut coating and the results are very promising. In fact, the results suggest that this new type of coating was very effective in controlling microbial growth in addition to being well accepted by consumers. Moreover, it adds value to chestnuts.

Another technology largely used in fresh fruits is modified atmosphere packaging (MAP), which is still under investigation in chestnuts. Panagou et al. (2006) observed that this technology combined with low temperature storage preserves chestnut quality, especially during long-term storage. Treatment with $CO_2$ under higher pressure has also been mentioned as a promising strategy for extending the shelf life of chestnuts (Navarro, 2012).

Chestnuts might also be affected by several fungi and insect worms, namely elephant weevil, *Balaninus elephas*, and chestnut tortrix, *Cydia splendana* (Chenlo et al., 2009), which are responsible for the rapid deterioration of nuts (Ertan et al., 2015) resulting in decreased value and consequently their acceptance by consumers. Gao et al. (2011) estimated the annual losses of chestnuts due to pests during storage as 35–50% of total production. In order to reverse the high perishability of chestnuts and to minimize post-harvest losses, different methodologies have been used. Traditionally, freezing, cold storage and drying are used to extend the shelf life of chestnuts (Gounga et al., 2008; Sakin-Yilmazer et al., 2014; Delgado et al., 2018) and to ensure year-round availability (Moreira et al., 2015; Lee et al., 2016b). However, such methodologies require proper storage conditions, namely a lower temperature, associated with adequate ventilation and relative air humidity in order to maintain the physical properties of fruits, such as colour or texture, and to preserve their nutritional and sensory properties (Panagou et al., 2006).

## 7 Chestnut co-products

The agricultural and industrial processing of chestnuts generates large quantities of waste material, mainly bur, shell, leaves and curing wastewater (Braga et al., 2015; Squillaci et al., 2018). The valorization of these products has been a new challenge for the chestnut value chain contributing to environmental (Pinto et al., 2017) and economic sustainability (Braga et al., 2015). In fact, such agro-industrial wastes, after valorization, can be used by other industries, such as pharmaceutical, food or cosmetics (Sapkota et al., 2010; Aires et al., 2016; Vella et al., 2018) due, in part, to the production of active molecules (Nazzaro et al., 2012; Squillaci et al., 2018).

Some examples of how the valorization of such co-products can be achieved are reported in the work of Aires et al. (2016) and Santos et al. (2016). The former shows how chestnut peels can be used to produce pomaces with high amounts of tannins, which can be used as phenol substitutes in the formulation of adhesives or as antioxidant supplements in food, cosmetic or pharmaceutical formulations. Santos et al. (2016) suggested the beneficial use of shell extracts on lettuce properties. According to these authors, the composting shell and seed coat or testa of the chestnut can potentially be used as growing media for lettuces as an alternative to non-renewable organic substrates as it results in increased yield and antioxidant capacity of lettuces. The use of rejected fruits by the industry, especially those of small size, can also be converted into different products (Delgado et al., 2017), making this crop even more profitable. In addition, application of chestnut co-products in the cosmetic industry is still an area under development. The growing demand for natural products that are environmentally friendly and are of benefit to health has caused an increased interest in chestnut extracts (Ribeiro et al., 2015), in part, due to its antioxidant properties (Sapkota et al., 2010). Almeida et al. (2015) indicated that the use of leaf extracts of chestnut in skincare products can contribute to the prevention and minimization of oxidative stress in the skin, which could otherwise result in photoageing, some dermatosis and cancer. Pinto et al. (2017) concluded that *C. sativa* bur extract could be a promising cosmetic ingredient in topical hydrogels. More recently, Squillaci et al. (2018)

suggested the beneficial use of chestnut shell extracts in cosmetic preparations as it showed anti-inflammatory activity, skin hydration capacity and the ability to protect collagen from degradation. Other parts of the fruit can also be used for other applications. For example, chestnut shell extracts can be used in the manufacture of shoe and sole leather due to the high level of tannins (Braga et al., 2015). Another example of alternative applications of chestnut shell extracts includes its use as a heavy metal adsorbent (Braga et al., 2015) or a source of fermentable sugars for biofuel production (Vella et al., 2018).

Although the main use of chestnuts is for human consumption, chestnuts are also being tested as a component of animal feed (Borges et al., 2008; Vasconcelos et al., 2010). De Jesús et al. (2016) proved that the incorporation of chestnuts into the diet of Celta pigs reduced the saturated and monounsaturated fatty acid content of such feed, enabling the production of healthier dry-cured meat products. Similar to these conclusions, Ferreira-Cardoso (2002) concluded that adding chestnuts to the diets of pigs had positive effects on the energy requirements of pigs and increased the sensorial taste of the meat and ham. Lee et al. (2016a) showed that the use of a small percentage of chestnuts (5%) in the feed could be useful for reducing the negative presence of tannins in the feed.

## 8 Health benefits of eating chestnuts

Chestnut consumption has been perceived as beneficial for health; however, nowadays there is a misleading association between nut intake and nut high fat content, which could contribute to weight gain (Vadivel et al., 2012). Chestnuts contain a low amount of fat, being a good alternative to other fat-rich nuts, without the weight gain effect but with the benefits of all the active compounds present in the fat fraction of this nut. The essential fatty acids found in chestnuts are linked to an array of biological processes (plasma lipid levels, cardiovascular and immune function, insulin action, and neuronal development and visual function) (Benatti et al., 2004) and their intake can result in decreased cholesterol levels and prevent coronary heart disease (CHD) (Garcia et al., 1992). Indeed, studies have shown that extracts of chestnut can actively reduce the oxidation of low-density lipoprotein (LDL), which is a key step in the pathological basis of cardiovascular disease (CVD) (Hansson, 2005). The intake of phytosterols and dietary fibre is also linked to the reduction of cardiovascular issues (Jesch and Carr, 2017). For phytosterols, an inverse dose-response relationship between their intake and LDL cholesterol concentration has been established (Ras et al., 2014), for an intake of up to 3 g/day, with mechanisms including inhibition of cholesterol ester hydrolysis, competition with cholesterol for solubilization or transport across the apical membrane of enterocytes (Jesch and Carr, 2017).

With regard to fibre, studies have confirmed the hypolipidaemic and cardioprotective effects in humans (Anderson et al., 2000; Naumann et al., 2006). Although the mechanism by which positive effects are achieved has not been elucidated, it may include interaction with and fermentation of fibre by gut bacteria, delayed gastric emptying, increased muscle glucose transporter type 4 expression and glucose uptake, decreased absorption of dietary fats and increased excretion of faecal cholesterol (Jesch and Carr, 2017). Dietary fibre has also been linked to beneficial effects for diabetes, cancer mechanisms and potential therapeutic agents in diseases such as various forms of colitis, antibiotic-associated diarrhoea and colon cancer (Prosky, 2000). Furthermore, it stimulates *Bifidobacterium* and *Lactobacillus* in the intestine (Blaiotta et al., 2013), which are also responsible for reducing serum total and LDL cholesterol (Jesch and Carr, 2017). Nut intake, including chestnuts, is linked to protection against type 2 diabetes and related metabolic syndromes, with those actions connected to decreases in oxidative damage and inflammatory biomarkers in blood lipids and the content of fatty acids, dietary fibre, vegetable proteins and polyphenols (Amarowicz et al., 2017). Chestnut ingestion also has positive effects on cancer prevention, due to its bioactive compound composition. For this cancer-preventive action, some major compounds are likely to contribute, such as fatty acids, especially omega-3 (Hardaman, 2002), but also vitamin C (Wenzel et al., 2004), vitamin E, and gallic and ellagic acids (Mertens-Talcott et al., 2006; Veluri et al., 2006; Geleijnse and Hollman, 2008). Compared with fresh nuts, ellagic acid can register a greater than twofold increase when chestnuts are boiled, with up to 12.2 mg/kg d.w. (Gonçalves et al., 2010) that can be transformed into urolithin metabolites by some species of intestinal bacteria, which play a major role in cardiovascular protection, and anti-inflammatory and anticarcinogenic properties (Selma et al., 2014). Vitamin C functions in the human body are also linked to bone, teeth and blood vessel formation, wound healing and is necessary for metabolic functions, activation of the B vitamin, folic acid, conversion of cholesterol to bile acids and conversion of tryptophan to serotonin (Chambial et al., 2013). Being a joint source of vitamins C and E, chestnuts are of paramount health importance. Vitamin E is a strong antioxidant, acting in the chain-breaking reactions during peroxidation of unsaturated lipids, LDL and other cell membrane components, reducing the risk of cancer and cardiovascular diseases (Vasconcelos et al., 2010). Chestnuts are also rich in B vitamins that, when deficient, can result in developmental abnormalities, haematologic status, iron absorption, gastrointestinal development (Powers, 2003), cognitive disorders and anaemia (Morris et al., 2007).

Furthermore, a well-balanced intake of minerals, that are almost entirely obtained from vegetables and fruits, is of great importance to human health. As mentioned before, chestnuts are a very good source of macro-elements, such as Ca, P, K, Mg and S, but also of micro-elements (Fe, Cu, Zn and Mn)

(Vasconcelos et al., 2010), that are linked to several health-related effects, and have been thoroughly reviewed by Martínez-Ballesta et al. (2010) and will be briefly detailed here. Calcium has been linked to reducing colon cancer and preventing the risk of pre-eclampsia, while suboptimal Ca intake can result in osteoporosis, hypercholesterolaemia and high blood pressure. Potassium is related to the transmittance of nerve impulses, heart muscle contraction and deficiency of this mineral can result in muscular difficulties, skin problems and alkalosis. Phosphorus, besides being essential for the production of ATP, GTP and CP, also regulates the activity of a number of proteins via phosphorylation reactions, and its deficiency results in bone pain, irregular breathing, fatigue, anxiety, numbness, skin sensitivity and changes in body weight. For magnesium, the described results of low intake are ageing and age-related disorders, such as diabetes, metabolic syndrome, hypertension and several cardiovascular conditions. Regarding the micro-elements, Fe deficiency is termed hypochromic anaemia, as this element is linked to the synthesis of haemoglobin and myoglobin. Copper, a mineral linked to enzyme function, can, when lacking in the body, result in anaemia, leukopenia and neuropenia, and skeletal disturbances. Zinc deficiency negatively affects the efficacy of the immune system, taste and smell senses, while impairing DNA synthesis, and causing hair loss and hypochromic anaemia. Finally, deficiencies in Mn intake lead to a reduction of cholesterol and red blood cells, while causing mucopolysaccharide abnormalities. Chestnuts have a major distinctive characteristic: they are gluten-free nuts and can therefore be used for the preparation of numerous products for gluten-sensitive, wheat allergy and celiac disease patients. In this way, these patients can benefit from all the health-related compounds present in chestnuts, without suffering from symptoms like diarrhoea, irritable bowel syndrome, malabsorption syndrome, weight loss and autoimmune diseases (Vasconcelos et al., 2010).

The chemical composition of chestnuts can change due to a number of factors, including genetic factors and environmental conditions (Míguelez et al., 2004; Pereira-Lorenzo et al., 2006; Borges et al., 2008; Poljak et al., 2016; Mert and Ertürk, 2017) suggesting different uses. For example, the presence or absence of some compounds can be of interest for processing purposes or health uses. Borges et al. (2008) in a study with eight sweet chestnut cultivars from three Portuguese areas identified two cultivars linked to health-promoting benefits, due to the presence of a high linoleic acid content (cv. Lada) or unsaturated fatty acids (cv. Aveleira).

Besides the nut itself, other chestnut organs or derived products have been referred to as having a beneficial health effect, either when consumed as-is or used as ingredients in other foods. Chestnut flowers have been referred to as one of the best sources of nectar and pollen for honeybees early in the summer (Kolayli et al., 2016). Pollen from chestnuts has also been studied and revealed interesting results, as it displays antioxidant and antimicrobial activity (Avşar et al.,

2016; Karadal et al., 2018) and exhibits good ability to protect hepatocytes from oxidative stress (Yıldız et al., 2013). In this study it was shown that the inclusion of chestnut pollen in the human diet reduces oxidative stress, with the pollen as a suitable alternative for the treatment of hepatocellular pathologies.

For a long time, honey has been associated with health-promoting effects due to high-quality natural compounds (Ajibola et al., 2012). Chestnut honey is considered one of the most delicious and is of high quality (Castro-Vázquez et al., 2010) due to its colour, aromatic physical properties (does not crystallize easily) and composition (particularly the natural volatile compounds related to the original nectar), when compared with blossom honey (Kolayli et al., 2016). Traditionally, it has been used in wound healing due to the high phenolic content and flavonoids (Alissandrakis et al., 2011; Meteoglu et al., 2015), and if impregnated with carboxymethyl cellulose sodium hydrogel paste can be a competitive candidate as a therapeutic dressing for diabetic ulcer wound healing, due to its potential to inhibit and kill bacteria (Park et al., 2017). Chestnut honey has also been used in the treatment of colds, reflux, gastritis and other disorders of the upper respiratory tract and gastric system (Kolayli et al., 2016). These effects are sustained by the strong antioxidant activity (Bertoncelj et al., 2007; Sarikaya et al., 2009; Kolayli et al., 2016; Karadal et al., 2018), which introduces the possibility of reducing health problems related to oxidation damage and associated diseases. Apart from the antioxidant activity, chestnut honey exhibits antimicrobial activity and can inhibit the growth of numerous bacteria and fungi (Kuncic et al., 2012; Karadal et al., 2018). Chestnut honey has also been reported to have antihyaluronidase and antiurease properties. However, excess hyaluronidase activity can cause hyaluronic acid deficiency, and its breakdown products can be harmful to the organism, with the inhibitors of this enzyme offering potential use for the treatment of several diseases, such as cancer and rheumatoid arthritis (Sunitha et al., 2013; Selenge et al., 2014). On the other hand, antiurease activity can be important, as excess urease activity is linked to the development of urolithiasis, pyelonephritis, hepatic encephalopathy, hepatic coma, urolithiasis and urinary catheter encrustation in humans and animals. Its inhibition is also important when dealing with *Helicobacter pylori* in gastric diseases (Kolayli et al., 2015). Chestnut honey differs significantly from floral honey, in phenolic composition, and in the recorded antioxidant and antimicrobial activity (Güneş et al., 2017). The major phenolic found in chestnut honeys is caffeic acid, followed by *p*-hydroxybenzoic acid or protocatechuic acid, depending of the origin of the honey, while floral honey contains higher amounts of protocatechuic acid. Chestnut honey has also been proven to have a higher amount of total phenolics when compared with floral honeys, as well as enhanced bioactive properties, namely an antioxidant effect, using the CHROMAC and ABTS methods, and antibacterial effects against *Staphylococcus aureus* and *Escherichia coli* (Güneş et al., 2017).

Propolis, another bee product that can originate in chestnuts, has garnered increased interest over the last few years, as it has beneficial properties for human health. As with other chestnut products, it exhibits antioxidant and antimicrobial activity (Sarikaya et al., 2009; Karadal et al., 2018). However, chestnut-based propolis also has an interesting effect against hereditary spherocytosis (HS), a type of anaemia characterized by microcytic and hyperchromic red cells, spherical in shape and without central pallor (Moreira et al., 2011). In this study, propolis extracts decreased the fragility of the erythrocyte membrane, with the effect related to the phenolic content of propolis. Furthermore, chestnut propolis also exhibits a monoamine oxidase inhibiting effect, an approach often used in the treatment of depressive disorders and some neurodegenerative illnesses, such as Parkinson's and Alzheimer's diseases, and may be effective in protecting humans against depression and similar diseases (Yıldız et al., 2014).

Extracts of the outer brown peel, leaf and catkins (male flowers) of chestnuts have shown an interesting ability to protect liposomes from peroxidation, as the phenolics of their extracts intercalate into the membranes to remove lipid peroxides (Živković et al., 2009b). Phenolic extracts prepared from chestnut leaf, catkin and spiny burs also prevented oxidative stress-induced β-cell (Rin-5F) death, a phenomenon with a central role in the onset of diabetes. The extracts increased cell viability by protecting DNA from oxidative damage and by enhancing the natural antioxidant system, while improving the cell's redox status, and inhibiting lipid peroxidation (Mujić et al., 2011). Chestnut catkins have been studied as an ingredient in Portuguese pastry and cheese, with the aim of creating functional foods. Carocho et al. (2015a,b) processed the Portuguese 'Serra da Estrela' cheese and Portuguese traditional cakes 'económicos', into functional products showing high phenolic content and higher antioxidant activity, which leads to an increase in consumer intake of the bioactive compounds present in chestnut flowers. Chestnut leaves have been traditionally used for teas in the treatment of coughs and diarrhoea (Calliste et al., 2005), lower back pain and stiff joints or muscles (Díaz-Reinoso et al., 2011; Reinoso et al., 2012) with their effect probably linked to the presence of phenolic compounds, namely ellagic acid and gallic acid derivatives (Živković et al., 2009c).

Bud extracts and bud preparations form a new category of plant products. Also based on the use of *Castanea* spp. as herbal medicines in homeopathy and in modern phytotherapy, old macerations of buds sampled from fertile long shoots (sleeping and breaking stages) from *C. sativa* and *C. crenata* were found to be a good source of vitamins, terpenes and polyphenolic compounds, which are responsible for the popularized effects on stagnant and vascular fluids or against recurrent cystitis and for its curative and restorative properties (Donno et al., 2014).

## 9 Future trends in research

The recent upward trend in chestnut production and consumption has been supported by recent research findings regarding its nutritional value and beneficial health effects, as well as improved knowledge and awareness of the stakeholders in the value chain that are willing to improve the use of either fresh or processed chestnuts. Although chestnut production is mainly focused on the production of fresh fruit and timber, the recent advances in food technology and health status have shown other neglected potentialities of this crop.

In the area of food science and technology, there are endless opportunities to explore other food and non-food products derived or obtained from chestnut production which will create added value in this value chain. Thus, further research should be aimed at developing novel products or new recipes from traditional chestnut products. For instance, honey and herbal teas from chestnut leaves, buds and catkins need further investigation, based on their potential health benefits and for added value in the chestnut value chain.

Concepts such as 'organic food', 'zero-kilometre food' or 'family farming' have an increasingly bigger role on the world's agricultural stage and could represent a big opportunity for producers to develop new market opportunities and expand the culture of this crop.

## 10 Acknowledgement

The authors would like to pay tribute to our colleague Richard Neil Bennett. A noted figure in chestnut research, Richard was prevented by a severe stroke from making what would have been a valuable contribution to this book.

## 11 References

Aires, A., Carvalho, R. and Saavedra, M. J. 2016. Valorization of solid wastes from chestnut industry processing: extraction and optimization of polyphenols, tannins and ellagitannins and its potential for adhesives, cosmetic and pharmaceutical industry. *Waste Manag.* 48, 457–64. doi:10.1016/j.wasman.2015.11.019.

Ajibola, A., Chamunorwa, J. P. and Erlwanger, K. H. 2012. Nutraceutical values of natural honey and its contribution to human health and wealth. *Nutr. Metab.* 9(1), 61. doi:10.1186/1743-7075-9-61.

Akbulut, M., Bozhuyuk, M. R., Ercisli, S., Skender, A. and Sorkheh, K. 2017. Chemical composition of seed propagated chestnut genotypes from Northeastern Turkey. *Not. Bot. Hort. Agrobot. Cluj-Napoca* 45(2), 425–30. doi:10.15835/nbha45210931.

Alasalvar, C. and Bolling, B. W. 2015. Review of nut phytochemicals, fat-soluble bioactives, antioxidant components and health effects. *Br. J. Nutr.* 113(S2), S68–78. doi:10.1017/S0007114514003729.

Alissandrakis, E., Tarantilis, P. A., Pappas, C., Harizanis, P. C. and Polissiou, M. 2011. Investigation of organic extractives from unifloral chestnut (*Castanea sativa* L.)

and eucalyptus (*Eucalyptus globulus* Labill) honey and flowers to identification of botanical marker compounds. *LWT Food Sci. Technol.* 44(4), 1042–51. doi:10.1016/j.lwt.2010.10.002.

Almeida, I. F., Valentão, P., Andrade, P. B., Seabra, R. M., Pereira, T. M., Amaral, M. H., Costa, P. C. and Bahia, M. F. 2008. In vivo skin irritation potential of a *Castanea sativa* (Chestnut) leaf extract, a putative natural antioxidant for topical application. *Basic Clin. Pharmacol. Toxicol.* 103(5), 461–7. doi:10.1111/j.1742-7843.2008.00301.x.

Almeida, I. F., Maleckova, J., Saffi, R., Monteiro, H., Goios, F., Amaral, M. H., Costa, P. C., Garrido, J., Silva, P., Pestana, N., et al. 2015. Characterization of an antioxidant surfactant-free topical formulation containing *Castanea sativa* leaf extract. *Drug Dev. Ind. Pharm. J.* 41(1), 148–55. doi:10.3109/03639045.2013.850712.

Amarowicz, R., Gong, Y. and Pegg, R. B. 2017. Recent advances in our knowledge of the biological properties of nuts. In: Ferreira, I. C. F. R., Morales, P. and Barros, L. (Eds), *Wild Plants, Mushrooms and Nuts: Functional Food Properties and Applications*. John Wiley and Sons Ltd, United Kingdom, pp. 377–409. Chapter 11.

Anderson, J. W., Davidson, M. H., Blonde, L., Brown, W. V., Howard, W. J., Ginsberg, H., Allgood, L. D. and Weingand, K. W. 2000. Long-term cholesterol-lowering effects of psyllium as an adjunct to diet therapy in the treatment of hypercholesterolemia. *Am. J. Clin. Nutr.* 71(6), 1433–8. doi:10.1093/ajcn/71.6.1433.

Antonio, A. L., Fernandes, Â., Barreira, J. C., Bento, A., Botelho, M. L. and Ferreira, I. C. 2011. Influence of gamma irradiation in the antioxidant potential of chestnuts (*Castanea sativa* Mill.) fruits and skins. *Food Chem. Toxicol.* 49(9), 1918–23. doi:10.1016/j.fct.2011.02.016.

Attanasio, G., Cinquanta, L., Albanese, D. and Matteo, M. D. 2004. Effect of drying temperature on physico-chemical properties of dried and rehydrated chestnuts (*Castanea sativa*). *Food Chem.* 88(4), 583–90. doi:10.1016/j.foodchem.2004.01.071.

Avşar, C., Özler, H., Berber, İ. and Civek, S. 2016. Phenolic composition, antimicrobial and antioxidant activity of *Castanea sativa* Mill. pollen grains from Black Sea region of Turkey. *Int. Food Res. J.* 23, 1711–6.

Barreira, J. C. M., Ferreira, I. C. F. R., Oliveira, M. B. P. P. and Pereira, J. A. 2008. Antioxidant activities of the extracts from chestnut flower, leaf, skins and fruit. *Food Chem.* 107(3), 1106–13. doi:10.1016/j.foodchem.2007.09.030.

Barreira, J. C., Casal, S., Ferreira, I. C., Oliveira, M. B. P. and Pereira, J. A. 2009. Nutritional, fatty acid and triacylglycerol profiles of *Castanea sativa* Mill. cultivars: a compositional and chemometric approach. *J. Agric. Food Chem.* 57(7), 2836–42. doi:10.1021/jf803754u.

Barreira, J. C., Ferreira, I. C., Oliveira, M. B. and Pereira, J. A. 2010. Antioxidant potential of chestnut (*Castanea sativa* L.) and almond (*Prunus dulcis* L.) by-products. *Food Sci. Technol. Int.* 16(3), 209–16. doi:10.1177/1082013209353983.

Barreira, J. C. M., Antonio, A. L., Günaydi, T., Alkan, H., Bento, A., Botelho, M. L. and Ferreira, I. C. F. R. 2012a. Chemometric characterization of gamma irradiated chestnuts from Turkey. *Radiat. Phys. Chem.* 81(9), 1520–4. doi:10.1016/j.radphyschem.2012.01.005.

Barreira, J. C., Casal, S., Ferreira, I. C., Peres, A. M., Pereira, J. A. and Oliveira, M. B. P. 2012b. Chemical characterization of chestnut cultivars from three consecutive years: chemometrics and contribution for authentication. *Food Chem. Toxicol.* 50(7), 2311–7. doi:10.1016/j.fct.2012.04.008.

Barros, A. I., Nunes, F. M., Gonçalves, B., Bennett, R. N. and Silva, A. P. 2011. Effect of cooking on total vitamin C contents and antioxidant activity of sweet chestnuts (*Castanea sativa* Mill.). *Food Chem.* 128(1), 165–72. doi:10.1016/j.foodchem.2011.03.013.

Basile, A., Sorbo, S., Giordano, S., Ricciardi, L., Ferrara, S., Montesano, D., Cobianchi, R. C., Vuotto, M. L. and Ferrara, L. 2000. Antibacterial and allelopathic activity of extract from *Castanea sativa* leaves. *Fitoterapia* 71(Suppl. 1), S110-6. doi:10.1016/S0367-326X(00)00185-4.

Belitz, H.-D. and Grosch, W. 1999. Minerals. In: Belitz, H.-D. and Grosch, W. (Eds), *Food Chemistry* (2nd edn.). Springer Verlag, Berlin, Germany, pp. 395–401. Chapter 7.

Bellini, E., Giordani, E., Marinelli, C., Migliorini, M. and Funghini, L. 2008. Marrone del Mugello PGI: nutritional and organoleptic quality of European chestnut (*Castanea sativa* Mill.). *IV International Chestnut Symposium*, Beijing, China, pp. 61-8. doi:10.17660/ActaHortic.2009.844.7.

Benatti, P., Peluso, G., Nicolai, R. and Calvani, M. 2004. Polyunsaturated fatty acids: biochemical, nutritional and epigenetic properties. *J. Am. Coll. Nutr.* 23(4), 281-302. doi:10.1080/07315724.2004.10719371.

Bento, A., Cabo, P. and Malheiro, R. 2016. Nuts: agricultural and economic importance worldwide. In: Ferreira, I. C. F. R., Morales, P. and Barros, L. (Eds), *Wild Plants, Mushrooms and Nuts: Functional Food Properties and Applications*. John Wiley and Sons Ltd, United Kingdom, pp. 353–76. Chapter 10.

Bertoncelj, J., Doberšek, U., Jamnik, M. and Golob, T. 2007. Evaluation of the phenolic content, antioxidant activity and colour of Slovenian honey. *Food Chem.* 105(2), 822-8. doi:10.1016/j.foodchem.2007.01.060.

Bettler, B., Kaupmann, K., Mosbacher, J. and Gassmann, M. 2004. Molecular structure and physiological functions of GABA(B) receptors. *Physiol. Rev.* 84(3), 835-67. doi:10.1152/physrev.00036.2003.

Blaiotta, G., La Gatta, B., Di Capua, M., Di Luccia, A., Coppola, R. and Aponte, M. 2013. Effect of chestnut extract and chestnut fiber on viability of potential probiotic *Lactobacillus* strains under gastrointestinal tract conditions. *Food Microbiol.* 36(2), 161-9. doi:10.1016/j.fm.2013.05.002.

Borges, O. P., Carvalho, J. S., Correia, P. R. and Silva, A. P. 2007. Lipid and fatty acid profiles of *Castanea sativa* Mill. chestnuts of 17 native Portuguese cultivars. *J. Food Compos. Anal.* 20, 80-9.

Borges, O., Gonçalves, B., de Carvalho, J. L. S., Correia, P. and Silva, A. P. 2008. Nutritional quality of chestnut (*Castanea sativa* Mill.) cultivars from Portugal. *Food Chem.* 106(3), 976-84. doi:10.1016/j.foodchem.2007.07.011.

Braga, N., Rodrigues, F. and Oliveira, M. B. 2015. *Castanea sativa* by-products: a review on added value and sustainable application. *Nat. Prod. Res.* 29(1), 1-18. doi:10.1080/14786419.2014.955488.

Brizi, C., Santulli, C., Micucci, M., Budriesi, R., Chiarini, A., Aldinucci, C. and Frosini, M. 2016. Neuroprotective effects of *Castanea sativa* Mill. Bark extract in human neuroblastoma cells subjected to oxidative stress. *J. Cell. Biochem.* 117(2), 510-20. doi:10.1002/jcb.25302.

Calliste, C. A., Trouillas, P., Allais, D. P. and Duroux, J. L. 2005. *Castanea sativa* Mill. leaves as new sources of natural antioxidant: an electronic spin resonance study. *J. Agric. Food Chem.* 53(2), 282-8. doi:10.1021/jf049341c.

Carocho, M., Antonio, A. L., Barros, L., Bento, A., Botelho, M. L., Kaluska, I. and Ferreira, I. C. 2012. Comparative effects of gamma and electron beam irradiation on the antioxidant potential of Portuguese chestnuts (*Castanea sativa* Mill.). *Food Chem. Toxicol.* 50(10), 3452-5. doi:10.1016/j.fct.2012.07.041.

Carocho, M., Barros, L., Antonio, A. L., Barreira, J. C. M., Bento, A., Kaluska, I. and Ferreira, I. C. F. R. 2013. Analysis of organic acids in electron beam irradiated chestnuts (*Castanea sativa* Mill.): effects of radiation dose and storage time. *Food Chem. Toxicol.* 55, 348–52. doi:10.1016/j.fct.2013.01.031.

Carocho, M., Antonio, A. L., Barreira, J. C. M., Rafalski, A., Bento, A. and Ferreira, I. C. F. R. 2014a. Validation of gamma and electron beam irradiation as alternative conservation technology for European chestnuts. *Food Bioprocess. Tech.* 7(7), 1917–27. doi:10.1007/s11947-013-1186-5.

Carocho, M., Calhelha, R. C., Queiroz, M. R. P., Bento, A., Morales, P., Soković, M. and Ferreira, I. C. F. R. 2014b. Infusions and decoctions of *Castanea sativa* flowers as effective antitumor and antimicrobial matrices. *Ind. Crops Prod.* 62, 42–6. doi:10.1016/j.indcrop.2014.08.016.

Carocho, M., Barreira, J. C., Bento, A., Morales, P. and Ferreira, I. C. 2014c. Chestnut flowers as functionalizing agents to enhance the antioxidant properties of highly appreciated traditional pastry. *Food Funct.* 5(11), 2989–95. doi:10.1039/c4fo00552j.

Carocho, M., Barros, L., Bento, A., Santos-Buelga, C., Morales, P. and Ferreira, I. C. 2014d. *Castanea sativa* Mill. flowers amongst the most powerful antioxidant matrices: a phytochemical approach in decoctions and infusions. *BioMed Res. Int.* 2014, 232956. doi:10.1155/2014/232956.

Carocho, M., Barreira, J. C. M., Antonio, A. L., Bento, A., Morales, P. and Ferreira, I. C. F. R. 2015a. The incorporation of plant materials in 'Serra da Estrela' cheese improves antioxidant activity without changing the fatty acid profile and visual appearance. *Eur. J. Lipid Sci. Technol.* 117(10), 1607–14. doi:10.1002/ejlt.201500018.

Carocho, M., Barreira, J. C. M., Barros, L., Bento, A., Cámara, M., Morales, P. and Ferreira, I. C. F. R. 2015b. Traditional pastry with chestnut flowers as natural ingredients: an approach of the effects on nutritional value and chemical composition. *J Food Compos. Anal.* 44, 93–101. doi:10.1016/j.jfca.2015.08.003.

Castro-Vázquez, L., Díaz-Maroto, M. C., De Torres, C. and Pérez-Coello, M. S. 2010. Effect of geographical origin on the chemical and sensory characteristics of chestnut honeys. *Food Res. Int.* 43(10), 2335–40. doi:10.1016/j.foodres.2010.07.007.

Chambial, S., Dwivedi, S., Shukla, K. K., John, P. J. and Sharma, P. 2013. Vitamin C in disease prevention and cure: an overview. *Indian J. Clin. Biochem.* 28(4), 314–28. doi:10.1007/s12291-013-0375-3.

Chenlo, F., Moreira, R., Chaguri, L. and Torres, M. D. 2009. Note. Sugar, moisture contents, and color of chestnuts during different storage regimes. *Food Sci. Technol. Int.* 15(2), 169–78. doi:10.1177/1082013208106321.

Chiarini, A., Micucci, M., Malaguti, M., Budriesi, R., Ioan, P., Lenzi, M., Fimognari, C., Gallina Toschi, T., Comandini, P. and Hrelia, S. 2013. Sweet chestnut (*Castanea sativa* Mill.) bark extract: cardiovascular activity and myocyte protection against oxidative damage. *Oxid. Med. Cell. Longev.* 2013, 471790. doi:10.1155/2013/471790.

Cho, G. S. and Jo, J. S. 2003. Screening of antimicrobial activity from *Castanea crenata* Sieb. et Zucc. leaves and flowers. I. Chemical compositions. *J. Korean Soc. Agric. Biotech.* 46, 257–61.

Comandini, P., Lerma-García, M. J., Simó-Alfonso, E. F. and Toschi, T. G. 2014. Tannin analysis of chestnut bark sample (*Castanea sativa* Mill.) by HPLC-DAD-MS. *Food Chem.* 157, 290–5. doi:10.1016/j.foodchem.2014.02.003.

Cristofori, V., Muganu, M., Graziosi, P., Bertazza, G. and Bignami, C. 2007. Comparison of nut traits and quality evaluation of chestnut (*Castanea sativa* Mill.) germplasm in Latium Region (Central Italy). *International Workshop on Chestnut Management in Mediterranean Countries - Problems and Prospects*, Bursa, Turkey, pp. 133–40. doi:10.17660/ActaHortic.2009.815.17.

Cruz, B. R., Abraão, A. S., Lemos, A. M. and Nunes, F. M. 2013. Chemical composition and functional properties of chestnut starch (*Castanea sativa* Mill). *Carbohyd. Polym.* 94(1), 594–602. doi:10.1016/j.carbpol.2012.12.060.

Dall'Asta, C., Cirlini, M., Morini, E., Rinaldi, M., Ganino, T. and Chiavaro, E. 2013. Effect of chestnut flour supplementation on physico-chemical properties and volatiles in bread making. *LWT Food Sci. Technol.* 53(1), 233–9. doi:10.1016/j.lwt.2013.02.025.

Das, A. K., Babylatha, R., Pavithra, A. S. and Khatoon, S. 2013. Thermal degradation of groundnut oil during continuous and intermittent frying. *J. Food Sci. Technol.* 50(6), 1186–92. doi:10.1007/s13197-011-0452-7.

De Jesús, M. C., Domínguez, R., Cantalapiedra, J., Iglesias, A. and Lorenzo, J. M. 2016. Effect of the amount of chestnuts in the diet of Celta pigs on the fatty acid profile of dry-cured lacon. *Grasas Aceites* 67(1), e119.

Delgado, T., Pereira, J. A., Ramalhosa, E. and Casal, S. 2016. Effect of hot air convective drying on the fatty acid and vitamin E composition of chestnut (*Castanea sativa* Mill.) slices. *Eur. Food Res. Technol.* 242(8), 1299–306. doi:10.1007/s00217-015-2633-5.

Delgado, T., Pereira, J. A., Ramalhosa, E. and Casal, S. 2017. Osmotic dehydration effects on major and minor components of chestnut (*Castanea sativa* Mill.) slices. *J. Food Sci. Technol.* 54(9), 2694–703. doi:10.1007/s13197-017-2706-5.

Delgado, T., Ramalhosa, E., Pereira, J. A. and Casal, S. 2018. Organic acid profile of chestnut (*Castanea sativa* Mill.) as affected by hot air convective drying. *Int. J. Food Prop.* 21(1), 557–65. doi:10.1080/10942912.2018.1454945.

Díaz-Reinoso, B., Moure, A., Domínguez, H. and Parajó, J. C. 2011. Membrane concentration of antioxidants from *Castanea sativa* leaves aqueous extracts. *Chem. Eng. J.* 175, 95–102. doi:10.1016/j.cej.2011.09.059.

Dimitri, G., Mastrocola, D., Sacchetti, G. and Neri, L. 2008. Chemical composition and functional properties of three sweet chestnut (*Castanea sativa* Mill.) ecotypes from Italy. *IV International Chestnut Symposium*, Beijing, China, pp. 41–6.

Dinis, L. T., Oliveira, M. M., Almeida, J., Costa, R., Gomes-Laranjo, J. and Peixoto, F. 2012a. Antioxidant activities of chestnut nut of *Castanea sativa* Mill. (cultivar 'Judia') as function of origin ecosystem. *Food Chem.* 132(1), 1–8. doi:10.1016/j.foodchem.2011.09.096.

Dinis, L. T., Peixoto, F., Ferreira-Cardoso, J. V., Morais, J. J. L., Borges, A. D. S., Nunes, F. M., Gomes-Laranjo, J. F., Costa, R. and Gomes-Laranjo, J. 2012b. Influence of the growing degree-days on chemical and technological properties of chestnut fruits (var. 'Judia'). *CyTA J. Food* 10(3), 216–24. doi:10.1080/19476337.2011.631713.

Donno, D., Beccaro, G. L., Masnari, F., Mellano, G. M., Cerutti, A. K. and Bounous, G. 2014. *Castanea* spp. as phytochemical source for herbal preparations. *Proceedings of the IInd European Congress on Chestnut. Acta Hortic.*, pp. 75–82. doi:10.17660/ActaHortic.2014.1043.8.

Ertan, E., Erdal, E., Alkan, G. and Algül, B. E. 2015. Effects of different postharvest storage methods on the quality parameters of chestnuts (*Castanea sativa* Mill.). *HortScience* 50(4), 577–81. doi:10.21273/HORTSCI.50.4.577.

Ertürk, U., Cevriye, M. and Arif, S. 2006. Chemical composition of fruits of some important chestnut cultivars. *Braz. Arch. Biot. Technol.* 49(2), 183–8. doi:10.1590/S1516-89132006000300000.

España, M. S. A., Rodríguez Galdón, B., Díaz Romero, C. and Rodríguez Rodríguez, E. 2011. Fatty acid profile in varieties of chestnut fruits from Tenerife (Spain) Perfil de ácidos grasos en variedades de castañas procedentes de Tenerife (España). *CyTA J. Food* 9(1), 77–81. doi:10.1080/19476331003686858.

FAO. 2017. Fruit and vegetables for health initiative – FAO. Available at: http://www.fao.org/3/a-i6807e.pdf.

FAOstat. 2018. Agriculture data (online). Available at: http://www.fao.org/faostat/en/#data/QC (accessed on 1 July 2018).

FAO/WHO. 2004. *Vitamin and Mineral Requirements in Human Nutrition* (2nd edn.). Bangkok, Thailand.

Fernandes, Â., Barreira, J. C., Antonio, A. L., Bento, A., Botelho, M. L. and Ferreira, I. C. 2011. Assessing the effects of gamma irradiation and storage time in energetic value and in major individual nutrients of chestnuts. *Food Chem. Toxicol.* 49(9), 2429–32. doi:10.1016/j.fct.2011.06.062.

Ferreira-Cardoso, J. V. 2002. Caracterização, valorização do fruto de *Castanea sativa* Mill.; Composição química, valor nutritivo e efeito da sua utilização na engorda de porcos Duroc. Doctoral thesis. Universidade de Trás-os-Montes e Alto Douro, Vila Real.

Gajcy, K., Lochyński, S. and Librowski, T. 2010. A role of GABA analogues in the treatment of neurological diseases. *Curr. Med. Chem.* 17(22), 2338–47. doi:10.2174/092986710791698549.

Galiñanes, C., Freire, M. S. and González-Álvarez, J. 2015. Antioxidant activity of phenolic extracts from chestnut fruit and forest industries residues. *Eur. J. Wood Prod.* 73(5), 651–9. doi:10.1007/s00107-015-0936-7.

Gao, H., Liu, S. and Shi, P. 2011. Progress of Chinese chestnut process technology and production model. *Food Sci.* 32, 141–3.

Garcia, J., Agar, I. and Streif, I. 1992. Fat content and fatty acid composition in individual seeds of pistachio varieties grown in Turkey. *Gartenbauwissenschaft* 57, 130–3.

Geleijnse, J. M. and Hollman, P. Ch. 2008. Flavonoids and cardiovascular health: which compounds, what mechanisms? *Am. J. Clin. Nutr.* 88, 12–3. doi:10.1093/ajcn/88.1.12.

Glushkova, M., Zhyanski, M. and Velinova, K. 2010. Nut quality assessment of chestnut cultivars from 'Ivanik' clone collection. *Forest Sci.* 1, 3–14.

Gonçalves, B., Borges, O., Costa, H. S., Bennett, R., Santos, M. and Silva, A. P. 2010. Metabolite composition of chestnut (*Castanea sativa* Mill.) upon cooking: proximate analysis, fibre, organic acids and phenolics. *Food Chem.* 122(1), 154–60. doi:10.1016/j.foodchem.2010.02.032.

Gonçalves, B., Borges, O., Rosa, E., Coutinho, J. and Silva, A. P. 2012. Effect of cooking on free amino acid and mineral profiles of sweet chestnut (*Castanea sativa* Mill.). *Fruits* 67(3), 201–14. doi:10.1051/fruits/2012013.

Gounga, M. E., Xu, S.-Y. and Wang, Z. 2007. Sensory attributes of freshly roasted and roasted freeze dried Chinese chestnut (*Castanea mollissima*) coated with whey protein isolate-pullulan edible coating. *Int. J. Agric. Res.* 2, 959–64.

Gounga, M. E., Xu, S. Y., Wang, Z. and Yang, W. G. 2008. Effect of whey protein isolate-pullulan edible coatings on the quality and shelf life of freshly roasted and freeze-dried Chinese chestnut. *J. Food Sci.* 73(4), E155–61. doi:10.1111/j.1750-3841.2008.00694.x.

Gounga, M. E., Omar, K. and Amadou, I. 2017. Volatile compounds analysis in chestnuts and chocolate-coated chestnuts using GC-MS. *Int. J. Curr. Res. Biosci. Plant Biol.* 5(3), 70–7.

Grdović, N., Dinić, S., Arambašić, J., Mihailović, M., Uskoković, A., Marković, J., Poznanović, G., Vidović, S., Zeković, Z., Mujić, A., et al. 2012. The protective effect of a mix of *Lactarius deterrimus* and *Castanea sativa* extracts on streptozotocin-induced oxidative stress and pancreatic β-cell death. *Br. J. Nutr.* 108(7), 1163–76. doi:10.1017/S0007114511006702.

Güneş, M. E., Şahin, S., Demir, C., Borum, E. and Tosunoğlu, A. 2017. Determination of phenolic compounds profile in chestnut and floral honeys and their antioxidant and antimicrobial activities. *J. Food Biochem.* 41(3), e12345. doi:10.1111/jfbc.12345.

Ham, J. S., Kim, H. Y. and Lim, S. T. 2015. Antioxidant and deodorizing activities of phenolic components in chestnut inner shell extracts. *Ind. Crops Prod.* 73, 99–105. doi:10.1016/j.indcrop.2015.04.017.

Hansson, G. K. 2005. Inflammation, atherosclerosis, coronary artery disease. *N. Engl. J. Med.* 352(16), 1685–95. doi:10.1056/NEJMra043430.

Hardaman, W. E. 2002. Omega-3 fatty acids to augment cancer therapy. *J. Nutr.* 132, 3508–12.

Hou, F., Shi, X., Li, Q., Xie, S., Wu, Y. and Ouyang, J. 2016. Nutritional quality of Chinese chestnut and effect of cooking on its bioactive compounds and antioxidant activity. *J. Food Process. Preserv.* 40(6), 1383–90. doi:10.1111/jfpp.12723.

Jeong, H. R., Jo, Y. N., Jeong, J. H., Jin, D. E., Song, B. G., Jin, Y. R., Kim, M., Lee, U. and Heo, H. 2012. Change in the chemical composition of chestnuts (*Castanea crenata*) from different periods. *Korean J. Food Sci. Technol.* 44(4), 393–400. doi:10.9721/KJFST.2012.44.4.393.

Jermini, M., Conedera, M., Sieber, T. N., Sassella, A., Schärer, H., Jelmini, G. and Hoehn, E. 2006. Influence of fruit treatments on perishability during cold storage of sweet chestnuts. *J. Sci. Food Agric.* 86(6), 877–85. doi:10.1002/jsfa.2428.

Jesch, E. D. and Carr, T. P. 2017. Food ingredients that inhibit cholesterol absorption. *Prev. Nutr. Food Sci.* 22(2), 67–80. doi:10.3746/pnf.2017.22.2.67.

Jukić, H., Jug, T., Nikolić, G., Trutić, N., Mujić, I., Rudić, D. and Zivković, J. 2013. Antioxidant and antibacterial properties of *Castanea sativa* Mill. catkins extracts. *II European Congress on Chestnut*, Debrecen, Hungary, pp. 167–73.

Kalogeropoulos, N., Chiou, A., Ioannou, M. S. and Karathanos, V. T. 2013. Nutritional evaluation and health promoting activities of nuts and seeds cultivated in Greece. *Int. J. Food Sci. Nutr.* 64(6), 757–67. doi:10.3109/09637486.2013.793298.

Karadal, F., Onmaz, N. E., Abay, S., Yildirim, Y., Al, S., Tatyuz, I. and Akcay, A. 2018. A study of antibacterial and antioxidant activities of bee products: propolis, pollen and honey samples. *Ethiop. J. Health Dev.* 32(2).

Kirbaslar, F. G., Türker, G., Özsoy-Günes, Z., Ünal, M., Dülger, B., Ertas, E. and Kizilkaya, B. 2012. Evaluation of fatty acid composition, antioxidant and antimicrobial activity, mineral composition and calorie values of some nuts and seeds from Turkey. *Rec. Nat. Prod.* 6(4), 339–49.

Kolayli, S., Sahin, H., Can, Z., Yildiz, O. and Sahin, K. 2015. Honey shows potent inhibitory activity against the bovine testes hyaluronidase. *J. Enzyme Inhib. Med. Chem.* 31(4), 599–602. doi:10.3109/14756366.2015.1054819.

Kolayli, S., Can, Z., Yildiz, O., Sahin, H. and Karaoglu, S. A. 2016. A comparative study of the antihyaluronidase, antiurease, antioxidant, antimicrobial and physicochemical properties of different unifloral degrees of chestnut (*Castanea sativa* Mill.) honeys. *J. Enzyme Inhib. Med. Chem.* 31(sup3), 96–104. doi:10.1080/14756366.2016.1209494.

Korel, F. and Balaban, M. 2008. Chemical composition and health aspects of chestnut (*Castanea* spp.). In: Alasalvar, C. and Shahidi, F. (Eds), *Tree Nuts: Composition, Phytochemicals and Health Effects*. CRC Press/Taylor and Francis Group, Boca Raton, FL, pp. 171–84. Chapter 11. doi:10.1201/9781420019391.ch11.

Küçük, M., Kolayli, S., Alpay Karaoğlu, Ş., Ulusoy, E., Baltaci, C. and Candan, F. 2007. Biological activities and chemical composition of three honeys of different types from Anatolia. *Food Chem.* 100(2), 526–34. doi:10.1016/j.foodchem.2005.10.010.

Kuncic, M. K., Jaklic, D., Lapanje, A. and Gunde-Cimerman, N. 2012. Antibacterial and antimycotic activities of Slovenian honeys. *Br. J. Biomed. Sci.* 69(4), 154–8.

Künsch, U., Schärer, H., Patrian, B., Hurter, J., Conedera, M., Sassella, A., Jermini, M. and Jelmini, G. 1999. Quality assessment of chestnut fruits. *Acta Hortic.* 494(494), 119–28. doi:10.17660/ActaHortic.1999.494.17.

Künsch, U., Schärer, H., Patrian, B., Höhn, E., Conedera, M., Sassella, A., Jermini, M. and Jelmini, G. 2001. Effects of roasting on chemical composition and quality of different chestnut (*Castanea sativa* Mill.) varieties. *J. Sci. Food Agric.* 81(11), 1106–12. doi:10.1002/jsfa.916.

Lau, A. and Tymianski, M. 2010. Glutamate receptors, neurotoxicity and neurodegeneration. *Pflugers Arch.* 460(2), 525–42. doi:10.1007/s00424-010-0809-1.

Lee, H. J., Choi, I. H., Kim, D. H., Amanullah, S. M. and Kim, S. C. 2016a. Nutritional characterization of tannin rich chestnut (*Castanea*) and its meal for pig. *J. Appl. Ani. Res.* 44(1), 258–62. doi:10.1080/09712119.2015.1031779.

Lee, U., Joo, S., Klopfenstein, N. B. and Kim, M.-S. 2016b. Efficacy of washing treatments in the reduction of postharvest decay of chestnuts (*Castanea crenata* 'Tsukuba') during storage. *Can. J. Plant Sci.* 96(1), 1–5. doi:10.1139/cjps-2015-0089.

Lenzi, M., Malaguti, M., Cocchi, V., Hrelia, S. and Hrelia, P. 2017. *Castanea sativa* Mill. bark extract exhibits chemopreventive properties triggering extrinsic apoptotic pathway in Jurkat cells. *BMC Complement. Altern. Med.* 17(1), 251. doi:10.1186/s12906-017-1756-6.

Li, Q., Shi, X., Zhao, Q., Cui, Y., Ouyang, J. and Xu, F. 2016. Effect of cooking methods on nutritional quality and volatile compounds of Chinese chestnut (*Castanea mollissima* Blume). *Food Chem.* 201, 80–6. doi:10.1016/j.foodchem.2016.01.068.

Lin, D. and Zhao, Y. 2007. Innovations in the development and application of edible coatings for fresh and minimally processed fruits and vegetables. *Compr. Rev. Food Sci. Food Saf.* 6(3), 60–75. doi:10.1111/j.1541-4337.2007.00018.x.

Liu, F., Chang, X., Hu, X., Brennan, C. S. and Guo, X. 2017. Effect of thermal processing on phenolic profiles and antioxidant activities in *Castanea mollissima*. *Int. J. Food Sci. Technol.* 52(2), 439–47. doi:10.1111/ijfs.13299.

Martínez-Ballesta, M. C., Dominguez-Perles, R., Moreno, D. A., Muries, B., Alcaraz-López, C., Bastías, E., García-Viguera, C. and Carvajal, M. 2010. Minerals in plant food: effect of agricultural practices and role in human health. A review. *Agron. Sustain. Dev.* 30(2), 295–309. doi:10.1051/agro/2009022.

McCarthy, M. A. and Meredith, F. I. 1988. Nutrient data on chestnuts consumed in the United States. *Econ. Bot.* 42(1), 29–36. doi:10.1007/BF02859026.

Mellano, M. G., Beccaro, G. L., Donno, D., Marinoni, D. T., Boccacci, P., Canterino, S., Cerutti, A. K. and Bounous, G. 2012. *Castanea* spp. biodiversity conservation: collection and characterization of the genetic diversity of an endangered species. *Genet. Resour. Crop Evol.* 59(8), 1727–41. doi:10.1007/s10722-012-9794-x.

Mencarelli, F. 2001. Postharvest handling and storage of chestnuts (Online). Available at: http://www.fao.org/docrep/006/ac645e/ac645e00.HTM (accessed on 7 August 2018).

Mert, C. and Ertürk, Ü. 2017. Chemical compositions and sugar profiles of consumed chestnut cultivars in the Marmara Region, Turkey. *Not. Bot. Hort. Agrobot. Cluj-Napoca* 45(1), 203–7. doi:10.15835/nbha45110729.

Mertens-Talcott, S. U., Lee, J. H., Percival, S. S. and Talcott, S. T. 2006. Induction of cell death in caco-2 human colon carcinoma cells by ellagic acid rich fractions from muscadine grapes (*Vitis rotundifolia*). *J. Agric. Food Chem.* 54(15), 5336–43. doi:10.1021/jf060563f.

Meteoglu, I., Kavas, N. C., Saricaoglu, M., Ilkaya, F., Guzel, H., Alici, O. and Ozturk, F. 2015. Chestnut honey and sherbet enhance the healing of burn wounds in rat model. *Clin. Investig. Med.* 38, E267–73.

Míguelez, J. D. L. M., Bernárdez, M. M. and Queijeiro, J. G. 2004. Composition of varieties of chestnuts from Galicia (Spain). *Food Chem.* 84(3), 401–4. doi:10.1016/S0308-8146(03)00249-8.

Milbury, P. E., Chen, C. Y., Dolnikowski, G. G. and Blumberg, J. B. 2006. Determination of flavonoids and phenolics and their distribution in almonds. *J. Agric. Food Chem.* 54(14), 5027–33. doi:10.1021/jf0603937.

Moreira, L. L., Dias, T., Dias, L. G., Rogão, M., Da Silva, J. P. and Estevinho, L. M. 2011. Propolis influence on erythrocyte membrane disorder (hereditary spherocytosis): a first approach. *Food Chem. Toxicol.* 49(2), 520–6. doi:10.1016/j.fct.2010.11.042.

Moreira, R., Chenlo, F., Torres, M. D., Rama, B. and Arufe, S. 2015. Air drying of chopped chestnuts at several conditions: effect on colour and chemical characteristics of chestnut flour. *Int. Food Res. J.* 22, 407–13.

Morris, M. S., Jacques, P. F., Rosenberg, I. H. and Selhub, J. 2007. Folate and vitamin B-12 status in relation to anemia, macrocytosis, and cognitive impairment in older Americans in the age of folic acid fortification. *Am. J. Clin. Nutr.* 85(1), 193–200. doi:10.1093/ajcn/85.1.193.

Mota, M., Pinto, T., Vilela, A., Marques, T., Borges, A., Caço, J., Ferreira-Cardoso, J., Raimundo, F. and Gomes-Laranjo, J. 2018. Irrigation positively affects the chestnut's quality: the chemical composition, fruit size and sensory attributes. *Sci. Hortic.* 238, 177–86. doi:10.1016/j.scienta.2018.04.047.

Mujić, I., Živković, J., Zeković, Z. P., Nikolić, G. M., Vidović, S. S., Prgomet, Z. and Dudaš, S. D. 2009. Capacity of extracts of *Castanea sativa* Mill. for removal of organic, hydrophilic radicals. *I European Congress on Chestnut-Castanea*, Cuneo, Italy, pp. 647–52. doi:10.17660/ActaHortic.2010.866.86.

Mujić, A., Grdović, N., Mujić, I., Mihailović, M., Živković, J., Poznanović, G. and Vidaković, M. 2011. Antioxidative effects of phenolic extracts from chestnut leaves, catkins and spiny burs in streptozotocin-treated rat pancreatic β-cells. *Food Chem.* 125(3), 841–9. doi:10.1016/j.foodchem.2010.08.068.

Muzzaffar, S., Maqbool, K., Wani, I. A., Masoodi, F. A. and Bhat, M. M. 2016. Physico-chemical characterization of sweet chestnut (*Castanea sativa* L.) starch grown in temperate climate of Kashmir Valley. *Acta Alim.* 45(2), 258–67. doi:10.1556/066.2016.45.2.13.

Naumann, E., van Rees, A. B., Onning, G., Oste, R., Wydra, M. and Mensink, R. P. 2006. β-glucan incorporated into a fruit drink effectively lowers serum LDL-cholesterol concentrations. *Am. J. Clin. Nutr.* 83(3), 601–5. doi:10.1093/ajcn.83.3.601.

Navarro, S. 2012. The use of modified and controlled atmospheres for the disinfestation of stored products. *J. Pest. Sci.* 85(3), 301–22. doi:10.1007/s10340-012-0424-3.

Nazzaro, M., Barbarisi, C., La Cara, F. and Volpe, M. G. 2011. Chemical and biochemical characterisation of an IGP ecotype chestnut subjected to different treatments. *Food Chem.* 128(4), 930–6. doi:10.1016/j.foodchem.2011.03.121.

Nazzaro, M., Mottola, M. V., La Cara, F., Del Monaco, G., Aquino, R. P. and Volpe, M. G. 2012. Extraction and characterization of biomolecules from agricultural wastes. *Chem. Eng. Trans.* 27, 331–6.

Neri, L., Dimitri, G. and Sacchetti, G. 2010. Chemical composition and antioxidant activity of cured chestnuts from three sweet chestnut (*Castanea sativa* Mill.) ecotypes from Italy. *J. Food Compos. Anal.* 23(1), 23–9. doi:10.1016/j.jfca.2009.03.002.

Noda, T., Takigawa, S., Matsuura-Endo, C., Suzuki, T., Hashimoto, N., Kottearachchi, N. S., Yamauchi, H. and Zaidul, I. S. 2008. Factors affecting the digestibility of raw and gelatinized potato starches. *Food Chem.* 110(2), 465–70. doi:10.1016/j.foodchem.2008.02.027.

Oh, S.-H., Moon, Y.-J. and Oh, C.-H. 2003. γ-aminobutyric acid content in selected uncooked foods. *Prev. Nutr. Food Sci.* 8, 75–8.

Okuda, T. 2005. Systematics and health effects of chemically distinct tannins in medicinal plants. *Phytochemistry* 66(17), 2012–31. doi:10.1016/j.phytochem.2005.04.023.

Otles, S. and Selek, I. 2012. Phenolic compounds and antioxidant activities of chestnut (*Castanea sativa* Mill.) fruits. *Qual. Ass. Saf. Crop* 4(4), 169–216.

Özel, H. B. 2015. The effects of origin difference on some chemical properties of the fruit of chestnut (*Castanea sativa* Mill.). *Rom. Biotechnol. Lett.* 20(6), 10969.

Öztürk, S., Çetin, M. and Çetin, M. 2010. Chemical composition and fatty acid contents of chestnuts grown in Bursa Province, Turkey. *As. J. Chem.* 22(3), 1845–52.

Panagou, E., Vekiari, S. A. and Mallidis, C. 2006. The effect of modified atmosphere packaging of chestnuts in suppressing fungal growth and related physicochemical changes during storage in retail packages at 0 and 8°C. *Adv. Hortic. Sci.* 20(1), 82–9.

Park, J. S., An, S. J., Jeong, S. I., Gwon, H. J., Lim, Y. M. and Nho, Y. C. 2017. Chestnut honey impregnated carboxymethyl cellulose hydrogel for diabetic ulcer healing. *Polymers* 9(7), 248.

Payne, J. A., Jaynes, R. A. and Kays, S. J. 1983. Chinese chestnut production in the United States: practice, problems, and possible solutions. *Econ. Bot.* 37(2), 187–200. doi:10.1007/BF02858784.

Pena-Mendez, E. M., Hernández-Suárez, M., Díaz-Romero, C. and Rodríguez-Rodríguez, E. 2008. Characterization of various chestnut cultivars by means of chemometrics approach. *Food Chem.* 107(1), 537–44.

Pereira-Lorenzo, S., Ramos-Cabrer, A. M., Díaz-Hernández, M. B., Ciordia-Ara, M. and Ríos-Mesa, D. 2006. Chemical composition of chestnut cultivars from Spain. *Sci. Hortic.* 107(3), 306–14. doi:10.1016/j.scienta.2005.08.008.

Pinto, D., Braga, N., Rodrigues, F. and Oliveira, M. B. P. P. 2017. *Castanea sativa* bur: an undervalued by-product but a promising cosmetic ingredient. *Cosmetics* 4(4), 50. doi:10.3390/cosmetics4040050.

Poljak, I., Vahčić, N., Gačić, M. and Idžojtić, M. 2016. Morphological characterization and chemical composition of fruits of the traditional Croatian chestnut variety 'Lovran Marron'. *Food Technol. Biotechnol.* 54(2), 189-99. doi:10.17113/ftb.54.02.16.4319.

Portela, E. M. A. C. and Louzada, J. L. P. 2012. Early diagnosis of boron deficiency in chestnut. *J. Plant Nut.* 35(2), 304–10. doi:10.1080/01904167.2012.636132.

Powers, H. J. 2003. Riboflavin (vitamin B-2) and health. *Am. J. Clin. Nutr.* 77(6), 1352-60. doi:10.1093/ajcn/77.6.1352.

Prosky, L. 2000. When is dietary fiber considered a functional food? *BioFactors* 12(1-4), 289-97. doi:10.1002/biof.5520120143.

Quave, C. L., Lyles, J. T., Kavanaugh, J. S., Nelson, K., Parlet, C. P., Krosbi, H. A., Heilmann, K. P. and Horswill, A. R. 2016. *Castanea sativa* (European chestnut) leaf extracts rich in ursene and oleanene derivatives block *Staphylococcus aureus* virulence and pathogenesis without detectable resistance. *PLoS ONE* 11(9), e0163655. doi:10.1371/journal.pone.0163655.

Rackham, R. 1994. Critical factors in establishing chestnuts in the Pacific Northwest. In: Double, M. L. and MacDonald, W. L. (Eds), *Proceedings of the International Chestnut Conference*, Morgantown, WV. West Virginia University Press, Morgantown, WV, pp. 193-6.

Ras, R. T., Geleijnse, J. M. and Trautwein, E. A. 2014. LDL-cholesterol-lowering effect of plant sterols and stanols across different dose ranges: a meta-analysis of randomised controlled studies. *Br. J. Nutr.* 112(2), 214-9. doi:10.1017/S0007114514000750.

Reinoso, B. D., Couto, D., Moure, A., Fernandes, E., Domínguez, H. and Parajó, J. C. 2012. Optimization of antioxidants - Extraction from *Castanea sativa* leaves. *Chem. Eng. J.* 203, 101-9. doi:10.1016/j.cej.2012.06.122.

Ribeiro, B., Rangel, J., Valentão, P., Andrade, P. B., Pereira, J. A., Bölke, H. and Seabra, R. M. 2007. Organic acids in two Portuguese chestnut (*Castanea sativa* Miller) varieties. *Food Chem.* 100(2), 504-8. doi:10.1016/j.foodchem.2005.09.073.

Ribeiro, A. S., Estanqueiro, M., Oliveira, M. B. and Sousa Lobo, J. M. 2015. Main benefits and applicability of plant extracts in skin care products. *Cosmetics* 2(2), 48-65. doi:10.3390/cosmetics2020048.

Ríos-Mesa, D., Pereira-Lorenzo, S., González-Díaz, A. J., Hernández-González, J. Z., González-Díaz, E. and Saúco, V. G. 2011. The status of chestnut cultivation and utilization in the Canary Islands. *Adv. Hortic. Sci.* 25(2), 90-8.

Ros, E. 2010. Health benefits of nut consumption. *Nutrients* 2(7), 652-82. doi:10.3390/nu2070652.

Sacchetti, G. and Pinnavaia, G. G. 2004. Compositional characteristics of some chestnut biotypes of Emiliano-Romagnolo Apennine. *III International Chestnut Congress*, Chaves, Portugal, pp. 241-6. doi:10.17660/ActaHortic.2005.693.29.

Sakin-Yilmazer, M., Dirim, S. N., Di Pinto, D. and Kaymak-Ertekin, F. 2014. Yoghurt with candied chestnut: freeze drying, physical, and rheological behaviour. *J. Food Sci. Technol.* 51(12), 3949-55. doi:10.1007/s13197-012-0890-x.

Sánchez-Monge, R., Blanco, C., López-Torrejón, G., Cumplido, J., Recas, M., Figueroa, J., Carrillo, T. and Salcedo, G. 2006. Differential allergen sensitization patterns in chestnut allergy with or without associated latex-fruit syndrome. *J. Allergy Clin. Immunol.* 118(3), 705-10. doi:10.1016/j.jaci.2006.04.058.

Santos, F. T., Goufo, P., Santos, C., Botelho, D., Fonseca, J., Queirós, A., Costa, M. S. S. M. and Trindade, H. 2016. Comparison of five agro-industrial waste-based composts as growing media for lettuce: effect on yield, phenolic compounds and vitamin C. *Food Chem.* 209, 293-301. doi:10.1016/j.foodchem.2016.04.087.

Sanz, M., Cadahía, E., Esteruelas, E., Muñoz, A. M., Simón, B. F., Hernández, T. and Estrella, I. 2010. Phenolic compounds in chestnut (*Castanea sativa* Mill.) heartwood. Effect of toasting at cooperage. *J. Agric. Food Chem.* 58(17), 9631-40. doi:10.1021/jf102718t.

Sapkota, K., Park, S. E., Kim, J. E., Kim, S., Choi, H. S., Chun, H. S. and Kim, S. J. 2010. Antioxidant and antimelanogenic properties of chestnut flower extract. *Biosci. Biotechnol. Biochem.* 74(8), 1527–33. doi:10.1271/bbb.100058.

Sarikaya, A. O., Ulusoy, E., Özturk, N., Tuncel, M. and Kolayli, S. 2009. Antioxidant activity and phenolic acid constituents of chestnut (*Castania sativa* Mill.) honey and propolis. *J. Food Biochem.* 33(4), 470–81. doi:10.1111/j.1745-4514.2009.00231.x.

Sathe, S. K., Monaghan, E. K., Kshirsagar, H. H. and Venkatachalam, M. 2009. Chemical composition of edible nut seeds and its implications in human health. In: Alasalvar, C. and Shahidi, F. (Eds), *Tree Nuts: Composition, Phytochemicals and Health Effects*. CRC Press Taylor and Francis Group, Boca Raton, FL, pp. 11–35. Chapter 2. doi:10.1201/9781420019391.ch2.

Sato, A., Tanaka, K., Takada, N., Sawamura, Y. and Hirabayashi, T. 2010. Comparison of phenolic content of easily removed pellicle of Japanese chestnut 'Porotan' with other Japanese and Chinese chestnut cultivars. *J. Jpn. Soc. Hortic. Sci.* 79(3), 258–62. doi:10.2503/jjshs1.79.258.

Sato, A., Truong, V.-D., Johanningsmeier, S. D., Reynolds, R., Pecota, K. V. and Yencho, G. C. 2017. Chemical constituents of sweet potato genotypes in relation to textural characteristics of processed French fries. *J. Food Sci.* 83(1), 60–73.

Selenge, E., Murata, T., Tanaka, S., Sasaki, K., Batkhuu, J. and Yoshizaki, F. 2014. Monoterpene glycosides, phenylpropanoids, and acacetin glycosides from *Dracocephalum foetidum*. *Phytochemistry* 101, 91–100. doi:10.1016/j.phytochem.2014.02.007.

Selma, M. V., Beltrán, D., García-Villalba, R., Espín, J. C. and Tomás-Barberán, F. A. 2014. Description of urolithin production capacity from ellagic acid of two human intestinal Gordonibacter species. *Food Funct.* 5(8), 1779–84. doi:10.1039/c4fo00092g.

Seo, D. J., Chung, M. J., Kim, D. J., You, J. K. and Choe, M. 2009. Nutritional constituent analysis of Korean chestnuts. *J. Korean Soc. Food Sci. Nutr.* 38(2), 166–76. doi:10.3746/jkfn.2009.38.2.166.

Seo, K. H., Lee, J. Y., Debnath, T., Kim, Y. M., Park, J. Y., Kim, Y. O., Park, S. J. and Lim, B. O. 2016. DNA protection and antioxidant potential of chestnut shell extracts. *J. Food Biochem.* 40(1), 20–30. doi:10.1111/jfbc.12179.

Sequeira, C. A., Ferreira-Cardoso, J. V. and Torres-Pereira, J. M. G. 2004. Effect of year and cultivar on chemical composition of chestnuts from northeastern Portugal. *III International Chestnut Congress*, Chaves, Portugal, pp. 271–8.

Silva, A. P., Santos-Ribeiro, R., Borges, O., Magalhães, B., Silva, M. E. and Gonçalves, B. 2001. Effects of roasting and boiling on the physical and mechanical properties of 11 Portuguese chestnut cultivars (*Castanea sativa* Mill.). *CyTA J. Food* 9(3), 214–9.

Silva, B. M., Andrade, P. B., Valentão, P., Ferreres, F., Seabra, R. M. and Ferreira, M. A. 2004. Quince (*Cydonia oblonga* Miller) fruit (pulp, peel, and seed) and jam: antioxidant activity. *J. Agric. Food Chem.* 52(15), 4705–12. doi:10.1021/jf040057v.

Silva, A. P., Oliveira, I., Silva, M. E., Guedes, C. M., Borges, O., Magalhães, B. and Gonçalves, B. 2016. Starch characterization in seven raw, boiled and roasted chestnuts (*Castanea sativa* Mill.) cultivars from Portugal. *J. Food Sci. Technol.* 53(1), 348–58. doi:10.1007/s13197-015-2047-1.

Sorice, A., Siano, F., Capone, F., Guerriero, E., Picariello, G., Budillon, A., Ciliberto, G., Paolucci, M., Constantini, S. and Volpe, M. G. 2016. Potential anticancer effects of polyphenols from chestnut shell extracts: modulation of cell growth, and cytokinomic and metabolomic profiles. *Molecules* 21(10), 1411. doi:10.3390/molecules21101411.

Squillaci, G., Apone, F., Sena, L. M., Carola, A., Tito, A., Bimonte, M., Lucia, A. D., Colucci, G., Cara, F. L. and Morana, A. 2018. Chestnut (*Castanea sativa* Mill.) industrial wastes as a valued bioresource for the production of active ingredients. *Proc. Biochem.* 64, 228–36. doi:10.1016/j.procbio.2017.09.017.

Sunitha, K., Suresh, P., Santhosh, M. S., Hemshekhar, M., Thushara, R. M., Marathe, G. K., Thirunavukkarasu, C., Kemparaju, K., Kumar, M. S. and Girish, K. S. 2013. Inhibition of hyaluronidase by N-acetyl cysteine and glutathione: role of thiol group in hyaluronan protection. *Int. J. Biol. Macromol.* 55, 39–46. doi:10.1016/j.ijbiomac.2012.12.047.

Tian, B., Tao, H., Hai-ying, Z., Hong-wei, L. and Li-ping, L. 2009. Changes in carbohydrate metabolism in coated chestnuts during storage. *Acta Hortic.* 844, 75–82.

Tsujita, T., Takaku, T. and Suzuki, T. 2008. Chestnut astringent skin extract, an α-amylase inhibitor, retards carbohydrate absorption in rats and humans. *J. Nutr. Sci. Vitaminol.* 54(1), 82–8. doi:10.3177/jnsv.54.82.

Vadivel, V., Kunyanga, C. N. and Biesalski, H. K. 2012. Health benefits of nut consumption with special reference to body weight control. *Nutrition* 28(11-12), 1089–97. doi:10.1016/j.nut.2012.01.004.

VALCAST project data. 2008. Projecto VALCAST II- Valorização da castanha portuguesa e opimização do seu processamento. Relatório Técnico Final. UTAD. Set 2008.

Vasconcelos, M. D. C. B. M., Bennett, R. N., Rosa, E. A. and Cardoso, J. V. F. 2007. Primary and secondary metabolite composition of kernels from three cultivars of Portuguese chestnut (*Castanea sativa* Mill.) at different stages of industrial transformation. *J. Agric. Food Chem.* 55(9), 3508–16. doi:10.1021/jf0629080.

Vasconcelos, M. C., Bennett, R., Rosa, E., Ferreira-Cardoso, J. V. and Nunes, F. 2009. Industrial processing of chestnut fruits (*Castanea sativa* Mill.) - Effects on nutrients and phytochemicals. *I European Congress on Chestnut-Castanea*, Cuneo, Italy, pp. 611–7.

Vasconcelos, M. C. B. M., Bennett, R. N., Rosa, E. A. and Ferreira-Cardoso, J. V. 2010. Composition of European chestnut (*Castanea sativa* Mill.) and association with health effects: fresh and processed products. *J. Sci. Food Agric.* 90(10), 1578–89. doi:10.1002/jsfa.4016.

Vaughan, J. G. and Geissler, C. A. 1997. *The New Oxford Book of Food Plants*. Oxford University Press, Oxford, UK.

Vella, F. M., Laratta, B., La Cara, F. and Morana, A. 2018. Recovery of bioactive molecules from chestnut (*Castanea sativa* Mill.) by-products through extraction by different solvents. *Nat. Prod. Res.* 32(9), 1022–32. doi:10.1080/14786419.2017.1378199.

Veluri, R., Singh, R. P., Liu, Z., Thompson, J. A., Agarwal, R. and Agarwal, C. 2006. Fractionation of grape seed extract and identification of gallic acid as one of the major active constituents causing growth inhibition and apoptotic death of DU145 human prostate carcinoma cells. *Carcinogenesis* 27(7), 1445–53. doi:10.1093/carcin/bgi347.

Vossen, P. 2000. Chestnut culture in California. Division of Agriculture and Natural Resources. Publ. 8010. University of California, Davis, CA, 18pp.

Wani, I. A., Hamid, H., Hamdani, A. M., Gani, A. and Ashwar, B. A. 2017. Physico-chemical, rheological and antioxidant properties of sweet chestnut (*Castanea sativa* Mill.) as affected by pan and microwave roasting. *J. Adv. Res.* 8(4), 399–405. doi:10.1016/j.jare.2017.05.005.

Wenzel, U., Nickel, A., Kuntz, S. and Daniel, H. 2004. Ascorbic acid suppresses drug-induced apoptosis in human colon cancer cells by scavenging mitochondrial superoxide anions. *Carcinogenesis* 25(5), 703–12. doi:10.1093/carcin/bgh079.

Woo, Y., Lee, H., Jeong, Y. S., Shin, G. Y., Oh, J. G., Kim, J. S. and Oh, J. 2017. Antioxidant potential of selected Korean edible plant extracts. *BioMed Res. Int.* 2017, Article ID 7695605. doi:10.1155/2017/7695605.

Yang, F., Liu, Q., Pan, S., Xu, C. and Xiong, Y. L. 2015. Chemical composition and quality traits of Chinese chestnuts (*Castanea mollissima*) produced in different ecological regions. *Food Biosci.* 11, 33–42. doi:10.1016/j.fbio.2015.04.004.

Yang, F., Huang, X., Zhang, C., Zhang, M., Huang, C. and Yang, H. 2018. Amino acid composition and nutritional value evaluation of Chinese chestnut (*Castanea mollissima* Blume) and its protein subunit. *RSC Adv.* 8(5), 2653–9. doi:10.1039/C7RA13007D.

Yıldız, O., Can, Z., Saral, O., Yuluğ, E., Öztürk, F., Aliyazıcıoğlu, R., Canpolat, S. and Kolaylı, S. 2013. Hepatoprotective potential of chestnut bee pollen on carbon tetrachloride-induced hepatic damages in rats. *Evid. Based Complement. Alternat. Med.* 2013, 461478. doi:10.1155/2013/461478.

Yıldız, O., Karahalil, F., Can, Z., Sahin, H. and Kolayli, S. 2014. Total monoamine oxidase (MAO) inhibition by chestnut honey, pollen and propolis. *J. Enzyme Inhib. Med. Chem.* 29(5), 690–4. doi:10.3109/14756366.2013.843171.

Zeković, Z., Mujić, I., Tumbas, V., Cvetković, D. and Spasojević, I. 2009. Antioxidant properties of phenolics in *Castanea sativa* Mill. Extracts. *Food Technol. Biotechnol.* 47(4), 421–7.

Zhao, S., Liu, J. Y., Chen, S. Y., Shi, L. L., Liu, Y. J. and Ma, C. 2011. Antioxidant potential of polyphenols and tannins from burs of *Castanea mollissima* Blume. *Molecules* 16(10), 8590–600. doi:10.3390/molecules16108590.

Zhu, F. 2016. Effect of processing on quality attributes of chestnut. *Food Bioprocess Tech.* 9(9), 1429–43. doi:10.1007/s11947-016-1749-3.

Živković, J., Mujić, I., Zeković, Z., Vidovic, S., Mujić, A. and Jokić, S. 2009a. Radical scavenging, antimicrobial activity and phenolic content of *Castanea sativa* extracts. *J. Cent. Eur. Agr.* 10(2), 175–82.

Živković, J., Zeković, Z., Mujić, I., Tumbas, V., Cvetković, D. and Spasojević, I. 2009b. Antioxidant properties of phenolics in *Castanea sativa* Mill. extracts. *Food Technol. Biotechnol.* 47(4), 421–7.

Živković, J., Zeković, Z., Mujić, I., Gođevac, D., Mojović, M., Mujić, A. and Spasojević, I. 2009c. EPR Spin-trapping and spin-probing spectroscopy in assessing antioxidant properties: example on extracts of catkin, leaves, and spiny burs of *Castanea sativa*. *Food Biophys.* 4(2), 126–33. doi:10.1007/s11483-009-9109-8.

Živković, J., Zeković, Z., Mujić, I., Vidović, S., Cvetković, D., Lepojević, Ž. and Trutić, N. 2010. Scavenging capacity of superoxide radical and screening of antimicrobial activity of *Castanea sativa* Mill. extracts. *Czech J. Food Sci.* 28(1), 61–8.

# Chapter 5

## Advances in understanding health benefits of pistachio

*Pablo Hernández-Alonso, Mònica Bulló and Jordi Salas-Salvadó, Universitat Rovira i Virgili, Spain*



### 1 Introduction

The *Pistacia* genus has been widely investigated due to its overall health benefits. It belongs to the genus of flowering plants from the family *Anacardiaceae* and contains about 20 species. Among them, *P. vera*, *P. atlantica*, *P. terebinthus*, *P. khinjuk* and *P. lentiscus* have been researched the most due to their traditional health-promoting effects. Parts of these plants have been used in traditional medicine for varying purposes, including as a tonic, antiseptic, antihypertensive and in the management of dental, gastrointestinal, liver, urinary tract and respiratory tract disorders (Rauf et al., 2017). However, scientific studies have also revealed a wide range of pharmacological uses from various parts of these species, such as antioxidant, antimicrobial, anti-inflammatory, antidiabetic, antitumour, antihyperlipidaemic, antiatherosclerotic and hepatoprotective activities as well as their beneficial effects in gastrointestinal disorders (Bozorgi et al., 2013).

This chapter focuses on the health-related effects of pistachio nuts, a kernel grown in *P. vera*. Pistachio is widely cultivated in the Mediterranean region, even though it probably originated in central and south-western Asia. Evidence of its consumption has been found in archaeological excavations, proving that it has long been associated with human activities (Salas-Salvadó et al., 2011). Pistachios are part of the nut family, an important food belonging to Mediterranean and other regional diets (Widmer et al., 2015). Over time, its consumption has varied. More than 30 years ago, the consumption of nuts was discouraged because of their high fat content. However, at the beginning of the 1990s, various randomized clinical trials (RCT) and animal experiments

http://dx.doi.org/10.19103/AS.2018.0042.06

started to demonstrate their potential beneficial effect on cardiovascular diseases (CVD) (Ros, 2010). In fact, certain review articles have addressed the beneficial effect of consuming nuts for overall cardiometabolic health (Carughi et al., 2016) or specifically to type 2 diabetes (T2D) (Hernández-Alonso et al., 2017b).

Nuts contain several macronutrients and micronutrients together with other important bioactive compounds that may, combined, contribute to modulate specific metabolic diseases such as hypercholesterolaemia, hypertension and T2D (reviewed in Ros, 2015). For this reason, the American Heart Association (AHA) (Lloyd-Jones et al., 2010; Stone et al., 2014), the Canadian Cardiovascular Society (Anderson et al., 2013) and the US Food and Drug Administration (US Food and Drug Administration, 2003) recommend consumption of nuts as part of a healthy diet, to prevent CVD. Epidemiological studies have also shown a significant link between nut consumption and a higher diet quality score or improved nutrient intake in children and adults (O'Neil et al., 2010, 2012). Nonetheless, the specific role of nuts in the development and progression of different diseases is still controversial.

This chapter explores the health benefits of pistachio consumption. It summarizes published clinical studies, *in vivo* and *in vitro* research, and reviews the potential mechanisms for understanding the beneficial role of pistachio intake on different diseases.

## 1.1 Production, import/export markets and consumption of pistachios

Domestic consumption of pistachios has varied widely over the last decade, as shown by the latest USDA report from 2016 to 2017 (Anon., 2017). Pistachio consumption increased by 28% between 2011–12 (523 521 tonnes (t)) and 2016-17 (672 427t). This coincided with a 62% increase in worldwide production of pistachios, from 480 095 t to 777 046 t over the same period.

Turkey is currently the highest consumer of pistachios (156 450 t), followed by the United States (121 652 t), China (116 100 t), the European Union (95 600 t) and Syria (54 200 t). However, the United States is by far the biggest producer (406 646 t) followed by Turkey (155 000 t) and Iran (153 000 t).In fact, production in the EU decreased by 34% during this same time period (from 12 700 to 8400 t). In 2011–12, the EU consumed 69 225 t, compared to 95 600 t in the last evaluated (2016–17) period - an increase of 38%. Over the same timeframe, US consumption of pistachios increased from 73 321 to 121 652 t (a 66% increase). Consumption of pistachios in Turkey and China increased by 83% and 66%, respectively, in the period evaluated, whereas pistachio consumption in Iran dropped by 26.5%. This difference between production and consumption has led to a major change in import and export markets around the globe. The United States has the highest export rate of pistachios (217 559 t, 59% of global exports), followed by Iran (144 300 t, 39% of global exports), Turkey (6600 t, 2% of global exports), the EU (1600 t, 0.4% of global exports) and Syria (500 t, 0.1% of global exports). The latter is the only country to have decreased its exports (by 33%) since 2011-12. The import market is dominated by China (116 100 t, 35% of global imports) and the EU (88 800 t, 27% of global imports), together representing more than 60% of overall imports. The United States has no import market, although it is a major pistachio supplier to the EU.

## 2 Nutritional composition of pistachios

Nuts are a rich source of fat, protein, carbohydrates and fibre. However, beyond macronutrient content, pistachios contain important bioactive compounds such as vitamins, minerals, phenolic compounds, carotenoids and phytosterols (USDA, 2015). Importantly, nutritional differences depend on the growing location of *P. vera*. The United States (California, Arizona, New Mexico), Turkey and Iran are the largest producers of pistachios, growing varieties that differ slightly in nutritional composition. Whereas US pistachios have fewer calories and contain higher amounts of lutein and zeaxanthin, Iranian pistachios are richer in linoleic acid and Turkish in calcium. Fatty acid composition and nutritional profile characteristics also depend on the climate in which the pistachios are grown (reviewed in Bulló et al., 2015). Moreover, as can be seen in Table 1, the concentration of some compounds change according to the type of dry roasting technique used - losing as much as 60% of lutein-zeaxanthin and 48.5% of vitamin A. However, their folate content remains unchanged, and phytosterols and total phenols can increase by 4.2% and 180.6% respectively. Roasting can increase total phenol content and antioxidant activity due to the release of bound phenolic compounds and formation of Maillard reaction products such as melanoidins (Chandrasekara and Shahidi, 2011).

### 2.1 Macronutrients

Compared to other nuts, pistachios have lower fat (mostly from poly- and monounsaturated fatty acids) and energy content. A dry-roasted serving (28.35 g) of pistachios contains 12.99 g of fat, 8.02 g of carbohydrate and 5.97 g of protein. They also contain 2.9 g of dietary fibre. Fat distribution in pistachios consist of 1.60 g of saturated fatty acids (SFA), 3.78 g of polyunsaturated fatty acids (PUFA) and 6.96 g of monounsaturated fatty acids (MUFA) (Table 1) (USDA, 2015). Within fatty acids, oleic and linoleic fatty acids, both recognized for their CV-preventive properties, represent more than 60% of the total fat content in pistachios (Gillingham et al., 2011). The amount of carbohydrate in pistachios,

**Table 1** Nutrient composition of raw and dry-roasted pistachio nut

| Nutrients | Raw pistachios (per 100 g) | Dry-roasted pistachios (per 100 g) | % of change from raw pistachio | Dry-roasted pistachios (per serving of 28.35 g) |
|---|---|---|---|---|
| *Energy and macronutrients* | | | | |
| Energy (Kcal) | 560 | 572 | | 162 |
| Water (g) | 4.37 | 1.85 | | 0.52 |
| Fat (g) | 45.32 | 45.82 | | 12.99 |
| SFA (g) | 5.91 | 5.65 | | 1.60 |
| MUFA (g) | 23.26 | 24.53 | | 6.96 |
| PUFA (g) | 14.38 | 13.35 | | 3.78 |
| Protein (g) | 20.16 | 21.05 | | 5.97 |
| Carbohydrates[a] (g) | 27.17 | 28.28 | | 8.02 |
| Fibre (g) | 10.6 | 10.3 | | 2.9 |
| *Micronutrients* | | | | |
| Ca (mg) | 105 | 107 | | 30 |
| Cu (mg) | 1.3 | 1.3 | | 0.37 |
| K (mg) | 1025 | 1007 | | 285 |
| Mg (mg) | 121 | 109 | | 31 |
| Mn (mg) | 1.2 | 1.2 | | 0.35 |
| Na (mg) | 1 | 6[d] | | 2 |
| P (mg) | 490 | 469 | | 133 |
| Se (µg) | 7 | 10 | | 2.8 |
| Zn (mg) | 2.2 | 2.3 | | 0.6 |
| *Essential amino acids* | | | | |
| Histidine (g) | 0.51 | 0.54 | | 0.15 |
| Isoleucine (g) | 0.92 | 0.96 | | 0.27 |
| Leucine (g) | 1.60 | 1.68 | | 0.48 |
| Lysine (g) | 1.14 | 1.19 | | 0.34 |
| Methionine (g) | 0.36 | 0.38 | | 0.11 |
| Phenylalanine (g) | 1.09 | 1.14 | | 0.32 |
| Threonine (g) | 0.68 | 0.71 | | 0.20 |
| Tryptophan (g) | 0.251 | 0.26 | | 0.07 |
| Valine (g) | 1.25 | 1.31 | | 0.37 |

*(Continued)*

**Table 1** *(Continued)*

| Nutrients | Raw pistachios (per 100 g) | Dry-roasted pistachios (per 100 g) | % of change from raw pistachio | Dry-roasted pistachios (per serving of 28.35 g) |
|---|---|---|---|---|
| *Non-essential amino acids with relevant content* | | | | |
| Glutamic acid (g) | 4.3 | 4.49 | | 1.27 |
| Arginine (g) | 2.13 | 2.23 | | 0.63 |
| *Vitamins and other bioactive molecules* | | | | |
| Lutein-Zeaxanthin (µg) | 2903 | 1160 | −60.04 | 329 |
| Thiamine (Vitamin $B_1$) (mg) | 0.87 | 0.70 | −19.54 | 0.20 |
| Niacin (Vitamin $B_3$) (mg) | 1.3 | 1.37 | 5.38 | 0.39 |
| Pyridoxine (Vitamin $B_6$) (mg) | 1.7 | 1.12 | −34.12 | 0.32 |
| Folate (µg) | 51 | 51 | 0.00 | 14 |
| Vitamin A (IU) | 516 | 266 | −48.45 | 75 |
| Vitamin K (µg) | NA | 13.2 | NA | 3.7 |
| Vitamin E (α-tocopherol) (mg) | 2.86 | 2.17 | −24.13 | 0.62 |
| γ-Tocopherol (mg) | 20.41 | 23.42 | 14.75 | 6.64 |
| β-Carotene (µg) | 305 | 159 | −47.87 | 45 |
| α-Carotene (µg) | 10 | 0 | −100.00 | 0 |
| Phytosterols[b] (mg) | 213 | 222 | 4.23 | 62.94 |
| Total phenols[c] (mg GAE) | 165 | 463 | 180.61 | 131.26 |

[a] Carbohydrates are calculated by difference.
[b] Phytosterols are the sum of stigmasterol, campesterol and β-sitosterol.
[c] Without skin. Total phenolic content in pistachio skin is 116.32 mg GAE/g fresh weight.
[d] Dry-roasted and salted increase this value up to 428 mg. Accession numbers for pistachio in the USDA National Nutrient Database: 12151 (raw), 12152 (dry-roasted, without salt added) and 12652 (dry-roasted, with salt added).
Abbreviations: GAE - gallic acid equivalent; IU - international units; MUFA - monounsaturated fatty acids; NA - not available; PUFA - polyunsaturated fatty acids; SFA - saturated fatty acids.
Nutrient information is taken from the US Department of Agriculture (USDA) National Nutrient Database for Standard Reference Release 28 (USDA, 2015). Total phenolic content is derived from Taş and Gökmen, 2017.

as in other nuts, is low to moderate, but pistachios contain 10% by weight of insoluble forms and 0.3% of soluble forms of fibre. Pistachios contain 2.9g of fibre per serving, which is 9.7% of men's and 11.6% of women's recommended dietary allowance (RDA). According to the USDA food composition tables, only almonds have similar amounts of fibre, at 13% of their weight (USDA, 2015). Epidemiological and clinical studies have consistently demonstrated that total fibre intake is inversely associated with weight gain (Ye et al., 2012), T2D (Kaczmarczyk et al., 2012), CVD (Anderson et al., 2000) and some types of cancer (Kaczmarczyk et al., 2012). Moreover, pistachios have a low glycaemic index (GI), which contributes to maintaining satiety for longer and lowering postprandial blood glucose concentrations (Gulati et al., 2014; Kendall et al., 2014).

Pistachios are also a good source of vegetable protein. They contain both essential (e.g. leucine, lysine) and non-essential amino acids, with a significant amount of glutamic acid (1.27 g per serving of dry-roasted pistachios) and arginine (0.63 g per serving of dry-roasted pistachios). Arginine is a conditionally indispensable amino acid, as a dietary source is required when endogenous synthesis cannot meet metabolic need (Trumbo et al., 2002). Arginine is a precursor to nitric oxide (NO), which is an endogenous vasodilator and an essential molecule in the cardiovascular system as it is a key regulator of vascular tone. In addition, due to its pro-oxidant capacity, NO is associated with numerous pathological conditions, such as hypertension, CVD and neurodegenerative disorders (Jaffrey and Snyder, 1995; Loscalzo and Welch, 1995). As NO synthase (NOS) inhibitors, based on arginine, are of particular interest for experimental as well as clinical applications (Víteček et al., 2012), pistachios could play a crucial role in preventing NOS-related diseases.

## 2.2 Vitamins and minerals

Compared to other nuts, pistachios are an excellent source of minerals such as potassium, phosphorus, magnesium, calcium and vitamins, including vitamin A, vitamin C and B group vitamins, with the exception of vitamin $B_{12}$ (cyanocobalamin) (Rothwell et al., 2012). In particular, pistachios contain relatively high amounts of thiamine (vitamin $B_1$), which is involved in intermediary carbohydrate metabolism. A serving of dry roasted pistachios contains 0.20mg of B1 (providing up to 16.7% and 18.2% of a man and woman's RDA respectively). The amount of pyridoxine (vitamin $B_6$) that is involved in the metabolism of amino acids and in the production of niacin is about 0.32 mg per serving basis of dry-roasted pistachios (30.8% of RDA for men and 23.5% for women). Finally, the amount of folate in pistachios provides approximately 3.5% of the RDA in both men and women. Among nuts, pistachios also stand out for high vitamin K content, with approximately 3.7 μg per serving (5.3% of the RDA). Beyond its role in bone metabolism (Bulló et al., 2011; Ibarrola-Jurado et al., 2012; Juanola-Falgarona et al., 2013), a higher dietary intake of vitamin K has been associated with a lower risk of several chronic diseases such as T2D (Bulló et al., 2011), cancer (Juanola-Falgarona et al., 2014; Nimptsch et al., 2010) and CVD (Juanola-Falgarona et al., 2014), thus increasing the potential health benefits of pistachio consumption. The beneficial role of pistachios in inflammatory-related diseases may also be explained by the relatively high amount of γ-tocopherol they contain (6.64 mg per serving basis) (Dietrich et al., 2006).

Pistachios are rich in several minerals, including copper (44.4% of RDA per serving basis), manganese (17.4% for men and 22.2% for women of RDA), phosphorus (13.6% of RDA), magnesium (8% for men and 10.5% for women of RDA) and potassium (7.7% for men and 10.4% for women of RDA). Because of their mineral profile, pistachios can play a beneficial role in blood pressure (BP) regulation and in bone-related diseases. Pistachios also contain significant amounts of zinc and selenium (Table 1), both of which have recognized antioxidant effects that can help the prevent CVD and some types of cancer (Hercberg et al., 2010; Huang et al., 2006).

## 2.3 Other bioactive compounds

Pistachios have the highest phytosterol content of all nuts, with 63 mg per serving, which include stigmasterol, campesterol and β-sistosterol. Phytosterols are structurally similar to cholesterol and several studies have demonstrated a dose-response reduction of cholesterol by phytosterols, even at levels similar to those found in plant-based diets that include pistachios (Ostlund et al., 2002). Although a product must provide 500 mg of phytosterols daily in order to take advantage of the FDA-approved health claim, the levels of phytosterols in pistachio nuts may be sufficient to play a synergistic role with unsaturated fatty acids and low SFA levels in order to maintain normal cholesterol levels.

Pistachios, pecans and walnuts are rich sources of phenolic compounds (e.g. anthocyanins, flavonoids, proanthocyanidins, flavonols, isoflavones) which are important as antioxidants and also for their chemopreventive, cardioprotective and vasoprotective properties (Bolling et al., 2011; Bulló et al., 2015). Phenolic compounds may have protective effects against diseases related to free radical overproduction, such as CVD and cancer (Hassellund et al., 2013). According to Tomaino et al. (2010) all phenolic groups found in pistachios are present in higher amounts in the skins compared with the seeds (Table 1). Pistachios are the only nut to contain anthocyanins, along with other phenolic compounds, in the skin. These phenolic compounds bind metals via *o*-diphenol groups, which is important in the inhibition of metal-induced lipid oxidation (Dixon et al., 2005). In studies using a simulated human digestion model, it was found that more than 90% of the polyphenols present in pistachios were released into the gastric compartment, with no noticeable difference between raw and roasted pistachios (Mandalari et al., 2013; Nikzadeh and Sedaghat, 2008). Importantly, polyphenols, tocopherols and xanthophylls from pistachios have proved to be rapidly accessible in the stomach, thus maximizing the possibility of absorption in the upper small intestine, and contributing to the beneficial relation between nut consumption and health-related outcomes (Mandalari et al., 2013). Due to their high phenolic content, pistachios are among the 50 foods with the highest antioxidant capacity (Halvorsen et al., 2006).

Pistachios are particularly rich in B-carotenes, which have been shown to prevent type 2 diabetes (Montonen et al., 2004; Sluijs et al., 2015). Furthermore, they are the only nut to contain significant amounts of lutein and zeaxanthin (USDA, 2015), which are two xanthophyll carotenoids responsible for the colour of pistachio nuts. Raw pistachios contain 2903 μg of lutein-zeaxanthin per 100 g, approximately 13 times as much as its nearest rival, hazelnuts, which contain only 92 μg (USDA, 2015). Carotenoid bioavailability depends on the source and interaction with other dietary components, and it has been shown that the interaction of β-carotene and lycopene with the lipid matrix increases the bioavailability of carotenoids (van het Hof et al., 2000).

Overall, pistachios contain a diverse and rich source of macronutrients, micronutrients, relevant amino acids and bioactive molecules, generally considered to have a major impact on health as will be covered in the next sections.

## 3 Health benefits of pistachio nut

The nut-health axis has been widely explored since the late 1980s. In particular, the inclusion of pistachios into different dietary patterns has been linked with improving parameters related to different cardiometabolic conditions. In this section, we summarize the outcomes that have been modulated by pistachio consumption (Table 2).

### *3.1 Classical risk factors for metabolic and cardiovascular disease*

#### *3.1.1 Insulin resistance, pre-diabetes and type 2 diabetes*

As we have explored, nuts are relatively low in carbohydrates (approximately 15% of total energy) with the exception of chestnut. Therefore, they have a lower impact on postprandial blood glucose levels and require less insulin secretion, thus favouring T2D control. T2D is one of the most common diseases worldwide, mainly due to an increase in the prevalence of obesity and physical inactivity. Moreover, T2D is a recognized risk factor for CVD and other chronic conditions and diseases, and is thus becoming a serious public health burden (Shaw et al., 2010; The Emerging Risk Factors Collaboration et al., 2010). Data from epidemiological studies suggest that the frequency of nut consumption is inversely related to an increased risk of T2D, mainly attributed to the fibre, healthy fats, antioxidants and anti-inflammatory compounds found in nuts (Damavandi et al., 2013; Jenkins et al., 2008; Kochar et al., 2010; Pan et al., 2013; Rajaram and Sabaté, 2006; Ros, 2010) in nuts.

Due to their low GI, pistachios may have a possible effect on reducing postprandial glycaemia and insulinaemia, especially when substituting carbohydrates, thereby potentially decreasing the risk of T2D and/or other metabolic conditions. The effect of pistachios, consumed alone or combined

**Table 2** Summary of human clinical trials assessing the health-related role of pistachios

| References | No. of subjects* (M/F) Type of subjects (age in years) | Study design (length of the intervention) | Control group | Intervention group(s) | Remarkable finding(s) |
|---|---|---|---|---|---|
| Edwards et al. (1999)[a] | 10 (4/6) Moderate HC (Range: 28–64) | Crossover (3 weeks/ period) | RD | 20% of energy in the form of pistachios (PD) | Significant decreases in TC, TC/HDLc and LDLc/HDLc ratio in PD compared with RD |
| Kocyigit et al. (2006)[a,b] | 44 (24/20) Healthy subjects (Range: 24–40) | Parallel (3 weeks of follow-up) | RD | 20% of energy in the form of pistachios (PD) | Significant decreases of TC, TC/ HDLc ratio and LDLc/HDLc ratio in PD compared with RD. Significant improvement in oxidative status |
| Sheridan et al. (2007)[a] | 15 (11/4) Moderate HC subjects (Mean ± SD: 60 ± 3) | Crossover (1 month/ period) | *Ad libitum* RD | *Ad libitum* diet plus pistachio supplementation (15% of energy; 57–85 g) | Significant decreases of TC/HDLc and LDLc/HDLc ratio, and increases of HDLc in PD compared with RD |
| Gebauer et al. (2008)[a] | 28 (10/18) Subjects with elevated LDLc (≥2.8 mmol/L) (Range: 35–61) | Crossover (1 month/ period) | LF-CD | PD1: 10% of energy in the form of pistachios PD2: 20% of energy in the form of pistachios | Both PD diets significantly decrease TC, LDLc and non-HDLc compared with CD |
| Kay et al. (2010)[c] | | | | | Both PD interventions significantly decreased ox-LDL and increased serum antioxidants (γ-tocopherol, lutein and β-carotene) |
| Holligan et al. (2014)[d] | | | | | A significant decrease in sdLDL levels was observed following both PD1 and PD2 *versus* CD |

*(Continued)*

**Table 2** *(Continued)*

| References | No. of subjects* (M/F) Type of subjects (age in years) | Study design (length of the intervention) | Control group | Intervention group(s) | Remarkable finding(s) |
|---|---|---|---|---|---|
| Li et al. (2010)[a,e] | 52<br>Healthy obese subjects<br>(Range: 20–65) | Parallel (3 months) | Energy restricted diet with 56 g/day (220 Kcal) of pretzels included | Energy restricted diet with 53 g/day (240 Kcal) of pistachios included (PD) | Significant reduction of BW and BMI in PD compared with CD. Decrease in TG in the pistachio group |
| Sari et al. (2010)[a,b,f,g] | 32 males<br>Healthy subjects<br>(Range: 21–24) | Sequential feeding trial (1 month on MD followed by 1 month on PD) | MD | Mediterranean-type diet with 20% of energy in form of pistachios (PD) | Significant decrease in FG, TC, TG, LDLc, TC/HDLc, LDLc/HDLc ratios and non-significant increase in HDLc, in PD compared with MD. Decrease in IL-6 concentrations and improvement of antioxidant capacity in PD vs. MD. PD improved endothelium-dependent vasodilation versus MD |
| Aldemir et al. (2011)[a,h] | 17 males<br>Erectile dysfunction individuals<br>(Range: 38–59) | Sequential feeding trial (3 weeks of intervention diet) | No CD | Pistachio diet: 100 g | Improvements of erectile function measured by IIEF-15 score after pistachio intervention. Lipid profile measured as TC, LDLc, TC/HDLc and LDLc/HDLc significantly decreased through intervention, and HDLc significantly increased |
| Baer et al. (2012)[i] | 16 (8/8)<br>Healthy subjects<br>(Range: 29–64) | Crossover (3 weeks/period) | Traditional American diet | PD1: CD + 42 g/day of pistachios | Pistachios contain significantly 5% less energy than the value calculated from the Atwater factors |

| References | No. of subjects* (M/F) Type of subjects (age in years) | Study design (length of the intervention) | Control group | Intervention group(s) | Remarkable finding(s) |
|---|---|---|---|---|---|
| Ukhanova et al. (2014)[j] | | | | PD2: CD + 84 g/day of pistachios | Pistachio consumption modulated gut microbiota. It included an increase in the number of potentially beneficial butyrate-producing bacteria and a decrease in the number of lactic acid bacteria |
| Wang et al. (2012)[e] | 90 (41/49) Subjects with MetS (Range: 25–65) | Parallel (3 months) | Nut-free AHA Step 1 diet | AHA Step 1 diet supplemented with: PD1: CD + 42 g/day of pistachios PD2: CD + 70 g/day of pistachios | Non-significant changes in BW, BMI or waist-to-hip ratio. Non-significant changes in BP, FG and blood lipids in PD1 or PD2 compared with CD |
| West et al. (2012)[f,k] | 28 (10/18) Subjects with dyslipidaemia (Range: 35–61) | Crossover (4 weeks/period) | LF-CD | PD1: CD + 10% of energy in form of pistachios PD2: CD + 20% of energy in form of pistachios | Significant reduction of SBP in PD1 vs. PD2. Significant decrease of peripheral vascular dilatation and heart rate in PD2 vs. CD |
| Gulati et al. (2014)[b,e,g] | 68 (37/31) Subjects with MetS (Mean ± SD: 42.5 ± 8.2) | Parallel (6 months) | Standard CD according to dietary guidelines for Asian Indians with exercise protocol | Standard diet with exercise protocol supplemented with unsalted pistachios (20% of daily energy) | Significant decrease in WC and trend towards reduction of subcutaneous adipose tissue (*P* = 0.07) in PD vs. CD. Significant reduction in levels of TC and LDLc. Significant decrease of FG but not in insulin levels and improvement of lipid profile in PD vs. CD. Moreover, CRP and TNF-α levels were reduced following nut diet versus nut-free diet |

*(Continued)*

**Table 2** *(Continued)*

| References | No. of subjects* (M/F) Type of subjects (age in years) | Study design (length of the intervention) | Control group | Intervention group(s) | Remarkable finding(s) |
|---|---|---|---|---|---|
| Hernández-Alonso et al. (2014)[g,l] | 54 (29/25) Subjects with pre-D (Mean (95% CI) 55 (53.4–56.8)) | Crossover (4 months/period) | Nut-free diet | Pistachio diet was supplemented with 57 g/day of pistachios (half roasted and salted) | FG, insulin, and HOMA-IR decreased significantly after the PD compared to CD. NS changes in BW, BP or lipid profile. Expression of IL-6 in lymphocytes was reduced during PD vs CD. However, circulating levels were unchanged |
| Hernández-Alonso et al. (2015)[d] | | | | | sLDL-P significantly decreased after PD compared to CD. The non-HDL-P significantly decreased under the PD compared to CD. Overall size of HDL-P significantly decreased in the PD |
| Hernández-Alonso et al. (2016)[m] | | | | | After the PD period, circulating levels of miR-192 and miR-375 were lower compared to CD period. Interestingly, changes in circulating miR-192 and miR-375 were positively correlated with FG, insulin and HOMA-IR |
| Hernández-Alonso et al. (2017)[n] | | | | | Three metabolites related with gut microbiota metabolism (i.e. hippurate, *p*-cresol sulphate and dimethylamine) were found decreased in PD compared with CD. Moreover, *cis*-aconitate (intermediate of the tricarboxylic acid (TCA)) was also found decreased following PD compared with CD. TMAO was found significantly reduced following PD |

| References | No. of subjects* (M/F) Type of subjects (age in years) | Study design (length of the intervention) | Control group | Intervention group(s) | Remarkable finding(s) |
|---|---|---|---|---|---|
| Parham et al. (2014)[l] | 44 (11/33) T2D subjects (Mean: 51) | Crossover (3 months/period) | Previous diet without pistachios | Two snacks of 25 g/day of pistachios | Marked decrease in $HbA_{1c}$ and FG concentrations in the pistachio diet group. No significant changes in BMI, BP, HOMA-IR and CRP |
| Nieman et al. (2014)[b,g,n,o,p] | 19 males Healthy subjects (Mean ± SEM: 38.0 ± 1.6) | Crossover (2 weeks/ period) | Nut-free diet | Pistachio supplementation (42.5 g/day) | Pistachio supplementation reduced 75-km cycling time trial performance. Pistachio supplementation increased post-exercise plasma levels of raffinose, sucrose and metabolites related to leukotoxic effects and oxidative stress |
| Kasliwal et al. (2015)[f] | 60 (46/14) Moderate HC subjects (Range: 25–60) | Parallel (3 months) | Lifestyle modification without nuts | Lifestyle modification plus 40 g/day of pistachios | Significant improvements in baPWV and BAFMD with pistachios supplementation versus no supplementation. Significant increase in HDLc and a reduction in LDLc, TC, TC/HDLc and FG following pistachio supplementation. NS differences in changes between interventions or following pistachios supplementation in BP, BW or WC |
| Sauder et al. (2014)[f,k] | 30 (15/15) T2D subjects (Mean ± SD: 56.1 ± 7.8) | Crossover (4 weeks/ period) | Nut-free diet | 20% of energy from pistachios | BP at rest and during stress did not differ between treatments. The PD significantly reduced total peripheral resistance, increased cardiac output and improved some measures of heart rate variability. Systolic ambulatory BP was significantly reduced following the PD, with the greatest reduction observed during sleep |

*(Continued)*

**Table 2** *(Continued)*

| References | No. of subjects* (M/F) Type of subjects (age in years) | Study design (length of the intervention) | Control group | Intervention group(s) | Remarkable finding(s) |
|---|---|---|---|---|---|
| Sauder et al. (2015)[a,f,g,l] | | | | | TC and TC/HDLc were significantly lower following the PD and TG were significantly reduced following the PD compared to the CD. NS treatment differences in FG and insulin. Changes in BW were not different between intervention periods |
| Assaf-Balut et al. (2017)[q] | 1000 females<br>Normoglycaemic pregnant women<br>(Mean ± SD: 33.0 ± 5.2) | Parallel (from 2-3 to 6-7 month of pregnancy) | Standard nut- and EVOO-free diet with limited fat intake | MD supplemented with EVOO and pistachios | The adjusted multivariable RR for GDM was 0.75 (95% CI: 0.57–0.98) intervention versus control. The MedDiet intervention also significantly improved the rates of insulin-treated GDM, prematurity and gestational weight gain, among other parameters |

Abbreviations: AHA, American Heart Association; AOP, antioxidant potential; BAFMD, brachial artery flow-mediated dilatations; baPWV, brachial-ankle pulse wave velocity; BMI, body mass index; CD, control diet; CI, confidence interval; EVOO, extra-virgin olive oil; FG, fasting glucose; GDM, gestational diabetes mellitus; HC, hypercholesterolaemic; HDLc, high-density lipoprotein cholesterol; $HbA_{1c}$, glycated haemoglobin; HOMA-IR, homeostatic model assessment of insulin resistance; IIEF-15, International Index of Erectile Function; IL-6, interleukin 6; LDLc, low-density lipoprotein cholesterol; LF, low-fat; MD, Mediterranean diet; MDA, malondialdehyde; MetS, metabolic syndrome; miR, microRNA; Ox-LDL, oxidized-low-density lipoprotein; PD, pistachio diet; RD, regular diet; RR, relative risk; SBP, systolic blood pressure; sdLDL, small and dense low-density lipoprotein; SEM, standard error of the mean; T2D, type 2 diabetes; TC, total cholesterol; TG, triglycerides; TMAO, trimethylamine N-oxide; TNF-α, tumour necrosis factor-α; VLDLc, very-low-density lipoprotein cholesterol.

* Number of subjects included in the main analysis. Articles with the same subjects of study describe secondary outcomes.

Main outcome(s) of each research article are as follows: [a] lipid profile; [b] oxidative stress/status; [c] antioxidant profile; [d] emerging markers of CVD (lipoprotein size); [e] body weight and/or body composition; [f] endothelial function; [g] inflammation; [h] erectile function; [i] Atwater factors; [j] gut microbiota; [k] blood pressure; [l] glucose/insulin metabolism/glycaemic control; [m] microRNAs; [n] metabolomics; [o] sport's performance; [p] immune dysfunction; [q] gestational diabetes mellitus.

with meals, on postprandial glycaemia has been also evaluated in two acute clinical trials (Kendall et al., 2011, 2014). Thus, whereas pistachios consumed alone had a minimal effect on postprandial glycaemia, the addition of pistachios (56 g) to foods with a high GI (pasta, parboiled rice and instant mashed potatoes) reduced, in a dose-dependent manner, the total postprandial glycaemic response by 20–30% (Kendall et al., 2011). In another acute crossover study conducted on 20 subjects with metabolic syndrome (MetS), 85.04 g of pistachios consumed with bread significantly reduced postprandial glycaemia and increased GLP-1 levels compared with bread alone, implying not only a benefit in postprandial levels of glucose but also for satiety status (Kendall et al., 2014).

A total of two studies have assessed – as a primary outcome – the role of pistachio consumption on the modulation of parameters related to glucose/insulin metabolism in subjects with T2D or pre-diabetes (pre-D) (Table 2). In a crossover study conducted on 48 diabetic patients who received 50 g/day of pistachios and a pistachio-free diet (three months/period), with an eight-week washout, a significant improvement in fasting glucose (FG) and glycated haemoglobin ($HbA_{1c}$) was observed during the period of pistachio consumption, while no changes in homeostatic model assessment of insulin resistance (HOMA-IR) were reported (Parham et al., 2014). To determine whether the beneficial effect of nut consumption on glucose and insulin metabolism could also be extended to pre-diabetic subjects, our group conducted a randomized crossover study. A total of 54 pre-diabetic subjects consumed a pistachio-supplemented diet (55 g pistachio/day) and a control diet (nut-free diet), each for four months, with a two-week washout period. We found a beneficial effect of pistachio intake on FG and insulin, which led to an improvement in HOMA-IR. Other cardiometabolic risk markers such as fibrinogen, oxidized-low-density lipoprotein (ox-LDL), platelet factor 4 and GLP-1 also improved during the consumption of pistachios (Hernández-Alonso et al., 2014). Moreover, one recent controlled trial, of pregnant women at eight to 12 gestational weeks, assessed the effect of a Mediterranean diet (MedDiet) supplemented with pistachios (25–30 g/day) and extra-virgin olive oil (EVOO; minimum 40 mL/day) versus a control standard diet (with limited fat intake) on gestational diabetes mellitus (GDM) incidence at 24–28 gestational weeks. The adjusted multivariable relative risk for GDM was 0.75 (95% confidence interval (CI): 0.57–0.98) in the intervention versus the control group. The MedDiet intervention also significantly improved the rates of insulin-treated GDM, prematurity and gestational weight gain, among other parameters, thus increasing the beneficial effect of pistachios in combination with EVOO in GDM prevention (Assaf-Balut et al., 2017).

In addition, five clinical studies have assessed the effect of pistachio supplementation on glucose concentrations as a secondary outcome, with contradictory results. In a controlled, crossover clinical trial, individuals were randomly assigned to a MedDiet or a MedDiet supplemented with pistachios for 20% of the energy intake, over a period of four weeks in each arm. During the intervention period, the participants showed a significant decrease in FG concentrations compared with the control period (Sari et al., 2010). A second parallel, randomized controlled study evaluated the effect of the AHA Step I diet supplemented with 42 g/day or 70 g/day of pistachios, compared with the effect of a control diet (non-supplemented AHA Step I), in Chinese subjects with MetS. After 12 weeks of intervention, no differences in FG or insulin were observed between groups; however, compared with baseline values, blood glucose was significantly increased in the control group at week 12 but not in the two pistachio groups (Wang et al., 2012). In a third parallel study, 60 MetS subjects were randomized to either an unsalted pistachio diet (20% energy) or a control diet for six months. A significant decrease in glucose blood levels was observed, but not in insulin. (Gulati et al., 2014). Sauder et al. studied the effects of the inclusion of pistachios for 20% of daily energy in a crossover study of 30 T2D subjects during a 4-week period. Levels of FG and insulin were not significantly modulated compared to the nut-free diet. However, they reported a reduction in fructosamine, a marker that reflects average blood glucose during the preceding two to three weeks (Sauder et al., 2015).

The most recent systematic review and meta-analysis designed to assess the link between nut consumption and risk of cancer and T2D was published in 2015. It included five studies that were linked to T2D. After pooling data from studies conducted across both genders, it was found that there was no statistically significant association between nut consumption and the risk of developing T2D (relative risk (RR): 0.98 (95% CI, 0.84–1.14)), even though the heterogeneity was significant (Pimpin et al., 2016). However, the latest results from a systematic review and meta-analysis after evaluating the effect of nuts on glycaemic control in diabetes showed significant reductions in $HbA_{1c}$ and FG, whereas no significant effects were observed in fasting insulin and HOMA-IR (Viguiliouk et al., 2014). All these findings therefore support pistachio inclusion in a healthy diet, reinforcing the need for longer, higher quality RCT evaluating the nut-T2D axis.

Despite positive results observed for glucose metabolism in fasting conditions or postprandial status, additional studies are necessary to evaluate the long-term effects of pistachio consumption on insulin resistance, secretion and diabetes control.

### 3.1.2 Dyslipidaemia

Lipid-related classical markers of CVD have been the main focus of research since the beginning of the 1990s due to the worldwide increase of dyslipidaemia.

Dyslipidaemia is a broad term that refers to a number of lipid disorders. Most (80%) lipid disorders are related to diet and lifestyle, although familial disorders (20%) are important as well (Smith, 2007).

The effect of pistachio consumption on classical lipid-related CV risk markers has been evaluated in six RCT as a primary outcome (Edwards et al., 1999; Gebauer et al., 2008; Kocyigit et al., 2006; Sari et al., 2010; Sauder et al., 2015; Sheridan et al., 2007) and in another four studies as a secondary outcome (Aldemir et al., 2011; Kasliwal et al., 2015; Li et al., 2010; Wang et al., 2012), providing between 10% and 20% of energy (or from 42 to 100 g/day) as pistachios versus diets avoiding the consumption of nuts (Table 2). In a total of five studies the authors found significant reductions of plasma total cholesterol (TC) concentrations in the pistachio-supplemented group (Aldemir et al., 2011; Edwards et al., 1999; Gebauer et al., 2008; Kocyigit et al., 2006; Sari et al., 2010), and in six of them, they found a significant reduction of the TC/high-density lipoprotein cholesterol (HDLc) ratio and LDLc/HDLc ratio (Aldemir et al., 2011; Edwards et al., 1999; Gebauer et al., 2008; Kocyigit et al., 2006; Sari et al., 2010; Sheridan et al., 2007). Moreover, LDLc concentrations were decreased in the pistachio-supplemented group in four studies (Aldemir et al., 2011; Gebauer et al., 2008; Kasliwal et al., 2015; Sari et al., 2010), whereas two studies reported no significant reductions in this recognized major cardiovascular risk factor (Edwards et al., 1999; Kocyigit et al., 2006), although the levels decreased but not significantly in Kocyigit et al. (2006). Only Wang et al. (2012), in a study on Chinese participants with MetS, found an increase in plasma LDLc levels after a 12-week period of dietary intervention with a normocaloric diet including different amounts of pistachios compared with the normocaloric diet alone. According to Wang et al., the nutrient content of the diet was insufficient to show changes in the secondary analyses of risk factors such as blood lipids. Notably, dietary intake was not controlled or reported, so it is difficult to ascertain the reason why LDLc increased in the high pistachio group (Wang et al., 2012). With respect to plasma HDLc, two studies found a significant increase in this parameter in those subjects supplemented with pistachios (Kasliwal et al., 2015; Sheridan et al., 2007).

In a pooled analysis of 25 intervention trials, participants who consumed an average of 67 g of nuts per day experienced a 5% decrease in TC, a nearly 7.5% decrease in LDLc and an 8% decrease in the LDLc/HDLc ratio. The effects of nut consumption were dose related, and different types of nuts had similar effects on blood lipid levels (Sabaté et al., 2010). In the latest systematic review, meta-analysis and dose-response studies composed of 61 trials (2582 subjects) have shown similar results. Interventions ranged from 3 to 26 weeks. Nut intake (per serving/day) lowered TC (−4.7 mg/dL; 95% CI: −5.3, −4.0 mg/dL), LDLc (−4.8 mg/dL; 95% CI: −5.5, −4.2 mg/dL), ApoB (−3.7 mg/dL; 95% CI: −5.2, −2.3 mg/dL) and triglycerides (−2.2 mg/dL; 95% CI: −3.8, −0.5 mg/dL).

The dose-response between nut intake and TC and LDLc was nonlinear, and stronger effects were observed for ≥60 g/day of nut consumption. The major factor determining the cholesterol-lowering effect appears to be nut dose rather than nut type. Importantly, these findings also highlighted the need to investigate potential stronger effects at high nut doses and among diabetic subjects (Del Gobbo et al., 2015).

In conclusion, evidence suggests that chronic pistachio consumption leads to an improvement in the lipid profile, which may reduce the risk of CVD. The main effects are related to a decrease in TC and LDLc, although beneficial improvements in HDLc have also been identified. Therefore, further research is needed in order to expand these results.

### *3.1.3 BP and endothelial function*

Worldwide, the prevalence of raised BP (systolic BP (SBP) ≥140 mmHg or diastolic BP (DBP) ≥90 mmHg) in adults (≥18) was 22.3% in 2014 and slightly decreased in 2015 to 22.1% (WHO, 2013). The most traditional and widely accepted view is that hypertension is a cause rather than a consequence of endothelial dysfunction (ED), although some authors suggest that the reverse may also be true. In fact, ED plays a role in the pathogenesis of atherosclerosis and might also increase the risk of IR and T2D (Tziomalos et al., 2010).

Different clinical trials have been designed to explore the effect of nuts on BP or endothelial function (EF). However, there is limited data from pistachio research. The role of pistachio consumption on BP has been assessed in subjects with dyslipidaemia, hypercholesterolaemia or T2D. However, no studies have specifically assessed BP modulation on subjects with pre- or hypertension. A beneficial effect of pistachios on BP was demonstrated in a randomized, crossover, clinical trial conducted on 28 dyslipidaemic individuals. Participants underwent one of three four-week interventions: a low-fat control diet, a diet in which 10% of total energy came in the form of pistachios, and a diet where 20% of energy came from pistachios. A dose-dependent reduction in systolic BP was observed in those subjects supplemented with pistachios, and a decrease in peripheral vascular dilation was observed with the highest doses of pistachios (West et al., 2012). Sauder et al. undertook a four-week study of T2D subjects, where 20% of energy in the diet came from pistachios versus 0% in the control group. They reported a reduction in total peripheral resistance, increase of cardiac output and an improvement of some measures of heart rate variability following pistachio intervention. Moreover, they found a reduction in ambulatory SBP in the PD (Sauder et al., 2014). Kasliwal et al. recently studied the effects of lifestyle supplementation plus 40 g/day of pistachios versus a nut-free diet in moderate hypercholesterolaemic subjects over a period of three months. Even though they reported insignificant changes in BP, they found an improvement in brachial-ankle pulse wave velocity (baPWV) and brachial artery flow-mediated dilatations (BAFMD) towards a better EF status (Kasliwal et al., 2015).

The BP lowering effects of pistachios have also been evaluated in three additional controlled feeding trials as a secondary outcome (Table 2) showing non-significant differences in changes in SBP or in DBP between those subjects supplemented with pistachios and those who did not receive supplementation (Sari et al., 2010; Sheridan et al., 2007; Wang et al., 2012). However, Sari et al. reported a significant improvement in the endothelium-dependent vasodilation following the PD compared with the Mediterranean diet (Sari et al., 2010).

In a recent meta-analysis comparing highest and lowest intakes of different food categories, an inverse association for the risk of hypertension was observed for 28 g nuts/day (RR: 0.70; 95% CI: 0.45, 1.08) (Schwingshackl et al., 2017). Importantly, results from the latest systematic review and meta-analysis of 21 RCT showed that nut consumption leads to a significant reduction in SBP in participants without T2D (mean difference (MD): −1.29; 95% CI: −2.35, −0.22; P = 0.02) but not in the total population. Subgroup analyses of different nut types suggest that pistachios, but not other nuts, significantly reduce SBP (MD: −1.82; 95% CI: −2.97, −0.67; P = 0.002) and DBP (MD: −0.80; 95% CI: −1.43, −0.17; P = 0.01). They also reported significant changes in DBP after the consumption of mixed nuts (Mohammadifard et al., 2015).

In conclusion, some evidence suggests that pistachios may reduce BP and improve different parameters of EF, which could contribute to a decrease in CV risk. However, due to the limited research, additional clinical trials should be conducted to contrast current knowledge.

### *3.1.4 Body weight modulation and satiety*

The number of people classified as overweight and obese is increasing dramatically. In 2016, 39% of adults (=18 years) were overweight and 13% were obese (World Health Organisation, 2016), thus new strategies for medical treatment and prevention are needed. Despite the fact that nuts, including pistachios, contain a significant amount of fat and are energy-dense foods, several epidemiologic studies have provided strong evidence that nut consumption is associated with neither body weight (BW) gain nor an increased risk of obesity (Bes-Rastrollo et al., 2007, 2009; Casas-Agustench et al., 2011; Flores-Mateo et al., 2013). The latest results from the European Prospective Investigation into Cancer and Nutrition (EPIC) have shown that, compared to non-consumers, subjects in the highest quartile of nut intake not only had 0.07 kg less BW gain over five years (95% CI: −0.12, −0.02), but also had 5% lower risk of becoming overweight (RR 0.95; 95% CI: 0.92, 0.98) or obese (RR 0.95; 95% CI: 0.90, 0.99) (Freisling et al., 2017).

Two studies have investigated the satiating properties of pistachios. Honselman et al. evaluated the impact on fullness and caloric intake of consuming in-shell pistachios or pistachio kernels. They conducted a randomized, crossover, controlled feeding trial including 140 university students aged 18–24 years old. Consumption of in-shell pistachios resulted in a lower calorie intake than kernel consumption (Honselman et al., 2011). In a second crossover feeding trial, involving 118 healthy individuals, the same authors demonstrated that the visual cue of the empty pistachio shells may have helped the participants to consume fewer pistachios and approximately 18% fewer calories (Kennedy-Hagan et al., 2011).

Different clinical trials have evaluated the effect of nuts on BW but only a few have been designed to assess the effect of pistachio consumption on BW as the main outcome (Table 2), including three randomized, controlled clinical trials by Gulati et al. (2014), Li et al. (2010) and Wang et al. (2012). During a three-month weight loss programme, using hypocaloric diets that provided 500 fewer Kcal than recommended, 70 overweight or obese individuals were randomly allocated to a pistachio diet group (consuming 53 g/day of pistachios) or to a pretzel-enriched diet group (consuming 56 g/day of salted pretzels). The pistachios or pretzels were consumed as an afternoon snack. During the intervention, the pistachio-supplemented group exhibited a significant reduction in BMI (−4.3% of the BMI), greater than that observed in the pretzel-diet group (−2% of the BMI) (Li et al., 2010). Similarly, Wang et al. assessed the effect of a three-month normocaloric diet supplemented, or not, with two different doses of pistachio nuts (70 or 42 g/day) on total BW maintenance in 90 subjects with MetS (Wang et al., 2012). The results showed that the consumption of any dosage of pistachios resulted in no significant changes in BMI or waist-to-hip ratio compared with the group following the AHA Step I recommendations. In 2014, a 24-week, randomized control trial, including 60 MetS subjects assigned randomly to either a pistachio (20% of total daily energy in the form of pistachios) or a control group for 6 months, did not find significant differences in BW. However, a significant decrease in waist circumference (WC) was observed in the pistachio group, compared with the control group, along with a trend towards reduced subcutaneous adipose tissue (Gulati et al., 2014).

In addition, nine randomized feeding trials evaluated the effect of eating pistachios on BW, WC and/or BMI as a secondary outcome. In all five studies, participants consumed at least 15% of total caloric intake in the form of pistachios; however, no significant effect on BW and/or BMI was observed compared with the control diet group (Edwards et al., 1999; Gebauer et al., 2008; Hernández-Alonso et al., 2014; Kasliwal et al., 2015; Kocyigit et al., 2006; Parham et al., 2014; Sari et al., 2010; Sauder et al., 2015; Sheridan et al., 2007). A review of the available data suggests that adding nuts to habitual diets of free-living individuals does not result in any appreciable increase in BW

(Flores-mateo et al., 2013; Li et al., 2010; Martínez-González and Bes-Rastrollo, 2011; Rajaram and Sabaté, 2006; Salas-Salvadó et al., 2008; Vadivel et al., 2012; Wang et al., 2012). Moreover, meta-analysis involving 33 clinical trials showed that a nut-enriched diet did not lead to increased BW, BMI or WC compared with control diets (Flores-Mateo et al., 2013).

Numerous biological mechanisms may explain the unexpected null effect of consuming pistachios, and other nuts, on adiposity. Nuts are rich in unsaturated fatty acids, and evidence suggests that MUFA and PUFA are more readily oxidized (Casas-Agustench et al., 2009; Reaven and Witztum, 1996) and have a greater thermogenic effect (Alper and Mattes, 2002) than SFA, which can result in decreased fat accumulation. In addition, there is much evidence to suggest that nuts have high satiating properties. Pistachios are energy dense and a good source of fibre, protein and unsaturated fats, which are dietary factors that increase satiety ratings. Therefore, pistachios may suppress hunger and subsequently reduce food intake (Astrup, 2005; Burton-Freeman, 2000; Hu, 2005). Furthermore, the physical structure of nuts may also contribute to their satiating effect, as they are crunchy and have to be mechanically reduced to particles small enough to swallow. It is thought that mastication may modify appetitive sensations as it activates mechanical, nutrient and sensory signalling systems (Cassady et al., 2009). Moreover, a small degree of fat malabsorption has been reported after nut intake, which is attributed to the fat being contained within walled cellular structures that are incompletely digested in the gut, an effect that could be exacerbated by insufficient mastication (Ellis et al., 2004). In fact, a crossover trial involving 16 healthy volunteers consuming pistachios (42 and 84 g/day) or a nut-free diet for three weeks, as part of a controlled diet, revealed that the metabolizable energy of pistachios, calculated from differences in faecal energy excretion during the different dietary regimes, is 5% less than the energy calculated by the Atwater general factors, implying that the energy from pistachios is not totally utilizable (Baer et al., 2012).

In summary, current data supports the fact that pistachio consumption does not lead to an increase in BW or other anthropometric parameters. However, in order to expand their beneficial role to BW control and loss, further research should be conducted pistachio consumption in *ad libitum* conditions.

## 3.2 Novel and emergent risk factors for metabolic and cardiovascular disease

Apart from the well-known risk factors for metabolic and CV diseases, novel and emergent parameters are being assessed due to their link with outcomes that are altered in the aforementioned diseases. This group includes different oxidant and antioxidant substances (e.g. ox-LDL and γ-tocopherol), the composition and size of lipoproteins, the nutrient-gene interaction and the modulation of certain metabolites and bacteria by nut consumption. In this section, we will summarize the current literature showing the effect of pistachio consumption on the modulation of these novel and emergent risk factors.

### 3.2.1 Oxidative stress

Pistachio antioxidants could play an important role in the control of oxidative stress. In 2006, Kocyigit et al. conducted a study involving 44 healthy men and women, who were randomized to a regular diet group or a pistachio diet group (accounting for 20% of their daily caloric intake in the form of pistachios) for three weeks. The participants consuming pistachios experienced an increase in blood antioxidant potential, determined by the production of thiobarbituric acid-reactive substances and decreased malondialdehyde levels, which is an important indicator of lipid peroxidation (Kocyigit et al., 2006). Kay et al. performed a crossover, randomized, controlled feeding trial involving 28 hypercholesterolaemic adults and showed that the consumption of diets containing 10% and 20% of energy from pistachios (32–63 and 63–126 g/day, respectively) led to increased antioxidant concentrations in serum, such as $\gamma$-tocopherol, lutein and $\beta$-carotene, whereas ox-LDL concentrations were decreased in comparison to the consumption of a control diet lacking pistachios (Kay et al., 2010). Finally, in a sequential feeding trial involving 32 healthy young men, Sari et al. determined the effect of a traditional MedDiet supplemented with pistachios by replacing the monounsaturated fat content, constituting approximately 20% of the daily caloric intake, for one month. They observed an increase in the total antioxidant status and superoxide dismutase, along with decreased inflammation and other oxidative markers (Sari et al., 2010).

### 3.2.2 Lipoprotein particle composition

Conventional CVD prevention strategies focus on decreasing LDLc concentrations in order to ameliorate the CV risk. Therefore, classical lipid profile assessment (i.e. LDLc, HDLc, TC, triglycerides) has been used in order to monitor the CV risk of the population including subjects with pre-D and T2D. However, an increasing amount of data suggests that preventive and therapeutic strategies should be addressed to monitor abnormalities on lipoprotein subfractions (small, medium, large) beyond classical lipid profile (Rizzo and Berneis, 2006). In fact, a novel procedure for quantifying plasma levels of lipoprotein subclass particles by proton nuclear magnetic resonance (NMR) spectroscopy was developed at the beginning of the century (Hu et al., 2002; Otvos, 2002). NMR spectroscopy provides concentrations of very low-density lipoprotein (VLDL), LDL and HDL subspecies simultaneously, and NMR-determined subclasses correspond well with those obtained with established methods (Hu et al., 2002). Few studies have evaluated the effect of nut consumption on the distribution and particle size of lipoprotein subclasses, which seems to be a better atherogenic indicator (German et al., 2006). The first study was conducted in a small number of adults ($n$ = 18) with combined hyperlipidaemia who consumed 48 g of walnuts for six weeks. Some fractions of cholesterol decreased, particularly in small-LDL-P and large-HDL-P, which suggested that the beneficial properties of nut intake on cardiovascular risk may be underlying by an additional mechanism (Almario et al., 2001). Concurring with these results, two crossover interventions with pistachio nuts showed a reduction in small-LDL (Holligan et al., 2014) together with HDL-P size (Hernández-Alonso et al., 2015), overall reinforcing these benefits in outcomes that may worsen pre-D and T2D status, and overall CVD.

### 3.2.3 Gene and microRNA modulation

The effect of diet on gene expression (i.e. nutrigenomics) is still in a developing stage. Gene expression after mixed nuts consumption has been previously reported (Ros et al., 2014) but our group was the first to assess changes in the leukocytes' expression of a set of genes after chronic pistachio consumption. In particular, a crossover study conducted in pre-diabetic subjects consuming pistachio for four months reported a downregulation of interleukin-6 (*IL-6*) and resistin (*RETN*) genes in leukocytes, whereas the pistachio diet led to a lower significant increase of solute carrier family 2 member 4 (*SLC2A4*, codifying GLUT4) (Hernández-Alonso et al., 2014). Consistent with this attenuation in the expression of glucose transporters, leukocytes significantly reduced the uptake of glucose following pistachio consumption, suggesting a decrease in the cell hyperactivity described in T2D. However, the modulatory role of nuts on the expression of genes related to inflammation, oxidative stress and glucose metabolism may also be mediated by the effect of nutrients on microRNAs (miRNAs). MiRNAs are RNA molecules that belong to a family of small non-coding RNAs with 20–25 nucleotides that post-transcriptionally and negatively regulate gene expression (Ambros, 2004; Ambros et al., 2003). Research into the nutritional modulation of miRNA is still in its infancy and few studies have assessed whether a specific dietary pattern, food, supplements or a particular nutrient can influence miRNA levels (Ross and Davis, 2014). Our group additionally revealed that pistachio consumption significantly diminished the levels of miR-375 and miR-192, both related to glucose and insulin metabolism (Hernández-Alonso et al., 2016a). Therefore, these results shed new light on the nutrient-gene-miRNA interactions in nuts. However, there is still much to discover about the exact mechanisms linking genes and miRNAs with disease status and the implication of including nuts as dietary components.

### 3.2.4 Modulation of metabolites and bacteria

Recent studies have pointed out that the gut microbiome might make an important contribution to the development of different diseases. Moreover, it is well established that long-term dietary patterns and shorter-term dietary variations influence gut microbiota composition (Conlon and Bird, 2014). In fact, many potential prebiotic components can be present in a particular food (e.g. fermentation of fibre from nuts to beneficial end products (e.g. butyric acid)) and may be associated with the transition to a healthier microbiota composition (Ansell et al., 2013; Ros, 2010). There is little information on the putative effect of nuts on microbiota modulation. Only one study has evaluated the effect of pistachio consumption on changes in microbiota. Ukhanova et al. carried out two separate crossover feeding trials with nuts and pistachios, and showed that they both significantly affected microbiota modulation (Ukhanova et al., 2014). However, a greater prebiotic effect was found after pistachio intake. In fact, pistachio increased the number of butyrate-producing bacteria, which are potentially beneficial, whereas the numbers of *Bifidobacteria* were not affected by the consumption of either nut (Ukhanova et al., 2014). Recently, dietary polyphenols have been found to be involved in gut microbiota dysbiosis processes, which may lead to a reduction of fat storage, inflammation and IR (reviewed in Anhê et al., 2015).

Very few studies have been made on metabolomic modulation after nut consumption. Of these, only two have focused on pistachio consumption. First in 2014, Nieman et al. showed a modulation in the levels of metabolites such as raffinose, (12*Z*)-9,10-dihydroxyoctadec-12-enoate, sucrose, together with some modulations of plasma amino acids and fatty acids following pistachio intake (Nieman et al., 2014). More recently, our group reported a modulation of gut-related metabolites and *cis*-aconitate, an intermediate of tricarboxylic acid, after consumption of pistachios for four months (Hernández-Alonso et al., 2017a).

Taken together, these results provide evidence of the beneficial effects of pistachios on metabolic diseases and CVD risk beyond the lipid lowering effect. Different novel and emergent risk factors of metabolic and CVD have been found to be improved following pistachio consumption. However, there is a huge need for research in this area in order to determine other molecules that may serve as biomarkers for disease.

## 3.3 In vitro and in vivo studies

Different species from the *Pistacia* genus have been examined in *in vitro* and *in vivo* studies due to their traditional and putative beneficial effects. However, in this section, we will focus on studies investigating the role of *P. vera* with a special emphasis on its kernel.

Non-edible parts extracted from *P. vera* such as its oleoresin have been shown to possess a manifest anti-inflammatory activity against carrageenan-induced hind paw oedema model in mice, without inducing any gastric damage at doses ranging from 250 to 500 mg/kg BW (Orhan et al., 2006). In fact, most of the *in vitro* research and studies conducted on animals suggest that the healthy properties of pistachios could be attributed partially to the content of the nuts' dietary antioxidants. First in 2007, Gentile et al. assessed the bioactive components in pistachios and their antioxidant activity (i.e. lag time oxidation of LDL), proving their beneficial role in oxidative-related impairments (Gentile et al., 2007). In 2012, they evaluated the effects of a hydrophilic extract (HPE) of *P. vera* on the production of reactive oxygen substances (ROS) in RAW 264.7 macrophage cells. A dose-dependent decrease of lipopolysaccharide (LPS)-induced ROS production was observed when cells were incubated with different concentrations of HPE, indicating proanthocyanidins as the bioactive responsible components of this effect (Gentile et al., 2012). Recently, Gentile et al. studied the effects of HPE and its polymeric proanthocyanidin fraction in a cell model that simulated some conditions of inflammatory bowel diseases, consisting of interleukin (IL)-1β-stimulated Caco-2 cells. They showed that HPE decreased prostaglandin (PG)E2 production, IL-6 and IL-8 release, and cyclooxygenase (COX)-2 expression. HPE also inhibited the increase in paracellular permeability and reduced NF-κB activation. These results suggest that pistachio nut components inhibit inflammatory response of intestinal epithelial cells *in vitro* and that intake of small amounts of pistachio nut can exert beneficial effects to gastrointestinal pathophysiology (Gentile et al., 2015).

Similarly, the incubation of RAW 264.7 murine macrophages with a pistachio oil extract (POE) for 24 h decreased some LPS-induced inflammatory markers such us Ifit-2, TNF-α and IL-6 (Zhang et al., 2010). This POE also reduced the Ifit-2, TNFα, IL-6 and IL-1β expression by 78%, 55%, 58% and 35%, respectively, in response to LPS stimulation of the same cells. Two studies on rats found increased antioxidant enzymatic activity in animals fed pistachios for eight weeks (Aksoy et al., 2007; Alturfan et al., 2009). In the first study, rats were divided into three groups of 12 animals and assigned to a control group fed a standard diet and two pistachio groups fed with a standard diet containing 20% or 40% of the energy in the form of pistachios. A significant increase in the activities of PON1 and arylesterase, both markers of antioxidant capacity, was shown in both groups supplemented with pistachios compared with the control group after 10 weeks' intervention (Aksoy et al., 2007). In the second study, rats were assigned to a control diet (standard commercial chow); a control diet supplemented with 1.26% of the total energy intake in the form of pistachios; a control diet with 1.63% of cholesterol, 0.41% cholic acid and 16.3% sunflower oil (hyperlipidaemic diet); or a hyperlipidaemic diet supplemented with 1.26% of the total energy intake in the form of pistachios. After eight weeks, rats fed with the hyperlipidaemic diet supplemented with pistachios had higher total antioxidant activity, determined by thiobarbituric acid-reactive substances, than rats fed with the hyperlipidaemic diet alone (Alturfan et al., 2009).

Polyphenol-rich extract from both skin and decorticated seeds possess good radical scavenger/antioxidant properties, as shown in a series of *in vitro* assays carried out by Martorana et al. Importantly, they reported that the skin extract is about ten times richer in phenols than the seed extract. Furthermore, both extracts reduce, when topically applied, UV-B-induced skin erythema in human volunteers. These findings suggest that extracts from both pistachios' skins and seeds could be successfully employed as photoprotective ingredients in topical cosmetic and pharmaceutical formulations (Martorana et al., 2013). Moreover, a beneficial cytoprotective activity of pistachio extracts (methanolic, water or ethyl acetate) against oxidative (ROS formation) and carbonyl stress has also been reported in a T2D model in hepatocytes from rats (Shahraki et al., 2014).

Bisignano et al. investigated the antimicrobial properties of polyphenol-rich fractions derived from raw shelled and roasted salted pistachios. Pistachio extracts were active against Gram-positive bacteria with a bactericidal effect observed against *Listeria monocytogenes*, *Staphylococcus aureus* and methicillin-resistant *Staphylococcus aureus* clinical isolates. As expected, they also reported that extracts from raw shelled pistachios are more active than those from roasted salted pistachios. The bactericidal activity of pistachio extracts could be used to help control the growth of some microorganisms in foods (i.e. improve safety) and may find application as a topical treatment for *S. aureus* (Bisignano et al., 2013). Grace et al. recently demonstrated the antioxidant and anti-inflammatory activity of roasted pistachios. They evaluated non-polar extracts from skin and nut kernel in 3T3L1 mouse adipocytes. Both extracts decreased lipid accumulation, with skin extract demonstrating the highest activity (17% inhibition). Radical scavenging assays indicated that all pistachio extracts significantly inhibited ROS, while skin non-polar extract was the most inhibiting to NO production in LPS-stimulated RAW 264.7 macrophages. Gene expression profiles associated with inflammation (IL-6, inducible NOS and COX2) were characterized in the LPS-stimulated RAW264.7 macrophages after treatment with pistachio extracts. Skin and kernel non-polar extracts were the most potent to inhibit the expression of COX2. The skin extract had the strongest effect in decreasing non-mitochondrial oxidative burst associated with inflammatory response in macrophages (Grace et al., 2016).

Recent research was conducted to investigate whether polyphenol extracts from natural raw shelled pistachios (NP) or roasted salted pistachio (RP) kernels have anti-inflammatory and antioxidant properties at lower doses than reported previously in *in vitro* and *in vivo* models. The monocyte/macrophage cell line J774 was used to assess the extent of protection by NP and RP against LPS-induced inflammation. Moreover, antioxidant activity of NP and RP was assessed in an *in vivo* rat model of paw oedema induced by carrageenan injection in the paw. Results from the *in vitro* study demonstrated that pre-treatment with NP and RP exerted a significant protection against LPS-induced inflammation. Western blot analysis showed that both NP and RP decreased the TNF-$\alpha$ and IL-1$\beta$ production in a dose-dependent way. These data demonstrated that, at lower doses, polyphenols present in pistachios possess antioxidant and anti-inflammatory properties. This may contribute towards a better understanding of the beneficial health effects associated with consumption of pistachios (Paterniti et al., 2017).

In summary, *in vitro* and *in vivo* studies have reported enhancement in different parameters with oxidation and inflammation being the most documented. However, further *in vitro* and *in vivo* studies should be conducted in order to improve parameters related to other CV and metabolic diseases.

## 4 Conclusion

Pistachios are a nutrient-dense nut with a healthy nutritional profile including fibre, healthy fat, phytosterols and antioxidant compounds, contributing to a reduced risk of different metabolic and CVDs. Growing evidence suggests that the inclusion of pistachios improves diet quality and provides several bioactive compounds with recognized properties for weight management, glycaemic control and vascular health. Despite further research still being needed in order to expand the current findings to other diseases and populations, current knowledge demonstrates the beneficial role of pistachios on maintaining and promoting health.

## 5 Where to look for further information

Different review articles have addressed the beneficial effects of pistachio consumption on different cardiometabolic conditions. For further information regarding specific health benefits of pistachio consumption and/or their nutrition value refer to these three latest reviews: Hernández-Alonso et al. (2016b), Bulló et al. (2015) and Dreher (2012).

## 6 References

Aksoy, N., Aksoy, M. and Bagci, C. (2007). Pistachio intake increases high density lipoprotein levels and inhibits low-density lipoprotein oxidation in rats. *The Tohoku Journal of Experimental Medicine* 212(1): 43–8.

Aldemir, M., Okulu, E., Neşelioğlu, S., et al. (2011). Pistachio diet improves erectile function parameters and serum lipid profiles in patients with erectile dysfunction. *International Journal of Impotence Research* 23(1): 32–8. DOI:10.1038/ijir.2010.33.

Almario, R. U., Vonghavaravat, V., Wong, R., et al. (2001). Effects of walnut consumption on plasma fatty acids and lipoproteins in combined hyperlipidemia. *The American Journal of Clinical Nutrition* 74(1): 72–9.

Alper, C. M. and Mattes, R. D. (2002). Effects of chronic peanut consumption on energy balance and hedonics. *International Journal of Obesity and Related Metabolic Disorders : Journal of the International Association for the Study of Obesity* 26(8): 1129–37. DOI:10.1038/sj.ijo.0802050.

Alturfan, A. A., Emekli-Alturfan, E. and Uslu, E. (2009). Consumption of pistachio nuts beneficially affected blood lipids and total antioxidant activity in rats fed a high-cholesterol diet. *Folia Biologica* 55: 132–6.

Ambros, V. (2004). The functions of animal microRNAs. *Nature* 431(7006): 350–5. DOI:10.1038/nature02871.

Ambros, V., Bartel, B., Bartel, D. P., et al. (2003). A uniform system for microRNA annotation. *RNA (New York, N.Y.)* 9(3): 277–9.

Anderson, J. W., Hanna, T. J., Peng, X., et al. (2000). Whole grain foods and heart disease risk. *Journal of the American College of Nutrition* 19(3 Suppl): 291S–9S.

Anderson, T. J., Grégoire, J., Hegele, R. A., et al. (2013). 2012 update of the Canadian Cardiovascular Society guidelines for the diagnosis and treatment of dyslipidemia for the prevention of cardiovascular disease in the adult. *The Canadian Journal of Cardiology* 29(2): 151–67. DOI:10.1016/j.cjca.2012.11.032.

Anhê, F. F., Varin, T. V., Le Barz, M., et al. (2015). Gut microbiota dysbiosis in obesity-linked metabolic diseases and prebiotic potential of polyphenol-rich extracts. *Current Obesity Reports* 4(4): 389–400. DOI:10.1007/s13679-015-0172-9.

Anon. (2017). World Production, Markets, and Trade Reports. Tree Nuts: World Markets and Trade. United States Department of Agriculture – Foreign Agricultural Service, https://www.fas.usda.gov/data/tree-nuts-world-markets-and-trade (accessed 4 July 2018).

Ansell, J., Parkar, S., Paturi, G., et al. (2013). Modification of the colonic microbiota. *Advances in Food and Nutrition Research* 68: 205–17. DOI:10.1016/B978-0-12-394294-4.00011-0.

Assaf-Balut, C., García de la Torre, N., Durán, A., et al. (2017). A Mediterranean diet with additional extra virgin olive oil and pistachios reduces the incidence of gestational diabetes mellitus (GDM): A randomized controlled trial: The St. Carlos GDM prevention study. *PloS ONE* 12(10): e0185873. DOI:10.1371/journal.pone.0185873.

Astrup, A. (2005). The satiating power of protein – a key to obesity prevention? *The American Journal of Clinical Nutrition* 82(1): 1–2.

Baer, D. J., Gebauer, S. K. and Novotny, J. A. (2012). Measured energy value of pistachios in the human diet. *The British Journal of Nutrition* 107(1): 120–5. DOI:10.1017/S0007114511002649.

Bes-Rastrollo, M., Sabaté, J., Gómez-Gracia, E., et al. (2007). Nut consumption and weight gain in a Mediterranean cohort: The SUN study. *Obesity (Silver Spring, Md.)* 15(1): 107–16. DOI:10.1038/oby.2007.507.

Bes-Rastrollo, M., Wedick, N. M., Martínez-González, M. A., et al. (2009). Prospective study of nut consumption, long-term weight change, and obesity risk in women. *The American Journal of Clinical Nutrition* 89(6): 1913–9. DOI:10.3945/ajcn.2008.27276.

Bisignano, C., Filocamo, A., Faulks, R. M., et al. (2013). In vitro antimicrobial activity of pistachio (*Pistacia vera* L.) polyphenols. *FEMS Microbiology Letters* 341(1): 62–7. DOI:10.1111/1574-6968.12091.

Bolling, B. W., Chen, C.-Y. O., McKay, D. L., et al. (2011). Tree nut phytochemicals: composition, antioxidant capacity, bioactivity, impact factors. A systematic review of almonds, Brazils, cashews, hazelnuts, macadamias, pecans, pine nuts, pistachios and walnuts. *Nutrition Research Reviews* 24(2): 244–75. DOI:10.1017/S095442241100014X.

Bozorgi, M., Memariani, Z., Mobli, M., et al. (2013). Five pistacia species (*P. vera*, *P. atlantica*, *P. terebinthus*, *P. khinjuk*, and *P. lentiscus*): A review of their traditional uses, phytochemistry, and pharmacology. *The Scientific World Journal*. DOI:10.1155/2013/219815.

Bulló, M., Estruch, R. and Salas-Salvadó, J. (2011). Dietary vitamin K intake is associated with bone quantitative ultrasound measurements but not with bone peripheral biochemical markers in elderly men and women. *Bone* 48(6): 1313–8. DOI:10.1016/j.bone.2011.03.767.

Bulló, M., Juanola-Falgarona, M., Hernández-Alonso, P., et al. (2015). Nutrition attributes and health effects of pistachio nuts. *British Journal of Nutrition* 113 Suppl 2: S79–S93. DOI:10.1017/S0007114514003250.

Burton-Freeman, B. (2000). Dietary fiber and energy regulation. *The Journal of Nutrition* 130(Suppl 2): 272S–5S.

Carughi, A., Feeney, M. J., Kris-Etherton, P., et al. (2016). Pairing nuts and dried fruit for cardiometabolic health. *Nutrition Journal* 15(1): 23. DOI:10.1186/s12937-016-0142-4.

Casas-Agustench, P., López-Uriarte, P., Bulló, M., et al. (2009). Acute effects of three high-fat meals with different fat saturations on energy expenditure, substrate oxidation and satiety. *Clinical Nutrition* 28(1): 39–45. DOI:10.1016/j.clnu.2008.10.008.

Casas-Agustench, P., Bulló, M., Ros, E., et al. (2011). Cross-sectional association of nut intake with adiposity in a Mediterranean population. *Nutrition, Metabolism, and Cardiovascular Diseases : NMCD* 21(7): 518–25. DOI:10.1016/j.numecd.2009.11.010.

Cassady, B. A., Hollis, J. H., Fulford, A. D., et al. (2009). Mastication of almonds: Effects of lipid bioaccessibility, appetite, and hormone response. *The American Journal of Clinical Nutrition* 89(3): 794–800. DOI:10.3945/ajcn.2008.26669.

Chandrasekara, N. and Shahidi, F. (2011). Effect of roasting on phenolic content and antioxidant activities of whole cashew nuts, kernels, and testa. *Journal of Agricultural and Food Chemistry* 59(9): 5006–14. DOI:10.1021/jf2000772.

Conlon, M. A. and Bird, A. R. (2014). The impact of diet and lifestyle on gut microbiota and human health. *Nutrients* 7(1): 17–44. DOI:10.3390/nu7010017.

Damavandi, R. D., Eghtesadi, S., Shidfar, F., et al. (2013). Effects of hazelnuts consumption on fasting blood sugar and lipoproteins in patients with type 2 diabetes. *Journal of Research in Medical Sciences : The Official Journal of Isfahan University of Medical Sciences* 18(4): 314–21.

Del Gobbo, L. C., Falk, M. C., Feldman, R., et al. (2015). Effects of tree nuts on blood lipids, apolipoproteins, and blood pressure: Systematic review, meta-analysis, and dose-response of 61 controlled intervention trials. *The American Journal of Clinical Nutrition* 102(6): 1347–56. DOI:10.3945/ajcn.115.110965.

Dietrich, M., Traber, M. G., Jacques, P. F., et al. (2006). Does gamma-tocopherol play a role in the primary prevention of heart disease and cancer? A review. *Journal of the American College of Nutrition* 25(4): 292–9.

Dixon, R. A., Xie, D.-Y. and Sharma, S. B. (2005). Proanthocyanidins – a final frontier in flavonoid research? *The New Phytologist* 165(1): 9–28. DOI:10.1111/j.1469-8137.2004.01217.x.

Dreher, M. L. (2012). Pistachio nuts: Composition and potential health benefits. *Nutrition Reviews* 70(4): 234–40.

Edwards, K., Kwaw, I., Matud, J., et al. (1999). Effect of pistachio nuts on serum lipid levels in patients with moderate hypercholesterolemia. *Journal of the American College of Nutrition* 18(3): 229–32.

Ellis, P. R., Kendall, C. W. C., Ren, Y., et al. (2004). Role of cell walls in the bioaccessibility of lipids in almond seeds. *The American Journal of Clinical Nutrition* 80(3): 604–13.

Flores-Mateo, G., Rojas-Rueda, D., Basora, J., et al. (2013). Nut intake and adiposity: Meta-analysis of clinical trials. *The American Journal of Clinical Nutrition* 97(6): 1346–55. DOI:10.3945/ajcn.111.031484.

Freisling, H., Noh, H., Slimani, N., et al. (2017). Nut intake and 5-year changes in body weight and obesity risk in adults: Results from the EPIC-PANACEA study. *European Journal of Nutrition*. DOI:10.1007/s00394-017-1513-0.

Gebauer, S. K., West, S. G., Kay, C. D., et al. (2008). Effects of pistachios on cardiovascular disease risk factors and potential mechanisms of action: A dose-response study. *The American Journal of Clinical Nutrition* 88(3): 651–9.

Gentile, C., Tesoriere, L., Butera, D., et al. (2007). Antioxidant activity of Sicilian pistachio (*Pistacia vera* L. Var. *Bronte*) nut extract and its bioactive components. *Journal of Agricultural and Food Chemistry* 55(3): 643–8. DOI:10.1021/jf062533i.

Gentile, C., Allegra, M., Angileri, F., et al. (2012). Polymeric proanthocyanidins from Sicilian pistachio (*Pistacia vera* L.) nut extract inhibit lipopolysaccharide-induced inflammatory response in RAW 264.7 cells. *European Journal of Nutrition* 51(3): 353–63. DOI:10.1007/s00394-011-0220-5.

Gentile, C., Perrone, A., Attanzio, A., et al. (2015). Sicilian pistachio (*Pistacia vera* L.) nut inhibits expression and release of inflammatory mediators and reverts the increase of paracellular permeability in IL-1β-exposed human intestinal epithelial cells. *European Journal of Nutrition* 54(5): 811–21. DOI:10.1007/s00394-014-0760-6.

German, J. B., Smilowitz, J. T. and Zivkovic, A. M. (2006). Lipoproteins: When size really matters. *Current Opinion in Colloid and Interface Science*. DOI:10.1016/j.cocis.2005.11.006.

Gillingham, L. G., Harris-Janz, S. and Jones, P. J. H. (2011). Dietary monounsaturated fatty acids are protective against metabolic syndrome and cardiovascular disease risk factors. *Lipids* 46(3): 209–28. DOI:10.1007/s11745-010-3524-y.

Grace, M. H., Esposito, D., Timmers, M. A., et al. (2016). In vitro lipolytic, antioxidant and anti-inflammatory activities of roasted pistachio kernel and skin constituents. *Food & Function* 7: 4285–98. DOI:10.1039/c6fo00867d.

Gulati, S., Misra, A., Pandey, R. M., et al. (2014). Effects of pistachio nuts on body composition, metabolic, inflammatory and oxidative stress parameters in Asian Indians with metabolic syndrome: A 24-wk, randomized control trial. *Nutrition* 30(2): 192–7. DOI:10.1016/j.nut.2013.08.005.

Halvorsen, B. L., Carlsen, M. H., Phillips, K. M., et al. (2006). Content of redox-active compounds (ie, antioxidants) in foods consumed in the United States. *The American Journal of Clinical Nutrition* 84(1): 95–135.

Hassellund, S. S., Flaa, A., Kjeldsen, S. E., et al. (2013). Effects of anthocyanins on cardiovascular risk factors and inflammation in pre-hypertensive men: A double-blind randomized placebo-controlled crossover study. *Journal of Human Hypertension* 27(2): 100–6. DOI:10.1038/jhh.2012.4.

Hercberg, S., Kesse-Guyot, E., Druesne-Pecollo, N., et al. (2010). Incidence of cancers, ischemic cardiovascular diseases and mortality during 5-year follow-up after stopping antioxidant vitamins and minerals supplements: A postintervention follow-up in the SU.VI.MAX Study. *International Journal of Cancer (Journal international du cancer)* 127(8): 1875–81. DOI:10.1002/ijc.25201.

Hernández-Alonso, P., Salas-Salvadó, J., Baldrich-Mora, M., et al. (2014). Beneficial effect of pistachio consumption on glucose metabolism, insulin resistance, inflammation, and related metabolic risk markers: A randomized clinical trial. *Diabetes Care* 37(11): 3098–105. DOI:10.2337/dc14-1431.

Hernández-Alonso, P., Salas-Salvadó, J., Baldrich-Mora, M., et al. (2015). Effect of pistachio consumption on plasma lipoprotein subclasses in pre-diabetic subjects. *Nutrition, Metabolism, and Cardiovascular Diseases : NMCD* 25(4): 396–402. DOI:10.1016/j.numecd.2015.01.013.

Hernández-Alonso, P., Giardina, S., Salas-Salvadó, J., et al. (2016a). Chronic pistachio intake modulates circulating microRNAs related to glucose metabolism and insulin resistance in prediabetic subjects. *European Journal of Nutrition*. DOI:10.1007/s00394-016-1262-5.

Hernández-Alonso, P., Bulló, M. and Salas-Salvadó, J. (2016b). Pistachios for health: What do we know about this multifaceted nut? *Nutrition Today* 51(3): 133–8.

Hernández-Alonso, P., Cañueto, D., Giardina, S., et al. (2017a). Effect of pistachio consumption on the modulation of urinary gut microbiota-related metabolites in pre-diabetic subjects. *The Journal of Nutritional Biochemistry*. DOI:10.1016/j.jnutbio.2017.04.002.

Hernández-Alonso, P., Camacho-Barcia, L., Bulló, M., et al. (2017b). Nuts and dried fruits: An update of their beneficial effects on type 2 diabetes. *Nutrients* 9(7). DOI:10.3390/nu9070673.

Holligan, S. D., West, S. G., Gebauer, S. K., et al. (2014). A moderate-fat diet containing pistachios improves emerging markers of cardiometabolic syndrome in healthy adults with elevated LDL levels. *The British Journal of Nutrition* 112(5): 744–52. DOI:10.1017/S0007114514001561.

Honselman, C. S., Painter, J. E., Kennedy-Hagan, K. J., et al. (2011). In-shell pistachio nuts reduce caloric intake compared to shelled nuts. *Appetite* 57(2): 414–7. DOI:10.1016/j.appet.2011.02.022.

Hu, F. B. (2005). Protein, body weight, and cardiovascular health. *The American Journal of Clinical Nutrition* 82(1): 242S–7S.

Hu, F. B., Stampfer, M. J., Haffner, S. M., et al. (2002). Elevated risk of cardiovascular disease prior to clinical diagnosis of type 2 diabetes. *Diabetes care* 25(7): 1129–34.

Huang, H.-Y., Caballero, B., Chang, S., et al. (2006). The efficacy and safety of multivitamin and mineral supplement use to prevent cancer and chronic disease in adults: A systematic review for a National Institutes of Health state-of-the-science conference. *Annals of Internal Medicine* 145(5): 372–85.

Ibarrola-Jurado, N., Salas-Salvadó, J., Martínez-González, M. A., et al. (2012). Dietary phylloquinone intake and risk of type 2 diabetes in elderly subjects at high risk of cardiovascular disease. *The American Journal of Clinical Nutrition* 96(5): 1113–8. DOI:10.3945/ajcn.111.033498.

Jaffrey, S. R. and Snyder, S. H. (1995). Nitric oxide: A neural messenger. *Annual Review of Cell and Developmental Biology* 11: 417–40. DOI:10.1146/annurev.cb.11.110195.002221.

Jenkins, D. J. A., Hu, F. B., Tapsell, L. C., et al. (2008). Possible benefit of nuts in type 2 diabetes. *The Journal of Nutrition* 138: 1752S–6S. DOI:138/9S-I/1752S [pii].

Juanola-Falgarona, M., Salas-Salvadó, J., Estruch, R., et al. (2013). Association between dietary phylloquinone intake and peripheral metabolic risk markers related to insulin resistance and diabetes in elderly subjects at high cardiovascular risk. *Cardiovascular Diabetology* 12(1): 7. DOI:10.1186/1475-2840-12-7.

Juanola-Falgarona, M., Salas-Salvadó, J., Martínez-González, M. Á., et al. (2014). Dietary intake of vitamin K is inversely associated with mortality risk. *The Journal of Nutrition*. DOI:10.3945/jn.113.187740.

Kaczmarczyk, M. M., Miller, M. J. and Freund, G. G. (2012). The health benefits of dietary fiber: Beyond the usual suspects of type 2 diabetes mellitus, cardiovascular disease and colon cancer. *Metabolism: Clinical and Experimental* 61(8): 1058–66. DOI:10.1016/j.metabol.2012.01.017.

Kasliwal, R. R., Bansal, M., Mehrotra, R., et al. (2015). Effect of pistachio nut consumption on endothelial function and arterial stiffness. *Nutrition (Burbank, Los Angeles County, Calif.)* 31(5): 678–85. DOI:10.1016/j.nut.2014.10.019.

Kay, C. D., Gebauer, S. K., West, S. G., et al. (2010). Pistachios increase serum antioxidants and lower serum oxidized-LDL in hypercholesterolemic adults. *The Journal of Nutrition* 140(6): 1093–8. DOI:10.3945/jn.109.117366.1093.

Kendall, C. W. C., Josse, A. R., Esfahani, A., et al. (2011). The impact of pistachio intake alone or in combination with high-carbohydrate foods on post-prandial glycemia. *European Journal of Clinical Nutrition* 65(6): 696–702. DOI:10.1038/ejcn.2011.12.

Kendall, C. W. C., West, S. G., Augustin, L. S., et al. (2014). Acute effects of pistachio consumption on glucose and insulin, satiety hormones and endothelial function in the metabolic syndrome. *European Journal of Clinical Nutrition*. DOI:10.1038/ejcn.2013.275.

Kennedy-Hagan, K., Painter, J. E., Honselman, C., et al. (2011). The effect of pistachio shells as a visual cue in reducing caloric consumption. *Appetite* 57(2): 418–20.

Kochar, J., Gaziano, J. M. and Djoussé, L. (2010). Nut consumption and risk of type II diabetes in the Physicians' Health Study. *European Journal of Clinical Nutrition* 64(1): 75–9. DOI:10.1038/ejcn.2009.121.

Kocyigit, A., Koylu, A. A. and Keles, H. (2006). Effects of pistachio nuts consumption on plasma lipid profile and oxidative status in healthy volunteers. *Nutrition, Metabolism, and Cardiovascular Diseases : NMCD* 16: 202–9. DOI:10.1016/j.numecd.2005.08.004.

Li, Z., Song, R., Nguyen, C., et al. (2010). Pistachio nuts reduce triglycerides and body weight by comparison to refined carbohydrate snack in obese subjects on a 12-week weight loss program. *Journal of the American College of Nutrition* 29: 198–203.

Lloyd-Jones, D. M., Hong, Y., Labarthe, D., et al. (2010). Defining and setting national goals for cardiovascular health promotion and disease reduction: the American Heart Association's strategic Impact Goal through 2020 and beyond. *Circulation* 121(4): 586–613. DOI:10.1161/CIRCULATIONAHA.109.192703.

Loscalzo, J. and Welch, G. (1995). Nitric oxide and its role in the cardiovascular system. *Progress in Cardiovascular Diseases* 38(2): 87–104. DOI:10.1016/S0033-0620(05)80001-5.

Mandalari, G., Bisignano, C., Filocamo, A., et al. (2013). Bioaccessibility of pistachio polyphenols, xanthophylls, and tocopherols during simulated human digestion. *Nutrition (Burbank, Los Angeles County, Calif.)* 29(1): 338–44. DOI:10.1016/j.nut.2012.08.004.

Martínez-González, M. A. and Bes-Rastrollo, M. (2011). Nut consumption, weight gain and obesity: Epidemiological evidence. *Nutrition, Metabolism, and Cardiovascular Diseases : NMCD* 21(Suppl 1): S40–5. DOI:10.1016/j.numecd.2010.11.005.

Martorana, M., Arcoraci, T., Rizza, L., et al. (2013). In vitro antioxidant and in vivo photoprotective effect of pistachio (*Pistacia vera* L., variety *Bronte*) seed and skin extracts. *Fitoterapia* 85: 41–8. DOI:10.1016/j.fitote.2012.12.032.

Mohammadifard, N., Salehi-Abarghouei, A., Salas-Salvadó, J., et al. (2015). The effect of tree nut, peanut, and soy nut consumption on blood pressure: A systematic review and meta-analysis of randomized controlled clinical trials. *The American Journal of Clinical Nutrition*. DOI:10.3945/ajcn.114.091595.

Montonen, J., Knekt, P., Jarvinen, R., et al. (2004). Dietary antioxidant intake and risk of type 2 diabetes. *Diabetes Care* 27(2): 362–6. DOI:10.2337/diacare.27.2.362.

Nieman, D. C., Scherr, J., Luo, B., et al. (2014). Influence of pistachios on performance and exercise-induced inflammation, oxidative stress, immune dysfunction, and metabolite shifts in cyclists: A randomized, crossover trial. *PLoS ONE* 9(11): e113725. DOI:10.1371/journal.pone.0113725.

Nikzadeh, V. and Sedaghat, N. (2008). Physical and sensory changes in pistachio nuts as affected by roasting temperature and storage. *American-Eurasian Journal of Agricultural & Environmental Science* 4(4): 478–83.

Nimptsch, K., Rohrmann, S., Kaaks, R., et al. (2010). Dietary vitamin K intake in relation to cancer incidence and mortality: Results from the Heidelberg cohort of the European Prospective Investigation into Cancer and Nutrition (EPIC-Heidelberg). *The American Journal of Clinical Nutrition* 91(5): 1348–58. DOI:10.3945/ajcn.2009.28691.

O'Neil, C. E., Keast, D. R., Fulgoni, V. L., et al. (2010). Tree nut consumption improves nutrient intake and diet quality in US adults: An analysis of national health and nutrition examination survey (NHANES) 1999–2004. *Asia Pacific Journal of Clinical Nutrition* 19(1): 142–50.

O'Neil, C. E., Keast, D. R., Nicklas, T. A., et al. (2012). Out-of-hand nut consumption is associated with improved nutrient intake and health risk markers in US children and adults: National Health and Nutrition Examination Survey 1999–2004. *Nutrition Research (New York, N. Y.)* 32(3): 185–94. DOI:10.1016/j.nutres.2012.01.005.

Orhan, I., Küpeli, E., Aslan, M., et al. (2006). Bioassay-guided evaluation of anti-inflammatory and antinociceptive activities of pistachio, *Pistacia vera* L. *Journal of Ethnopharmacology* 105(1–2): 235–40. DOI:10.1016/j.jep.2005.10.023.

Ostlund, R. E., Racette, S. B. and Stenson, W. F. (2002). Effects of trace components of dietary fat on cholesterol metabolism: Phytosterols, oxysterols, and squalene. *Nutrition Reviews* 60(11): 349–59.

Otvos, J. D. (2002). Measurement of lipoprotein subclass profiles by nuclear magnetic resonance spectroscopy. *Clinical Laboratory* 48(3–4): 171–80.

Pan, A., Sun, Q., Manson, J. E., et al. (2013). Walnut consumption is associated with lower risk of type 2 diabetes in women. *The Journal of Nutrition* 143(4): 512–8. DOI:10.3945/jn.112.172171.

Parham, M., Heidari, S., Khorramirad, A., et al. (2014). Effects of pistachio nut supplementation on blood glucose in patients with type 2 diabetes: A randomized crossover trial. *The Review of Diabetic Studies : RDS* 11(2): 190–6. DOI:10.1900/RDS.2014.11.190.

Paterniti, I., Impellizzeri, D., Cordaro, M., et al. (2017). The anti-inflammatory and antioxidant potential of pistachios (*Pistacia vera* L.) *In Vitro* and *In Vivo*. *Nutrients* 9(8): 915. DOI:10.3390/nu9080915.

Pimpin, L., Wu, J. H. Y., Haskelberg, H., et al. (2016). Is butter back? A systematic review and meta-analysis of butter consumption and risk of cardiovascular disease, diabetes, and total mortality. *PLoS ONE* 11(6): e0158118. DOI:10.1371/journal.pone.0158118.

Rajaram, S. and Sabaté, J. (2006). Nuts, body weight and insulin resistance. *The British Journal of Nutrition* 96(Suppl 2): S79–86.

Rauf, A., Patel, S., Uddin, G., et al. (2017). Phytochemical, ethnomedicinal uses and pharmacological profile of genus Pistacia. *Biomedicine & Pharmacotherapy* 86: 393–404. DOI:10.1016/j.biopha.2016.12.017.

Reaven, P. D. and Witztum, J. L. (1996). Oxidized low density lipoproteins in atherogenesis: role of dietary modification. *Annual Review of Nutrition* 16: 51–71. DOI:10.1146/annurev.nu.16.070196.000411.

Rizzo, M. and Berneis, K. (2006). Low-density lipoprotein size and cardiovascular risk assessment. *QJM* 99(1): 1–14. DOI:10.1093/qjmed/hci154.

Ros, E. (2010). Health benefits of nut consumption. *Nutrients* 2: 652–82. DOI:10.3390/nu2070652.

Ros, E. (2015). Nuts and CVD. *The British Journal of Nutrition* 113 Suppl 2: S111–20. DOI:10.1017/S0007114514003924.

Ros, E., Martínez-González, M. A., Estruch, R., et al. (2014). Mediterranean diet and cardiovascular health: Teachings of the PREDIMED study. *Advances in Nutrition (Bethesda, Md.)* 5(3): 330S–6S. DOI:10.3945/an.113.005389.

Ross, S. A. and Davis, C. D. (2014). The emerging role of microRNAs and nutrition in modulating health and disease. *Annual Review of Nutrition* 34(1): 305–36. DOI:10.1146/annurev-nutr-071813-105729.

Rothwell, J. A., Urpi-Sarda, M., Boto-Ordoñez, M., et al. (2012). Phenol-Explorer 2.0: A major update of the Phenol-Explorer database integrating data on polyphenol metabolism and pharmacokinetics in humans and experimental animals. *Database : The Journal of Biological Databases and Curation*. DOI:10.1093/database/bas031.

Sabaté, J., Oda, K. and Ros, E. (2010). Nut consumption and blood lipid levels: A pooled analysis of 25 intervention trials. *Archives of Internal Medicine* 170: 821–7. DOI:10.1001/archinternmed.2010.79.

Salas-Salvadó, J., Fernández-Ballart, J., Ros, E., et al. (2008). Effect of a Mediterranean diet supplemented with nuts on metabolic syndrome status: one-year results of the PREDIMED randomized trial. *Archives of Internal Medicine* 168: 2449–58. DOI:10.1001/archinte.168.22.2449.

Salas-Salvadó, J., Casas-Agustench, P. and Salas-Huetos, A. (2011). Cultural and historical aspects of Mediterranean nuts with emphasis on their attributed healthy and nutritional properties. *Nutrition, Metabolism, and Cardiovascular Diseases : NMCD* 21 Suppl 1: S1–6. DOI:10.1016/j.numecd.2010.10.013.

Sari, I., Baltaci, Y., Bagci, C., et al. (2010). Effect of pistachio diet on lipid parameters, endothelial function, inflammation, and oxidative status: A prospective study.

*Nutrition (Burbank, Los Angeles County, Calif.)* 26: 399–404. DOI:10.1016/j.nut.2009.05.023.

Sauder, K. A., McCrea, C. E., Ulbrecht, J. S., et al. (2014). Pistachio nut consumption modifies systemic hemodynamics, increases heart rate variability, and reduces ambulatory blood pressure in well-controlled type 2 diabetes: A randomized trial. *Journal of the American Heart Association* 3(4). DOI:10.1161/JAHA.114.000873.

Sauder, K. A., McCrea, C. E., Ulbrecht, J. S., et al. (2015). Effects of pistachios on the lipid/lipoprotein profile, glycemic control, inflammation, and endothelial function in type 2 diabetes: A randomized trial. *Metabolism* 64(11): 1521–9. DOI:10.1016/j.metabol.2015.07.021.

Schwingshackl, L., Schwedhelm, C., Hoffmann, G., et al. (2017). Food groups and risk of hypertension: A systematic review and dose-response meta-analysis of prospective studies. *Advances in Nutrition (Bethesda, Md.)* 8(6): 793–803. DOI:10.3945/an.117.017178.

Shahraki, J., Zareh, M., Kamalinejad, M., et al. (2014). Cytoprotective effects of hydrophilic and lipophilic extracts of pistacia vera against oxidative versus carbonyl stress in rat hepatocytes. *Iranian Journal of Pharmaceutical Research : IJPR* 13(4): 1263–77.

Shaw, J. E., Sicree, R. A. and Zimmet, P. Z. (2010). Global estimates of the prevalence of diabetes for 2010 and 2030. *Diabetes Research and Clinical Practice* 87: 4–14. DOI:10.1016/j.diabres.2009.10.007.

Sheridan, M. J., Cooper, J. N., Erario, M., et al. (2007). Pistachio nut consumption and serum lipid levels. *Journal of the American College of Nutrition* 26(2): 141–8.

Sluijs, I., Cadier, E., Beulens, J. W. J., et al. (2015). Dietary intake of carotenoids and risk of type 2 diabetes. *Nutrition, Metabolism and Cardiovascular Diseases* 25(4): 376–81. DOI:http://dx.doi.org/10.1016/j.numecd.2014.12.008.

Smith, D. G. (2007). Epidemiology of dyslipidemia and economic burden on the healthcare system. *The American Journal of Managed Care* 13 Suppl 3: S68–71.

Stone, N. J., Robinson, J. G., Lichtenstein, A. H., et al. (2014). 2013 ACC/AHA guideline on the treatment of blood cholesterol to reduce atherosclerotic cardiovascular risk in adults: A report of the American College of Cardiology/American Heart Association Task Force on practice guidelines. *Circulation* 129(25 Suppl 2): S1–S45. DOI:10.1161/01.cir.0000437738.63853.7a.

Taş, N. G. and Gökmen, V. (2017). Phenolic compounds in natural and roasted nuts and their skins: A brief review. *Current Opinion in Food Science* 14: 103–9.

The Emerging Risk Factors Collaboration, Sarwar, N., Gao, P., et al. (2010). Diabetes mellitus, fasting blood glucose concentration, and risk of vascular disease: A collaborative meta-analysis of 102 prospective studies. *The Lancet* 375(9733): 2215–22. DOI:10.1016/S0140-6736(10)60484-9.

Tomaino, A., Martorana, M., Arcoraci, T., et al. (2010). Antioxidant activity and phenolic profile of pistachio (*Pistacia vera* L., variety *Bronte*) seeds and skins. *Biochimie* 92: 1115–22. DOI:10.1016/j.biochi.2010.03.027.

Trumbo, P., Schlicker, S., Yates, A. A., et al. (2002). Dietary reference intakes for energy, carbohydrate, fiber, fat, fatty acids, cholesterol, protein and amino acids. *Journal of the American Dietetic Association* 102(11): 1621–30. DOI:10.1016/S0002-8223(02)90346-9.

Tziomalos, K., Athyros, V. G., Karagiannis, A., et al. (2010). Endothelial dysfunction in metabolic syndrome: Prevalence, pathogenesis and management. *Nutrition, Metabolism and Cardiovascular Diseases*. DOI:10.1016/j.numecd.2009.08.006.

Ukhanova, M., Wang, X., Baer, D. J., et al. (2014). Effects of almond and pistachio consumption on gut microbiota composition in a randomised cross-over human feeding study. *The British Journal of Nutrition* 111(12): 2146–52. DOI:10.1017/S0007114514000385.

USDA (2015). National Nutrient Database. Available at: http://www.ars.usda.gov/ba/bhnrc/ndl (accessed 11 December 2017).

US Food and Drug Administration (2003). *Qualified Health Claims: Letter of Enforcement Discretion - Nuts and Coronary Heart Disease*. Rockville, MD: Center for Food Safety and Applied Nutrition.

Vadivel, V., Kunyanga, C. N. and Biesalski, H. K. (2012). Health benefits of nut consumption with special reference to body weight control. *Nutrition*. DOI:10.1016/j.nut.2012.01.004.

van het Hof, K. H., West, C. E., Weststrate, J. A., et al. (2000). Dietary factors that affect the bioavailability of carotenoids. *Journal of Nutrition* 130(3): 503–6.

Viguiliouk, E., Kendall, C. W. C., Blanco Mejia, S., et al. (2014). Effect of tree nuts on glycemic control in diabetes: A systematic review and meta-analysis of randomized controlled dietary trials. *PLoS ONE* 9(7): e103376. DOI:10.1371/journal.pone.0103376.

Víteček, J., Lojek, A., Valacchi, G., et al. (2012). Arginine-based inhibitors of nitric oxide synthase: Therapeutic potential and challenges. *Mediators of Inflammation* 2012: 318087. DOI:10.1155/2012/318087.

Wang, X., Li, Z., Liu, Y., et al. (2012). Effects of pistachios on body weight in Chinese subjects with metabolic syndrome. *Nutrition Journal* 11(1): 20. DOI:10.1186/1475-2891-11-20.

West, S. G., Gebauer, S. K., Kay, C. D., et al. (2012). Diets containing pistachios reduce systolic blood pressure and peripheral vascular responses to stress in adults with dyslipidemia. *Hypertension* 60(1): 58–63. DOI:10.1161/HYPERTENSIONAHA.111.182147.

WHO (2013). Global action plan for the prevention and control of noncommunicable diseases 2013-2020. *World Health Organization*: 102. DOI:978 92 4 1506236.

Widmer, R. J., Flammer, A. J., Lerman, L. O., et al. (2015). The Mediterranean diet, its components, and cardiovascular disease. *American Journal of Medicine* 128(3): 229–38. DOI:10.1016/j.amjmed.2014.10.014.

World Health Organisation (2016). WHO | Obesity and overweight. *WHO*. DOI:/entity/mediacentre/factsheets/fs311/en/index.html.

Ye, E. Q., Chacko, S. A., Chou, E. L., et al. (2012). Greater whole-grain intake is associated with lower risk of type 2 diabetes, cardiovascular disease, and weight gain. *The Journal of Nutrition* 142(7): 1304–13. DOI:10.3945/jn.111.155325.

Zhang, J., Kris-Etherton, P. M., Thompson, J. T., et al. (2010). Effect of pistachio oil on gene expression of IFN-induced protein with tetratricopeptide repeats 2: A biomarker of inflammatory response. *Molecular Nutrition & Food Research* 54 Suppl 1: S83–92. DOI:10.1002/mnfr.200900244

# Chapter 6

# Advances in detecting tree nut allergens

*Stef J. Koppelman and Melanie L. Downs, University of Nebraska-Lincoln, USA*



## 1 Introduction

Food allergy remains a major problem affecting consumers. Together with peanut, tree nuts belong to a small group of allergens responsible for the great majority of food allergies (Venter and Arshad 2011). It should be noted that peanut is a groundnut rather than a tree nut. Although peanut allergy is clinically important, peanuts are not considered in this chapter.

European and US regulations are in place to regulate labelling of tree nut-containing food products (EU 2011; FDA 2018), and most countries have adopted similar regulations (Gendel 2012). The main tree nuts producing allergic reactions are (FSA 2015):

- Walnut (*Juglans regia*)
- Pecan (*Carya Illinoinensis*)
- Cashew (*Anacardium occidentale*)
- Pistachio (*Pistacia vera*)
- Hazelnut (*Corylus avellana*)
- Almond (*Prunus dulcis*)
- Brazil nut (*Bertholletia excelsa*)
- Macadamia nut (*Macadamia ternifolia*)

http://dx.doi.org/10.19103/AS.2018.0042.08

This list is valid for most developed countries with some local variations. As an example, hazelnut is more commonly used in Europe than in the United States, resulting in a lower importance of hazelnut as culprit of tree nut allergic reactions in the United States (Andorf et al. 2017). It is important to note that some nuts are botanically related and share common epitopes resulting in sensitivity to both nuts, even while sensitization may have originated from only one type. This applies to walnut and pecan as well as cashew and pistachio (Goetz et al. 2005).

The problem of allergic reaction is not just consumption of the nuts themselves but a range of products containing nuts as ingredients. These include (FSA 2015; Knol and Wickman 2016):

- Praline (hazelnut)
- Marzipan (almond)
- Frangipane (almond)
- Nougat (sugar paste made with hazelnuts)
- Nut butters and spreads
- Pesto sauce (almond)
- Nut flavourings
- Nut oils (e.g. walnut oil in salad dressings)
- Nut milk (e.g. almond milk)
- Breakfast cereals

Highly refined plant oils may be free of protein and consequently free of allergens (Peeters et al. 2004). However, crude oils or gourmet oils used for preparing foods may contain nut proteins (Teuber et al. 1997), making it impossible to say that nut oils, in general, are allergen-free.

These nuts produce two main types of allergic reactions (Stiefel et al. 2017):

- Primary nut allergy: characterized as an immunoglobulin E (IgE)-mediated reaction to storage proteins in nuts
- Pollen food syndrome (PFS), also known as oral allergy syndrome: a distinct disorder with IgE to pollen-related allergens, similar to hay fever, which can be triggered by nuts

As well as safety, it is important to note that detection is also important in quality management in checking ingredient composition or potential adulteration, for example. This chapter begins by providing an overview of tree nut allergies and their prevalence. It then discusses allergenic proteins and thresholds. Finally, it reviews both general issues in developing effective detection methods and the range of techniques available for particular tree nuts.

## 2 Overview of tree nut allergies

The immunological mechanisms and clinical manifestation of tree nut allergy are discussed elsewhere in this book. This section provides a brief overview of tree nut allergies, focussing on the most important tree nuts from the consumer and food industry point of view. A meta-analysis by McWilliam et al. (2015) of over 30 studies from Europe and the United States suggests that the overall prevalence of tree nut allergy may be as high as 1.6%, with some evidence of an increase from low levels in the 1990s (NAS 2017). Estimates for different regions and countries vary. Tree nut allergy (excluding peanuts) may affect around 1% of population in the United States (Sicherer et al. 2010). It has been estimated that nut allergy (including peanuts) affects up to 2% of children and 0.5% of adults in the United Kingdom (Stiefel et al. 2017). Studies of tree nut allergies in the United Kingdom suggest a prevalence amongst children and teenagers ranging from 0.7% to as high as 2.2%, depending on age (Pereira et al. 2005; Venter et al. 2008). The prevalence of PFS related to hazelnut may be as high as 4.6% (Osterballe et al. 2005).

There are significant variations between different types of nut. As an example, the prevalence of allergy has been estimated at 0.12% for walnut and 0.36% for hazelnut in a UK population (Venter et al. 2016), while in the United States walnut allergy is more prevalent than hazelnut allergy (Andorf et al. 2017). There are also regional variations in the relative importance of individual tree nut allergies. The most common allergy reported in European studies was hazelnut (at around 17% of all tree nut allergies) whereas walnut and cashew (at up to 30% each) were most common in the United States (McWilliam et al. 2015). Variations in the prevalence of particular tree nut allergies between different European counties are discussed in EFSA (2014).

A further complication is cross-reactivity among tree nuts as a result of similarities in amino acid sequence and structure between storage proteins in different types of tree nut (Roux et al. 2003; Knol and Wickman 2016; Smeekens et al. 2018). Research suggests that allergic subjects have a just under 40% chance of being allergic to two or more species of tree nut (Sicherer 2001; Brough et al. 2015). Strongly cross-reactive groups include walnut, pecan and hazelnut, as well as cashew and pistachio (Goetz et al. 2005). Allergic symptoms range from mild (PFS) to severe, including anaphylaxis (Bock et al. 2001). Tree nuts, together with peanut and shellfish are the most common triggers of food-induced anaphylaxis in adults (Järvinen 2011).

## 3 Important tree nut allergens

An allergic reaction stems from an abnormal immune response to specific antigens such as proteins in foods (Taylor and Hefle 2001). Allergic reactions to tree nuts have been characterized as a type 1 immunoglobulin E (IgE)-mediated reaction to storage proteins in nuts (Stiefel et al. 2017). Storage proteins can be classified according to their structural and functional properties. The majority of tree nut allergens belong to storage proteins and examples are conglutins or 2S albumins, vicilins or 7S globulins, legumins or 11S globulins, and non-specific lipid transfer proteins (nsLTPs) (Breiteneder and Radauer 2004; Breiteneder, 2016). While for some tree nuts these individual allergen have been identified and characterized, for some other three nuts the allergens have only poorly been described even while it is evident that they can trigger severe allergic reactions. Table 1 provides an overview of the allergens for the most important tree nuts. Allergen names are given by the WHO/IUIS Allergen Nomenclature Sub-Committee (http://www.allergen.org), endorsed by the European Food Safety Authority (EFSA 2014).

## 4 Thresholds for allergic patients and the required sensitivity of detection methods

A key issue is the thresholds for these allergens, that is the maximum amount of an allergenic food that can be tolerated without producing an adverse reaction (Taylor et al. 2004). Studies have shown a lowest observed adverse effect level (LOAEL) as high as 32 mg of hazelnut protein though other studies have shown reactions to as little as 1 mg (Ortolani et al. 2000; Wensing et al. 2002; Hansen et al. 2003). A further complication is the heat-labile nature of some allergens, particularly pollen-related allergens such as Cor a 1, which means they are affected by heat treatments such as drying, roasting or baking. Studies have, for example, suggested higher thresholds for roasted compared to raw hazelnuts in individuals with PFS (Wensing et al. 2002; Hansen et al. 2003; Worm et al. 2009). In comparison to pollen-related allergens, however, seed storage protein allergens are generally much more heat stable. A meta-analysis of 30 studies showed a modest effect of thermal processing on seed storage proteins in hazelnut, almond, cashew, Brazil nut, walnut, pecan and pistachio (Masthoff et al. 2013). LOAEL have not been established for all tree nuts commonly consumed and quality of data on thresholds that is available varies between studies. A different way to consider thresholds is the dose to which 5% of the population reacts ($ED_{05}$). $ED_{05}$ is used for risk assessment studies. For walnut, cashew and hazelnut, the $ED_{05}$ have been modelled based on threshold distributions in European populations (Blom et al. 2013; Blankestijn et al. 2017), and were about 4, 7 and 0.3 mg nut protein, respectively.

Government agencies have suggested a range of LOAELs for tree nut allergens from 0.02 to 7.5 mg of protein (FDA 2006). While clinical exposures are expressed in doses (e.g. mg), allergen levels in food are typically measured in concentrations (e.g. mg/kg or parts per million (ppm)). A quantitative risk

**Table 1** Allergen names and WHO/IUIS Allergen nomenclature

| Common name | Botanical name | Conglutin 2S albumin | Viciline 7S globulin | Legumin 11S gobulin | Bet v 1-homologue PR-10 | LTP | Key references |
|---|---|---|---|---|---|---|---|
| Almond | *Prunus dulcis* | D | D | Pru du 6 | Pru du 4 | Pru du 3 | Costa et al. (2012) |
| Brazil nut | *Bertholletia excelsa* | Ber e 1 | ~ | Ber e 2 | ~ | ~ | Alcocer et al. (2012), Guo et al. (2007) |
| Cashew | *Anacardium occidentale* | Ana o 3 | Ana o 1 | Ana o 2 | ~ | ~ | Mendes et al. (2016) |
| Coconut | *Cocos nucifera* | ~ | Coc n 1 | D | ~ | ~ | Teuber and Peterson (1999) |
| Hazelnut | *Corylus avellana* | Cor a 14 | Cor a 11 | Cor a 9 | Cor a 1 | Cor a 8 | Costa et al. (2016) |
| Macadamia | *Macadamia ternifolia* | S | ~ | S | ~ | ~ | Herbst et al. (2010), De Knop et al. (2010) |
| Pecan | *Carya Illinoinensis* | Car i 1 | Car i 2 | Car i 4 | ~ | ~ | Sharma et al. (2011a,b), Zhang et al. (2016) |
| Pine nut | *Pinus pinea* | Pin p 1 | D | D | D | | Cabanillas and Novak (2015) |
| Pistachio | *Pistacia vera* | Pis v 1 | Pis v 3 | Pis v 11/Pis v 5 | ~ | ~ | Ahn et al. (2009) |
| Walnut | *Juglans regia (Juglans nigra)* | Jug r 1 (Jug n 1) | Jug r 2 (Jug n 2) | Jug r 4 | Jug r 5 | Jug r 3/Jug r 8 | Costa et al. (2014) |

WHO/IUIS Allergen nomenclature database: www.allergen.org.
D: described but no official name given.
S: Suggested that this is an allergen, but not confirmed by identification such as mass spectrometry.
~: Not described or suggested.

assessment model to relate an individual's threshold to risk of experiencing an allergic reaction to packaged food products was recently described for peanut (Baumert et al. 2018). While such work has not been done for tree nuts yet, elements of the model can be used to determine what lower limit of detection food allergen detection methods should have. If one consumes a single food item of 1000 g (1 kg; a conservatively high consumption amount), and the individual's threshold is 1 mg of protein (a conservatively low threshold illustrating a highly sensitive individual), a contamination level of 1 mg/kg or 1 ppm protein would be a relevant lower limit of detection. Because the protein content of tree nuts is between 15% and 25%, this 1 ppm protein corresponds to 4–6 ppm tree nut residue. This is in line with earlier postulated limits for food allergen detection of 10 ppm or less (Taylor and Nordlee 1995; Koppelman 2006).

## 5 Detection methods: technical aspects

In addition to the required sensitivity mentioned in the previous section, several other key challenges in analysis need to be addressed (Scheibe et al. 2001):

- Recovery of analytes from food matrices
- Impact of processing (e.g. heat treatment) on reactivity
- Cross-reactivity in detection of different allergens

Key requirements for any analytical method include:

- Specificity: the ability to detect particular allergens
- Sensitivity: the ability to detect low levels of an allergen
- Quantitative ability: to measure amounts of an allergen
- Flexibility: the ability to deal with a range of foods and types of sample
- Reliability: the ability to produce consistent, reproducible results
- Ease of use

Methods that have been shown to detect food allergens reliably use immunological techniques such as the enzyme-linked immunosorbent assay (ELISA) (Krska et al. 2003; Poms et al. 2004; Schubert-Ullrich et al. 2009; Fielder et al. 2010). ELISA has been found to be both specific and sensitive and can be used for most food-related analytical samples (Blais and Philippe 2003; Akkerdaas et al. 2002a). Limitations include matrix effects, insufficient extraction of protein, cross-reactivity and insufficient reproducibility (Garber and Perry 2010; EFSA 2014). Immunoassays can also present as lateral flow device (Stephane et al. 2002). These are highly convenient in use but are usually not quantitative. Such lateral flow assays are mainly used for screening purposes and positive samples can be reanalysed by ELISA to obtain a quantitative result. ELISA and lateral flow devices are the most commonly used methods for routine testing.

There is growing interest in the use of DNA-based detection, based on polymerase chain reaction (PCR) techniques to complement immunological assays (Popping et al. 2005). DNA is generally more stable than proteins and thus suitable for analysis of processed foods (EFSA 2014). PCR is an indirect method in that it identifies a marker molecule rather than the allergenic fraction. Whilst highly sensitive, the indirect nature of the technique increases the risk of false positive or negative results (Holzhauser et al. 2000). An advantage of PCR over immunoassays is that multiplex assays are relatively easy to develop and Engel et al. (2009) have described a method for simultaneous detection of different tree nuts, peanut and sesame. Several PCR methods are currently available for routine testing.

Biosensors have also been investigated for their ability to detect food allergens using signals from antibody, oligonucleotide probe or aptamer binding. Unlike the colorimetric or fluorescent signals generated in traditional ELISA and PCR methods, biosensors utilize optical, electrochemical or electromechanical signals to detect target binding (Alves et al. 2016). While biosensor methods have potential advantages in terms of real-time monitoring, they still face the challenges associated with adequate extraction, matrix effects and processing-induced changes to analytical targets that impact ELISA and PCR methods. Several allergen biosensors have been reported in the literature, but none are commercially available at the current time, substantially limiting the technology's application and impact in the food industry.

## 6 Detection methods: methods for particular allergens

Because of its widespread use in food processing, a range of detection methods for hazelnut have been developed. An early paper compared specificity and sensitivity of various immunoassays utilizing animal antibodies and serum from hazelnut allergic patients (Koppelman et al. 1999). While both platforms resulted in comparable sensitivity, immunoassays based on animal antibodies are currently preferred for ethical and safety reasons. Since then, many other tests for hazelnut and other tree nuts have been described. Such methods show good sensitivity and are not affected by processing steps such as roasting, though some are affected by the Maillard reaction (Blais and Philippe 2001; Akkerdaas et al. 2000b; Holzhauser et al. 2002; Ben Rejeb et al. 2003; Garber and Perry 2010; Trashin et al. 2011; Cucu et al. 2012). Detection methods for almond show good sensitivity and selectivity, as well as ability to deal with processing effects, though there is some cross-reactivity with other nut traces such as walnut and macadamia (Hlywka et al. 2000; Roux et al. 2001; Garber and Perry 2010; Costa et al. 2012).

Detection techniques for Brazil nut have good sensitivity and low cross-reactivity (Blais et al. 2002; Clemente et al. 2004; Sharma et al. 2009; de la Cruz et al. 2013). Methods for walnut and cashew also show good levels of recovery and sensitivity but with some cross-reactivity (Niemann and Hefle 2003; Wei et al. 2003; Doi et al. 2008; Niemann et al. 2009; Sakai et al. 2010; Gaskin and Taylor 2011). Methods for pecan also suffer from cross-reactivity (Venkatachalem et al. 2006; Polenta et al. 2010). As has been noted, cross-reactivity is a particular problem for tree nut that are closely related such as walnut and pecan or cashew and pistachio. Ben Rejeb et al. (2005) have developed a multiresidue enzyme immunoassay for the simultaneous detection of hazelnut, almond, cashew and Brazil nut. This can be used as screening method and positive samples can be reanalysed to identify and quantify the particular nut(s) present.

By choosing the DNA targets well, PCR methods can be highly specific, even for tree nut that are closely related such as walnut and pecan, and cashew and pistachio (Engel et al. 2009). PCR-based methods have been developed for most commonly consumed tree nuts and show good sensitivities in the lower ppm range (reviewed by Sharma et al. 2017).

In addition to various assays developed by the research community, industry has developed immunoassays and PCR-based methods for the most commonly used tree nuts. Most of these assays are available as a test kit for identification (lateral flow immunoassay and PCR) or as quantitation assay (ELISA and PCR). Some assays that screen for mixed nuts are available as well. Table 2 summarizes the sensitivities of the assays from a European, US and Australian manufacturer. Table 2 is not exhaustive and there are other test manufacturers providing similar tests though often with a more limited portfolio.

All these commercially available tests are in the lower ppm range allowing adequate detection of traces of tree nut allergen. However, because extraction of the analyte from the matrix remains as a critical step that may have different efficiencies for different food matrices, it is recommended to investigate recovery of extraction for individual food matrices (Ben Rejeb et al. 2003). It may turn out that one assay is suitable for a cookie-like matrix, while another assay is more suited for a chocolate-like matrix.

## 7 Detection methods: emerging techniques

Due to some of the limitations encountered with immunoassay methods, such as matrix and processing effects on recovery and the possibility for unwanted antibody cross-reactivity, alternative methods are currently under research and development. One such emerging technique is characterization, detection and quantification of allergens using mass spectrometry. Mass spectrometry

**Table 2** Sensitivities of commercially available immunoassays and PCR, with LLOQ in ppm (LLOD in ppm)

| | Manufacturer 1 (Neogen) | | Manufacturer 2 (Romer) | | Manufacturer 3 (ELISA systems) | | Manufacturer 4 (rBioPharm) | | | Manufacturer 5 (3M/Elution) | | Manufacturer 6 (BioFront) | Manufacturer 7 (BioTeCon/ ImmunoLab) | |
|---|---|---|---|---|---|---|---|---|---|---|---|---|---|---|
| Optional row | ELISA | LFD | ELISA | LFD | ELISA | LFD | ELISA | LFD | PCR | ELISA* | LFD* | ELISA | ELISA | PCR |
| Almond | 2.5 | (5) | 0.4 (0.2) | ? | 0.5 | - | 2.5 (1.2) | (1) | - | 1 | 2 | 1 (0.15) | (0.2) | - |
| Brazil nut | - | - | - | - | - | - | | (1) | (0.4) | 1 | - | 1 (0.14) | (0.2) | - |
| Cashew | - | - | 2 (0.2) | ? | - | - | 2.5 (0.9) | (1) | 1 (0.4) | 0.9 | 2 | 1 (0.12) | (0.2) | - |
| Coconut | - | - | - | - | - | - | - | - | - | 2 | 2 | 1 (0.13) | (0.4) | - |
| Hazelnut | 2.5 | (5–10) | 1 (0.3) | ? | 0.5 | - | 2.5 (0.19) | (1) | - | 1 | 2 | 1 (0.04) | (0.3) | 0.8 (0.1) |
| Macadamia | - | - | 1 (1) | ? | - | - | 1 (0.38) | (1) | (0.4) | 0.3 | - | 2 (0.13) | (0.1) | - |
| Pecan | - | - | - | - | - | - | - | - | (4) | 0.7 | 2 | 1 (0.17) | (0.2) | - |
| Pine nut | - | - | - | - | - | - | - | - | - | 1.5 | - | 2 (0.24) | - | - |
| Pistachio | - | - | 1 (0.13) | ? | - | - | - | (1) | 1 (0.4) | 1 | 2 | 1 (0.12) | (0.13) | - |
| Walnut | 2.5 | - | 2 (0.25) | ? | - | - | - | (10) | 1 (0.4) | 2 | 2 | 1 (0.22) | (0.35) | 0.8 (0.1) |
| 'Tree nuts' | - | (5–10)** | - | - | - | - | - | - | (1)*** | - | - | - | - | - |

-: Not available.
?: Available, but LLOQ or LLOD not given.
*: Results in ppm protein, not ppm total nut as other kits do.
**: Nut mix: almond, hazelnut, pecan, walnut, cashew, pistachio.
***: Nut mix: peanut, hazelnut, walnut.

methods are potentially well-suited for allergen analysis because they directly detect markers of the food component of concern (proteins), they are highly specific to the protein sequence level, they are compatible with more rigorous extraction conditions and they can be multiplexed to detect several allergen targets in one assay.

As is the case with the majority of current proteomics applications, mass spectrometry analysis of food allergens is primarily conducted utilizing bottom-up proteomics workflows. Bottom-up proteomics methods analyse peptides derived from proteins of interest, rather than intact proteins, and are reviewed extensively elsewhere (Aebersold and Mann 2016; Yates et al. 2009). The fundamental steps in this type of analysis are the following: extraction of proteins from the source of interest, reduction and alkylation of cysteine residues to eliminate disulphide bonds, digestion of proteins into peptides of suitable size (typically utilizing trypsin), separation of the resulting peptides by liquid chromatography and analysis of the eluting peptides by mass spectrometry (Aebersold and Mann, 2016; Yates et al. 2009). While some exceptions do exist, tandem mass spectrometry is used in nearly all recent instances of food allergen analysis.

Mass spectrometers can conduct analysis of peptides in two main ways, utilizing either discovery or targeted modes. In discovery mode, the instrument analyses peptides without any predetermined targets, with the goal of identifying as many of the peptides present in a sample as possible. Untargeted methods can be used for characterization of allergens when the amino acid sequences are not entirely known, or when food processing may have altered certain amino acids (Nitride et al. 2013; Downs et al. 2016). In contrast, analyses conducted in targeted mode only measure predetermined peptides of interest. Depending on the type of instrument, targeted methods can be either selected reaction monitoring (SRM, also known as multiple reaction monitoring or MRM) or parallel reaction monitoring (PRM) methods (Peterson et al. 2012; Picotti and Aebersold 2012). In both method types, the MS instrument monitors a specific combination of a precursor and a fragment ion, known as a transition, to indicate the presence of a target peptide. In order to enhance specificity, multiple transitions are typically monitored for each peptide. Both SRM and PRM methods can be used for quantification by utilizing heavy stable isotope labelled peptide internal standards and/or external calibration curves (Barnidge et al. 2003; Gerber et al. 2003; Peterson et al. 2012). An advantage of reference peptides over antibodies (used for immunoassays) is that peptides can be manufactured chemically, thereby minimizing differences between lots.

The use of mass spectrometry methods for the detection and quantification of food allergens, including tree nuts, presents both benefits and challenges. MS methods can be multiplexed to detect and quantify several protein targets in a single analysis, which may be particularly helpful in the case of tree nuts. In addition, unlike many immunoassays, MS methods do not rely on structure for protein detection and therefore can detect proteins with conformations that have been altered by various types of food processing, as long as those proteins can be extracted and digested. Sample preparation allows for denaturing conditions which means that allergens can be released from various types of food matrices even after food processing conditions such as heating. Similarly, as detection by MS is based upon protein sequence, there are opportunities for MS methods to have a higher level of specificity than methods relying on detection of protein conformation. However, this reliance on protein sequence for identification is also one of the critical challenges for applying MS methods to tree nut detection. First, for some tree nuts there are currently very few published protein sequences. For example there are less than 100 protein sequences each for pecan (18), cashew (95), Brazil nut (95) and macadamia nut (99), compared with the 45 744 sequences now available from walnut following the publication of a complete genome and accompanying proteome (The UniProt Consortium 2017). The lack of sufficient protein sequence information for the different tree nuts can affect both the ability to select high-quality peptide targets for each tree nut and to determine whether peptides are unique to the target tree nut of interest. The challenge of selecting unique peptides for tree nuts is further exacerbated by the fact that the close phylogenetic relationships among tree nuts and between tree nuts and other foods leads to cases of highly homologous protein sequences. In the case of the currently published sequences for 11S globulins, for example, walnut and pecan sequences are up to 94.5% identical, cashew and pistachio are up to 78.8% identical, and almond and peach are up to 97.8% identical. For both untargeted and targeted methods, the levels of sequence similarity among tree nuts and other foods have been shown to cause specificity challenges (Ansari et al. 2012; Garber et al. 2016).

In the past decade, several authors have published targeted methods for the detection of different tree nuts by mass spectrometry. In some cases, these methods have been multiplexed methods that have included food allergen targets in addition to certain tree nuts, while in other instances the methods have focused specifically on tree nuts. As seen in Table 3, several methods have been published for hazelnut (11), almond (7), cashew (7), walnut (7) and pistachio (5), with fewer publications including other tree nuts, such as pecan (3), Brazil nut (2), pine nut (2), chestnut (1), coconut (1) and macadamia nut (1). Some of these methods have been qualitative in nature (Ansari et al. 2012; Heick et al. 2011; Korte et al. 2016; New et al. 2018; Sealey-Voyksner et al. 2016), with others providing quantitative data through the use of external standard curves (Bignardi et al. 2013; Gu et al. 2018; Korte and Brockmeyer 2016; Pilolli et al. 2017, 2018; Planque et al. 2017).

**Table 3** Mass spectrometry methods for tree nut detection and quantification

| | | | Method reference | | | | | | | | | | | | | | | | |
|---|---|---|---|---|---|---|---|---|---|---|---|---|---|---|---|---|---|---|---|
| | | | Bignaridi et al. (2013) | | Gu et al. (2018) | Heick et al. (2011) | Korte et al. (2016a) | | | Korte et al. (2016b) | Pilolli et al. (2018) | Pilolli et al. (2018) | Planque et al. (2017) | New et al. (2018) | | | Ansari et al. (2012) | New et al. (2018) | Sealey-Voyksner et al. (2016) |
| Tree nut | Protein class | IUIS Allergen name | Baked | Chocolate | Chocolate | Baked | Baked | Ice cream | Chocolate | Ice cream | Baked | Baked | Multiple* | Baked | Ice cream | Chocolate | Peptide targets | Peptide targets | Peptide targets |
| Almond | 11S | Pru d 6 | 3.1 (0.9) | 30 (9) | 6.1 (1.9) | (3) | 1 (1) | 1 (0.3) | 1 (1) | (1.8) | | | 23.6 | | | | | | x |
| Brazil nut | 2S | Ber e1 | | | | | | | | | | | | | | | | x | x |
| Cashew | 11S | Ana o 2 | 1.6 (0.5) | 50 (15) | 12.6 (3.8) | | 0.3 (0.3) | 1 (0.3) | 1 (1) | (4.3) | | | 13.7 | | | | | x | x |
| | 2S | Ana o 3 | | | | | | | | | | | 13.7 | | | | | | x |
| | 7S | Ana o 1 | | | | | | | | | | | | | | | | | |
| Chestnut | Chitinase | ND | | | | | | | | | | | | | | | | | x |
| | HSP | ND | | | | | | | | | | | | | | | | | x |
| Coconut | Oleosin | ND | | | | | | | | | | | | | | | | | x |
| Hazelnut | 11S | Cor a 9 | 4.5 (1.3) | 49 (14) | 11.4 (3.3) | (5) | 3 (1) | 3 (1) | 3 (3) | (4) | 20 (7) | 80 (20) | 16.7 | 3 (0.25) | 10 (2) | 30 (10) | x | | x |
| | 7S | Cor a 11 | | | | | | | | | | | | | | | x | | |
| | LTP | Cor a 8 | | | | | | | | | | | | | | | x | | |
| | N/A | N/A | | | | | | | | | | | | | | | | x | |

| | | | | | | | | | | | | | |
|---|---|---|---|---|---|---|---|---|---|---|---|---|---|
| Macadamia | De novo | De novo | | | | | | | | | | | x |
| Pecan | 7S | Car i 2 | | | | | | | | | 54.5 | x | x |
| Pine nut | 11S | ND | | | | | | | | | | | x |
| | 7S | ND | | | | | | | | | | x | |
| Pistachio | 11S | Pis v 11<br>Pis v 5 | | | 6.4<br>(2.0) | | 1 (0.3) | 3 (1) | 3 (3) | (5.1) | 12.4 | | x |
| | 2S | Pis v 1 | | | | | | | | | 12.4 | | x |
| | 7S | Pis v 3 | | | | | | | | | | | |
| Walnut | 11S | Jug r 4 | 2.6<br>(0.8) | 18 (5) | | | (3) | (3) | 10 (3) | (21.4) | | | |
| | 2S | Jug r 1 | | | | (70) | | | | (21.4) | 32.8 | | x |
| | 7S | Jug r 2 | | | 13.1<br>(3.9) | | 3 (1) | 3 (3) | (10) | (5.7) | 65.7 | | |
| Walnut + Pecan | 7S | Jug r 2 + Car i 2 | | | | | | | | | 16.4 | | |

For each method, matrix and tree nut combination, the most sensitive peptide LOQ (LOD) values are given in ppm tree nut (mg tree nut/kg matrix). Where original reference gives values in ppm total protein, values have been converted to ppm nut using the USDA nutrient database reference values. Baked: includes cookies and bread; Chocolate: includes dark and milk chocolate. x: For methods without defined LOQ or LOD values, x indicates peptide targets from protein class were included in the method. N/A: Peptide sequence not given.

*: Multiple matrices: chocolate, cookie, ice cream and tomato sauce.

ND: Allergen not defined in IUIS database.

De novo: Target peptide sequence determined through de novo MS analysis, not present in databases.

A number of similarities can be observed among the methods, starting with the targets selected for detection. In many of the methods, the peptide targets selected were from 11S globulin proteins from each of the tree nuts. The selection of peptides from similar proteins by different method developers could be due to a number of factors, including that 11S globulin proteins tend to be among the most abundant protein groups in tree nuts and that the 11S globulins are relatively large, in comparison to other seed storage proteins, and would therefore generate a larger number of possible tryptic peptides. In addition to selecting peptides from similar proteins, the methods also frequently selected the same peptide targets, and in some cases the same transitions were also selected. While the largest number of peptide targets were from 11S globulin proteins, some targets were also selected from other seed storage proteins (7S globulins and 2S albumins) as well as other protein classes (e.g. oleosins) for particular tree nuts.

Despite the overall similarities in the peptide targets and transitions used in the various tree nut methods, the performance of the methods varied considerably with respect to the limits of detection and quantification for the tree nuts. Depending on the type of nut, the target peptide, the instrument platform and the matrix in which the tree nut was detected, reported limits of quantification range from 0.05 ppm total cashew protein in bread to 25 ppm total almond or pecan protein in chocolate (Korte and Brockmeyer 2016; Planque et al. 2017). Even for a single peptide used by multiple method developers, such as ADIYTPEVGR from cashew, reported limits of quantification vary widely (e.g. 0.05–2.0 ppm total cashew protein in ice cream) (Korte and Brockmeyer 2016; Planque et al. 2017). Given the wide ranges in detection and quantification of tree nuts under different conditions, end users will likely need to conduct validation studies for the intended tree nut target in the matrix of concern prior to effective use of these types of MS methods.

These techniques are still relatively new and require expensive equipment as well as highly trained operators. Mass spectrometry-based methods are also being developed for pharmaceutical proteins and validated for complex matrices such as serum (Russo et al. 2017). It is expected that this development will pave the way for using mass spectrometry-based method in the food industry as well, similar to other QC methods that have originated from the pharmaceutical industry.

## 8 Conclusion and future trends

As with other allergenic foods, the presence of undeclared tree nuts is a critical food safety hazard of concern for the food industry and regulatory authorities. Effective food allergen management strategies employed by the food industry incorporate data from analytical methods for allergen control validation and verification. Currently, the market for assays to detect and quantify tree nut allergens in food products and environmental samples is dominated by immunoassays and PCR-based methods and, for both these platforms, reagents and ready-to-use test kits are commercially available. However, these methods have their limitations, as previously described. Although immunoassays and PCR-based methods are currently the preferred options to detect and quantify amounts of tree nuts in food products, mass spectrometry-based methods will become more important in the near future. This may enable testing for multiple tree nuts in one assay, providing a powerful approach for QC purposes in the food industry. Due to their speed and ease of use, however, immunoassay-based methods may remain the primary analytical choice for many applications in the near-term, with mass spectrometry analyses serving as robust confirmatory methods.

## 9 Where to look for further information

### 9.1 Regulations for allergen labelling

Regulations for allergen labelling for Europe and the United States can be found in the following guidelines:

- Europe: Regulation (EU) No 1169/2011 of the European Parliament and of the Council of 25 October 2011
- USA: www.fda.gov/Food/GuidanceRegulation/GuidanceDocumentsRegulatoryInformation/Allergens/default.htm
- Canada: https://www.canada.ca/en/health-canada/services/food-nutrition/food-labelling/allergen-labelling.html
- Australia and New Zealand: http://www.foodstandards.gov.au/industry/labelling/Pages/default.aspx
- An overview of regulations for other countries can be found at: https://farrp.unl.edu/IRChart

### 9.2 Ongoing research to establish thresholds for food allergens in food allergic individuals

The Dutch research institute TNO and the Food Allergy Research and Resource Program (FARRP) of the University of Nebraska-Lincoln conducted a project to collect information on food allergen thresholds including tree nuts (https://farrp.unl.edu/thresholds-for-allergenic-foods).

### 9.3 Tree nut allergen testing services and risk assessment

Allergen detection services are provided by several laboratories and a local service provider is often used. Risk assessment services related to allergen contamination and support for the design of allergen control plans for the food industry are provided by Food Allergy Research and Resource Program (FARRP) of the University of Nebraska-Lincoln (https://farrp.unl.edu/allergen-control-food-industry).

The University of Manchester (UK) coordinated a project funded by the European Commission on integrated approaches to food allergen and allergy risk management (iFAAM) where not only European countries participated but also the United States and Australia. The final report of iFAAM project was recently delivered (https://cordis.europa.eu/result/rcn/208641_en.html).

Workshops on food allergen management for the US food industry are regularly organized by the Food Allergy Research and Resource Program (FARRP) of the University of Nebraska-Lincoln. This programme also hosts webinars and other outreach activities on food allergen management (https://farrp.unl.edu/farrp-sponsored-workshop-series).

### 9.4 Food allergen detection method harmonization and validation

AOAC International currently has efforts underway to develop standardized requirements for evaluating the performance of ELISA-based methods for food allergen detection, including tree nuts. The standard requirements will form a framework, under which allergen detection and quantification methods can progress towards official method status (www.aoac.org).

## 10 References


Aebersold, R. and Mann, M. (2016). Mass-spectrometric exploration of proteome structure and function. *Nature*, 537(7620): 347–55.

Ahn, K., Bardina, L., Grishina, G., Beyer, K. and Sampson, H. A. (2009). Identification of two pistachio allergens, Pis v 1 and Pis v 2, belonging to the 2S albumin and 11S globulin family. *Clin. Exp. Allergy*, 39(6): 926–34.

Akkerdaas, J., Wensing, M., Knulst, A., Hefle, S., Aalberse, R. and Van Ree, R. (2002a). Sensitive and specific measurement of crude and processed hazelnut protein by a sandwich-type ELISA. *J. Allergy Clin. Immunol.*, 109: 302–7.

Akkerdaas, J., Shcocker, F., Alcocer, M., Schilte, P., Knulst, A., Hefle, S., et al. (2002b). Detection and immunological characterisation of hazelnut 2S albumin. *J. Allergy Clin. Immunol.*, 109: 328–35.

Alcocer, M., Rundqvist, L. and Larsson, G. 2012. Ber e 1 protein: The versatile major allergen from Brazil nut seeds. *Biotechnol. Lett.*, 34(4): 597–610.

Alves, R. C., Barroso, M. F., González-García, M. B., Oliveira, M. B. P. P. and Delerue-Matos, C. (2016). New trends in food allergens detection: Toward biosensing strategies. *Crit. Rev. Food Sci. Nutr.*, 56(14), 2304–19.

Andorf, S., Borres, M. P., Block, W., Tupa, D., Bollyky, J. B., Sampath, V., et al. (2017). Association of clinical reactivity with sensitization to allergen components in multifood-allergic children. *J. Allergy Clin. Immunol. Pract.*, 5(5): 1325–34.

Ansari, P., Stoppacher, N. and Baumgartner, S. (2012). Marker peptide selection for the determination of hazelnut by LC-MS/MS and occurrence in other nuts. *Anal. Bioanal. Chem.*, 402(8), 2607–15.

Barnidge, D. R., Dratz, E. A., Martin, T., Bonilla, L. E., Moran, L. B. and Lindall, A. (2003). Absolute quantification of the g protein-coupled receptor rhodopsin by LC/MS/MS using proteolysis product peptides and synthetic peptide standards. *Anal. Chem.*, 75(3): 445–51.

Baumert, J. L., Taylor, S. L. and Koppelman, S. J. (2018). Quantitative assessment of the safety benefits associated with increasing clinical peanut thresholds through immunotherapy. *J Allergy Clin. Immunol. Pract.*, (2): 457–65.

Ben Rejeb, S., Abbott, M., Davies, D., Querry, J., Cleroux, C., Streng, C., et al. (2003). Immunochemical-based method for detection of hazelnut proteins in processed foods. *J. AOAC Int.*, 86(3): 557–63.

Ben Rejeb, S., Abbott, M., Davies, D., Cleroux, C. and Delahaut, P. (2005). Multi-allergen screening immunoassay for the detection of protein markers of peanut and four tree nuts in chocolate. *Food Addit. Contam.*, 22: 709–15.

Bignardi, C., Mattarozzi, M., Penna, A., Sidoli, S., Elviri, L., Careri, M., et al. (2013). A rapid size-exclusion solid-phase extraction step for enhanced sensitivity in multi-allergen determination in dark chocolate and biscuits by liquid chromatography-tandem mass spectrometry. *Food Anal. Method.*, 6(4): 1144–52.

Blais, B. and Philippe, L. (2001). Detection of hazelnut proteins in foods by enzyme immunoassay using egg yolk antibodies. *J. Food Prot.*, 64: 895–98.

Blais, B. and Philippe, L. (2003). Multiplex enzyme immunoassay system for the simultaneous detection of multiple allergens in food. *Food Control*, 14: 43–7.

Blais, B., Omar, M. and Philippe, L. (2002). Detection of Brazil nut proteins in food by enzyme immunoassay. *Food Agric. Immunol.*, 14: 163–8.

Blankestijn, M. A., Remington, B. C., Houben, G. F., Baumert, J. L., Knulst, A. C., Blom W. M., et al. 2017. Threshold dose distribution in walnut allergy. *J. Allergy Clin. Immunol. Pract.*, 5(2): 376–80.

Blom, W. M., Vlieg-Boerstra, B. J., Kruizinga, A. G., van der Heide, S., Houben, G. F. and Dubois, A. E. 2013. Threshold dose distributions for 5 major allergenic foods in children. *J. Allergy Clin. Immunol.*, 131(1): 172–9.

Bock, S., Munoz-Furlong, A. and Sampson, H. (2001). Fatalities due to anaphylactic reactions to foods. *J. Allergy Clin. Immunol.*, 107: 191–3.

Breiteneder, H. (2016). Allergen families and databases. In: Matricardi, P., Kleine-Tebbe, J., Hoffmann, H., et al. (Eds), *EAACI Molecular Allergology User's Guide*. European Academy of Allergy and Clinical Immunology (EAACI), Zurich, Switzerland. http://www.eaaci.org/documents/MAUG_book.pdf.

Breiteneder, H. and Radauer, C. (2004). A classification of food plant allergens. *J. Allergy Clin. Immunol.*, 113: 821–30.

Brough, H., Turner, P., Wright, T., Fox, A. T., Taylor, S. L., Warner, J. O., et al. (2015). Dietary management of peanut and tree nut allergy: What exactly should patients avoid? *Clin. Exp. Allergy*, 45: 859–71.

Cabanillas, B. and Novak, N. (2015). Allergic reactions to pine nut: A review. *J. Investig. Allergol. Clin. Immunol.*, 25(5): 329–33.

Clemente, A., Chambers, S., Lodi, F., Nicoletti, C. and Brett, G. (2004). Use of the indirect competitive ELISA for the detection of Brazil nut in food products. *Food Control*, 15: 65–9.

Costa, J., Mafra, I. Carrapatoso, I. and Oliveira, M. (2012). Almond allergens: Molecular characterization, detection and clinical relevance. *J. Agric. Food Chem.*, 60: 1337–49.

Costa, J., Carrapatso, I., Oliveria, M. and Mafra, I. (2014). Walnut allergens: Molecular characterization, detection and clinical relevance. *Clin. Exp. Allergy*, 44: 319–41.

Costa, J., Mafra, I., Carrapatoso, I. and Oliveira, M. B. (2016). Hazelnut allergens: Molecular characterization, detection, and clinical relevance. *Crit. Rev. Food Sci. Nutr.*, 56(15): 2579–605.

Cucu, T., Devreese, B., Trashin, S., Kerkaert, B., Rogge, M. and De Meulenaer, B. (2012). Detection of hazelnut in foods using ELISA: Challenges related to detectability in processed foods. *J. AOAC Int.*, 95: 149–56.

De Knop, K. J., Hagendorens, M. M., Bridts, C. H., Stevens, W. J. and Ebo, D. G. (2010). Macadamia nut allergy: 2 case reports and a review of the literature. *Acta Clin. Belg.*, 65(2): 129–32

De la Cruz, S., Lopez-Calleja, I, Alcocer, M., Gonzalez, I., Martin, R. and García, T. (2013). Selection of recombinant antibodies by phage display technology and application for the detection of allergenic Brazil nut in processed foods. *J. Agric. Food Chem.*, 61: 10310–19.

Doi, H., Touhata, Y., Shibata, H., Sakai, S., Urisu, A., Akiyama, H., et al. (2008). Reliable enzyme-linked immunosorbent assay for the determination of walnut protein in processed foods. *J. Agric. Food Chem.*, 56: 7625–30.

Downs M. L., Baumert J. L., Taylor S. L. and Mills E. N. (2016). Mass spectrometric analysis of allergens in roasted walnuts. *J. Proteomics*, 142: 62–9.

EFSA (2014). Scientific Opinion on the evaluation of allergenic foods and food ingredients for labelling purposes. *EFSA Journal*, 12(11): 1-286. http://onlinelibrary.wiley.com/doi/10.2903/j.efsa.2014.3894/epdf.

Engel, K. H., Demmel, A., Ehlert, A., Hupfer, C. and Busch, U. (2009). Simultaneous detection of DNA from ten food allergens by ligation-dependent probe amplification. *Food Addit. Contam.*, 26(4): 409–18.

EU (2011). *Regulation (EU) No 1169/2011 of the European Parliament and of the Council of 25 October 2011*. https://eur-lex.europa.eu/legal-content/EN/TXT/?uri=uriserv:OJ.L_.2011.304.01.0018.01.ENG&toc=OJ:L:2011:304:TOC.

FDA (2006). *Approaches to Establishing Thresholds for Major Food Allergens and for Gluten in Food*. Food and Drug Administration (FDA). https://www.fda.gov/downloads/Food/GuidanceRegulation/UCM316632.pdf.

FDA (2018). *Food Allergens and Gluten-Free Guidance Documents and Regulatory Information*. Food and Drug Administration (FDA). www.fda.gov/Food/GuidanceRegulation/GuidanceDocumentsRegulatoryInformation/Allergens/default.htm.

Fielder, R., Higgs, W. and Barden K. (2010). Nut allergen detection. In: Popping, B., Diaz-Amigo, C. and Hoenicke, K. (Eds), *Molecular, Biological and Immunological Techniques and Applications for Food Chemists*. Hohn Wiley and Sons, New York, NY, pp. 377–406.

FSA (2015). *Guidance on Allergy Management and Consumer Information: Best Practice Guide on Managing Food Allergens with Particular Reference to Avoiding Cross-contamination and Using Appropriate Advisory Labelling*. Food Standards Agency, London, UK https://www.food.gov.uk/sites/default/files/multimedia/pdfs/maycontainguide.pdf.

Garber, E. and Perry, J. (2010). Detection of hazelnuts and almonds using commercial ELISA kits. *Anal. Bioanal. Chem.*, 396: 1939–45.

Garber, E. A. E., Parker, C. H., Handy, S. M., Cho, C. Y., Panda, R., Samadpour, M., Reynaud, D. H. and Ziobro, G. C. (2016). Presence of undeclared food allergens in cumin: The need for multiplex methods. *J. Agric. Food Chem.*, 64(5): 1202–11.

Gaskin, F. and Taylor, S. (2011). Sandwich enzyme-linked immunosorbent assay (ELISA) for detection of cashew nut in food. *J. Food Sci.*, 76: 218–26.

Gendel, S. M. (2012). Comparison of international food allergen labeling regulations. *Regul. Toxicol. Pharmacol.*, 63(2): 279–85.

Gerber, S. A., Rush, J., Stemman, O., Kirschner, M. W. and Gygi, S. P. (2003). Absolute quantification of proteins and phosphoproteins from cell lysates by tandem MS. *Proc. Natl. Acad. Sci.*, 100(12): 6940–5.

Goetz, D. W., Whisman, B. A. and Goetz, A. D. 2005. Cross-reactivity among edible nuts: Double immunodiffusion, crossed immunoelectrophoresis, and human specific igE serologic surveys. *Ann. Allergy Asthma Immunol.*, 95(1): 45–52.

Gu, S., Chen, N., Zhou, Y., Zhao, C., Zhan, L., Qu, L., Cao, C., Han, L., Deng, X., Ding, T., Song, C. and Ding, Y. (2018). A rapid solid-phase extraction combined with liquid chromatography-tandem mass spectrometry for simultaneous screening of multiple allergens in chocolates. *Food Control*, 84: 89–96.

Guo, F., Jin, T., Howard, A. and Zhang, Y. Z. 2007. Purification, crystallization and initial crystallographic characterization of brazil-nut allergen Ber e 2. *Acta Crystallogr. Sect. F Struct. Biol. Cryst. Commun.*, 63(Pt 11): 976–9.

Hansen, K., Ballmer-Weber, B., Luttkopf, D., Skov, P., Wuthrich, B., Bindslev-Jensen, C., et al. (2003). Roated hazelnuts – allergenic activity evaluated by a double-blind placebo-controlled food challenge. *Allergy*, 58(2): 132–8.

Heick, J., Fischer, M. and Pöpping, B. (2011). First screening method for the simultaneous detection of seven allergens by liquid chromatography mass spectrometry. *J. Chromatogr. A*, 1218(7), 938–43.

Herbst, R. A., Wahl, R. and Frosch, P. J. 2010. Specific IgE reactivity and identification of potential allergens in macadamia allergy. *J. Eur. Acad. Dermatol. Venereol.*, 24(11): 1361–3.

Hlywka, J., Hefle, S. and Taylor, S. (2000). A sandwich enzyme-linked immunosorbent assay for detection of almonds in foods. *J. Food Prot.*, 63(2): 252–7.

Holzhauser, T., Wangorsch, A. and Vieths, S. (2000). Polymerase chain reaction (PCR) for detection of potentially allergenic residues in complex food matrices. *Eur. Food Res. Technol.*, 211: 360–5.

Holzhauser, T., Stephan O. and Vieths, S. (2002). Detection of potentially allergenic hazelnut residues in food: A comparative study with DNA pcr-ELISA and protein-sandwich ELISA. *J. Agric. Food Chem.*, 50: 5808–15.

Järvinen K. M. (2011). Food-induced anaphylaxis. *Curr. Opin. Allergy Clin. Immunol*, 11(3): 255–61.

Knol, E. and Wickman, M. (2016). Tree nut and seed allergy. In: Matricardi, P., Kleine-Tebbe, J., Hoffmann, H., et al. (Eds), *EAACI Molecular Allergology User's Guide*. European Academy of Allergy and Clinical Immunology (EAACI), Zurich, Switzerland. http://www.eaaci.org/documents/MAUG_book.pdf.

Koppelman, S. (2006). Detecting tree nuts and seeds in food. In: Koppelman, S. and Hefle, S. (Eds), *Detecting Allergens in Food*. Woodhead Publishing Limited, Cambridge, UK.

Koppelman, S. J., Knulst, A. C., Koers, W. J., Penninks, A. H., Peppelman, H., Vlooswijk, R., et al. (1999). Comparison of different immunochemical methods for the detection and quantification of hazelnut proteins in food products. *J. Immunol. Method.*, 229(1-2): 107-20.

Korte, R. and Brockmeyer, J. (2016). MRM3-based LC-MS multi-method for the detection and quantification of nut allergens. *Anal. Bioanal. Chem.*, 408(27): 7845-55.

Korte, R., Lepski, S. and Brockmeyer, J. (2016). Comprehensive peptide marker identification for the detection of multiple nut allergens using a non-targeted LC-HRMS multi-method. *Anal. Bioanal. Chem.*, 408(12): 3059-69.

Krska, R., Welzig, E. and Baumgartner, S. (2003). Immunological detection of allergenic proteins in food. *Anal. Bioanal. Chem.*, 378(1): 63-5.

Masthoff, L., Hoff, R., Verhoeckx, K. C., van Os-Medendorp, H., Michelsen-Huisman, A., Baumert, J. L., et al. (2013). A systematic review of the effect of thermal processing on the allergenicity of tree nuts. *Allergy*, 68: 983-93.

McWilliam, V., Koplin, J., Lodge, C., Tang, M., Dharmage, S. and Allen, K. (2015). The prevalence of tree nut allergy: A systematic review. *Curr. Allergy Asthma Rep.*, 15(9): 54-62.

Mendes, C., Costa, J., Vicente, A. A., Oliveira, M. B. and Mafra, I. 2016. Cashew nut allergy: Clinical relevance and allergen characterisation. *Clin. Rev. Allergy Immunol.* 1 September 2016 [Epub ahead of print] Review. PMID: 27585580.

NAS (2017). *Finding a Path to Safety in Food Allergy: Assessment of the Global Burden, Causes, Prevention, Management and Public Policy*. National Academies of Sciences, Engineering and Medicine: The National Academies Press, Washington DC.

New, L. S., Schreiber, A., Stahl-Zeng, J. and Liu, H.-F. (2018). Simultaneous analysis of multiple allergens in food products by LC-MS/MS. *J. AOAC Int.*, 101(1): 132-45.

Niemann, L. and Hefle, S. (2003). Validated ELISA for detection of undeclared walnut residues in food. *J. Allergy Clin. Immunol.*, 111: 248-52.

Niemann, L., Taylor, S. and Hefle, S. (2009). Detection of walnut residues in food using an enzyme-linked immunosorbent assay. *J. Food Sci.*, 74: 51-7.

Nitride, C., Mamone, G., Picariello, G., Mills, C., Nocerino, R., Canani, R. B. and Ferranti, P. (2013). Proteomic and immunological characterization of a new food allergen from hazelnut (*Corylus avellana*). *J. Proteomics*, 86: 16-26.

Ortolani, C., Ballmer-Weber, B, Hansen, K., Ispano, M., Wuthrich, B., Bindslev-Jensen, C., et al. (2000). Hazelnut allergy: A double-blind placebo-controlled food challenge multi-center study. *J. Allergy Clin. Immunol.*, 105: 577-81.

Osterballe, M., Hansen, T., Mortz, C., Host, A. and Bindslev-Jensen, C. (2005). The prevalence of food hypersensitivity in an unselected population of children and adults. *Pediatr. Allergy Immunol.*, 16: 567-73.

Peeters, K. A., Knulst, A. C., Rynja, F. J., Bruijnzeel-Koomen, C. A. and Koppelman, S. J. (2004). Peanut allergy: Sensitization by peanut oil-containing local therapeutics seems unlikely. *J. Allergy Clin. Immunol.*, 113(5): 1000-1.

Pereira, B., Venter, C., Grundy J., Clayton, C., Arshad, S. and Dean, T. (2005). Prevalence of sensitization to food allergens, reported adverse reaction to foods, food avoidance, and food hypersensitivity among teenagers. *J. Allergy Clin. Immunol.*, 116: 884-92.

Peterson, A. C., Russell, J. D., Bailey, D. J., Westphall, M. S. and Coon, J. J. (2012). Parallel reaction monitoring for high resolution and high mass accuracy quantitative, targeted proteomics. *Mol. Cell. Proteomics*, 11(11): 1475-88.

Picotti, P. and Aebersold, R. (2012). Selected reaction monitoring-based proteomics: Workflows, potential, pitfalls and future directions. *Nat. Meth.*, 9(6): 555-66.

Pilolli, R., De Angelis, E. and Monaci, L. (2017). Streamlining the analytical workflow for multiplex MS/MS allergen detection in processed foods. *Food Chem.*, 221: 1747–53.

Pilolli, R., De Angelis, E. and Monaci, L. (2018). In house validation of a high resolution mass spectrometry Orbitrap-based method for multiple allergen detection in a processed model food. *Anal. Bioanal. Chem.*, 410(22): 5653–62.

Planque, M., Arnould, T., Dieu, M., Delahaut, P., Renard, P. and Gillard, N. (2017). Liquid chromatography coupled to tandem mass spectrometry for detecting ten allergens in complex and incurred foodstuffs. *J. Chromatogr. A*, 1530: 138–51.

Polenta, G., Godefroy-Benrejeb, S., Delahaut, P., Weber, D. and Abbott, M. (2010). Development of a competitive ELISA for the detection of pecan traces in food. *Food Anal. Method.*, 3: 375–81.

Poms, R., Klein, C. and Anklam, E. (2004). Methods for allergen analysis in food: A review. *Food Addit. Contam.*, 21(1): 1–31.

Popping, B., Schneed, K., Koelln, S., Pardigol, A., Dan, A. and Pinero, D. (2005). Detection of allergens in food: ELISA, PCR and multi-allergen screening assays. In: Amerongen, A., Barug, D. and Lauwaars, M. (Eds), *Rapid Methods for Biological and Chemical Contaminants in Food and Feed*. Wageningen Academic Publishers, Wageningen, The Netherlands.

Roux, K., Teuber, S., Robotham, J. and Sathe S. (2001). Detection and stability of almond allergens in food. *J. Agric. Food Chem.*, 49: 2131–36.

Roux, K., Teuber S. and Sathe, S. (2003). Tree nut allergens. *Int. Arch. Allergy Immunol.*, 131(4): 234–44.

Russo, R., Rega, C., Caporale, A., Tonon, G., Scaramuzza, S., Selis, F., et al. (2017). Ultra-performance liquid chromatography/multiple reaction monitoring mass spectrometry quantification of trastuzumab in human serum by selective monitoring of a specific peptide marker from the antibody complementarity-determining regions. *Rapid Commun. Mass Spectrom.*, 31(14): 1184–92.

Sakai, S., Adachi, R., Adiyama, H., Teshima, R., Doi, H. and Shibata, H. (2010). Determination of walnut protein in processed foods by enzyme-linked immunosorbent assay. *J. AOAC Int.*, 93: 1255–61.

Scheibe, B., Weiss, W., Rueff, F., Przybilla, B. and Gorg, A. (2001). Detection of trace amounts of hidden allergens: Hazelnut and almond proteins in chocolate. *J. Chromatogr. B Biomed. Sci. Appl.*, 756: 229–37.

Schubert-Ullrich, P., Rudolf, J., Ansari, P., Galler, B., Fuhrer, M., Molinelli, A., et al. (2009). Commercialised rapid immunoanalytical tests for the determination of allergenic food proteins: An overview. *Anal. Bioanal. Chem.*, 395: 69–81.

Sealey-Voyksner, J., Zweigenbaum, J. and Voyksner, R. (2016). Discovery of highly conserved unique peanut and tree nut peptides by LC-MS/MS for multi-allergen detection. *Food Chem.*, 194, 201–11.

Sharma, G. M., Roux, K. and Sathe, S. (2009). A sensitive and robust competitive enzyme-linked immunosorbent assay for Brazil nut detection. *J. Agric. Food Chem.*, 57: 769–76.

Sharma, G. M., Irsigler, A., Dhanarajan, P., Ayuso, R., Bardina, L., Sampson, H. A., et al. (2011a). Cloning and characterization of 2S albumin, Car i 1, a major allergen in pecan. *J. Agric. Food Chem.* 2011 Apr 27;59(8): 4130–9.

Sharma, G. M., Irsigler, A., Dhanarajan, P., Ayuso, R., Bardina, L., Sampson, H. A., et al. (2011b). Cloning and characterization of an 11S legumin, Car i 4, a major allergen in pecan. *J. Agric. Food Chem.*, 59(17): 9542–52

Sharma, G. M., Khuda, S. E., Parker, C. H., Eischeid, A. C. and Pereira, M. (2017). Detection of allergen markers in food: Analytical methods. Food and Drug Administration Papers. 6.

Sicherer, S. (2001). Clinical implications of cross-reactive food allergens. *J. Allergy Clin. Immunol.*, 108: 881–90.

Sicherer, S., Munoz-Furlong, A., Godbold, J. and Sampson, H. (2010). US prevalence of self-reported peanut, tree nut and sesame allergy: 11-year follow-up. *J. Allergy Clin. Immunol.*, 125: 1322–6.

Smeekens, J. M., Bagley, K. and Kulis, M. 2018. Tree nut allergies: Allergen homology, cross-reactivity, and implications for therapy. *Clin. Exp. Allergy* 27 April 2018. [Epub ahead of print] Review.

Stephan, O., Moller, N., Lehmann, S., Holzhauser, T. and Vieths, S. (2002). Development and validation of two dipstick type immunoassays for determination of trace amounts of peanut and hazelnut in processed foods. *Eur. Food Res. Technol.*, 215: 431–6.

Stiefel, G., Anagnostou, K., Boyle, R., Braithwaite, N., Ewan, P., Fox, A. T., et al. (2017). BSACI guideline for the diagnosis and management of peanut and tree nut allergy. *Clin. Exp. Allergy*, 47: 719–39.

Taylor, S. L. and Hefle, S. (2001). Ingredient and labelling issues associated with allergenic foods. *Allergy*, 56(Suppl. 67): 64–9.

Taylor, S. L. and Nordlee, J. A. (1995). Detection of food allergens. *Food Technol.*, 5, 231.

Taylor, S. L., Hefle, S., Bindslev-Jensen, C., Atkins, F., Andre, C., Bruijnzeel-Koomen, C., et al. (2004). A consensus protocol for the determination of the threshold doses for allergenic foods: How much is too much? *Clin. Exp. Allergy*, 34: 689–95.

Teuber, S. S. and Peterson, W. R. (1999). Systemic allergic reaction to coconut (*Cocos nucifera*) in 2 subjects with hypersensitivity to tree nut and demonstration of cross-reactivity to legumin-like seed storage proteins: New coconut and walnut food allergens. *J. Allergy Clin. Immunol.*, 103(6): 1180–5.

Teuber, S. S., Brown, R. L. and Haapanen, L. A. (1997). Allergenicity of gourmet nut oils processed by different methods. *J. Allergy Clin. Immunol.*, 99(4): 502–7.

The UniProt Consortium (2017). UniProt: The universal protein knowledgebase. *Nucleic Acids Res.*, 45(D1): D158–69.

Trashin, A., Cucu, T., Devreese, B. Adriaens, A. and De Meulenaer, B. (2011). Development of a highly-sensitive and robust Cor a 9-specific enzyme-linked immunosorbent assay for the detection of hazelnut traces. *Anal. Chim. Acta*, 708: 116–22.

Venkatachalem, M., Teuber, S., Peterson, W., Roux, K. and Sathe, S. (2006). Antigenic stability of pecan proteins. *J. Agric. Food Chem.*, 54: 1449–58.

Venter, C. and Arshad, S. (2011). Epidemiology of food allergy. *Pediatr. Clin. North Am.*, 58(2): 3276–49.

Venter, C., Pereira, B., Voigt, K., Grundy, J., Clayton, C. B., Higgins, B., et al. (2008). Prevalence and cumulative incidence of food hypersensitivity in the first 3 years of life. *Allergy*, 63: 354–9.

Venter, C., Patil, V., Grundy J., Glasbey, G., Twiselton, R., Arshad, S. H., et al. (2016). Prevalence and cumulative incidence of food hyper-sensitivity in the first 10 years of life. *Pediatr. Allergy Immunol.*, 27: 452–8.

Wei, Y., Sathe, S., Teuber, S. and Roux, K. (2003). A sensitive sandwich ELISA for the detection of trace amounts of cashew nuts in foods. *J. Agric. Food Chem.*, 51: 3215–21.

Wensing, M., Pennnks, A., Hefle, S., Akkerdas, J., van Ree, R., Koppelman, S. J., et al. (2002). The range of minimum provoking doses in hazelnut-allergenic patients as determined by double-blind placebo-controlled food challenges. *Clin. Exp. Allergy*, 32: 1757–62.

Worm, M., Hompes, S., Fiedler, E., Illner, A., Zuberbier, T. and Vieths, S. (2009). Impact of native, heat-processed and encapsulated hazelnuts on the allergic response in hazelnut-allergic patients. *Clin. Exp. Allergy*, 39: 159–66.

Yates, J. R., Ruse, C. I., and Nakorchevsky, A. (2009). Proteomics by mass spectrometry: Approaches, advances, and applications. *Annu. Rev. Biomed. Eng.*, 11(1): 49–79.

Zhang, Y., Lee, B., Du, W. X., Lyu, S. C., Nadeau, K. C., Grauke L. J., et al. (2016). Identification and characterization of a new pecan [*Carya illinoinensis* (Wangenh.) K. Koch] allergen, Car i 2. *J. Agric. Food Chem.*, 64(20): 4146–51.

# Chapter 7

## Integrated disease management in tree nut cultivation

*Andrea Vannini and Carmen Morales-Rodríguez, Department for Innovation in Biological, Agro-food and Forest systems (DIBAF) - University of Tuscia, Italy*



### 1 Introduction

Integrated disease management and use of protocols and products that minimize the environmental impact are important aspects of fruit cultivation. Tree nut cultivations are challenged by several pests and pathogens both native and exotic, which have a significant impact on the economic sustainability and quality of production. Remediation to such problems was found in the selection and use of resistant varieties or hybrids, including for hazelnut in the United States against the Eastern filbert blight, and for chestnut against chestnut blight and Chinese gall wasp. In the recent past and still at the present in some countries, the use of chemicals represents the easiest and cheapest method to control pests such as weevils for chestnut or wood and fruit insects affecting hazelnut. In this regard, an integrated approach has been commonly introduced that provides an indication of the most appropriate time of the treatment based on host phenology, pest biological cycle and climatic parameters. However, the massive use of chemicals in tree nut plantations, often located in the proximity of natural or urban areas, legitimately raises concerns for environmental sustainability and human safety. As an example, chestnut groves in European

http://dx.doi.org/10.19103/AS.2018.0042.25

countries such as Italy, Greece and Turkey are typical agroforestry productions where most of the orchards are derived from high stands or coppices. As a consequence, they are scattered within large natural forest areas where the use of pesticide would have a tremendous impact on biodiversity. In such context, the integrated disease management should aim to eliminate the use of chemicals, to introduce protocols based on the employment, where necessary, of eco-friendly molecules and to adopt preventive measures in an area-wide integrated pest management.

In this chapter, an approach is proposed for chestnut cultivation and, specifically, for the management and mitigation of damage caused by ink disease, one of the most significant diseases threatening chestnut cultivation worldwide.

Ink disease of chestnut represents one of the first examples of invasive new pathogens challenging a tree considered both a forest and fruit tree species. The first record of the disease was in Portugal in 1838, although it is believed to have been present in Spain since 1726 (Crandall, 1950). It has since been reported in many European countries, including Spain, Italy, France and the United Kingdom on sweet chestnut (*Castanea sativa*), and in the United States on American chestnut (*C. dentata*) and other chestnut species (Crandall et al., 1945). Cases of ink disease have been recorded in Australia affecting *C. sativa* plantations (Washington, 1983). Two species of *Phytophthora* are responsible for ink disease, namely *P.* × *cambivora* and *P. cinnamomi*. Ink disease is one of the most destructive diseases of chestnut. It causes root and/or collar rot of trees and seedlings in nurseries, plantations and forests (Vannini and Vettraino, 2001). Disease symptoms include crown decline and the presence of flame-shaped necrosis in the bark tissues at the collar (Fig. 1). Tree crown appears

**Figure 1** Symptoms of ink disease on adult orchard tree: (a) crown decline with reduced leaf size and discolouration; (b) flame-shaped bark necrosis at the collar. Source: photos Andrea Vannini.

lighter and paler and gradually declines. This can lead to tree death in only few years depending on its age and size.

## 2 Area-wide integrated management of ink disease

Integrated management of ink disease is difficult and must take into consideration several variables and factors associated with its spread, incidence and severity. Such variables can be considered within a general context of area-wide and landscape heterogeneity. Physical and human infrastructures such as the existing network of natural drainage and roads offer preferential corridors of spread and dispersal of *Phytophthora*. Other variables include climatic factors that influence the prevalence of one of the two *Phytophthora* species over the other, cultural practices and host resistance. Also relevant is the pressure exerted by activities related to specific land uses, such as the presence of nurseries or other type of plantation in the neighbouring areas that could act as a bridge for host shift and the introduction of the pathogens in disease-free areas. Last, but not the least, the quality and level of monitoring of the territory could alter the application efficacy of control measures.

## 3 Monitoring and GIS mapping

In fact, the first step for an area-wide integrated ink disease management (AWIIM) is an accurate monitoring activity that integrates proximal and remote sensing, and geographic information system (GIS) applications. The efficiency of remote sensing and GIS in identifying diseased foci within chestnut districts has been demonstrated in Italy through several experimental activities (Vannini et al., 2005; Papale et al., 2006; Natili, 2010). In Fig. 2 the potential to highlight infection foci with images acquired by the airborne multispectral camera is clearly shown.

The overlap in a GIS of multispectral images with informative layers on land use and landscape heterogeneity provides a detailed and updated scenario of the distribution of the disease and its association with territorial factors,

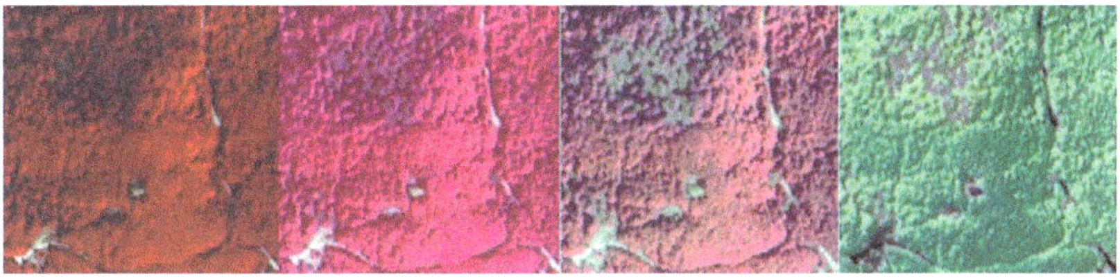

**Figure 2** Ink disease infection foci evidenced with false colour images after elaboration of a multispectral acquisition. The grey spot indicates declining chestnut crowns. Source: modified from Papale et al. (2006).

improving the planning and localization of area-wide control measures. In this regard, the strict association of ink disease with the network of natural drainage and roads crossing or bordering the orchards is well known, which has been demonstrated via statistical analysis (Vannini et al., 2010; Natili, 2010).

## 4 Landscape features

*Phytophthora* inoculum is easily dispersed along roads because of the superficial flow of water and the passive transport of infected soil particles by vehicles and pedestrians. Similarly, *Phytophthora* inoculum is also transported through water streams, rivers, creeks and natural drainage. Figures 3 and 4 show the distribution of ink disease foci as a function of the distance from roads and natural drainage respectively. Thus, in the context of AWIIM, sanitary measures must be considered in order to avoid the uncontrolled movement of inoculum from infected to disease-free areas. To achieve this, Vannini et al. (2010) suggested the following precautions and practices: (1) the restriction of the transit of humans and machines on infected groves and along rural roads crossing or bordering the groves, at least during flooding periods; (2) the disinfection of tyres and shoes when exiting infected areas; (3) the management of superficial water flows and rural roads by removing water in an artificial drainage net to channel the infected water outside the chestnut area. Measures as listed in points (1) and (2) are cost-effective, but their successful application depends on education and awareness among the chestnut stakeholders and

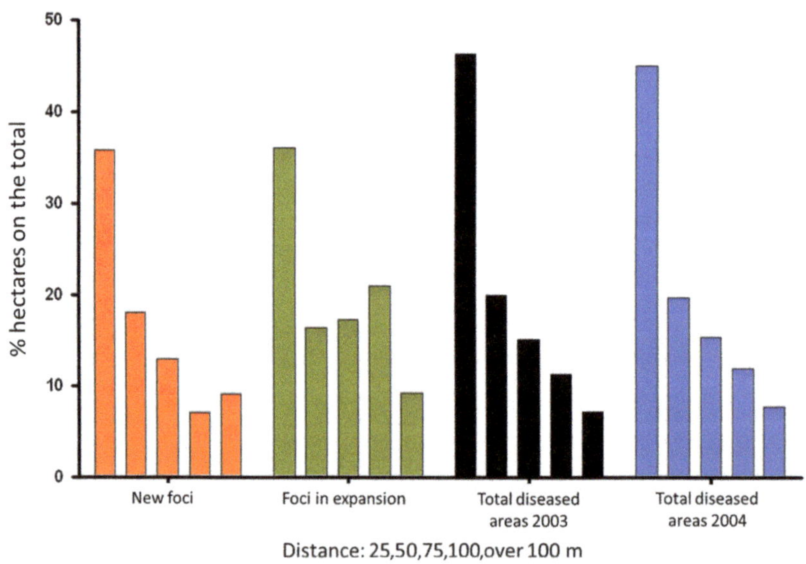


**Figure 3** Area of ink disease foci as in function of the distance from roads. Source: modified from Papale et al. (2006).

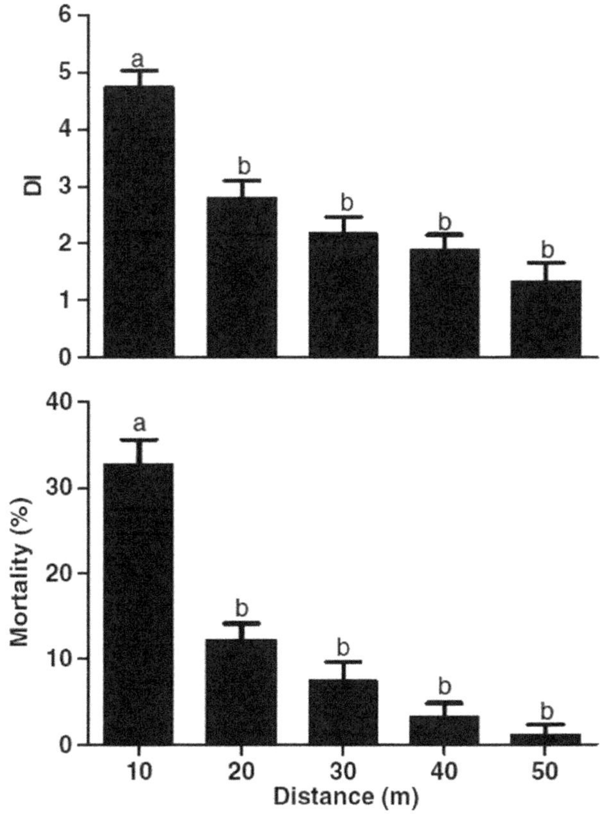


**Figure 4** Disease index (DI) and mortality as a function of the distance of trees from natural drainage. Different letters indicate significant differences according to Tukey's multiple comparison test (DI) and Dunn's post-test (non-parametric) (tree mortality); vertical bar represent SD. Source: modified from Vannini et al. (2010).

private citizens. The latter measure is costly and not easy to implement in large infection foci.

## 5 Tree and soil treatments

The second step of AWIIM is the increase of the orchard's resilience, and the reduction of the *Phytophthora* inoculum in soil and in the rhizosphere of the infected trees. Vannini et al. (2009, 2010) proposed a protocol including the use of chemicals, such as metalaxyl and copper, to reduce the inoculum at rhizosphere level and within the artificial drainage network. These products proved to be highly efficient in reducing the resident inoculum in the soil (Vannini et al., 2010). The effect of metalaxyl on the control of ink disease of chestnut was previously reported by Washington (1983) and other authors

(Vannini and Vettraino, 2004). Wicks (1988) demonstrated the efficacy of metalaxyl field application in trenches around and close to the trunk to control *P. cambivora* crown rot of almond. Metalaxyl efficiently reduces the inoculum density, although a relevant risk exists in inducing resistance in *Phytophthora* populations (Erwin and Ribeiro, 1996). The use of copper compounds, particularly Bordeaux mixture, to control ink disease has a long history (Vannini and Vettraino, 2004) and the sensitivity of *Phytophthora* to copper compounds is widely reported (Coelho et al., 2005). However, the use of these products is nowadays restricted in semi-natural ecosystems in most countries because of their long persistence in the environment, and possible contamination of water tables. More recently, Morales-Rodríguez et al. (2016) demonstrated the efficacy of biofumigation with a commercial pellet of *Brassica carinata* BioFence® (Triumph Italia SPA) in reducing the inoculum of *Phytophthora cinnamomi* during 'in vivo' experiments. Biofumigation has the potential to be a valid substitute for conventional pesticides in controlling oomycetes in the soil. It refers to the suppression of soil-borne pests and pathogens by biocidal compounds released by *Brassicaceae* species when glucosinolates (GLs) in their tissues are hydrolysed by the myrosinase enzyme. Hydrolysis products of GLs, in particular, the isothiocyanates (ITCs), are known to have broad biocidal activity, including insecticidal, nematocidal, fungicidal, antibiotic and phytotoxic effects (Morales-Rodríguez et al., 2016). Data on large-scale biofumigation in chestnut orchards is still lacking, but the results obtained in restricted areas or microcosmos are extremely encouraging, with a reduction of the inoculum in the rhizosphere of the host plant down to one-tenth of the controls (Morales-Rodríguez et al., 2016). Furthermore, BioFence® is completely eco-friendly with no impact on biodiversity. An additional factor to consider within an AWIIM is the increase of resilience of the orchards. Among the eco-friendly compounds, K-phosphite proved to be the most effective in inducing resistance in plants against *Phytophthora* spp. In the late 1980s, phosphite was demonstrated to be effective in the control of *P. cinnamomi* in agriculture (Cohen and Coffey, 1986) and in native ecosystems using foliar applications (Shearer and Fairman, 1991). Stem injection of phosphite has previously been used to protect chestnut against *Phytophthora* root and collar rot with encouraging results (Lim, 1993). The long-lasting action of phosphite (4–5 years) after one injection makes this treatment a practical option for the control of ink disease spread (Shearer et al., 2006). Vannini et al. (2009) and Vettraino et al. (2010) carried out applications of K-phosphite in ink disease-affected orchards through stem injection to trees (Fig. 5). In particular, both healthy looking trees and diseased trees were treated. A noticeable improvement to the crown aspect of treated trees was obtained in the following season (Vannini et al., 2009). The treatment was effective in reducing the increment of severity and in stopping the disease spread, although there was no variation in incidence (Vettraino et al., 2010). Figure 6 shows the

**Figure 5** Chestnut trees under treatment with K-phosphite by stem injection. Photo: Andrea Vannini.

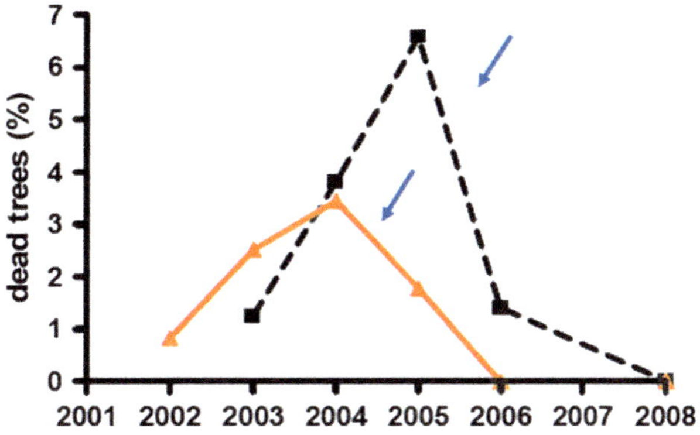


**Figure 6** Percentage of dead trees affected by ink disease in the period 2002-08 in two areas (hyphen and straight lines) that received the integrated disease control protocol. Arrows indicate the year of application of the protocol. Source: modified from Vannini et al. (2009).

results (in terms of tree mortality) of the application of protocol proposed by Vannini et al. (2009), including the management of drainage treatments with metalaxyl and copper, and stem injection of K-phosphite.

## 6 Resistant varieties

A further aspect to consider in AWIIM is the host resistance/susceptibility to ink disease. Asiatic species are traditionally considered resistant to ink disease. In 1952, a breeding programme was started in France to select interspecific hybrids between Asiatic and European chestnuts, to be used as ink disease-resistant rootstocks. Clones with a high tolerance to *P. cinnamomi* were selected as rootstocks for high-quality cultivars. The most common of these are 'Marsol' (CA07), 'Maraval' (CA74), 'Ferosacre' (CA90), 'Marigoule' (CA15) and 'Marlhac' (CA118) (Salesses et al., 1993). These hybrids are *C. crenata* × *C. sativa* or the reciprocal. Additional clones, mostly hybrid, have been selected showing resistance to ink disease in Spain (Miranda-Fontaíña et al., 2007), France and Portugal (Vannini and Vettraino, 2004). Hybrid clones resistant to ink disease, such as '125', '7521' and '90025', have been reported as being excellent for timber production due to their vigour. These resistant clones can be used as a rootstock to establish new plantations with high-value fruit varieties, both hybrid or pure *C. sativa*. With this approach, a limiting factor is the risk of genetic incompatibility specifically to *C. sativa* varieties. However, some of the obtained hybrids have good compatibility with cultivars if used as rootstocks, that is 'HS' (Avanzato, 2009).

## 7 Quality and certification of propagation material

Quality checks and inspection of nursery productions are part of the AWIIM. It is well understood that most of the epidemic diseases in plant pathology have been caused by non-native invasive pathogens introduced into a new environment via the trading of plants and their commodities (Santini et al., 2013). The most recent invasion to affect European chestnut was the Chinese gall wasp, which was introduced in Europe and, previously, in the United States, through uninspected propagation material. Historically *Cryphonectria parasitica* and most probably the ink disease pathogens, *P. cinnamomi* and *P. cambivora*, were also globally dispersed through propagation material. Jung et al. (2016) carried out an analysis of the incidence of *Phytophthora* spp. in 732 European nurseries producing forest transplants, larger specimen trees, landscape plants and ornamentals, plus 2525 areas in which trees and shrubs were planted. Forty-nine *Phytophthora* taxa were recorded in 670 nurseries (91.5%); within these nurseries, 1614 of 1992 nursery stands (81.0%) were infested with *Phytophthora* spp., although most affected plants appeared healthy. In forest and landscape plantings, 56 *Phytophthora* taxa were recovered from 1667 of 2525 tested sites (66.0%). Based on these findings, it is evident that the ink disease pathogens, both being polyphagous species, can be introduced unnoticed in disease-free areas directly from the nursery with chestnut propagation material, or through host shift from neighbouring gardens and plantations. Therefore, a strict control of propagation material produced in nurseries providing chestnut plant material for planting is necessary. Since symptoms of the disease might not be visible on plants for planting, and because of chemical treatment provided in the nurseries, alternative methods of detection must be employed. These include the use of sentinel susceptible species or highly specific and sensitive molecular detection methods. Furthermore, nurseries providing chestnut propagation material should agree with best-practice protocols, including the traceability of the propagation material, the banishing of the use of systemic pesticides with fungi-static activity and the use of pathogen-free irrigation water. In a wider context, these sanitary measures should be used even for propagation material of other species recognized as hosts of the two *Phytophthora*s, particularly those commonly used as ornamentals in gardens or for productive plantation, in areas bordering the chestnut cultivation sites.

## 8 Conclusions

Large-scale integrated disease control is a complex system taking into account variables both strictly related to the orchard and others typical of the neighbouring landscape context. The studies carried out in the chestnut orchards aiming to control ink disease demonstrated the combined efficacy of different measures targeting specific phases of the invasion process. Furthermore, the advancement in knowledge and the increased availability of environmentally sustainable products support the design of eco-friendly protocols to replace conventional pesticides and enhance the resilience of the orchards. The AWIIM model can be taken as an example allowing integrated pest management to be applied to other tree nut cultivations.

## 9 Where to look for further information

Centres of expertise on large area use of phosphite treatments to control *Phytophthora* spp.:

Centre for Phytophthora Science & Management (CPSM) - Murdoch University (AU) - http://cpsm-phytophthora.org/phytophthora.php

UC Berkeley Forest Pathology and Mycology Lab - https://nature.berkeley.edu/matteolab/

Laboratory of Plant Protection of the Department for Innovation in Biological, Agro-food and Forest systems (DIBAF) - University of Tuscia, Viterbo (Italy) - http://www.unitus.it/en/dipartimento/dibaf

Centres of expertise for treatments with *Brassica*-based products:

Council for Agricultural Research and Economics - Research Centre for Industrial Crops (CRA-CIN), Via di Corticella, 133, 40128 Bologna, Italy - http://sito.entecra.it/portale/cra_dati_istituto.php?lingua=EN&id=206&id_gruppo=&tipo=&tipo_org=&sezione=&opz_menu=&access_flag=0

## 10 References


Avanzato, D. 2009. Following chestnut footprints (*Castanea* spp.) cultivation and culture, folklore and history, traditions and uses. *Scripta Horticulturae* N. 9. International Society for Horticultural Science.

Coelho, V., Coutinho, S. and Gouveia, M. E. 2005. Sensitivity to copper and phosphite of *Phytophthora* species associated with ink disease of chestnut. *Acta Hortic.* 693(693), 641–4. doi:10.17660/ActaHortic.2005.693.86.

Cohen, Y. and Coffey, M. D. 1986. Systemic fungicides and the control of oomycetes. *Annu. Rev. Phytopathol.* 24(1), 311–38. doi:10.1146/annurev.py.24.090186.001523.

Crandall, B. S. 1950. The distribution and significance of the chestnut root rot *Phytophthoras*, *P. cinnamomi* and *P. cambivora*. *Plant Dis. Rep.* 6(34), 194–6.

Crandall, B. S., Gravatt, G. F. and Ryan, M. M. 1945. Root disease of *Castanea* species and some coniferous and broadleaf nursery stocks, caused by *Phytophthora cinnamomi*. *Phytopathology* 35, 162–80.

Erwin, D. C. and Ribeiro, O. K. 1996. *Phytophthora Diseases Worldwide*. APS Press, St. Paul, MN.

Jung, T., Orlikowski, L., Henricot, B., Abad-Campo, P., Aday, A. G., Aguín Casal, O., Bakonyi, J., Cacciola, S. O., Cech, T., Chavarriaga, D., et al. 2016. Widespread *Phytophthora* infestations in European nurseries put forest, semi-natural and horticultural ecosystems at high risk of *Phytophthora* diseases. *For. Path.* 46(2), 134–63. doi:10.1111/efp.12239.

Lim, T. M. 1993. Trunk injection with phosphorous acid for controlling *Phytophthora* on chestnuts: early promising results. *Proceedings of the 9th Biennial Conference of the Australasian Plant Pathology Society*, Hobart, Tasmania, 4–8 July. Australasian Plant Pathology Society, Hobart, Australia, p. 25. Abstract.

Miranda-Fontaíña, M. E., Fernández-López, J., Vettraino, A. M. and Vannini, A. 2007. Resistance of *Castanea* clones to *Phytophthora cinnamomi*: testing and genetic control. *Silvae Genet.* 56(1–6), 11–21. doi:10.1515/sg-2007-0002.

Morales-Rodríguez, C., Vettraino, A. M. and Vannini, A. 2016. Efficacy of biofumigation with *Brassica carinata* commercial pellets (BioFence®) to control vegetative and reproductive structures of *Phytophthora cinnamomi*. *Plant Dis.* 100(2), 324–30. doi:10.1094/PDIS-03-15-0245-RE.

Natili, G. 2010. Remote sensing, GIS and geostatistic techniques for mapping and epidemiology study of ink disease. PhD Thesis. University of Tuscia, Viterbo, Italy.

Papale, D., Belli, C., Gioli, B., Noce, S., Valentini, R. and Vannini, A. 2006. Nuovi sistemi di telerilevamento aereo multispettrale ad alta risoluzione: esempi di applicazione

nell'agricoltura di precisione. *Atti del Workshop 'il Telerilevamento per un'agricoltura sostenibile'*, 20 aprile 2006, Piacenza.
Salesses, G., Chapa, J. and Chazerans, P. 1993. The chestnut in France - Cultivars - Breeding programs. *Proceedings of the International Congress on Chestnut*, Spoleto, Italy, 20-23 October, pp. 331-7.
Santini, A., Ghelardini, L., De Pace, C., Desprez-Loustau, M. L., Capretti, P., Chandelier, A., Cech, T., Chira, D., Diamandis, S., Gaitniekis, T., et al. 2013. Biogeographical patterns and determinants of invasion by forest pathogens in Europe. *New Phytol.* 197(1), 238-50. doi:10.1111/j.1469-8137.2012.04364.x.
Shearer, B. L. and Fairman, R. G. 1991. Control of *Phytophthora cinnamomi* species in native communities with phosphorous acid. In: *Proceedings of the Conservation Biology in Australia and Oceania Conference*. University of Queensland, Brisbane, p. 72.
Shearer, B. L., Fairman, R. G. and Grant, M. 2006. Effective concentration of phosphite in controlling *Phytophthora cinnamomi* following stem injection of *Banksia* species and *Eucalyptus marginata*. *Forest Pathol.* 36, 119-35. doi:10.1111/j.1439-0329.2006.00440.x.
Vannini, A. and Vettraino, A. M. 2001. Ink disease in chestnuts: impact on the European chestnut. *For. Snow Landsc. Res.* 76(3), 345-50.
Vannini, A. and Vettraino, A. M. 2004. Epidemiology and control of the main diseases of sweet chestnut. *Inf. Fitopat.* 54, 20-4.
Vannini, A., Vettraino, A. M., Fabi, A., Montaghi, A., Valentini, R. and Belli, C. 2005. Monitoring ink disease of chestnut with the airborne multispectral system A.S.P.I.S. *Acta Hortic.* 693(693), 529-34. doi:10.17660/ActaHortic.2005.693.68.
Vannini, A., Franceschini, S., Vuono, G., Natili, G., Paganini, R. and Vettraino, A. M. 2009. Integrated control protocol to mitigate and eradicate ink disease in chestnut orchards. *Acta Hortic.* 844(844), 461-4. doi:10.17660/ActaHortic.2009.844.65.
Vannini, A., Natili, G., Anselmi, N., Montaghi, A. and Vettraino, A. M. 2010. Distribution and gradient analysis of Ink disease in chestnut forests. *Forest Pathol.* 40(2), 73-86. doi:10.1111/j.1439-0329.2009.00609.x.
Vettraino, A. M., Franceschini, S., Natili, G., Paganini, R., Vuono, G., Alicicco, D. and Vannini, A. 2010. Integrated control protocol (ICP) of Ink disease of chestnut in Central Italy: principles and future perspectives. *Acta Hortic.* 866, 425-30.
Washington, W. S. 1983. The occurrence and control of *Phytophthora* root rot and crown rot of chestnut in Victoria. *Aust. Plant Pathol.* 12, 57-8.
Wicks, T. J. 1988. Effect of metalaxyl on the control of *Phytophthora* crown rot of almonds. *Aust. J. Exp. Agric.* 28(4), 547-52. doi:10.1071/EA9880547.

# Chapter 8

# Tree nut cultivation, ecosystem services, biodiversity and conservation

*Gabriele Beccaro, Marta De Biaggi, Chiara Ferracini and Simona Bonelli, University of Turin, Italy*



## 1 Agro-ecosystem services and dis-services in tree nut orchards

Since the 1992 Rio Convention on Biological Diversity (CBD), growing awareness and concerns have arisen about the impact of human activities on the environment, to the point that biodiversity conservation has gradually gained relevance in international and national agendas.

However, agriculture, forestry, fisheries, water management, extraction industries and energy production still largely shape the fate of the planet's biodiversity, exerting direct pressures such as land-use change, pollution and climate change (Kok et al., 2018). Agricultural production, in particular, is a very competitive sector, with yield being a great concern for growers who are at the same time under the pressing demand for high-quality products, and tree cultivation is no exception. As an example, conventional agriculture is still leading to the intensification of production techniques, especially of pesticide use in orchards worldwide (Demestihas et al., 2017). Nevertheless, the increased use of pesticides is proving to be cost-ineffective and unsustainable given the negative effects of these practices on the environment and on human health (Berg and Tam, 2018; Demestihas et al., 2017; Simon et al., 2010).

http://dx.doi.org/10.19103/AS.2018.0042.27

The growing concern about the overuse of agricultural inputs requires new priorities in agriculture, including tree nut production, to sustain acceptable levels of production whilst preserving natural resources. As stated in the CBD's Strategic Plan for 2011–20, it is now clear that 'biological diversity underpins ecosystem functioning and the provision of ecosystem services essential for human well-being' (CBD, 2010).

The 'ecosystem services' concept appeared in the scientific literature in the mid-1960s and early 1970s (Helliwell, 1969; Odum and Odum, 1972), and defined by Daily (1997) as 'the conditions and processes through which natural ecosystems, and the species that make them up, sustain and fulfil human life'. Ecosystem services need to be classified to allow comparisons and potential trade-offs between individual services (Wallace, 2007). Since 2005, the concept has gained broader attention in the sustainability debate, as the United Nations' Millennium Ecosystem Assessment included ecosystem services firmly on the policy agenda, and it now represents the basis for new strategies for the sustainable use and conservation of ecosystems and their contributions to human well-being (MA, 2005; Gómez-Baggethun et al., 2010).

Several international initiatives have been undertaken to describe and evaluate the full range of services provided by Nature, in order to define an effective framework for natural resource decisions. The ecosystem services may be classified into four main categories (MA, 2005):

- supporting services of soil formation, photosynthesis, biodiversity and nutrient cycling;
- provisioning services of fresh water, food, fibre and fuel production;
- regulating services concerning the buffering and filtering of water, carbon and gasses;
- providing cultural services of heritage, recreation and spiritual values.

In spite of the increasing categorisation of the different services provided by natural capital, there is general consensus on the intimate interconnection of certain ecological functions and associated ecosystem services. Research efforts have been focusing on the development of dynamic models to highlight the interdependencies among ecosystem functions, services and values (Haines-Young and Potschin, 2018), trying to avoid the potential changes in the logic of conservation from ethical obligation to economic self-interest, as a consequence of market-based mechanisms (Mancini et al., 2018). Despite the fact that the concept of ecosystem services has not yet featured much in research on orchard cultivation, some conclusions can be drawn for the sustainable management of tree nut cultivation.

The management of agricultural ecosystems aims at optimising the provisioning ecosystem services of food, fibre and fuel, while depending upon

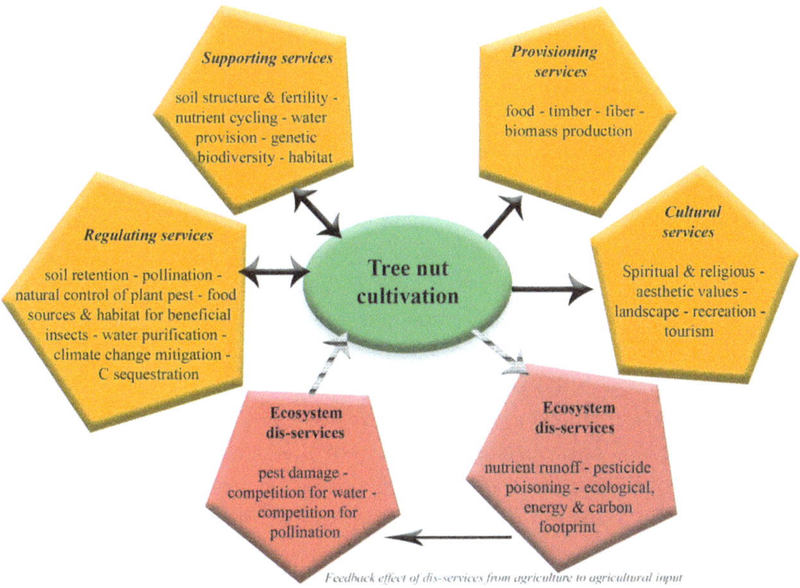


**Figure 1** Tree nut ecosystem services and dis-services to and from agriculture. Black arrows indicate services, whereas dashed grey arrows indicate dis-services. Source: modified from Zhang et al. (2007).

a large spectrum of supporting and regulating services, such as pollination and soil fertility, which ultimately determine the underlying agricultural ecosystems' biophysical capacity (Zhang et al., 2007). It is important to note that, like other tree fruit species, tree nut ecosystems provide and depend on a range of ecosystem services, but that they are also vulnerable to a range of ecosystem 'dis-services' such as pests, as shown in Fig. 1 (Zhang et al., 2007). It is clear that the management of multiple ecosystem services and dis-services (and their relative flow to and from agriculture) represents a great challenge and opportunity for agro-ecosystems (Zhang et al., 2007) such as tree nut cultivation.

The following sections will present the main ecosystem services and dis-services associated with tree nut cultivations, with examples and case studies in order to give an overview on the complex management of intertwined biodiversity conservation and production issues.

## 2 Regulating and supporting services

As orchard ecosystems, tree nut cultivation can provide potential ecosystem services affecting soil health (Montanaro et al., 2017). Soils are a complex system deeply interconnected with ecosystem processes (Adhikari and Hartemink, 2016). Soil-based physiochemical and biological processes result in a number services such as supply of nutrients, water storage, biodiversity and climate regulation. These are significantly affected by by soil management practices (Montanaro et al., 2017).

The potential C sequestration by tree nut agro-ecosystems and the related regulating services (e.g. greenhouse gas mitigation) are receiving increasing attention, in particular, in areas where orchards are highly concentrated. Previous research on carbon sequestration rates in orchards such as olive, apple, peach and kiwifruit showed similar ranges to those of forests, from 2.4 to 12.5 t C/ha/year (Montanaro et al., 2017; Clothier et al., 2016), in particular, during their first years of life (Demestihas et al., 2017). Clothier et al. (2016) showed that kiwifruit orchards enhance soil 'growing', extending the zone of carbon down through the soil profile. Moreover, pruning practices intensively used in orchards, increase photosynthesis rates (Demestihas et al., 2017). The implementation of appropriate management practices (e.g. minimum tillage input of crop residues, dead wood recycling) may increase worldwide carbon sequestration by between 400 and 800 Mt C/year in agricultural soils (IPCC, 2003).

In addition, growing evidence shows that orchard management may significantly influence the ability of fruit tree ecosystems to increase C stocks leading to a better provisioning service. The valorisation of potentially recyclable elements (e.g. pruning woods, mowed grasses and senescent leaves as cover crops) may increase fruits and nuts production by up to 30–50% compared to a conventional field (tillage, removal of pruning residuals) (Montanaro et al., 2017). However, mouse populations may increase in untilled orchards, and for this reason appropriate adjustments are required to help prevent damage (Rutter and Shepard, 2002).

Orchard management practices also strongly influence soil nitrogen availability for nitrogen uptake by plants, a key element for soil fertility. Previous studies on apple orchards showed that the adequate use of ground covers, cover crops and composts significantly increase abundance and activity of microorganisms (bacteria, fungi, actinomycetes) that enable mineralisation of organic matter and humification, therefore improving the immobilisation of nitrogen (Tagliavini et al., 2007; Demestihas et al., 2017). In addition, leaf and pruned wood restitution to soils enables closing the carbon and nitrogen biogeochemical cycles, therefore maintaining soil fertility (Demestihas et al., 2017). The maintenance of soil fertility reduces the need to rely on synthetic nitrogen fertiliser inputs, therefore mitigating nitrous oxide ($N_2O$) emissions, a greenhouse gas of which 86% is emitted by agriculture (CITEPA, 2018). Studies on mycoflora (parasitic, mycorrhizal, saprotrophic and wood decay) in chestnut ecosystems pointed out the primary role of the type of forest management, and the presence of branches, twigs, leaves or large wood on the site influences the number of wood decomposers, which also represent an economically valuable food resource (Laganà et al., 1999; Diamandis and Perlerou, 2001).

Soil structure maintenance plays a crucial role in the overall soil fertility, determining the farming location and the quality and quantity of agricultural output (Zhang et al., 2007). Many tree nut orchards are established in hilly or mountainous areas (hazelnut, walnut, chestnut, almond). Soil erosion reduces water-holding capacity and induces several dis-services (e.g. nutrient and biodiversity loss), impoverishing the topsoil and reducing orchard productivity (Montanaro et al., 2017). Estimated mean rainfall erosion (3.5 Mg/ha/year in orchards, winter and spring crops) together with tillage operations can bring erosion losses beyond the tolerable erosion level (38.8 Mg/ha/year) (Montanaro et al., 2017). Minimising soil erosion through adequate management, such as vegetation or litter layers, conservation tillage or 'limited deficit irrigation' techniques, have proven to preserve soil functions and minimise the loss of productivity induced by erosion (Montanaro et al., 2017). Chestnut cultivations provide valuable services as the orchard and rooting structure provide long-term hydrogeological and environmental protection (Bounous, 2014)

Tree productivity is closely tied to water status, therefore water-regulating services corresponding to water flow and hydrological cycle maintenance are crucial for agriculture production. As water scarcity is increasing in several areas of the planet and progressively aggravated by climate change, innovative strategies are being studied and applied to reduce water consumption with minimum impact on nut crop productivity. For instance, deficit irrigation strategies are applied in pistachio and almond production in Mediterranean agrosystems where production intensification is critically increasing irrigation water demand (López-López et al., 2018), also using plant-based stress sensors to provide the relevant information (sap flow, stem diameter) for use in irrigation scheduling (Galindo et al., 2017; Goldhamer and Fereres, 2004).

## 3 Tree nut cultivation and biodiversity

The particular features of tree nut systems determine their potential contribution to biodiversity: the permanency of the system (often standing longer than other major tree crops), the frequent extensive status of the cultivations, multi-strata design within and in the surrounding areas and pest management (Simon et al., 2010). Indeed increased landscape complexity, generally associated with improved food and habitat availability for animal species as compared to monoculture landscapes, is correlated with diversity and abundance of natural enemies, and with enhanced pest control in some cases (Zhang et al., 2007).

The perennial character of tree nut orchards enhances the stability of the system and its resilience, maintaining food-webs over the years. The persistence of multiple suitable habitats, also provided by the nut trees themselves, contributes to the presence and development of a diversified arthropod community including herbivore, predators, scavengers, parasitoids, but also their natural predators such as birds and bats (Simon et al., 2010).

The multi-strata design of orchards includes arboreal strata for individual trees as well as grass and other types of cover at ground level. Such structural complexity creates high habitat diversity with a range of food sources and means of shelter, thereby increasing animal diversity and abundance.

A mosaic of crop and non-crop habitats is crucial for the maintenance of vascular plant species and structural diversity and as a consequence for the conservation of habitats for terrestrial wildlife species in agricultural landscapes (Sullivan and Sullivan, 2006). In agricultural landscape the habitat heterogeneity could be ensured only by a non-intensive agriculture and pastures. The environmental heterogeneity is positively correlated with biodiversity. This hypothesis is accumulating consensus year by year. Recently Monck-Whipp et al. (2018) found strong evidence that bat activity increases with farmland heterogeneity. Among agrosystems, permanent crop systems could play a crucial role for wildlife conservation if a sustainable agricultural scheme (with low input of pesticides) is followed.

Permanent crop systems such as olive groves, oak tree systems, vineyards, fruit and nut tree orchards, represent some of the most important semi-natural farmlands in Europe. For this reason they are considered the most multifunctional high nature value (HNV) farming systems in Europe and their conservation should be made central to agricultural and rural policies (Kizos et al., 2012). Obviously these crops must follow three criteria to be considered as HNV: low intensity of land use, presence of semi-natural vegetation and features, and presence of a landscape mosaic.

The positive effect of the habitat heterogeneity was recently pointed out when explaining carabid richness in farmlands. According to Fusser et al. (2018) the carabid community in organically managed fields increased with increasing landscape heterogeneity, whereas this effect was absent in conventional fields. The authors collected 43 carabid species in a system of fields situated in 18 different landscapes in an intensively cultivated area in Germany (Fusser et al., 2018).

Several studies found a positive correlation between habitat heterogeneity/diversity and animal species diversity as assessed by Tews et al. (2004), who screened 85 publications in the period 1960–2003. According to the 'habitat heterogeneity hypothesis', authors found in 85% of the reviewed publications a positive effect on animal diversity for arthropods, birds, mammals, amphibians and reptiles. The vegetation structure in particular is responsible for bird species diversity especially in forests (Tews et al., 2004).

Heterogeneity at farmland landscape level is strongly related to farmland biodiversity and, as a consequence, its decline is related to farming practices change. Benton et al. (2003) pointed out that agricultural intensification causes the replacement of habitat heterogeneity with habitat homogeneity. That could affect animals, both at individual survival and population dynamics (e.g. encouraging or depressing dispersal) increasing local extinction probability (Benton et al., 2003).

Planted (multispecies) hedgerows and natural habitats surrounding orchards improve plant and animal diversity, favouring the formation of diverse habitats and ecosystems (Simon et al., 2010). Alomar et al. (2018) observed a positive effect of local (flower abundance and diversity in the groundcover of almond fields) and natural habitat surrounding the fields on the presence of pollinators and almond production.

Traditional land-use systems, as opposed to modern conventional heavily mechanised agriculture (Bignal et al., 1995), are typically low input and low output per hectare, optimising resource use and minimising risks through polyculture and other spatially and temporally plant diversified environments applied to the whole land (Baiamonte et al., 2015).

The presence of single trees, hedgerows and uncultivated areas within and surrounding the orchards is significantly positively correlated with the total and beneficial arthropod diversity, incorporating important microhabitats for both plants and animals (Baiamonte et al., 2015; Simon et al., 2010). Baiamonte et al. (2015) highlight the importance of structural diversity in mixed orchards in Sicily that comprise different tree species (almond, pear, apple, fig, apricot, cherry, azarole, medlar etc.) resulting in diverse mosaic-type habitats. These orchards not only provide a habitat for a wide range of plant and animal species, but also represent a valuable agro-biodiversity reservoir.

## *3.1 Biodiversity conservation in chestnut cultivations*

Chestnut stands (orchards and coppices) are among the most typical elements of the southern European mountain landscape and a protected habitat (9260 *Castanea sativa* woods) according to the European Union (Directive 92/43/EEC). Although the associated fauna has been poorly investigated, several studies have pointed out the importance for biodiversity conservation of this permanent crop. This anthropogenic landscape could host a large portion of biodiversity and even rare and endangered species.

Among the endangered species inhabiting chestnut one of the more emblematic species is *Osmoderma eremita* Scopoli (Coleoptera, Cenoniidae), an umbrella species for the preservation of the entire invertebrate community associated with hollow trees, protected by Habitat Directive in Europe (Directive 92/43/EEC). This coleopteran species mainly inhabits trunk hollows containing large amounts of wood mould (Antonsson et al., 2005). According to Antonsson et al. (2005) in Europe the survival of *O. eremita* is linked to the restoration and maintenance of habitats related to traditional agricultural landscapes, and of course to the preservation of remnants of natural forest with old trees.

Obrist et al. (2011) studied bat fauna in orchards, and in particular in chestnut orchards, in the southern Swiss Alps. The authors detected 12 bat species in foraging activity, and observed a clear choice of bats for managed chestnut compared to abandoned ones underlining the importance to maintain an open orchard for the conservation of endangered bat species.

Spada et al. (2008) investigated *Nyctalus leisleri* Kuhl (Chiroptera, Vespertilionidae) in the same area. They observed that bats roosted mainly in big and live chestnut trees, with large diameter and no vegetation near the entrance. Roost trees were located closer to streams, in woodlands with a higher percentage of sweet chestnut trees and a lower tree density.

In the centre of China (Hechuan District Chongqing), Ren and Chen (2009) conducted the most comprehensive study to date published on the arthropod fauna associated with chestnut orchards. The researchers collected 12 orders, 63 families and 136 species of arthropods in the years 2001–04.

In 2008 and 2009 in northeastern Portugal spider communities associated with the canopies of chestnut trees were studied to determine the best soil management practices. Three chestnut groves each subject to a different agricultural practice (grazed, tilled or untilled) were sampled. Over the 2 years of this study, 15 families of spiders were found in the canopies of the chestnut trees – Philodromidae, Araneidae, Thomisidae, Lyniphiidae, Dictynidae and Theridiidae were the most abundant families (Benhadi-Marin et al., 2013).

Greco et al. (2016), investigating chestnut cultivation systems in southern Italy, revealed the role of the forest system for Lepidoptera conservation, and in particular for moths. Forest moth species have important functional roles as selective herbivores, pollinators, detritivores and prey for birds. On the other hand, chestnut woods provide a wide range of habitats and niches for several Lepidoptera species with different ecological needs. The authors sampled 328 moth species, five of which are endemic to Italy, and pointed out the importance of the management of chestnut forests, since in this kind of habitat the diversity of Lepidoptera ranges from specialists, that live in both coppice clearing and canopy, to widespread species. The studied system is complex, with different successional stages of chestnut forest. According to the authors, while mature coppices host the highest number of species, orchards for fruit production host moth communities similar to those present in just harvested coppices, therefore creating a permanent habitat for heliophilous and thermophilous species. The maintenance of mosaic-like forests ensures the presence of different habitats for moths with different ecological needs.

Furthermore, soil fungi are not only essential for sustaining the ecological integrity and productivity of chestnut orchards, as discussed in Section 2, but they may also represent an important income as some fungal species produce edible sporocarps. *Boletus edulis, B. aestivalis, B. aereus, B. pinophilus, B. fragrans, Russula cyanoxantha* and *Lepista nuda, Macrolepiota procera, Cantharellus cibarius* are among the most marketable edible mushrooms (Martins et al., 2011; Metaxas, 2013).

### 3.2 Crucial issues for animal diversity: cultivation intensification and abandonment

Both cultivation intensification and abandonment may cause dramatic loss of natural biodiversity. This is particularly well-demonstrated for invertebrates with scarce dispersal abilities.

Chestnut woods as an anthropogenic landscape require specific measures to address preservation and to support genetic variability necessary to guide their evolutionary trend (Mellano et al., 2012). Chestnut orchards represent ecosystems characterised by specific types of flora and fauna, many of which are endangered, which should justify their inclusion among the 'habitats' that need protection. Many of these systems are not profitable today compared to more intensive systems, but their value for nature conservation is extremely high.

For these reasons in the Southern Alps and the Apennines, where chestnut woods still dominate the low and middle mountain landscape, a large amount of chestnut orchards have disappeared and been replaced by mixed woodlands with different percentages of chestnut, beech and hornbeam. Pezzi et al. (2011) evaluated this phenomenon in a portion of Tuscany Apennine. The authors calculated the abandonment of farming activities in 267 ha and a parallel decrease in the number of residents associated with a degradation of natural and cultural landscape.

Obrist et al. (2011) evaluated the effect of abandonment of silvo-pastoral practices in chestnut orchards on the associated fauna, clearly demonstrating the negative effect of unmanaged orchards that cause a decline in the number of species richness and bat foraging activity. The authors explain how the dense vegetation structure of unmanaged chestnut woods negatively affects bat foraging activity despite prey availability. Hunting manoeuvres are limited, especially in bat species with low flight manoeuvrability, which are 14 times more common in managed orchards.

## 4 Pest management and animal diversity in tree nut orchards

Particular attention must be given to the ecosystem services and dis-services provided by animal species in orchards and in the surrounding landscape, depending on the species and the season.

Pest management strategies have traditionally been developed at the field scale and have often relied on the use of broad-spectrum pesticides. In today's socio-economic context, developing environmentally friendly pest management strategies has become an urgent matter as the intensive use of agrochemical inputs is recognised to have detrimental impacts on the environment. Together with soil organisms, insects, spiders and birds may provide pest control and crop pollination services in orchards, and are extremely valuable if sustainable agriculture management techniques are applied. Indeed, crop pollination is the best-known ecological service performed by insects, with an estimated value of €153 billion (Demestihas et al., 2017).

Many agro-ecosystems are, in fact, unfavourable environments for beneficial insects due to high levels of disturbance. The manipulation of the landscape and habitat structures is thus a major concern in order to influence the dynamics of both insect pests and their natural enemies, enhancing conservation biological control in agricultural systems and promoting beneficial organisms that deliver important ecosystem services at the landscape scale.

There is growing evidence showing that more complex or diversified landscapes that have high proportions of non-crop habitat such as forests, field margins and wetlands enhance natural enemy abundance and diversity in different environments (Morandin et al., 2014). In agro-ecosystems, biodiversity has a key role in pest control and a high diversity of natural enemies should offer more 'solutions' for controlling pests in different seasons or environmental conditions (Altieri, 1999).

The goal of habitat management is to create a suitable ecological infrastructure within the agricultural landscape to provide resources such as food for adult natural enemies, alternative prey or hosts, and shelter from adverse conditions (Landis et al., 2000). Fragmentation of habitats has a greater effect on the abundance and species richness of higher trophic levels, and parasitoids in general (Wilby and Thomas, 2002).

Multifunctional landscapes, which promote biodiversity and provide favourable conditions for agriculture based on ecological principles, may contribute to the development of productive and sustainable agricultural systems (Bianchi et al., 2006). Non-crop habitats can therefore enhance the abundance and diversity of natural enemy species in the agricultural landscape. However, depending on their vegetation composition, non-crop habitats can also stimulate potential pest species (Bianchi et al., 2006). Understanding the determinants of insect–plant interactions and pest population dynamics is a prerequisite for developing innovative crop protection strategies and increasing the sustainability of agro-ecosystems (Rusch et al., 2013).

## 4.1 Insect pest species management and conservation of tree nuts

In recent years, the introduction of invasive alien species (IAS) has led to changes in the structure and composition of ecosystems with consequent significant detrimental impacts to biodiversity. Expanding populations of exotic invasive species are considered to be the main direct drivers of biodiversity loss across the globe, disrupting ecosystems in their structure and composition directly through displacement and extinction of native species, even if for the vast majority of exotic species no quantitative information is available to assess the magnitude of the effects of such introductions (Jeschke et al., 2014).

The European chestnut (*Castanea sativa* Mill.) is a multipurpose tree species valued for nuts, timber, tannins, hydrogeological protection and landscape. Invasive pathogens and pests of chestnut have significantly impacted the rural economy, leading to a severe decline in coppices and orchards. Among the main chestnut-feeding pests in Europe, the chestnut leafroller (*Pammene fasciana* L.), the chestnut tortrix (*Cydia splendana* Hübner) and the beech tortrix (*C. fagiglandana* Zeller; Lepidoptera, Tortricidae) may cause significant damage in Europe. In Southern Europe, these species cause major damage to sweet chestnut (*C. sativa*), but other common hosts are oaks (*Quercus* spp.) and beech (*Fagus sylvatica* L.). Furthermore, the chestnut weevil (*Curculio elephas* Gyllenhal; Coleoptera, Curculionidae) represents an important pest of the European chestnut, attacking both chestnuts and acorns (*Quercus* spp.).

However, the most threatening insect pest affecting chestnut in the last decade is surely represented by the Asian chestnut gall wasp (ACGW), *Dryocosmus kuriphilus* Yasumatsu (Hymenoptera, Cynipidae), which was first reported in Europe (Italy) in 2002 (Brussino et al., 2002). This exotic pest, native to China, is considered as one of the major pests attacking chestnut trees (*Castanea* spp.) worldwide, and its infestation may cause a strong reduction of the photosynthetic activity and nut production, with yield losses as high as 80% (Battisti et al., 2014). ACGW galls are known to support species richness, closed communities of inquilines and parasitoids that have become a model system in community ecology (Stone et al., 2002). In China, its native range, ACGW is limited by a complex of natural enemies. By contrast, in the countries where ACGW was accidentally introduced (Japan, Korea, the United States, Europe), the native parasitoids did not effectively control the pest, and the Chinese parasitoid *Torymus sinensis* Kamijo (Hymenoptera, Torymidae) was released, also due to the ineffectiveness of both chemical control and ACGW-resistant chestnuts (Moriya et al., 2003). This parasitoid proved to be an outstanding biocontrol agent (BCA) for managing this exotic invasive pest in Italy, and classical biological control using *T. sinensis* was also performed in Croatia, France, Hungary, Portugal, Slovenia, Spain and Turkey (Ferracini et al., 2018).

Although not effective as BCAs, several native generalist wasp species, belonging to cynipids which produce galls on oaks in particular, have been observed alongside the gall wasp. Recently, the agroforestry system (e.g. oak density and type of chestnut formation) was displayed as a possible determinant of richness and density of the native parasitoid community. In fact, in relation to habitat variability, chestnut growing in mixed forest (namely chestnut coppice producing small nuts and usually managed for wood production, surrounded by various woody broadleaf species) revealed a higher number of associated specimens than chestnut cultivated in orchard (Ferracini et al., 2018).

Furthermore, woody and herbaceous vegetation may also act as sources of pollen and nectar, which are essential prerequisites for many insects. For example, the longevity and fecundity of parasitoids are substantially increased when nectar sources are available. Most commonly, nectar- or pollen-rich plants are used with the former benefiting parasitoids and the latter benefiting hoverflies (Gurr et al., 2003), with a marked effect on the effectiveness of natural enemies controlling pest insect populations. That is why a promotion of flowering in chestnut orchards may also benefit the released *T. sinensis*.

Different bug species belonging to the families Coreidae and Pentatomidae (Hemiptera, Heteroptera) are serious economic pests of many food crops and ornamental plants around the world. Included among the food crop hosts are hard-shelled nuts such as pistachio, pecan, macadamia and hazelnut. Currently, stink bug pests such as *Gonocerus acuteangulatus* (Goeze) (Hemiptera, Coreidae), *Nezara viridula* (L.) and *Palomena prasina* (L.) (Hemiptera, Pentatomidae) are the primary insect pests of cultivated hazelnuts (*Corylus avellana* L.) in Europe and Turkey. Adults and nymphs can cause different types of damage in relation to the period when they feed on the nuts (e.g. seed abortion, early nut drop, whitish or brownish spots, spongy tissue and surface depression), being responsible for economic damage to hazelnut in the main producing areas of Europe and Turkey (Bosco et al., 2018; Hedstrom et al., 2014). Furthermore, *G. acuteangulatus*, and the western conifer seed bug (*Leptoglossus occidentalis* Heidemann; Heteroptera, Coreidae) have reported on pistachio (*Pistacia vera* L. (Anacardiaceae)), as well (Moraglio et al., 2014; Uyemoto et al., 1986). Also in this case, the abundance and protection of beneficial organisms and a proper management of the hazelnut agro-ecosystem, with the presence of favourite bushes that could influence the bug behaviour, highlight the importance in developing alternative and environmentally friendly control techniques (Moraglio et al., 2014).

Besides the well-known hazelnut bugs, the recent invasion of the exotic brown marmorated stink bug (*Halyomorpha halys* Stål; Hemiptera, Pentatomidae) represents a serious threat to hazelnut in Europe and Turkey, as already observed in North America, where this species has also been producing economic losses in hazelnut crops (Bosco et al., 2018). This pest, native to Eastern Asia, is a highly polyphagous pentatomid (>120 different host plants), spreading rapidly worldwide, notably through human-mediated activities. In Italy, after the first detection in 2012, it has rapidly spread and economic damage has been reported in commercial fruit orchards. Damage is caused by feeding of the adults and nymphs on fruits, buds, leaves and stems of their host plants, which include many economically important field and tree crops, vegetables, ornamentals, herbaceous perennials, shrubs and forest trees (Haye and Weber, 2017). To date, chemical control has been the most widely used strategy for managing *H. halys* in the northeastern United States; however, the most effective insecticides are generally broad-spectrum in their activity, thus not compatible with most IPM systems, and disruptive to natural enemies (Haye et al., 2015). A more environmentally friendly and self-sustaining control method could involve the use of biological control agents, but research to evaluate the most promising candidate is still ongoing. A recent paper by Maistrello et al. (2017) showed that hedges are an important reservoir for both adults and nymphs throughout the season, likely because in this area they find a suitable assortment of fruits that ripen at different times. Recent studies on the influence of natural habitats surrounding almond cultivations demonstrated that almond production was positively influenced by flower abundance in the fields that increased the diversity of wild pollinators (Alomar et al., 2018; Schaeckermann et al., 2015).

Habitat management can therefore favour development and survival of beneficial insects, providing a starting point for the development of specific strategies for its sustainable management.

Increasing numbers of hazelnut growers throughout the world are adopting the IPM approach and its future looks promising. Even so, considerable work still needs to be done in order to develop and implement IPM. For instance, many beneficial insects and mites also live in hazelnut orchards (Ecevit et al., 1996; Özman and Cobanoglu, 2001). In particular, *Kampimodromus aberrans* (Oud.) (Acari: Phytoseiidae) and *Verticillium lecanii* (Zimm.) Viegas (Fungi: Hyphomycetes) show potential as BCAs for *Phytoptus avellanae* Nal. and *Cecidophyopsis vermiformis* Nal. (Acarina: Eryiophyoidea) (Özman-Sullivan and Akca, 2005). To complement the use of sulphur, it is also recommended to remove big buds in winter, as this largely removes the sources of infestation for the following spring. So, big buds should never be burned to ensure that useful insects continue to live. Removing them is enough (Özman-Sullivan and Akca, 2005).

## 5 Cultural services of tree nut cultivation

Tree nuts are widely appreciated worldwide and represent an important part of the agronomic production in several countries such as China, Turkey and Italy. Moreover, the domestication of indigenous nut trees in poor African countries is starting to have a significant impact on rural development (Asaah et al., 2011).

The dissemination of these commodities began in prehistoric times following the trade routes of emerging civilisations, thus becoming intimately connected to local cultural traditions. As an example, almond's widespread desirability and easy transportability appear to have made it an important commodity in prehistoric trade from Central Asia westward to the Mediterranean. Moreover, almond's horticultural characteristics (e.g. early flowering) contributed to the development of culinary practices and associated rich folklore traditions (Gradziel, 2011). Walnut types also spread in all directions from Afghanistan along the Silk Road (Avanzato et al., 2014), becoming one of the most economically important tree nuts in the world. *C. sativa* monumental trees over 1000 years old can be found in Turkey, where the use of chestnut trees and fruits is deeply rooted in the country's tradition (Topçu et al., 2009).

In relation to their long history, and for their economical and botanical importance, tree nut cultivations have a rich cultural heritage, providing a wide range of benefits such as aesthetic, recreational and spiritual experiences, contributing to physical and psychological well-being. The cultural services associated with tree nut forests and orchards depend on the management of finite resources, biodiversity and ecosystem dynamics.

Cultural and biological interactions over the centuries have contributed to the development of agricultural landscapes of global significance, associated with local cultures and contributing to the identity of communities (Daniel et al., 2012; Luo et al., 2009). Moreover, the interaction between cultural values and ecosystems may impact positively biodiversity protection and improve local communities' support for protected area management (Higgins-Zogib et al., 2010). As an example, Sicilian traditional agro-ecosystems are configured in a heterogeneous mosaic that includes significant elements of naturalness sprawl interspersed in cultivated areas (e.g. hazelnut, almond, chestnut, pistachio), contributing to the ecological connectivity of agricultural landscapes and to biodiversity preservation (Baiamonte et al., 2015).

Improving farmers' perceptions of ecosystem services, within agricultural landscapes, might help overcome the several economic evaluation constraints that derive from the intangible nature of the majority of the benefits provided by cultural services (Montanaro et al., 2017). As an example, the dense walnut forests covering mountains in Eastern Afghanistan need protection from severe over-grazing and illegal logging not only for their landscape values, but also for their precious genetic resources that maintain the status of commercial cultivars (Avanzato et al., 2014).

## 6 Conclusions

There are some efforts being made to assess multiple ecosystem services in tree nut agro-ecosystems. Models can be adapted to analyse the relationships between agricultural management, ecosystem functions and ecosystem services and dis-services, and to simulate different management scenarios. The STICS model (Brisson et al., 2009), for instance, is a widely used dynamic model that has been adapted to simulate apple orchards (Demestihas et al., 2018), and could also be applied to tree nut cultivations.

However, despite efforts dedicated to understanding the biophysical aspects of ecosystem services related to tree nut cultivation systems, policies should incorporate the value of ecosystem services into land management decisions integrating different sectors and international institutions (MA, 2005).

## 7 Where to look for further information

As discussed in this chapter, a standard introduction to the ecosystem services concept is the Millennium Ecosystem Assessment (MA), www.millenniumassessment.org, convened by the then-United Nations Secretary-General Kofi Annan in 2000. Its objective is to assess the consequences of ecosystem change for human well-being and the scientific basis for action needed to enhance the conservation and sustainable use of those systems and their contribution to human well-being. The technical volumes and the synthesis reports are downloadable and provide a state-of-the-art scientific appraisal of the condition and trends in the world's ecosystems.

Together with MA, the Common International Classification of Ecosystem Services (CICES), www.cices.eu, is an important source of information on the functioning of ecosystem services based on systematic approaches. Developed from the work on environmental accounting undertaken by the European Environment Agency (EEA), CICES supports their contribution to the revision of the System of Environmental-Economic Accounting (SEEA), which is currently being led by the United Nations Statistical Division (UNSD).

Scientific information on ecosystem services is rather scattered in various journals; however, a growing number of specialized journals is gaining importance. Examples of scientific journals where updated information can be found are:

- *Ecosystem Services*, Elsevier B.V., www.journals.elsevier.com/ecosystem-services
- *Ecological Economics*, Elsevier B.V., www.journals.elsevier.com/ecological-economics

- *Ecological Modelling*, Elsevier B.V., www.journals.elsevier.com/ecological-modelling
- *Current Opinion in Environmental Sustainability*, Elsevier B.V., www.journals.elsevier.com/current-opinion-in-environmental-sustainability/

Within the framework of the European Union Common Agricultural Policy, https://ec.europa.eu/agriculture/envir/biodiv_en, are addressed the preservation of habitats and biodiversity by specific rural development measures, targeted toward the preservation of habitats and biodiversity. A stronger link is now forming to advance the promise of ecosystem services provision from European agroforestry in decision making including various actors, stakeholders, and institutions.

## 8 References

Adhikari, K. and Hartemink, A. E. 2016. Linking soils to ecosystem services – a global review. *Geoderma* 262, 101–11. doi:10.1016/j.geoderma.2015.08.009.

Alomar, D., González-Estévez, M. A., Traveset, A. and Lázaro, A. 2018. The intertwined effects of natural vegetation, local flower community, and pollinator diversity on the production of almond trees. *Agric. Ecosyst. Environ.* 264, 34–43. doi:10.1016/j.agee.2018.05.004.

Altieri, M. A. 1999. The ecological role of biodiversity in agroecosystems. In: Paoletti, M. G. (Ed.), *Invertebrate Biodiversity as Bioindicators of Sustainable Landscapes*. Elsevier, pp. 19–31.

Antonsson, K., Audisio, P., Ballerio, A., Carpaneto, G. M., Chobot, K., Gjurasin, B., Hanssen, O., Huijbregts, H., Lakatos, F. and Martin, O. 2005. *Osmoderma eremita* (Coleoptera, Scarabaeidae, Cetoniinae) in Europe. *Anim. Biodiv. Conserv.* 28(1), 1–44.

Asaah, E. K., Tchoundjeu, Z., Leakey, R. R. B., Takousting, B., Njong, J. and Edang, I. 2011. Trees, agroforestry and multifunctional agriculture in Cameroon. *Int. J. Agr. Sustain.* 9(1), 110–9. doi:10.3763/ijas.2010.0553.

Avanzato, D., McGranahan, G. H., Vahdati, K., Botu, M., Iannamico, L. and Assche, J. van 2014. Following walnut footprints (*Juglans regia* L.): cultivation and culture, folklore and history, traditions and uses. *Scr. Hortic.* 17, 442.

Baiamonte, G., Domina, G., Raimondo, F. M. and Bazan, G. 2015. Agricultural landscapes and biodiversity conservation: a case study in Sicily (Italy). *Biodivers. Conserv.* 24(13), 3201–16. doi:10.1007/s10531-015-0950-4.

Battisti, A., Benvegnù, I., Colombari, F. and Haack, R. A. 2014. Invasion by the chestnut gall wasp in Italy causes significant yield loss in *Castanea sativa* nut production. *Agric. For. Entomol.* 16(1), 75–9. doi:10.1111/afe.12036.

Benhadi-Marin, J., Pereira, J. A., Barrientos, J.-A., Bento, A. and Santos, S. A. P. 2013. Araneae communities associated with the canopies of chestnut trees in the northeastern part of Portugal: the influence of soil management practices. *Eur. J. Entomol.* 110(3), 501–8. doi:10.14411/eje.2013.066.

Benton, T. G., Vickery, J. A. and Wilson, J. D. 2003. Farmland biodiversity: is habitat heterogeneity the key? *Trends Ecol. Evol.* 18(4), 182–8. doi:10.1016/S0169-5347(03)00011-9.

Berg, H. and Tam, N. T. 2018. Decreased use of pesticides for increased yields of rice and fish-options for sustainable food production in the Mekong Delta. *Sci. Total Environ.* 619-620, 319–27. doi:10.1016/j.scitotenv.2017.11.062.

Bianchi, F. J., Booij, C. J. and Tscharntke, T. 2006. Sustainable pest regulation in agricultural landscapes: a review on landscape composition, biodiversity and natural pest control. *Proc. R. Soc. Lond. B Biol. Sci.* 273(1595), 1715–27. doi:10.1098/rspb.2006.3530.

Bignal, E. M., McCracken, D. I. and Corrie, H. 1995. Defining European low-intensity farming systems: the nature of farming. In: McCracken, D. I., Bignal, E. M. and Wenlock, S. E. (Eds), *Farming on the Edge: the Nature of Traditional Farmland in Europe*. Joint Nature Conservation Committee, Peterborough, pp. 29–37.

Bosco, L., Moraglio, S. T. and Tavella, L. 2018. Halyomorpha halys, a serious threat for hazelnut in newly invaded areas. *J. Pest. Sci.* 91(2), 661–70. doi:10.1007/s10340-017-0937-x.

Bounous, G. 2014. *Il Castagno: Risorsa Multifunzionale in Italia e nel Mondo*. Edagricole, Bologna, Italy.

Brisson, N., Launay, M., Mary, B., and Beaudoin, N.2009. *Conceptual Basis, Formalisations and Parameterization of the STICS Crop Model*. Editions Quae, Paris.

Brussino, G., Bosio, G., Baudino, M., Giordano, R., Ramello, F. and Melika, G.2002. Pericoloso insetto esotico per il castagno europeo. *Inf. Agrar.* 58(37), 59–62.

CBD. 2010. The strategic plan for biodiversity 2011–2020 and the Aichi biodiversity targets. UNEP/CBD/COP/DEC/X/2, Nagoya, Aichi Prefecture, Japan.

CITEPA. 2018. Rapport National d'Inventaire pour la France au titre de la Convention cadre des Nations Unies sur les Changements Climatiques et du Protocole de Kyoto (CCNUCC). Available at: https://www.citepa.org/fr/activites/inventaires-des-emissions/ccnucc.

Clothier, B., Green, S., Müller, K., Gentile, R. M., Mason, K., Herth, I. and Holmes, A. 2016. Orchard natural capital supplying valuable ecosystem services. *Acta Hort.* 1130, 1–10.

Daily, G. 1997. *Nature's Services: Societal Dependence on Natural Ecosystems*. Island Press, Washington DC.

Daniel, T. C., Muhar, A., Arnberger, A., Aznar, O., Boyd, J. W., Chan, K. M., Costanza, R., Elmqvist, T., Flint, C. G., Gobster, P. H., et al. 2012. Contributions of cultural services to the ecosystem services agenda. *Proc. Natl. Acad. Sci. U.S.A.* 109(23), 8812–9. doi:10.1073/pnas.1114773109.

Demestihas, C., Plénet, D., Génard, M., Raynal, C. and Lescourret, F. 2017. Ecosystem services in orchards. A review. *Agron. Sustain. Dev.* 37(2), 12. doi:10.1007/s13593-017-0422-1.

Demestihas, C., Plénet, D., Génard, M., Garcia de Cortazar-Atauri, I., Launay, M., Ripoche, D., Beaudoin, N., Simon, S., Charreyron, M., Raynal, C., et al. 2018. Analyzing ecosystem services in apple orchards using the STICS model. *Eur. J. Agron.* 94, 108–19. doi:10.1016/j.eja.2018.01.009.

Diamandis, S. and Perlerou, C. 2001. The mycoflora of the chestnut ecosystems in Greece. *Forest Snow Landsc. Res.* 76(3), 499–504.

Ecevit, O., Tuncer, C., Özman, S., Mennan, S. and Akca, I. 1996. Natural enemies found in Black Sea Region hazelnut orchards and their potential in biological control. In: *Proceedings of the Symposium: Hazelnut and Other Nutrition Crops*. Turkey, pp. 293–307.

Ferracini, C., Bertolino, S., Bernardo, U., Bonsignore, C. P., Faccoli, M., Ferrari, E., Lupi, D., Maini, S., Mazzon, L. and Nugnes, F.2018. Do *Torymus sinensis* (Hymenoptera:

Torymidae) and agroforestry system affect native parasitoids associated with the Asian chestnut gall wasp? *Biol. Control* 121, 36–43.
Fusser, M. S., Holland, J. M., Jeanneret, P., Pfister, S. C., Entling, M. H. and Schirmel, J. 2018. Interactive effects of local and landscape factors on farmland carabids. *Agric. Forest Entomol.* 20(4), 549–57. doi:10.1111/afe.12288.
Galindo, A., Collado-González, J., Griñán, I., Corell, M., Centeno, A., Martín-Palomo, M. J., Girón, I. F., Rodríguez, P., Cruz, Z. N. and Memmi, H. 2017. Deficit irrigation and emerging fruit crops as a strategy to save water in Mediterranean semiarid agrosystems. *Agric. Water Manag.* 202, 311–24.
Goldhamer, D. A. and Fereres, E. 2004. Irrigation scheduling of almond trees with trunk diameter sensors. *Irrig. Sci.* 23(1), 11–9. doi:10.1007/s00271-003-0088-0.
Gómez-Baggethun, E., De Groot, R., Lomas, P. L. and Montes, C. 2010. The history of ecosystem services in economic theory and practice: from early notions to markets and payment schemes. *Ecol.Econ.* 69(6), 1209–18. doi:10.1016/j.ecolecon.2009.11.007.
Gradziel, T. M. 2011. 2 Origin and dissemination of almond. *Hortic. Rev.* 38, 23.
Greco, S., Infusino, M., Bernardini, V., Turco, R. and Scalercio, S. 2016. Timber vs. fruit Production: partitioning of moth diversity within a mosaic-like chestnut forest. *Contemp. Eng. Sci.* 9(28), 1387–402. doi:10.12988/ces.2016.68141.
Gurr, G. M., Wratten, S. D. and Luna, J. M. 2003. Multi-function agricultural biodiversity: pest management and other benefits. *Basic Appl. Ecol.* 4(2), 107–16. doi:10.1078/1439-1791-00122.
Haines-Young, R. and Potschin, M. B. 2018. Common international classification of ecosystem services (CICES) V5.1 and guidance on the application of the revised structure. Available at: www.cices.eu.
Haye, T. and Weber, D. C. 2017. Special issue on the brown marmorated stink bug, *Halyomorpha halys*: an emerging pest of global concern. *J. Pest. Sci.* 90(4), 987–8. doi:10.1007/s10340-017-0897-1.
Haye, T., Fischer, S., Zhang, J. and Gariepy, T. 2015. Can native egg parasitoids adopt the invasive brown marmorated stink bug, Halyomorpha halys (Heteroptera: Pentatomidae), in Europe? *J. Pest. Sci.* 88(4), 693–705. doi:10.1007/s10340-015-0671-1.
Hedstrom, C. S., Shearer, P. W., Miller, J. C. and Walton, V. M. 2014. The effects of kernel feeding by *Halyomorpha halys* (Hemiptera: Pentatomidae) on commercial hazelnuts. *J. Econ. Entomol.* 107(5), 1858–65. doi:10.1603/EC14263.
Helliwell, D. R. 1969. Valuation of wildlife resources. *Reg. Stud.* 3(1), 41–7. doi:10.1080/09595236900185051.
Higgins-Zogib, L., Dudley, N., Mallarach, J.-M. and Mansourian, S. 2010. Beyond belief: linking faiths and protected areas to support biodiversity conservation. In: Stolton, S. and Dudley, N. (Eds), *Arguments for Protected Areas: Multiple Benefits for Conservation and Use*. Earthscan Publications, London, pp. 145–58.
IPCC. 2003. *Intergovernmental Panel on Climate Change. Good Practice Guidance for Land Use, Land-Use Change and Forestry*. Institute for Global Environmental Strategies (IGES), Hayama.
Jeschke, J. M., Bacher, S., Blackburn, T. M., Dick, J. T., Essl, F., Evans, T., Gaertner, M., Hulme, P. E., Kühn, I., Mrugała, A., et al. 2014. Defining the impact of non-native species. *Conserv. Biol.* 28(5), 1188–94. doi:10.1111/cobi.12299.
Kizos, T., Plieninger, T., Schaich, H. and Petit, C. 2012. HNV permanent crops: olives, oaks, vines, fruit and nut trees. In: Oppermann, R., Beaufoy, G. and Jones, G. (Eds),

*High Nature Value Farming in Europe - 35 European Countries, Experiences and Perspectives*. Verlag Regionalkultur, Ubstadt-Weiher, pp. 70–84.

Kok, M. T. J., Alkemade, R., Bakkenes, M., van Eerdt, M., Janse, J., Mandryk, M., Kram, T., Lazarova, T., Meijer, J., van Oorschot, M., et al. 2018. Pathways for agriculture and forestry to contribute to terrestrial biodiversity conservation: a global scenario-study. 221, 137–50. doi:10.1016/j.biocon.2018.03.003.

Laganà, A., Loppi, S. and De Dominicis, V. 1999. Relationship between environmental factors and the proportions of fungal trophic groups in forest ecosystems of the central Mediterranean area. *For. Ecol. Manag.* 124(2-3), 145–51. doi:10.1016/S0378-1127(99)00061-4.

Landis, D. A., Wratten, S. D. and Gurr, G. M. 2000. Habitat management to conserve natural enemies of arthropod pests in agriculture. *Annu. Rev. Entomol.* 45(1), 175-201. doi:10.1146/annurev.ento.45.1.175.

López-López, M., Espadador, M., Testi, L., Lorite, I. J., Orgaz, F. and Fereres, E. 2018. Water use of irrigated almond trees when subjected to water deficits. *Agric. Water Manag.* 195, 84–93. doi:10.1016/j.agwat.2017.10.001.

Luo, Y., Liu, J. and Zhang, D. 2009. Role of traditional beliefs of Baima Tibetans in biodiversity conservation in China. *For. Ecol. Manag.* 257(10), 1995–2001. doi:10.1016/j.foreco.2009.01.001.

MA. 2005. *Ecosystems and Human Well-Being: Synthesis*. Island Press, Washington DC.

Maistrello, L., Vaccari, G., Caruso, S., Costi, E., Bortolini, S., Macavei, L., Foca, G., Ulrici, A., Bortolotti, P. P., Nannini, R., et al. 2017. Monitoring of the invasive *Halyomorpha halys*, a new key pest of fruit orchards in northern Italy. *J. Pest. Sci.* 90(4), 1231–44. doi:10.1007/s10340-017-0896-2.

Mancini, M. S., Galli, A., Coscieme, L., Niccolucci, V., Lin, D., Pulselli, F. M., Bastianoni, S. and Marchettini, N. 2018. Exploring ecosystem services assessment through ecological footprint accounting. *Ecosyst. Serv.* 30, 228–35. doi:10.1016/j.ecoser.2018.01.010.

Martins, A., Marques, G., Borges, O., Portela, E., Lousada, J., Raimundo, F. and Madeira, M. 2011. Management of chestnut plantations for a multifunctional land use under Mediterranean conditions: effects on productivity and sustainability. *Agroforest Syst.* 81(2), 175–89. doi:10.1007/s10457-010-9355-2.

Mellano, M. G., Beccaro, G. L., Donno, D., Marinoni, D. T., Boccacci, P., Canterino, S., Cerutti, A. K. and Bounous, G. 2012. *Castanea* spp. biodiversity conservation: collection and characterization of the genetic diversity of an endangered species. *Genet. Resour. Crop Evol.* 59(8), 1727–41. doi:10.1007/s10722-012-9794-x.

Metaxas, A. M. 2013. *Chestnut (Castanea spp.) Cultivar Evaluation for Commercial Chestnut Production in*. Hamilton, County, TN.

Monck-Whipp, L., Martin, A. E., Francis, C. M. and Fahrig, L. 2018. Farmland heterogeneity benefits bats in agricultural landscapes. *Agric. Ecosyst. Environ.* 253, 131-9. doi:10.1016/j.agee.2017.11.001.

Montanaro, G., Xiloyannis, C., Nuzzo, V. and Dichio, B. 2017. Orchard management, soil organic carbon and ecosystem services in Mediterranean fruit tree crops. *Sci. Hort.* 217, 92–101. doi:10.1016/j.scienta.2017.01.012.

Moraglio, S. T., Ingegno, B. L. and Tavella, L. 2014. Studies on semiochemicals produced by *Gonocerus acuteangulatus* for their use as control methods in hazelnut orchards. *Acta Hortic.* 1052(1052), 281–7. doi:10.17660/ActaHortic.2014.1052.38.

Morandin, L. A., Long, R. F. and Kremen, C. 2014. Hedgerows enhance beneficial insects on adjacent tomato fields in an intensive agricultural landscape. *Agric. Ecosyst. Environ.* 189, 164–70. doi:10.1016/j.agee.2014.03.030.

Moriya, S., Shiga, M. and Adachi, I. 2003. Classical biological control of the chestnut gall wasp in Japan. In: *Proceedings of the 1st International Symposium on Biological Control of Arthropods*. USDA Forest Service, Washington, pp. 407–15.

Obrist, M. K., Rathey, E., Bontadina, F., Martinoli, A., Conedera, M., Christe, P. and Moretti, M. 2011. Response of bat species to sylvo-pastoral abandonment. *For. Ecol. Manag.* 261(3), 789–98. doi:10.1016/j.foreco.2010.12.010.

Odum, E. P. and Odum, H. T. 1972. Natural areas as necessary components of man's total environment. In: *Transactions of the 37th North American Wildlife and Natural Resources Conference*. Wildlife Management Institute, Washington DC, vol. 37, pp. 178–89.

Özman, S. and Cobanoglu, S. 2001. Current status of hazelnut mites in Turkey. *Acta Hortic.* 556, 479–87.

Özman-Sullivan, S. K. and Akca, I. 2005. Efficiency of pesticides against big bud mites. *Phytoptus avellanae* Nal. and *Cecidophyopsis vermiformis* Nal.(Acarina: Eriophyoidea)] on hazelnut. *Acta Hortic.* 686(686), 393–400. doi:10.17660/ActaHortic.2005.686.54.

Pezzi, G., Maresi, G., Conedera, M. and Ferrari, C. 2011. Woody species composition of chestnut stands in the Northern Apennines: the result of 200 years of changes in land use. *Landsc. Ecol.* 26(10), 1463–76. doi:10.1007/s10980-011-9661-8.

Ren, S. and Chen, B. 2009. Study of species biodiversity of Arthropod community in chestnut orchard. *J. Chongqing Norm. Univ. Nat. Sci.* 3: 006.

Rusch, A., Valantin-Morison, M., Sarthou, J. P. and Roger-Estrade, J. 2013. Effect of crop management and landscape context on insect pest populations and crop damage. *Agric. Ecosyst. Environ.* 166, 118–25. doi:10.1016/j.agee.2011.05.004.

Rutter, P. and Shepard, M. 2002. *Hybrid Hazelnut Handbook.* (1st edn.). Badgersett Research Corporation, Canton, MN. Available at: http://www.badgersett.com/HazHandbook1.html.

Schaeckermann, J., Pufal, G., Mandelik, Y. and Klein, A. 2015. Agro-ecosystem services and dis-services in almond orchards are differentially influenced by the surrounding landscape. *Ecol. Entomol.* 40(S1), 12–21. doi:10.1111/een.12244.

Simon, S., Bouvier, J.-C., Debras, J.-F. and Sauphanor, B. 2010. Biodiversity and pest management in orchard systems. A review. *Agron. Sustain. Dev.* 30(1), 139–52. doi:10.1051/agro/2009013.

Spada, M., Szentkuti, S., Zambelli, N., Mattei-Roesli, M., Moretti, M., Bontadina, F., Arlettaz, R., Tosi, G. and Martinoli, A. 2008. Roost selection by non-breeding Leisler's bats (*Nyctalus leisleri*) in montane woodlands: implications for habitat management. *Acta Chiropt* 10(1), 81–8. doi:10.3161/150811008X331117.

Stone, G. N., Schönrogge, K., Atkinson, R. J., Bellido, D. and Pujade-Villar, J. 2002. The population biology of oak gall wasps (Hymenoptera: Cynipidae). *Annu. Rev. Entomol.* 47(1), 633–68. doi:10.1146/annurev.ento.47.091201.145247.

Sullivan, T. P. and Sullivan, D. S. 2006. Plant and small mammal diversity in orchard versus non-crop habitats. *Agric. Ecosyst. Environ.* 116(3-4), 235–43. doi:10.1016/j.agee.2006.02.010.

Tagliavini, M., Tonon, G., Scandellari, F., Quiñones, A., Palmieri, S., Menarbin, G., Gioacchini, P. and Masia, A. 2007. Nutrient recycling during the decomposition of apple leaves (Malus domestica) and mowed grasses in an orchard. *Agric. Ecosyst. Environ.* 118(1-4), 191-200. doi:10.1016/j.agee.2006.05.018.

Tews, J., Brose, U., Grimm, V., Tielbörger, K., Wichmann, M. C., Schwager, M. and Jeltsch, F. 2004. Animal species diversity driven by habitat heterogeneity/ diversity: the importance of keystone structures. *J. Biogeogr.* 31(1), 79-92. doi:10.1046/j.0305-0270.2003.00994.x.

Topçu, K., Serdar, U., Öztürk, A. and Genç, M. 2009. A chestnut monument tree in Sinop: Padisah Kestanesi. *Acta Hortic.* 815, 163-70.

Uyemoto, J. K., Ogawa, J. M., Rice, R. E., Teranishi, H. R., Bostock, R. M. and Pemberton, W. M. 1986. Role of several true bugs (Hemiptera) on incidence and seasonal development of pistachio fruit epicarp lesion disorder. *J. Econ. Entomol.* 79(2), 395-9. doi:10.1093/jee/79.2.395.

Wallace, K. J. 2007. Classification of ecosystem services: problems and solutions. *Biol. Conserv.* 139(3-4), 235-46. doi:10.1016/j.biocon.2007.07.015.

Wilby, A. and Thomas, M. B. 2002. Are the ecological concepts of assembly and function of biodiversity useful frameworks for understanding natural pest control? *Agric. For. Entomol.* 4(4), 237-43. doi:10.1046/j.1461-9563.2002.00165.x.

Zhang, W., Ricketts, T. H., Kremen, C., Carney, K. and Swinton, S. M. 2007. Ecosystem services and dis-services to agriculture. *Ecol. Econ.* 64(2), 253-60. doi:10.1016/j. ecolecon.2007.02.024.

# Part 2

## Improving individual types of tree nuts

# Chapter 9

## Towards sustainable production of walnut (*Juglans regia* L.)

*Annarita Marrano and David Neale, University of California-Davis, USA*



## 1 Introduction

A walnut is defined as the nut of any tree of the genus *Juglans*, which encompasses more than 20 species and belongs to the botanical family of *Juglandaceae*, order *Fagales* (McGranahan and Leslie, 2009). All species included in the genus *Juglans* produce edible nuts, but the Persian or English walnut (*J. regia* L.) is the only species widely cultivated for nut production and will be the primary focus of this chapter.

The evolutionary history of walnut has been shaped by the complex interaction of human and biogeographical forces (Pollegioni et al., 2017). It is supposed that the Persian walnut was domesticated in Central Asia, specifically in the foothills of the Western Himalayas from the Kashmir region to Tajikistan and Kyrgyzstan (Zeven and Zhukovskiĭ, 1975). Due to human migration through the Silk Road and the Persian Royal Road, walnut spread across Asia, particularly to Uzbekistan, north Iran, the Caucasus region and the eastern region of Turkey, and then east into Northern India, Nepal and Western China (Bernard et al., 2018). Even though palynological data support the presence of *J. regia* in Europe and the Balkans during the Last Glacial Period (Carrion and Sanchez-Gomez, 1992), how the Persian walnut reached Europe is still controversial. Recently, Pollegioni et al. (2017) confirmed the presence of glacial refugia of walnut in the eastern Mediterranean by integrating fossil, historical and genetic data. Moreover, the analysis of

http://dx.doi.org/10.19103/AS.2018.0042.12

genetic structure reported by Pollegioni et al. (2017) indicated southeastern Europe as a contact zone between the Balkans and western Asia, where walnut populations accumulated high genetic diversity and admixture. From the Balkans, *J. regia* expanded throughout Europe starting from 2500 BP as a consequence of the widespread use of walnut by Greek and Roman civilizations (Mercuri et al., 2013). Finally, during the Roman Empire between the first and second centuries, walnut cultivation reached north-central Europe (Bakels and Jacomet, 2003), from where it was brought to America in the sixteenth century by the earliest settlers. The name 'English' walnut was chosen by the American colonists to distinguish it from the native American black walnut (*J. nigra*).

Currently, Persian walnuts are grown in a wide area extending from North, Central and South America to Europe and North Africa, through the Caucasus to China and Oceania (FAOSTAT data, 2016). Over 3.7 million metric tons (MT) of walnut (in-shell) were produced in 2016, with China leading the world production, followed by the United States, Iran, Turkey and Ukraine (FAOSTAT statistics, 2016; Fig. 1). The United States dominates the global export of walnut with 460 101 in-shell MT in 2016, followed by Chile (115 000 MT), Ukraine (56 100 MT), Moldova (34 000 MT) and China (19 000 MT) (USDA-Foreign Agricultural Service).

Overall, the global production of walnut has grown by almost 250% over the last 20 years (FAOSTAT statistics, 2016). This enormous growth is mainly attributed to an increase in regular consumption of walnut due to its numerous benefits to human health (Costa et al., 2014). Walnut kernels

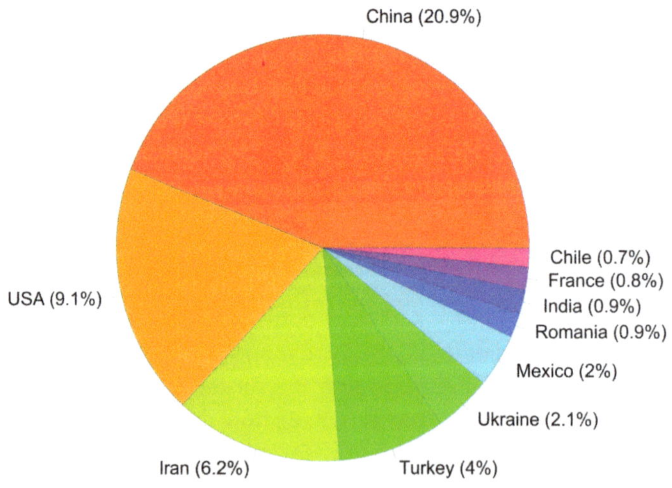


**Figure 1** Top ten producers of walnut in the world. For each country, the percentage (by in-shell tons) of the global walnut production is shown in brackets (FAOSTAT data, 2016).

are a nutrient-dense food due to their content of fats, proteins, vitamins and minerals (Martínez et al., 2010). Oils are the most prominent nutrient in walnut, especially the amount of polyunsaturated fatty acids (up to 78% of total fatty acid content), mainly linoleic acid and alpha-linolenic acid (ALA) (Taha, 2011). ALA is a plant-based essential omega-3 fatty acid and has been associated with anti-inflammatory and antiatherogenic effects as well as to a 10% risk reduction for coronary heart disease (CHD) (Wu et al., 2014). In addition to a favourable lipid profile, walnut is also superior to other nuts for its polyphenolic composition with the highest antioxidant efficacy (Vinson and Cai, 2012). Also, walnut consumption has been associated with a lower risk of type 2 diabetes (Pan et al., 2013) and the maintenance of brain health with age (Poulose et al., 2014).

Even though the interest in walnut can be dated back to the Romans, the genetic improvement of walnut started only in the twentieth century (McGranahan and Leslie, 2012). The most important walnut breeding programmes worldwide are located in California, France and China. They all aim to have a high yield of kernels with high visual and taste qualities, ease of shell cracking, well-adapted flowering and harvest date, and resistance to the major walnut diseases. Recently, Bernard et al. (2018) reported a comprehensive overview of the past and current goals of commercial walnut breeding programmes around the world. In addition, walnut breeding programmes are also facing the challenge of increasing production sustainably. There are potentially different practices to make walnut cultivation ecologically and economically friendly, such as reducing input consumption (i.e. water and chemicals), increasing and preserving crop diversity, and developing new cultivars with genetic resistance to biotic and abiotic stresses (Wezel et al., 2014). The primary goal of this chapter is to review recent efforts and achievements of walnut breeding programmes to address sustainable walnut production.

## 2 Sustainable solutions for disease control in walnut

One of the major challenges for walnut growers is to control new and recurrent pest and pathogen-related epidemics. To date, the application of chemicals has been the primary strategy for disease prevention and control. However, the increasing demand for sustainable agriculture requires research for alternative and environmentally friendly solutions. Walnut breeding programmes worldwide have devoted some effort in this direction, by (1) screening *in-situ* and *ex-situ* walnut collections to identify innate disease resistances (Frutos, 2010), (2) investing in molecular biology to understand the molecular and cellular bases of host-pathogen interactions (Romero-Suarez et al., 2012) and (3) investigating the effectiveness of environmentally neutral forms of chemical control (Horibe et al., 2017).

## 2.1 Cherry leafroll virus

An important disease of Persian walnut is blackline disease, caused by cherry leafroll virus (CLRV; Mircetich and Rowhani, 1984). CLRV is pollen-borne, and infection occurs at flowering. Infected English walnut trees are usually symptomless unless they are grafted onto resistant rootstocks. A hypersensitive reaction (HR) occurs when the virus reaches the graft union between the scion and a resistant rootstock, resulting in scion necrosis (McGranahan et al., 1997). Therefore, blackline disease is a peculiar case where resistance is undesirable in the rootstock.

In 1982, the Walnut Improvement Program (WIP) of University of California, Davis (UC Davis) initiated a backcross breeding programme with two main objectives: (1) development of vigorous rootstocks with a symptomless response to infection by CLRV and (2) the introgression of resistance to CLRV from the Northern California black walnut into English walnut, so avoiding the occurrence of HR and necrosis at the grafting union (McGranahan et al., 1997). The first backcross generation (BC1) of walnut hybrids ((*J. hindsii* × *J. regia*) × *J. regia*) segregating for hypersensitivity to CLRV was used to perform a bulk segregant analysis and identify randomly amplified polymorphic DNA (RAPD) markers linked to blackline resistance genes (Woeste et al., 1996). Two RAPD fragments were associated with the hypersensitivity to CLRV, inherited as a single dominant trait. One of the two RAPD markers (RAPD marker OP-K15$_{720}$) was cloned, sequenced and converted into a sequence characterized amplified region (SCAR) marker. This locus is still used today to screen for hypersensitivity to CLRV in the backcross programme for selecting the most promising individuals as parents for future backcross generations. In 2017, the WIP began to evaluate 90 BC5 seedlings for virus resistance as well as nut quality, harvest date and yield (Leslie et al., 2018). Also, a new procedure for screening the CLRV resistance in the greenhouse has been developed to avoid the standard and time-consuming field-testing procedure (Leslie et al., 2018).

One of the main drawbacks of the CLRV resistance introgression into English walnut from *J. hindsii* is that unfavourable nut-related characteristics of black walnuts, such as thick shells and packing tissue, are retained in the BC progeny. This makes it necessary to perform several generations of backcrosses to reduce the *J. hindsii* genetic background, and obtain individuals resistant to CLRV and with the necessary nut qualities for commercial walnut production. Gene silencing through an RNA interference- (RNAi-) mediated approach is an alternative approach to traditional breeding for the control of blackline disease. It would allow generating new rootstocks with the HR gene silenced and, therefore, tolerant to the CRLV. Sudarshana et al. (2011) focused on the identification of walnut genome sequences homologous to well-characterized *R* genes in other species. The gene silencing approach could also be a suitable approach to target CLRV with RNAi via a modified inter-stock grafted between scion and rootstock. In this way, it would be possible to prevent the CLRV infection from reaching the rootstock and, therefore, triggering blackline (Leslie et al., 2016). In this effort, double-stranded RNA (dsRNA) from W8-CLRV-inoculated *Chenopodium quinoa* plants was extracted for whole genome deep sequencing (Sudarshana et al., 2012). A hairpin DNA construct was engineered and inserted into competent *Agrobacterium tumefaciens* cells to genetically transform English walnut inter-stocks and target the genomic RNAs of CLRV (Lynn, 2015).

The production of English walnut cultivars on their own roots is another possible solution to avoid blackline disease (Vahdati et al., 2004). To date, several studies have been carried out to compare field performances between self-rooted and grafted walnut cultivars. Hasey et al. (2001) reported the comparison of micropropagated self-rooted 'Chandler' trees with those conventionally grafted onto seedling Paradox rootstock. Own-rooted 'Chandler' trees outperformed the grafted ones in vigour and yield for five years. Likewise, López (2001) described the field behaviour of 20 000 micropropagated trees of five commercial varieties ('Serr', 'Chandler', 'Vina', 'Sunland' and 'Sorrento'). Excellent vigour and root system were observed in *in vitro* propagated cultivars, without any delay in nut production. This behaviour was very homogeneous over the years, supporting the idea that *in vitro* propagation can assure more genetic uniformity than grafting (McGranahan and Leslie, 1988). Rovira and Aletà (2010) compared self-rooted and grafted trees of the English cultivars 'Chandler' and 'Vina' in North Spain. Both cultivars showed important differences in phenology between self-rooted and grafted materials. In particular, *in vitro* trees leafed and bloomed out one week later, which can be an advantage in some areas with high risk of spring frost. In addition, no differences were observed in fruit production between the two tested methods of propagation after ten years. Overall, all these findings demonstrate that the production of English walnut cultivars on their own roots is a promising strategy to avoid expensive and time-consuming grafting protocols as well as to provide genetically uniform material without the risk of blackline disease.

## 2.2 Crown gall disease

Crown gall disease is caused by *Agrobacterium tumefaciens* (Smith and Townsend, 1907) and can severely limit the growth and productivity of infected trees. During *A. tumefaciens* infection, the *Agrobacterium*-transferred DNA (T-DNA) is excised from the tumour-inducing plasmid and inserted into the plant genome. The gall structure arises from the expression of T-DNA oncogenes and the biosynthesis of new phytohormone pathway constituents that alter the normal hormone sensitivity of plant cells. Two well-characterized T-DNA oncogenes required for normal tumour development are the tryptophan monooxygenase (*iaaM*) and isopentenyl transferase (*ipt*) genes, involved in the biosynthesis of auxins and cytokinins (Escobar et al., 2001). Common agronomic practices to prevent *A. tumefaciens* infection are fumigation, phytosanitation and biocontrol. However, the effectiveness of those measures has not been consistent in English walnut. Escobar et al. (2002) developed transgenic lines of cultivar 'Chandler' (*J. regia*) recalcitrant to gall formation, by expressing inducers of post-transcriptional gene silencing homologous to *iaaM* and *ipt* and, therefore, activating the degradation of the corresponding oncogene mRNA transcripts in planta. Since 'Chandler' is not used as a rootstock in the United States, transgenic 'Paradox' lines resistant to crown gall were produced subsequently using the same gene silencing strategy. The resulting disease-resistant rootstocks have not been released for growers, due to regulatory and consumer acceptance issues (Leslie and McGranahan, 2014). Another approach for the development of crown gall resistant walnuts is to screen *Juglans* germplasm collections for innate resistance to *A. tumefaciens*. Several open-pollinated progenies of *J. microcarpa, J. major* and *Pterocarya* spp. held at the *Juglans* germplasm collection of the USDA-ARS National Clonal Germplasm Repository have shown high levels of crown gall resistance (Kluepfel et al., 2013).

## 2.3 Walnut blight

Walnut blight is another important bacterial disease that negatively affects productivity and nut quality. Its causal agent is *Xanthomonas arboricola* pv. *Juglandis* (*Xaj*), which attacks leaves, flowers, fruits, buds and young twigs. Lesions begin as small dark-brown spots with a yellowish halo, followed by the appearance of large areas of dark dead tissue once the infection spreads (Belisario and Zoina, 1995). Wet weather and early flowering are the most important factors that increase *Xaj* infection risk and severity (Aletà et al., 2001). Moreover, the bacterium overwinters in infected twigs, buds and fruits, extending the host susceptibility period and, therefore, making disease control more complicated. Arquero et al. (2006) observed that the incidence of walnut blight varied between the different areas of the canopy, with the low external part showing the highest percentage of affected fruits. This result can be explained by the fact that, in general, high relative humidity and low solar radiation prevail in the low external area of the canopy. Moreover, in the same study a correlation between the presence of aphids (*Cromaphis juglandicola*) and the development of walnut blight has been noticed in leaves.

The disease is commonly controlled using preventive spraying with copper bactericides. An efficient and sustainable way to avoid walnut blight disease would be to identify resistant individuals. The evaluation of an *ex-situ* collection of *J. regia* revealed segregation for *Xaj* tolerance in natural conditions (Frutos, 2010). This result suggests that anatomical differences in the epidermis of young twigs, leaves and flowers may be correlated with the tolerance to *Xaj* infection. Aletà et al. (2001) evaluated blight susceptibility on nuts and leaves in a collection of walnut seedling trees from Spanish wild populations, in relation to five commercial cultivars ('Adams-10', 'Amigo', 'Chandler', 'Franquette' and 'Vina'). In general, the Spanish walnut selections showed less blight susceptibility in nuts than leaves. The genotype 'MBLu-20' stood out for its low susceptibility in all organs analysed, comparable to that of 'Adams-10', used as a reference for blight tolerance. Likewise, Botu et al. (2010) evaluated eleven Romanian cultivars together with foreign varieties for several traits, including fruit bearing and resistance to walnut blight. The Romanian cultivars showed fewer fruits infected by *Xaj* than the varieties introduced from other countries, such as USA and France. Moreover, the UC Davis WIP released the cultivar 'Gillet', selected for its low blight incidence combined with high percentage of light kernels and yield, good vigour and large nuts (McGranahan and Leslie, 2006). Another solution for walnut blight control is the selection of late blooming varieties that can escape the favourable environmental conditions for *Xaj* infection during early spring (Germain, 1997). Also, a new research project began in mid-2017 in France to investigate the phenotypic and genetic diversity associated with *Xaj* tolerance in local germplasm resources (Bernard et al., 2018).

The identification of both early biomarkers of the disease and alternative sustainable formulations to copper-based biocides is another strategy to prevent *Xaj* epidemics. Solar et al. (2012a) studied the endogenous phenolic profile of healthy fruits sampled from infected trees of cvs 'Franquette', 'Cisco' and 'Sampion' with different levels of susceptibility to infection by *Xaj*. They observed a negative correlation between the total amount of phenolic compounds and blight severity in all cultivars, which indicates the role of phenolics in fruit-bacteria interactions. In particular, juglone and gallic acid seemed to be the main phenolic compounds involved in the defence mechanism of walnut against bacterial blight. Recently, Horibe et al. (2017) observed that the activity of the two polyphenol oxidase (PPO) genes in walnut (*JrPPO1* and *JrPPO2*; Martínez-García et al., 2016) was positively correlated with the susceptibility to *Xaj*. Therefore, they proposed both PPO genes as biomarkers for host susceptibility.

In the attempt to find alternative products for walnut blight control, Solar et al. (2012b) applied prohexadione-Ca (ProCa) in an adult walnut orchard of cv. 'Franquette' to test its ability to reduce severity of *Xaj* infection, and its effect on the phenolic content of walnut fruits. ProCa significantly reduced walnut blight severity regardless of its concentration and first application date. Also, treated fruits showed higher total phenolic content, more hydroxycinnamic acids and flavanols, suggesting a potential role of ProCa in triggering the defence mechanisms of walnut through the regulation of phenolic synthesis. At the same time, Horibe et al. (2017) validated the effectiveness of the biological ε-polylysine (EPL) to kill *Xaj* strain 417. EPL is an antimicrobial peptide produced by natural fermentation and used as a preservative in some products. The validated capacity of EPL to kill a copper-resistant strain of *Xaj* is a promising result towards the complete replacement of toxic chemicals for walnut blight control.

## *2.4* Phytophthora *spp. and nematode-related diseases*

*Phytophthora* crown and root rot (PCRR) is one of the most severe diseases of Persian walnut worldwide (Browne et al., 2011). More than ten species of *Phytophthora* have been identified as PCRR-causing agents, but *Phytophthora cinnamomi* and *P. citricola* are especially aggressive (Mircetich and Matheron, 1983). Belisario et al. (2006) isolated *Phytophthora* species from roots, flame-shaped lesions on trunks and soil collected from declining trees in English walnut plantations in Italy and Southwestern France. Six different species of *Phytophthora* (*P. cactorum, P. cambivora, P. cinnamomi, P. citricola, P. cryptogea* and *P. nicotianae*) were identified, with *P. cinnamomi* being the most widespread and consistently associated with dead walnut trees. PCRR is also widely spread in Iran, where it is called 'black water' by the local farmers. Banihashemi and Ghaderi (2006) reported the isolation of several *Phytophthora* species from symptomatic walnut trees and soil in Southern Iran, including *P. cactorum*, *P. citricola* and *P. citrophora* as the most abundant. Careful soil water management can minimize the incidence and severity of PCRR, but rootstocks resistant to the causal agents are necessary. Indeed, most of the current commercial rootstocks used for walnut cultivation in California are susceptible to *P. cinnamomi* and *P. citricola*. Browne et al. (2011) evaluated seedlings from seven open-pollinated selections of Chinese wingnut (*Pterocarya stenoptera*) for resistance to *P. cinnamomi* and *P. citricola*. Although all tested seed families were highly resistant to both pathogens, they were not graft-compatible with popular walnut cultivars. In 2010, the UC Davis WIP released the rootstock cultivar 'RX1' (*J. microcarpa* × *J. regia*), selected for its resistance to PCRR in early screening in germplasm resources (McGranahan et al., 2010a). Browne et al. (2015) recently confirmed the resistance of 'RX1' to *P. cinnamomi* in a commercial walnut orchard infected by the pathogen.

Lesion nematodes (*Pratylenchus vulnus*) are other soil-borne pathogens of Persian walnut. They feed on young walnut roots, seriously compromising their development. Since the past practice to fumigate nurseries and new orchards before planting is not sustainable, the identification of a genetic solution to this pest is desirable. An in-depth screening for nematode resistance in a broad germplasm collection, including seedlings from commercial nursery rootstock sources, diverse *Juglans* species of the USDA NCGR collection and Paradox hybrids generated by the WIP, identified a *J. hindsii* × *J. regia* 'VX211' hybrid as tolerant to lesion nematode. 'VX211' has shown very vigorous growth under nematode feeding pressure and was released as a commercial rootstock in 2010 (McGranahan et al., 2010b).

## 3 Adaptation to environmental changes in walnut

Since the 1950s the entire globe has experienced surface warming and extreme climate events, including a decrease in cold temperature extremes and an increase in the number of massive precipitation events in some regions. The projections for future climate change are for further warming and long-lasting changes in all the climate. Such scenarios will undoubtedly affect agriculture, including increased water deficit, freeze-thaw events and the expansion of pest ranges (Luedeling et al., 2011). Accordingly, recent studies in walnut have focused on physiological responses to environmental factors, notably water, temperature and nutrients. Gauthier and Jacobs (2011) summarized the studies carried out in the last 35–60 years regarding walnut stress physiology, emphasizing how there is still uncertainty regarding the magnitude of potential effects of climate change on walnut. The first signal of adaptation to global warming is the change of the timing of phenological events during the plant life cycle. The general observation is that the majority of spring plant phenological events are trending earlier across Europe, North America and Asia (Menzel et al., 2006; Schwartz et al., 2006; Ma and Zhou, 2012). Črepinšek et al. (2009) studied the influence of increasing winter and spring temperature on bud break in Slovenia. In particular, they analysed the bud-breaking date of the cultivars 'Franquette' and 'G-139' over two time periods: 1984–90 (P1) and 2000–6 (P2). A shift of 3 days (d) earlier for 'G-139' and 7 d earlier for 'Franquette' was observed during P2, when the mean air temperature from January to April increased by 0.9°C compared to P1. Therefore, they concluded that the average bud-breaking day could be advanced by as much as 4 weeks by 2060, with a resulting increase of frost risk. However, this forward march of spring phenological events is not universal: some plant species are responding more to lack of winter chill than increased spring temperature. Pope et al. (2013) used a 60-year data set for timing the vegetative bud break and male flowering of the cultivar 'Payne'. The date of male flowering moved forward in spring, with satisfied chilling requirements. The vegetative bud break changed trends during the 60 years: before 1994, leaf-out timing advanced with warmer spring temperatures and an adequate chilling. Around 1994 chilling became suboptimal, delaying the leaf-out date. Therefore, as global temperature rises, winter chill decreases and more spring heat will be required for bud break. These findings indicate that certain walnut growing regions could become unsuitable for walnut growth, whereas areas currently considered too cold might present the climate requirements for walnut cultivation (Gauthier and Jacobs, 2011).

A temporary but still beneficial solution could be to select cultivars adapted to local climate conditions or with a timing of phenological events suitable for specific climates. Thus, very early cultivars such as 'Serr' could be a good selection for cultivation in mild Mediterranean climates, while late varieties like 'Franquette' should be preferred in mountainous areas, where spring frost is common at the end of April. Rouskas and Zakynthinos (2001) evaluated seventy walnut selections from Greek seedling populations adapted to local conditions. Based on five years observations on bud breaking, they identified 16 selections very early, 27 early, 23 medium early and 4 medium late. This material could be useful as gene pool for walnut cultivation in Greece and in those areas with a similar climate diversity. Likewise, domestic (i.e. 'Maraş -12', 'Maraş -18', 'Şebin' and 'Şen 2') and foreign (i.e. 'Fernor', 'Chandler', 'Fernette', 'Howard') cultivars were evaluated in the climate conditions of two Turkish cities, Bursa (Ertürk et al., 2014) and Niksar (Akça et al., 2014). In both studies, the domestic varieties presented early leafing time, which makes them susceptible to spring frosts. On the contrary, the foreign cultivars showed late leafing time and seem to be more indicated for cultivation in those regions. Moreover, Cosmulescu et al. (2010) studied the influence of environmental factors on flowering and pollination of six Romanian walnut cultivars ('Jupâneşti', 'Germisara', 'Roxana', 'Mihaela', 'Geoagiu 65', 'Orăştie') in Oltenia, aiming to identify genotypes with biological requirements closely similar to the ecological factors of the region.

### 3.1 Frost tolerance

Another consequence of climate change on walnut production may be an increased risk of frost damage. Walnut is a deciduous tree and undergoes winter dormancy (Warren, 1998). The endodormancy release starts when chilling requirements are satisfied, preventing trees from initiating growth during transient warm events. Buds subsequently break when heat requirements (ecodormancy) have been completed in early spring. This annual cycle achieves *frost avoidance*. In Persian walnut, buds enter into the dormant state during November and release from dormancy during March to April after fulfilment of the chilling requirements (Fig. 2) (Gholizadeh et al., 2017). *Cold acclimation* is another mechanism to avoid freezing damage that involves major changes in the biochemical and physiological state of the plant to improve low-temperature stress tolerance (Preston and Sandve, 2013). Dormancy and cold acclimation are driven by the same environmental signals: short photoperiod and low temperatures. Thus, while endodormancy occurs during winter, trees are also cold acclimating. Both dormancy and cold acclimation are

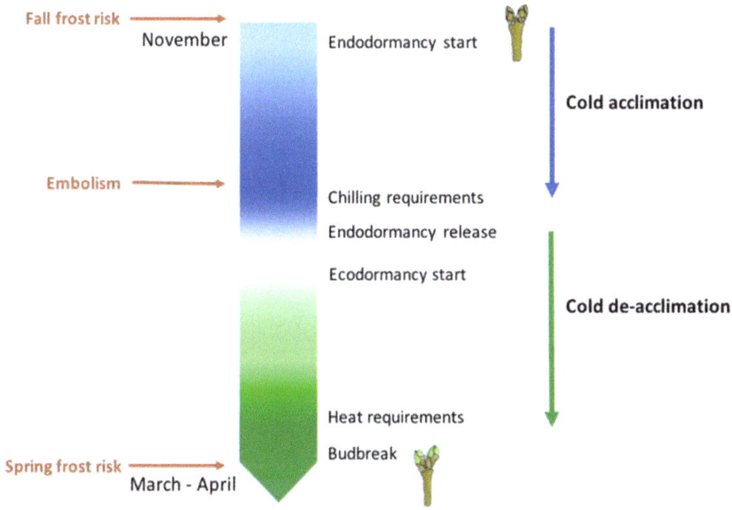


**Figure 2** Phenological events and temperature acclimation from endodormancy to bud break in walnut. The timing of frost and embolism risks are also indicated.

heritable as variation exists among cultivars. Charrier et al. (2011) evaluated the degree to which differences in dormancy release and cold acclimation were related to environmental and genetic effects by comparing walnut trees growing in common garden conditions. In particular, they measured budburst date, chilling and heat requirements of four nut-producing *J. regia* cultivars ('Chandler', 'Franquette,' 'Lara' and 'Serr') and two wood-producing hybrids (*J. regia* × *J. nigra*) for two winters in lowland and mountain locations. A strong genotypic effect on frost resistance was observed only after endodormancy release, explaining the differences in bud break date among walnut trees. In particular, earlier genotypes were de-acclimating sooner and grew faster than later ones. These findings highlight the increase of freeze damage risk in earlier cultivars, mostly through previous exposure of buds and new leaves. The same research group compared different organs and tissues of young potted and mature orchard English walnut trees for seasonal changes in frost hardiness during different years (Charrier et al., 2013). They observed cold acclimation and de-acclimation in all aboveground tissues, with buds being the most sensitive to freezing and bark the most resistant. Moreover, Charrier et al. (2013) investigated the relationship between frost hardiness and two physiological parameters, namely water and soluble carbohydrate (glucose, fructose, sucrose: GFS) contents. Indeed, during the process of frost acclimation, the water content of plant tissues decreases, due to the gradual reduction of water supply via the root system and ongoing water evaporation in aerial parts (Poirier et al., 2010). Trees simultaneously synthesize soluble compounds, such as proline and GFS. These metabolites act as water substitutes, maintaining a hydrophilic environment for all types of biological molecules and lowering the freezing point of cytosol. Based on the best model for frost hardiness of Charrier et al. (2013), cold hardiness is due to the interactive effect of water content and GFS. Poirier et al. (2006) compared the cold hardiness of seven walnut fruit cultivars ('Chandler', 'Franquette', 'Fernor', 'Fernette', 'Lara', 'Pedro', 'Serr') and the hybrid *J. regia* × *J. nigra* NG38 in Southern Central France. Freezing tolerance was significantly different between winter and the other seasons, reaching its minimum during bud breaking and fall beginning. Overall, NG38 appeared to be the most hardy cultivar. However, the maximum GFS concentration was not strictly correlated with cold resistance in autumn, where other factors are likely involved. In addition to changes in water content and GFS concentrations, Aslamarz and Vahdati (2010) observed differences in leaf stomatal density between hardy, semi-hardy and non-hardy walnut genotypes. In particular, the density of stomata was lower in hardy walnuts than in non-hardy genotypes. This relationship between cold hardiness and stomata density can be explained by the fact that, while water movements in the stem stops when the temperature drops below 0°C, water continues to evaporate in the aerial part through the stomata, increasing the risk of freezing injuries. Therefore, similarly to drought stress, stomata density might be used to predict hardiness in walnut.

The assessment of genetic variability and heritability of frost hardiness is crucial for future breeding for adaptive traits. Aslamarz et al. (2010a) evaluated cold hardiness of four walnut cultivars ('Serr', 'Pedro', 'Hartley' and 'Lara') and four native Iranian genotypes ('$Z_{63}$', '$Z_{53}$', '$Z_{30}$' and '$Z_{67}$') using thermal analysis and freezing technique. They categorized cultivars and genotypes in three groups based on the degree of hardiness: hardy ('Lara' and 'Pedro'), semi-hardy ('Hartley', '$Z_{63}$', '$Z_{53}$' and '$Z_{67}$') and sensitive ('Serr' and '$Z_{30}$'). The same walnut trees were then assessed for chilling and heat requirements (Aslamarz et al., 2010b). As for cold hardiness, cultivars and genotypes were classified in three categories based on chilling requirements: low ('Serr' and '$Z_{30}$'), medium ('$Z_{53}$', '$Z_{67}$', 'Lara' and 'Pedro') and high requirements ('Hartley' and '$Z_{63}$'). Aletà et al. (2014) assessed the response to sub-zero temperatures in popular commercial cultivars of English walnut ('Chandler', 'Franquette', 'Howard', 'Fernette', 'Serr', 'Fernor' and 'Tulare'). Regarding winter frost resistance, 'Chandler' stood out from the other varieties. However, when considering the susceptibility to sudden autumn frost, a different cultivar ranking was obtained, with 'Serr' and 'Fernette' being less damaged than 'Chandler', 'Franquette' and 'Tulare'. Guàrdia et al. (2016) carried out a progeny test of Persian walnut for autumn and winter frost hardiness. The heritability of autumn frost hardiness was 0.68, and that for winter frost hardiness 0.77. Such high heritability values indicate promising selection and breeding for these traits in walnut.

### 3.2 Drought tolerance

An effect of climate change might be reduced water availability for a crop such as walnut that requires large amounts of water for optimum growth and productivity (Fulton and Buchner, 2006). Water stress in walnut is expressed as leaf abscission, reduced shoot growth, sunburn and, eventually, tree decline and death. Regarding drought adaptation, walnut is a *drought avoider*: under high leaf-to-air vapour pressure deficit (VPD) or low leaf water potential, walnut's close stomata preventing xylem cavitation (Cochard et al., 2002). Moreover, in response to drought stress trees accumulate several organic compounds, such as proline, sugars and polyols. Naser et al. (2010) induced water stress in Persian walnut genotypes using polyethylene glycol-6000 to study the mechanisms involved in drought tolerance. They determined the amount of proline and sugar accumulated at different water potentials in the seedlings of four *J. regia* genotypes ('Lara', 'Panegine$_{20}$', 'Serr' and 'Chandler'). A reduction in seed germination under drought stress was observed in all genotypes. However, cultivars 'Panegine$_{20}$' and 'Chandler' showed the highest drought tolerance, correlated with an increased accumulation of proline and soluble sugars, and a decreased starch content. Hence, Naser et al. (2010) proposed proline and soluble sugars concentrations as biochemical markers of drought stress level in walnut. Likewise, Scartazza et al. (2001) monitored photosynthesis, growth and storage of accumulated carbohydrate during the growing season of 2-year-old seedlings of the cultivar 'Lozeronne', subjected to water shortage from August till the end of the vegetative season. The drought treatment induced a decrease of photosynthesis, which consequently caused a reduction in carbohydrate accumulation and growth. More importantly, the effects of drought lasted till the beginning of the following vegetative season, even after the recovery of an optimal water status. Due to a reduced amount of storage carbohydrate, plants subjected to summer drought showed earlier sprouting and lower growth rate of spring shoot. Vahdati (2014) reported a wide overview of the morphological, physiological and molecular responses to drought and salt stress in walnut. In particular, after collecting genotypes tolerant to abiotic stresses around the world, he observed differences in shoot growth, seed germination, stomatal density and mineral composition between tolerant and sensitive accessions.

An alternative and inexpensive way to screen for drought tolerance in walnut is to assess the carbon isotope composition of leaf tissue as an indirect index of water use efficiency (WUE). Plants discriminate against $^{13}C$ during stomatal diffusion and photosynthesis. However, if the $CO_2$ assimilation increases or the stomatal conductance ($g_s$) decreases, a deficit in available intracellular $CO_2$ occurs, and the plant is forced to reduce its discrimination against $^{13}C$. Since the intrinsic WUE can be expressed as the ratio of assimilation to stomatal conductance or transpiration ($A/g_s$), carbon isotope discrimination

($\Delta^{13}C$) is linked to WUE through the effect of A or $g_s$ on the intracellular $CO_2$ concentrations: as WUE increases due to stomatal closure (decrease $g_s$) or an increase in A, intercellular $CO_2$ and $\Delta^{13}C$ decrease. Therefore, low values of $\Delta^{13}C$ indicate high WUE (Farquhar et al., 1989). Aletà et al. (2009) evaluated the genetic variation in leaf $\Delta^{13}C$ for 22 *J. regia* open-pollinated progenies originating from seven regions in Spain and investigated whether such variation was also correlated with climatic indicators of native habitats. Genetic variation in carbon isotope composition was observed between families, with the most water use efficient genotypes coming from drought-prone areas in Spain with low rainfall. Moreover, this study emphasized the existence of a trade-off between growth and drought tolerance in *J. regia*.

Famula et al. (2018) characterized carbon isotope discrimination and foliar nitrogen concentration in 60 individuals of the UC Davis WIP. A Bayesian analysis utilizing genomic relationships revealed 'Chandler' as the least water use efficient, and 'Solano' and 'Tulare', two recently released cultivars, as the highest ranked genotypes for WUE. Also, an association mapping test was performed using over 126K single nucleotide polymorphisms (SNPs), identifying four loci associated with $\Delta^{13}C$. Those molecular markers were in linkage disequilibrium with genes involved in the response to abiotic stresses. In addition, M. Arab (pers. comm.) utilized the new Axiom® Walnut700K SNP array (Affymetrix, Santa Clara, CA) (Marrano et al., 2018) to genotype a population of 95 walnut trees collected from hot and dry provinces of Iran. Population structure analysis identified five genetic subgroups defined by the geographical origin, which are quite different in terms of natural aridity. Offspring of the 95 genotyped trees have been extensively phenotyped for morpho-physiological and biochemical drought tolerance-related traits, revealing 15 genotypes with an enhanced ability to survive under water deficit conditions. The 95 trees will be used to carry out marker–trait association analysis for drought tolerance.

## 4 Yield

Agriculture can be sustainable not only from an ecological and environmental perspective but also regarding improvements in human life (Giovannoni and Fabietti, 2013). Due to the caloric and nutrient contents of walnut, it would be essential to increase its accessibility to all humanity. A higher and more stable yield is among the major objectives of any walnut breeding programme. In Persian walnut, the yield is correlated with three factors: the number of pistillate flowers produced, the per cent set and the size of nuts (Forde, 1975). The number of pistillate flowers produced depends on the number of reproductive buds each year and the number of flowers per shoot. Walnut trees can produce flowers on shoots from (1) terminal buds (*terminal bearing;* Fig. 3), (2) terminal and subterminal buds (*intermediate bearing*) or (3) terminal, subterminal and

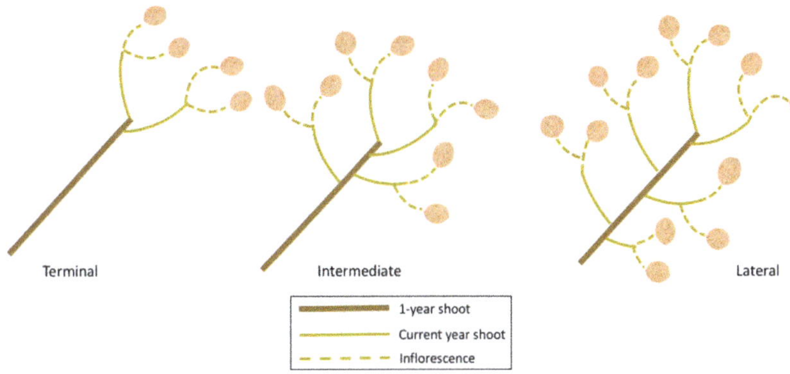


**Figure 3** Types of fruit-bearing in walnut (*J. regia* L.)

axillary buds (*lateral bearing*) (Solar and Štampar, 2003). Thus, young trees of lateral bearing cultivars, producing nuts along the entire length of one-year-old shoots, have a much higher yield (Germain, 1997). Lateral bearing trees are most commonly found in Western China and areas of Central Asia, but occasionally they have been identified elsewhere (Germain, 1997). Lateral bearing has been the focus of most walnut breeding programmes across the world. In California, the first cultivar selected for this trait was 'Payne', intensively used afterward as a source of lateral fruitfulness in breeding programmes (Tulecke and McGranahan, 1994). Germain (1990) tried to combine by crossing the late leafing date, low blight susceptibility and high kernel quality of the French cultivar 'Franquette' with the lateral bearing of Californian varieties. Although 32 hybrids from a population of 720 seedlings showed a combination of the desired characteristics, this breeding attempt demonstrated how lateral fruit-bearing is negatively correlated with late leafing and blight resistance. In Turkey, the cultivar 'Şebin', whose later buds on shoots are fruitful for about 43%, has become really popular due to its high kernel ratio and excellent yield, together with high nut quality and precocity (Akça, 2014). However, in some Turkish orchards established with 'Şebin', an elevated ratio of female flower abortions (87.8%) and, therefore, low yield have been observed. The application of aminoethoxyvinylglycine hydrochloride (AVG) combined with gibberellic acid (GA3) raised considerably the fruit set in those problematic orchards (Akça et al., 2012). Later bearing genotypes have been also selected and released as cultivars in Ukraine, with particular reference to the three varieties 'Bukovyns'ky', 'Prydnistrovs'ky' and 'Chernivets'ky', which are characterized by high yield, early fruit bearing, tolerance to diseases and excellent nut quality.

Botu et al. (2014) investigated the influence of genotype (G), environment (E) and genotype-environment interaction (GxE) on yield in 18 walnut cultivars coming from Romania (10), France (4) and the United States (4). For the terminal bearing varieties, 43.9% of the yield was influenced by G, while 24.4% depended on environmental effects and 12.2% on GxE. On the contrary, for the lateral bearing cultivars, G accounted for 42.2% of the yield, E for 29.1% and GxE for 13.8%. The highest values of yield caused by genotype were observed for 'Valcor' and 'Jupâneşti' among the terminal bearing cultivars, and for 'Vina', 'Ferjean' and 'Hartley' for the lateral ones. Recently, Martínez-García et al. (2017) used historical pedigree and heirloom phenotypic data from the UC Davis WIP to estimate breeding values and genetic parameters for both yield and lateral bearing. Breeding value is defined as the additive genetic merit of an individual, in other words, the fraction of individual's genetic makeup that contributes to the phenotypic value of a trait in its progeny (Wilson et al., 2010). Yield showed a lower value of both repeatability (0.82) and heritability (0.54) than lateral bearing (0.96 and 0.75, respectively). Therefore, non-genetic effects explained a high percentage of the total phenotypic variance for yield, which is less repeatable over multiple seasons than the lateral bearing. By ranking the individuals based on their breeding values, Martínez-García et al. (2017) selected twenty top individuals for yield and lateral bearing separately, which may be used as parents in future breeding cycles. The efficiency of selection for yield and lateral bearing can be enhanced with the identification of molecular markers tightly linked to underlying genes (Xu and Crouch, 2008). Dvorak et al. (2016) have started, recently, to fine-map the *Lateral-Bearing*-1 (*LB1*) gene in an $F_2$ 'Chandler' × 'Chandler' progeny at the UC Davis WIP. This project is the first step towards marker-assisted selection (MAS) for lateral bearing in walnut, which can save time and resources in comparison with classical phenotype-based selection (Collard and Mackill, 2008).

The fruiting behaviour of walnut affects nut weight and homogeneity over the life of a tree (Lauri et al., 2001). Terminal bearing cultivars are known to produce nuts of regular weight from year to year, while in lateral bearing cultivars nut weight and yield decrease as the tree ages, along with an increase in nut weight heterogeneity. Kelc et al. (2007) investigated the relationship between nut weight and the morphometric features of one-year-old and current year shoots in four genotypes with different fruiting habits: 'Franquette' (terminal bearing), 'Zdole-59' (intermediate bearing), 'Lara' (lateral bearing and densely branched) and 'Fernor' (lateral bearing and sparsely branched). They observed that large and heavy nuts could only be present under conditions of long and thick floriferous shoots. This can be explained by the high levels of water and nutrient transport, gas exchange and carbohydrate resources present in such shoots (Kelc et al., 2007). Moreover, the number and type of morphological characteristics developed on vigorous shoots depend on the tree fruiting behaviour. In the case of lateral bearing cultivars, a one-year shoot will bear at least two current year fruiting shoots (Fig. 3), while in cultivars with a terminal habit, each one-year shoot will bear only one inflorescence at a terminal position. This study showed that only in the terminal bearing cultivar 'Franquette' nut weight was affected by the number of nuts on current year shoots. In lateral bearing cultivars, the correlation between the number of nuts on current year shoots and nut weight was also negative but not significant. Finally, Kelc et al. (2007) concluded that in terminal cultivars, nut weight could be predicted by one-year shoot diameter, while in lateral bearing cultivars with densely branched shoots, the length and the number of nuts of current year shoots can be most useful for nut weight predictions.

Valdebenito et al. (2017) confirmed the strong positive correlation between the morphological characteristics of the shoot and the number and size of nuts. They studied mono-cyclic shoots located in three canopy zones (upper, lower light-exposed and lower-shaded) of mature 'Chandler' trees in three different orchards in Chile. It was observed that the number and size of fruits depended on shoot diameter and length, as well as leaf area. In particular, longer and thicker shoots with a greater leaf area will produce more fruits but with a smaller nut diameter. Moreover, the position of the shoot within the canopy seemed to be correlated with the number and size of fruits produced: shoots located in the outer and upper zones of the tree produce more and larger nuts than shoots located in the inner part of the canopy. The higher crop load of the upper and lower-exposed shoots is related to higher exposure and interception of light, as already proposed by Lampinen et al. (2014).

## 5 Future trends and conclusion

During the last two centuries, a dramatic loss in crop genetic diversity has occurred, dangerously shrinking the genetic pool available for both natural selection and plant breeders (Esquinas-Alcazar, 2005). Exploration and conservation of existing genetic diversity are crucial to deal with environmental and demographic changes. In walnut, many studies have been conducted on the genetic diversity preserved in natural populations throughout the global walnut distribution area (Table 1). These studies have revealed a considerable diversity in wild *J. regia*. Eastern and Southeastern Europe showed the highest value of genetic diversity, most likely resulting from the genetic admixture of Asian and European gene pools during millennia of human trade across these regions (Ebrahimi et al., 2016; Christopoulos et al., 2010). Moreover, *J. regia* populations from Iran were particularly rich in unique alleles suggesting that urgent conservation actions are necessary for this area. Interesting morphological and phenological variation was also observed in almost all the natural collections analysed, notably for fruit-bearing habit and nut-related characteristics. These findings stimulate investment in more efforts in exploring natural resources and developing *ex-situ* and *in-situ* conservation programmes to avoid genetic erosion in walnut.

**Table 1** List of the most recent reports about genetic and phenotypic diversity within natural populations of *J. regia*

| Geographical origin | Molecular markers used | Traits evaluated | References |
|---|---|---|---|
| Uzbekistan | SSRs | Nut size, weight and composition | Mapelli et al. (2014) |
| Greece | Inter-Simple Sequence Repeat (ISSR) | Growth habit, fruit-bearing, tree vigour, bud break, nut weight, kernel percentage and colour | Christopoulos et al. (2010) |
| Algeria, Germany, Hungary, Iran, Iraq, Libya, Morocco, The Netherlands, Slovakia, Tunisia, Turkey, Greece, Italy, Spain | SSRs | - | Ebrahimi et al. (2016) |
| Kyrgyzstan, Uzbekistan, China, Pakistan, Tajikistan, Iran, Turkey, Georgia | SSRs | - | Pollegioni et al. (2017) |
| Romania | RAPDs, SSRs | 24 morphological traits (i.e. kernel colour, bud break period, bearing type, yield, nut weight and size) | Pope et al. (2013) |
| Iran | SSRs | Tree vigour, growth habit, budburst, dichogamy, bearing type, maturity time, nut weight, kernel weight and colour | Vahdati et al. (2014) |
| China | EST-SSRs | - | Han et al. (2016) |
| Midwestern USA, Hungary, Slovakia, Italy | SSRs | - | Ebrahimi et al. (2017) |
| Turkey | - | Nut-related traits, bearing type, vegetative and reproductive bud break, tree vigour, growth habit, disease tolerance (i.e. walnut blight) | Akça and Ozongun (2004) |
| Iran | - | Phenological and nut-related traits (i.e. bud break, kernel weight, shell thickness), yield and fruit-bearing type | Arzani et al. (2008) |
| India | - | Fruit-bearing habit, kernel quality and harvest date | Nainwal and Nainwal (2015) |

Microsatellites (SSR) have been the most widely used marker for studying genetic diversity in walnut. Woeste et al. (2002) first developed thirty SSR loci from three black walnut genotypes, offering to the walnut research community an easy and cost-effective tool to characterize genetic diversity in germplasm collections as well as to verify the identity of important cultivars. More recently, the construction of two large-insert BAC libraries of the cultivar 'Chandler', and the identification of more than 48K high-quality BAC End Sequences (BESs) provided a new resource for studying diversity (Wu et al., 2012). The BESs have been used to identify SNPs and develop the first 6K SNP Infinium iSelect BeadChip (Illumina®, San Diego) for walnut. This SNP array has been used to genotype a large walnut population of 425 $F_1$ trees from a cross of the English cultivars 'Chandler' × 'Idaho', and construct a dense genetic map as the first prerequisite for obtaining a walnut BAC clone-based physical map (Luo et al., 2015).

Genetic resources in walnut reached a new level with the release of the first version of the *J. regia* genome sequence by Martínez-García et al. (2016). More than 32K genes were predicted in the walnut genome (667 Mb), providing insights into the genetics of important walnut phenotypes, such as the phenolic composition of nuts. The availability of a walnut reference genome changes the approach that can be used for studying intraspecific variation, by offering the opportunity to discover any allelic variant at each genic locus. Recently, Stevens et al. (2018) sequenced the whole genome of the founders of the UC Davis WIP, originally imported from different parts of the world (i.e. France, China, Afghanistan). The alignment of the sequence reads onto the walnut reference genome allowed the identification of 17.8M SNPs, which were later filtered based on quality-related parameters to build the novel Walnut 700K Axiom® SNP array (Marrano et al., 2018). The development of this genome-based tool opens a new phase in walnut genetics, where classical phenotype-based breeding evolves into MAS. This novel genomic tool will enable large-scale genotyping of walnut collections and, thus, the identification of the genetic bases of important traits in walnuts, such as yield and WUE. It will also facilitate obtaining an overview of the genetic diversity across the genome, tracking those regions most differentiated among walnut accessions. Moreover, genome-wide markers can be used to predict the overall performance of individuals in walnut breeding populations using genome selection (GS) (Desta and Ortiz, 2014). GS can accelerate breeding cycles by capturing the effects of all major and minor quantitative trait loci for target traits (Poland, 2015).

Epigenetics is another subject that will gain interest in future research on walnut. Epigenetics is the combination of all molecular mechanisms that regulate plant gene expression and modulates phenotypes in response to environmental stimuli (Álvarez-Venegas and De-la-Peña, 2016). DNA methylation, dynamic changes in chromatin properties and the biogenesis of small RNAs are among the major epigenetic mechanisms. They can create new heritable genetic variants (epialleles) so that they can work as a powerful source of diversity for important agronomical traits, such as crop production (Rodriguez Lopez and Wilkinson, 2015) and stress adaptation (Mirouze and Paszkowski, 2011). Therefore, it will be essential to understanding the contribution of epigenetic mechanisms to the phenotypic diversity and plasticity in walnut, and how its exploitation can benefit walnut breeding programmes towards sustainable agriculture.

## 6 Where to look for further information

Among the numerous resources on walnut history, cultivation and breeding, we recommend the book chapter 'Walnuts' of Harold I. Forde in *Advances in Fruit Breeding* (1975) and the recent review, 'Walnut: past and future of genetic improvement' of Bernard et al. (2018). The former covers the history and main objectives of walnut genetic improvement, while the latter provides an exhaustive overview of the latest achievements of molecular and 'omics' studies in walnut, together with a standard introduction of the main walnut breeding programmes in the world.

Detailed descriptions of walnut cultivation in each country are also reported in *Following Walnut Footprints (Juglans regia* L.*) Cultivation and Culture, Folklore and History, Traditions and Uses* published in *Scripta Horticulturae* in 2014.

Moreover, updates on walnut cultivation and research reports can be found at:

- The Food and Agriculture Organization of the United Nations (FAOSTAT) web site (http://www.fao.org/faostat/en/#data)
- The Walnut Research Report Database of University of California, Division of Agriculture and Natural Resources (http://ucanr.edu/sites/cawalnut/)
- The Acta Horticulturae for Workgroup Walnuts of the International Society for Horticultural Science (https://www.ishs.org/acta-horticulturae/F27)

## 7 References

Akça, Y. (2014), '"Şebin" Persian walnut', *Acta Horticulturae*, 1050, pp. 139–42. doi:10.17660/ActaHortic.2014.1050.16.

Akça, Y. and Ozongun, S. (2004), 'Selection of late leafing, late flowering, laterally fruitful walnut (*Juglans regia*) types in Turkey', *New Zealand Journal of Crop and Horticultural Science*, 32, pp. 337–42.

Akça, Y., Mustafa, O., Umran, E., et al. (2012), 'The effects of Avg and Ga(3) treatments on pistillate (Female) flower abortion in "Sebin" walnut cultivar', *Acta Scientiarum Polonorum-Hortorum Cultus*, 11(4), pp. 179–85.

Akça, Y., Ünal, B., Çelik, M., et al. (2014), 'Comparison of some promising Turkish and foreign walnut cultivars', *Acta Horticulturae*, 1050, pp. 143-50. doi:10.17660/ActaHortic.2014.1050.17.

Aletà, N., Ninot, A., Moragrega, C., et al. (2001), 'Blight sensitivity of Spanish selections of *J. regia*', *Acta Horticulturae*, 544, pp. 353-62.

Aletà, N., Vilanova, A., Diaz, R., et al. (2009), 'Genetic variation for carbon isotope composition in *Juglans regia* L.: Relationships with growth, phenology and climate of origin', *Annals of Forest Science*, 66, p. 413. doi:10.1051/forest/2009021.

Aletà, N., Vilanova, A., Tomàs, E., et al. (2014), 'Frost resistance in seven commercial walnut cultivars', *Acta Horticulturae*, 1050, pp. 389-94.

Álvarez-Venegas, R. and De-la-Peña, C. (2016), 'Editorial: Recent advances of epigenetics in crop biotechnology', *Frontiers in Plant Science*, 7, pp. 1-3. doi:10.3389/fpls.2016.00413.

Arquero, O., Lovera, M., Rodriguez, R., et al. (2006), 'Walnut blight (*Xanthomonas arboricola* pv. *juglandis*): Factors influencing disease incidence', *Acta Horticulturae*, 705, pp. 443-6.

Arzani, K., Mansouri-Ardakan, H., Vezvaei, A., et al. (2008), 'Morphological variation among persian walnut (*Juglans regia*) genotypes from central Iran', *New Zealand Journal of Crop and Horticultural Science*, 36, pp. 159-68. doi:10.1080/01140670809510232.

Aslamarz, A. A. and Vahdati, K. (2010), 'Stomatal density and ion leakage as indicators of cold hardiness in Walnut', *Acta Horticulturae*, 861, pp. 321-4. doi:10.17660/ActaHortic.2010.861.44.

Aslamarz, A. A., Vahdati, K., Rahemi, M., et al. (2010a), 'Cold-hardiness evaluation of Persian walnut by thermal analysis and freezing technique', *Acta Horticulturae*, 861, pp. 269-72. doi:10.17660/ActaHortic.2010.861.35.

Aslamarz, A. A., Vahdati, K., Rahemi, M., et al. (2010b), 'Evaluation of chilling-heat requirements of some Persian walnut cultivars', *Acta Horticulturae*, 861, pp. 317-20. doi:10.17660/ActaHortic.2010.861.43.

Bakels, C. and Jacomet, S. (2003), 'Access to luxury foods in Central Europe during the Roman period: The archaeobotanical evidence', *World Archaeology*, 34(3), pp. 542-57. doi:10.1080/0043824021000026503.

Banihashemi, Z. and Ghaderi, F. (2006), 'Phytophthora species associated with crown and root rot of walnut in southern Iran', *Proceedings of the Fifth International Walnut Symposium*, 705, pp. 425-8.

Belisario, A. and Zoina, A. (1995), 'Occurrence of Persian (English) walnut blight and its control in the nursery', *European Journal of Forest Pathology*, 25(4), pp. 224-31. doi:10.1111/j.1439-0329.1995.tb01006.x.

Belisario, A., Maccaroni, M., Vettraino, A. M., et al. (2006), 'Phytophthora species associated with decline and death of English walnut in Italy and France', *Acta Horticulturae*, 705, pp. 401-7.

Bernard, A., Lheureux, F. and Dirlewanger, E. (2018), 'Walnut: Past and future of genetic improvement', *Tree Genetics and Genomes*, 14(1). doi:10.1007/s11295-017-1214-0.

Botu, M., Tudor, M. and Papachatzis, A. (2010), 'Evaluation of some walnut cultivars with different bearing habits in the ecological conditions of Oltenia – Romania', *Acta Horticulturae*, 861, pp. 119-26.

Botu, M., Achim, G., Cosmulescu, S., et al. (2014), 'The influence of ecological conditions and genotype on walnut yield north of Oltenia - Romania', *Acta Horticulturae*, 1050, pp. 271-6. doi:10.17660/ActaHortic.2014.1050.36.

Browne, G. T., Grant, J. A., Schmidt, L. S., et al. (2011), 'Resistance to *Phytophthora* and graft compatibility with Persian walnut among selections of Chinese wingnut', *HortScience*, 46(3), pp. 371–6.

Browne, G. T., Leslie, C. A., Grant, J. A., et al. (2015), 'Resistance to species of *Phytophthora* identified among clones of *Juglans microcarpa* × *J. regia*', *HortScience*, 50(8), pp. 1136–42.

Carrion, J. S. and Sanchez-Gomez, P. (1992), 'Palynological data in support of the survival of walnut (*Juglans regia* L.) in the western Mediterranean area during last glacial times', *Journal of Biogeography*, 19(6), p. 623. doi:10.2307/2845705.

Charrier, G., Bonhomme, M., Lacointe, A., et al. (2011), 'Are budburst dates, dormancy and cold acclimation in walnut trees (*Juglans regia* L.) under mainly genotypic or environmental control?' *International Journal of Biometeorology*, 55(6), pp. 763–74. doi:10.1007/s00484-011-0470-1.

Charrier, G., Poirier, M., Bonhomme, M., et al. (2013), 'Frost hardiness in walnut trees (*Juglans regia* L.): How to link physiology and modelling?' *Tree Physiology*, 33(11), pp. 1229–41. doi:10.1093/treephys/tpt090.

Christopoulos, M. V., Rouskasb, D., Tsantilia, E., et al. (2010), 'Germplasm diversity and genetic relationships among walnut (*Juglans regia* L.) cultivars and Greek local selections revealed by Inter-Simple Sequence Repeat (ISSR) markers', *Scientia Horticulturae*, 125(4), pp. 584–92. doi:10.1016/j.scienta.2010.05.006.

Cochard, H., Coll, L., Le Roux, X., et al. (2002), 'Unraveling the effects of plant hydraulics on stomatal closure during water stress in walnut', *Plant Physiology*, 128(1), pp. 282–90. doi:10.1104/pp.010400.

Collard, B. C. and Mackill, D. J. (2008), 'Marker-assisted selection: An approach for precision plant breeding in the twenty-first century', *Philosophical Transactions of the Royal Society B: Biological Sciences*, 363(1491), pp. 557–72. doi:10.1098/rstb.2007.2170.

Cosmulescu, S., Baciu, A., Botu, M., et al. (2010), 'Environmental factors' influence on walnut flowering', *Acta Horticulturae*, 861, pp. 83–8.

Costa, J., Carrapatoso, I., Oliveira, M., et al. (2014), 'Walnut allergens: Molecular characterization, detection and clinical relevance', *Clinical and Experimental Allergy*, 44(3), pp. 319–41. doi:10.1111/cea.12267.

Črepinšek, Z., Solar, M., Štampar, F., et al. (2009), 'Shifts in walnut (*Juglans regia* L.) phenology due to increasing temperatures in Slovenia', *Journal of Horticultural Science and Biotechnology*, 84(1), pp. 59–64. doi:10.1080/14620316.2009.11512480.

Desta, Z. A. and Ortiz, R. (2014), 'Genomic selection: Genome-wide prediction in plant improvement', *Trends in Plant Science*, 19(9), pp. 592–601. doi:10.1016/j.tplants.2014.05.006.

Dvorak, J., Aradhya, M., Leslie, C., et al. (2016), '2015 Discovery of the causative mutation of the lateral bearing phenotype in walnut', *Walnut Research Report 2015 - California Walnut Board* (year 3), pp. 55–8.

Ebrahimi, A., Zarei, A., Lawson, S., et al. (2016), 'Genetic diversity and genetic structure of Persian walnut (*Juglans regia*) accessions from 14 European, African, and Asian countries using SSR markers', *Tree Genetics & Genomes*, 12(6), p. 114. doi:10.1007/s11295-016-1075-y.

Ebrahimi, A., Zarei, A. K., McKenna, J. R., et al. (2017), 'Genetic diversity of Persian walnut (*Juglans regia*) in the cold-temperate zone of the United States and Europe', *Scientia Horticulturae*, 220, pp. 36–41. doi:10.1016/j.scienta.2017.03.030.

Ertürk, U., Mert, C., Soylu, A., et al. (2014), 'Evaluation of some domestic and foreign walnut cultivars in the conditions of Bursa, Turkey', *Acta Horticulturae*, 1050, pp. 123-9.

Escobar, M. A., Civerolo, E. L., Summerfelt, K. R., et al. (2001), 'RNAi-mediated oncogene silencing confers resistance to crown gall tumorigenesis', *Proceedings of the National Academy of Sciences*, 98(23), pp. 13437-42. doi:10.1073/pnas.241276898.

Escobar, M. A., Leslie, C. A., McGranahan, G. H., et al. (2002), 'Silencing crown gall disease in walnut (*Juglans regia* L.)', *Plant Science*, 163(3), pp. 591-7. doi:10.1016/S0168-9452(02)00164-4.

Esquinas-Alcazar, J. (2005), 'Protecting crop genetic diversity for food security: Political, ethical, and technical challenges', *Nature Reviews Genetics*, 6, pp. 946-53.

Famula, R. A., Eckert, A. and Wegrzyn, J., et al. (2018), 'Association genetics of carbon isotope discrimination and leaf morphology in a breeding population of *Juglans regia* L.', *Tree Genetics & Genomes*, 107(2), 105-14.

Farquhar, G. D., Ehleringer, J. R. and Hubick, K. T. (1989), 'Carbon isotope discrimination and photosynthesis', *Annual Review of Plant Physiology and Plant Molecular Biology*, 40(1), pp. 503-37. doi:10.1146/annurev.pp.40.060189.002443.

Forde, H. I. (1975), 'Walnuts', *in* Janick, J. and Moore, J. N. (Eds), *Advances in Fruit Breeding*, Purdue University Press, West Lafayette, IN, pp. 439-55.

Frutos, D. (2010), 'Bacterial diseases of walnut and hazelnut and genetic resources', *Journal of Plant Pathology*, 92, pp. S1.79-S1.85.

Fulton, A. and Buchner, R. (2006), 'The effect of water stress on walnut trees growth, productivity and economics', UC Farm Advisors Draft Publication, University of California, Tehama County.

Gauthier, M. M. and Jacobs, D. F. (2011), 'Walnut (*Juglans* spp.) ecophysiology in response to environmental stresses and potential acclimation to climate change', *Annals of Forest Science*, 68(8), pp. 1277-90. doi:10.1007/s13595-011-0135-6.

Germain, E. (1990), 'Inheritance of late leafing and lateral bud fruitfulness in walnut (*Juglans regia* L.). Phenotypic correlations among some traits of the trees.', *Acta Horticulturae*, 284, pp. 125-33.

Germain, E. (1997), 'Genetic improvement of the persian walnut (*Juglans regia* L.)', *Acta Horticulturae*, 442, 21-32. Available at: http://www.actahort.org/books/442/442_2.htm.

Gholizadeh, J., Sadeghipour, H. R., Abdolzadeh, A., et al. (2017), 'Redox rather than carbohydrate metabolism differentiates endodormant lateral buds in walnut cultivars with contrasting chilling requirements', *Scientia Horticulturae*, 225, pp. 29-37. doi:10.1016/j.scienta.2017.06.058.

Giovannoni, E. and Fabietti, G. (2013), 'What is sustainability? A review of the concept and its applications', *in* Busco, C., Frigo, M. L., Riccaboni, A., et al. (Eds), *Integrated Reporting: Concepts and Cases that Redefine Corporate Accountability*, Springer, Cham, Switzerland, pp. 21-40. doi:10.1007/978-3-319-02168-3.

Guàrdia, M., Charrier, G., Vilanova, A., et al. (2016), 'Genetics of frost hardiness in *Juglans regia* L. and relationship with growth and phenology', *Tree Genetics and Genomes*, 12(5), p. 83. doi:10.1007/s11295-016-1038-3.

Han, H., Woeste, K. E., Hu, Y., et al. (2016), 'Genetic diversity and population structure of common walnut (*Juglans regia*) in China based on EST-SSRs and the nuclear gene phenylalanine ammonia-lyase (PAL)', *Tree Genetics & Genomes*, 12(111), p. 2. doi:10.1007/s11295-016-1064-1.

Hasey, J., Westerdahl, B. B., Micke, W. C., et al. (2001), 'Yield performance of own-rooted 'Chandler' versus 'Chandler' walnut on Paradox rootstock', *Acta Horticulturae*, 544, pp. 489–93.

Horibe, T., Sagawa, C., Jacobson, A., et al. (2017), Developing a sustainable management strategy to control walnut blight, *Walnut Research Report - UC Walnut Research Conference*.

Kelc, D., Štampar, F. and Solar, A. (2007), 'Fruiting behaviour of walnut trees influences the relationship between the morphometric traits of the parent wood and nut weight', *Journal of Horticultural Science and Biotechnology*, 82(3), pp. 439–45. doi:10.1080/14620316.2007.11512256.

Kluepfel, D., Aradhya, M., McClean, A., et al. (2013), 'Identification and characterization of crown gall resistant *Juglans Germplasm*', *Walnut Research Report 2012 - California Walnut Board*, 1, pp. 413–18.

Lampinen, B., Upadhyaya, S., Udompetaikul, V., et al. (2014), 'A second generation mobile platform for assessing midday canopy photosynthetically active radiation interception in orchard systems', *Acta Horticulturae*, 1058, pp. 105–12.

Lauri, P. E., Delort, F., Germain, E., et al. (2001), 'Factors affecting nut weight in walnut (*Juglans regia* L.) - an analysis of genotypes with contrasting branching patterns P.E.', *Acta Horticulturae*, 544, pp. 265–73.

Leslie, C. A. and McGranahan, G. H. (2014), 'The California Walnut Improvement Program: Scion breeding and rootstock development', *Acta Horticulturae*, (1050), pp. 81–8. doi:10.17660/ActaHortic.2014.1050.9.

Leslie, C., McGranahan, G., Hackett, W., et al. (2016), 'Walnut improvement program 2015', *Walnut Research Report 2015 - California Walnut Board*, pp. 1–38.

Leslie, C., Brown, P. J. McGranahan, G. H., et al. (2018), 'Walnut improvement program 2017', *Walnut Research Report 2017 - California Walnut Board*, pp. 11–42.

López, J. M. (2001), 'Field behaviour of self-rooted walnut trees of different cultivars produced by tissue culture and planted in Murcia (Spain)', *Acta Horticulturae*, 544, pp. 543–6. doi:10.17660/ActaHortic.2001.544.76.

Luedeling, E., Steinmann, K. P., Zhang, M., et al. (2011), 'Climate change effects on walnut pests in California', *Global Change Biology*, 17(1), pp. 228-38. doi:10.1111/j.1365-2486.2010.02227.x.

Luo, M.-C., You, F. M., Li, P., et al. (2015), 'Synteny analysis in Rosids with a walnut physical map reveals slow genome evolution in long-lived woody perennials.', *BMC Genomics*, 16(1), p. 707. doi:10.1186/s12864-015-1906-5.

Lynn, N. (2015), *Cherry Leaf Roll Virus and Blackline Disease in Walnut*, University of California Davis.

Ma, T. and Zhou, C. (2012), 'Climate-associated changes in spring plant phenology in China', *International Journal of Biometeorology*, 56(2), pp. 269–75. doi:10.1007/s00484-011-0428-3.

Mapelli, S., Bertani, A., Malvolti, M. E., et al. (2014), 'Resources of *Juglans Regia* in Uzbekistan: A valuable step along the Silk Road', *Acta Horticulturae*, 1050, pp. 55–62. doi:10.17660/ActaHortic.2014.1032.7.

Marrano, A., Martínez-García, P. J., Bianco, L., et al. (2018), 'Development and validation of a high-density 700K SNP genotyping array for walnut (*Juglans regia*)' (*In preparation*).

Martínez, M. L., Martínez, M. L., Labuckas, D. O., et al. (2010), 'Walnut (*Juglans regia* L.): Genetic resources, chemistry, by-products', *Journal of the Science of Food and Agriculture*, 90(12), pp. 1959–67. doi:10.1002/jsfa.4059.

Martínez-García, P. J., Crepeau, M. W., Puiu, D., et al. (2016), 'The walnut (*Juglans regia*) genome sequence reveals diversity in genes coding for the biosynthesis of non-structural polyphenols', *Plant Journal*, 87(5), pp. 507–32. doi:10.1111/tpj.13207.

Martínez-García, P. J., Famula, R. A., Leslie, C., et al. (2017), 'Predicting breeding values and genetic components using generalized linear mixed models for categorical and continuous traits in walnut (*Juglans regia*)', *Tree Genetics & Genomes*, 13(5), p. 109. doi:10.1007/s11295-017-1187-z.

McGranahan, G. and Leslie, C. A. (1988), 'In vitro propagation of mature Persian walnut cultivars', *HortScience*, 23(1), p. 220.

McGranahan, G. and Leslie, C. A. (2006), 'Walnut tree named "Gillet"', US Patent No. 17,135.

McGranahan, G. and Leslie, C. A. (2009), 'Breeding walnuts (*Juglans Regia*)', *in* Jain, S. M. and Priyadarshan, P. M. (Eds), *Breeding Plantation Tree Crops: Temperate Species*, Springer-Verlag, New York, NY, pp. 249–72. doi:10.1007/978-0-387-71203-1.

McGranahan, G. H. and Leslie, C. (2012), 'Walnut', *in* Badenes, M. L. and Byrne, D. H. (Eds), *Fruit Breeding*, Springer-Verlag, New York, NY, pp. 827–46. doi:10.1007/978-1-4419-0763-9.

McGranahan, G. H., Leslie, C. A. and Woeste, K. E. (1997), 'Backcross breeding walnuts for resistance to the cherry leafroll virus', *Acta Horticulturae*, 422, pp. 121–7.

McGranahan, G., Leslie, C., Hackett, W., et al. (2010a), 'Walnut rootstock "RX1"', UC Patent No. 20,649.

McGranahan, G., Leslie, C., Hackett, W., et al. (2010b), 'Walnut rootstock "VX211"', UC Patent No. 21,179.

Menzel, A., Sparks, T. H., Estrella, N., et al. (2006), 'European phenological response to climate change matches the warming pattern', *Global Change Biology*, 12(10), pp. 1969–76. doi:10.1111/j.1365-2486.2006.01193.x.

Mercuri, A. M., Bandini Mazzanti, M., Florenzano, A., et al. (2013), 'Olea, Juglans and Castanea: The OJC group as pollen evidence of the development of human-induced environments in the Italian peninsula', *Quaternary International*, 303, pp. 24-42. doi:10.1016/j.quaint.2013.01.005.

Mircetich, S. and Matheron, E. (1983), 'Phytophthora root and crown rot of walnut trees', *Phytopathology*, 73, p. 1487–8. doi:10.1094/Phyto-73-1481.

Mircetich, S. M. and Rowhani, A. (1984), 'The relationship of cherry leafroll virus and blackline disease of English walnut trees.', *Phytopathology*, 74.

Mirouze, M. and Paszkowski, J. (2011), 'Epigenetic contribution to stress adaptation in plants', *Current Opinion in Plant Biology*, 14(3), pp. 267–74. doi:10.1016/j.pbi.2011.03.004.

Nainwal, N. C. and Nainwal, K. (2015), 'Lateral bearing walnut cultivars and improved postharvest practices for promotion of export in India', *Acta Horticulturae*, 1088, pp. 435–8. doi:10.17660/ActaHortic.2015.1088.76.

Naser, L., Vahdati, K., Kholdebarin, B., et al. (2010), 'Soluble sugars and proline accumulation play a role as effective indices for drought tolerance screening in Persian walnut (*Juglans regia* L.) during germination', *Fruits*, 65(2), pp. 97–112. doi:10.1051/fruits/20010005.

Pan, A., Sun, Q. and Manson, J. (2013), 'Walnut consumption is associated with lower risk of type 2 diabetes in women', *The Journal of Nutrition*, 143, pp. 512–18. doi:10.3945/jn.112.172171.questionnaires.

Poirier, M., Bodet, C., Ploquin, S., et al. (2006), 'Walnut cultivar performance of cold resistance in south central France', *Acta Horticulturae*, 705, pp. 281–5.

Poirier, M., Lacointe, A. and Améglio, T. (2010), 'A semi-physiological model of cold hardening and dehardening in walnut stem', *Tree Physiology*, 30(12), pp. 1555–69. doi:10.1093/treephys/tpq087.

Poland, J. (2015), 'Breeding-assisted genomics', *Current Opinion in Plant Biology*, 24, pp. 119–24. doi:10.1016/j.pbi.2015.02.009.

Pollegioni, P., Woeste, K., Chiocchini, F., et al. (2017), 'Rethinking the history of common walnut (*Juglans regia* L.) in Europe: Its origins and human interactions', *PLoS ONE*, 12(3), pp. 1–24. doi:10.1371/journal.pone.0172541.

Pope, K. S., Dose, V., Da Silva, D., et al. (2013), 'Detecting nonlinear response of spring phenology to climate change by Bayesian analysis', *Global Change Biology*, 19(5), pp. 1518–25. doi:10.1111/gcb.12130.

Poulose, S. M., Miller, M. G. and Shukitt-Hale, B. (2014), 'Role of walnuts in maintaining brain health with age', *The Journal of Nutrition*, 144(4), p. 561S–6S. doi:10.3945/jn.113.184838.

Preston, J. C. and Sandve, S. R. (2013), 'Adaptation to seasonality and the winter freeze', *Frontiers in Plant Science*, 4, pp. 1–18. doi:10.3389/fpls.2013.00167.

Rodriguez Lopez, C. M. and Wilkinson, M. J. (2015), 'Epi-fingerprinting and epi-interventions for improved crop production and food quality', *Frontiers in Plant Science*, 6, p. 397. doi:10.3389/fpls.2015.00397.

Romero-Suarez, S., Jordan, B. and Heinemann, J. A. (2012), 'Isolation and characterization of bacteriophages infecting *Xanthomonas arboricola* pv. *juglandis*, the causal agent of walnut blight disease', *World Journal of Microbiology and Biotechnology*, 28(5), pp. 1917–27. doi:10.1007/s11274-011-0992-z.

Rouskas, D. and Zakynthinos, G. (2001), 'Preliminary evaluation of seventy walnut (*Juglans regia* L.) seedlings selections in Greece.', *Acta Horticulturae*, 544, pp. 61–72.

Rovira, M. and Aletà, N. (2010), 'Productive behavior of self-rooted and grafted plants in Persian walnut', *Acta Horticulturae*, 861, pp. 215–20. doi:10.17660/ActaHortic.2010.861.28.

Scartazza, A., Proietti, S., Moscatello, S., et al. (2001), 'Effect of water shortage on photosynthesis, growth and storage carbohydrate accumulation in walnut (*Juglans regia* L.)', *Acta Horticulturae*, 544, pp. 227–32. Available at: http://agris.fao.org/agris-search/search.do?recordID=US201301003682.

Schwartz, M. D., Ahas, R. and Aasa, A. (2006), 'Onset of spring starting earlier across the Northern Hemisphere', *Global Change Biology*, 12(2), pp. 343–51. doi:10.1111/j.1365-2486.2005.01097.x.

Smith, E. F. and Townsend, C. O. (1907), 'A plant-tumor of bacterial origin', *Science*, 25(643), pp. 671–3. doi:10.1126/science.25.643.671.

Solar, A. and Štampar, F. (2003), 'Genotypic differences in branching pattern and fruiting habit in common walnut (*Juglans regia* L.)', *Annals of Botany*, 92(2), pp. 317–25. doi:10.1093/aob/mcg137.

Solar, A., Jakopic, J., Veberic, R., et al. (2012a), 'Correlations between *Xanthomonas arboricola* pv. *juglandis* severity and endogenous juglone and phenolic acids in walnut', *Journal of Plant Pathology*, 94(1), pp. 229–35. doi:10.4454/jpp.fa.2012.013.

Solar, A., Jakopic, J., Nour, V., et al. (2012b), 'Prohexadione-CA induces reduction in bacterial blight severity and alteration in phenolic content in walnuts', *Journal of Plant Pathology*, 94, pp. S1.47–S1.52. doi:10.4454/jpp.v94i1sup.008.

Stevens, K. A., Woeste, K., Chakraborty, S., et al. (2018), 'Genomic variation among and within six *Juglans* species', *G3: Genes|Genomes|Genetics*, 8, pp. 1–37. doi:10.1534/g3.118.200030.

Sudarshana, M., Mahmood, N. and Rowhani, A. (2011), 'Characterization of r genes involved in resistance to cherry leaf roll virus in Paradox hybrids', *Walnut Research Report 2010 - California Walnut Board*, pp. 245–51.

Sudarshana, M., Lynn, N. and Rowhani, A. (2012), 'Genomic and horticultural approaches to manage blackline disease in walnuts', *Walnut Research Report 2011 - California Walnut Board*, pp. 323–7.

Taha, N. A. (2011), 'Utility and importance of walnut, *Juglans regia* Linn: A review', *African Journal of Microbiology Research*, 5(32), pp. 5796–805. doi:10.5897/AJMR11.610.

Tulecke, W. and McGranahan, G. (1994), The Walnut Germplasm Collection of the University of California, Davis. A Description of the Collection and a History of the Breeding Program of Eugene F. Serr and Harold I. Forde. Report No. 13. Genetic Resources Conservation Program, Division of Agriculture and Natural Resources, University of California, Davis, CA.

Vahdati, K. (2014), 'Walnut tolerance to abiotic stresses: Approaches and prospects', *Acta Horticulturae*, 1050, pp. 399–406. doi:10.17660/ActaHortic.2014.1050.56.

Vahdati, K., Leslie, C., Zamani, Z., et al. (2004), 'Rooting and acclimatization of in vitro-grown shoots from mature trees of three Persian walnut cultivars', *HortScience*, 39(2), pp. 324–7.

Valdebenito, D., Farías, D., Oyanedel, E., et al. (2017), 'The morphology of a Walnut (*Juglans regia* L.) shoot is affected by its position in the canopy and correlated to the number and size of its fruits', *Scientia Horticulturae*, 220, pp. 303–9. doi:10.1016/j.scienta.2017.03.046.

Vinson, J. A. and Cai, Y. (2012), 'Nuts, especially walnuts, have both antioxidant quantity and efficacy and exhibit significant potential health benefits', *Food & Function*, 3(2), pp. 134–40. doi:10.1039/C2FO10152A.

Warren, G. J. (1998), 'Cold stress: Manipulating freezing tolerance in plants', *Current Biology*, 8(15), pp. R514–16. doi:10.1016/S0960-9822(07)00335-1.

Wezel, A., Casagrande, M., Celette, F., et al. (2014), 'Agroecological practices for sustainable agriculture. A review', *Agronomy for Sustainable Development*, 34(1), pp. 1–20. doi:10.1007/s13593-013-0180-7.

Wilson, A. J., Réale, D., Clements, M. N., et al. (2010), 'An ecologist's guide to the animal model', *Journal of Animal Ecology*, 79(1), pp. 13–26. doi:10.1111/j.1365-2656.2009.01639.x.

Woeste, K., McGranahan, G. and Bernatzky, R. (1996), 'The identification and characterization of a genetic marker linked to hypersensitivity to the cherry leafroll virus in walnut.', *Molecular Breeding*, 2(3), pp. 261–6.

Woeste, K., Burns, R., Rhodes, O., et al. (2002), 'Thirty polymorphic nuclear microsatellite loci from black walnut', *The Journal of Heredity*, 93(1), pp. 58–60.

Wu, J., Gu, Y. Q., Hu, Y., et al. (2012), 'Characterizing the walnut genome through analyses of BAC end sequences', *Plant Molecular Biology*, 78(1–2), pp. 95–107. doi:10.1007/s11103-011-9849-y.

Wu, L., Piotrowski, K., Rau, T., et al. (2014), 'Walnut-enriched diet reduces fasting non-hdl-cholesterol and apolipoprotein b in healthy caucasian subjects: A randomized controlled cross-over clinical trial', *Metabolism: Clinical and Experimental*, 63(3), pp. 382–91. doi:10.1016/j.metabol.2013.11.005.

Xu, Y. and Crouch, J. H. (2008), 'Marker-assisted selection in plant breeding: From publications to practice', *Crop Science*, 48(2), pp. 391–407. doi:10.2135/cropsci2007.04.0191.

Zeven, A. and Zhukovskiĭ, P. M. (1975), *Dictionary of Cultivated Plants and their Centres of Diversity, Excluding Ornamentals, Forest Trees, and Lower Plants*, Centre for Agricultural Publishing and Documentation, Wageningen. Available at: https://core.ac.uk/download/pdf/29387092.pdf

# Chapter 10

## Advances in cultivation of walnuts

*D. L. McNeil, University of Tasmania, Australia*



## 1 Introduction

Walnuts are a major world crop which have had a considerable increase in production in recent years. Global production statistics have been given in detail in McNeil (2010) and thus this chapter will concentrate on updating the situation. Global production has increased by 4.14 fold between 1985 and 2014 (Table 1).[1] Most of that increase (Fig. 1) has come from expanded areas in less developed traditional growing regions (e.g. China) as well as in new developed (e.g. Central America, Australia) and new less developed (e.g. Chile to reach 109000 tonnes in 2018; FAS, 2017) growing regions. The nature of development has been diverse among regions and consequently the issues facing each region are diverse. As examples, the traditional major growing area of the west coast of the United States has not had large area increases with most of its production increase arising from yield increases. California along with Oceania, Central and South America all have production systems based on large intensively managed orchards, mechanization, Mediterranean climate, irrigated, high-yielding US varieties, well-developed cool storage and processing facilities targeting high-quality export products (UC Davis, 2017). In addition to this UC Davis web extension site there are numerous written extension manuals supporting these growers (McNeil et al., 2014; McNeil and Evans, 2014; Adem, 2009, 2010; Ramos, 1997). Over this expansionary phase

1 These data are from FAOSTAT (2017). The US FAS (2017) production data for China and other developing areas are often much lower (e.g. China 710 000 tonnes, 2014 compared with FAO 1 600 000). The differences are largely accountable due to differences over official estimates and nuts that do not enter commerce.

http://dx.doi.org/10.19103/AS.2018.0042.18

**Table 1** Walnut (in-shell equivalent) global production characteristics and changes between 1985 and 2014 by region

| | Proportional increase 2014/1985 | | | % world production | |
|---|---|---|---|---|---|
| Area | Area | Production | Yield | 1985 | 2014 |
| World | 2.55 | 4.14 | 1.63 | 1.000 | 1.000 |
| China | 8.01 | 13.14 | 1.64 | 0.146 | 0.463 |
| Asia minus China | 2.25 | 4.20 | 1.87 | 0.220 | 0.223 |
| Northern America | 1.62 | 2.61 | 1.61 | 0.238 | 0.150 |
| Central America | 15.26 | 21.24 | 1.39 | 0.007 | 0.036 |
| South America | 3.40 | 4.35 | 1.28 | 0.021 | 0.022 |
| Eastern Europe | 0.66 | 1.19 | 1.80 | 0.184 | 0.053 |
| Southern Europe | 0.74 | 0.83 | 1.12 | 0.124 | 0.025 |
| Western Europe | 0.93 | 1.22 | 1.31 | 0.058 | 0.017 |
| Northern Africa[a] | 2.34 | 18.41 | 7.87 | 0.002 | 0.011 |
| Oceania | - | 27.62 | - | 0.000 | 0.001 |

[a] Northern and total Africa are the same.
*Source*: FAO (2017).

(1985–2014) European production areas have tended to decline with increases in yields balancing out reducing areas (Fig. 1, Table 1), thus maintaining production at fairly steady levels resulting in significantly reduced proportion

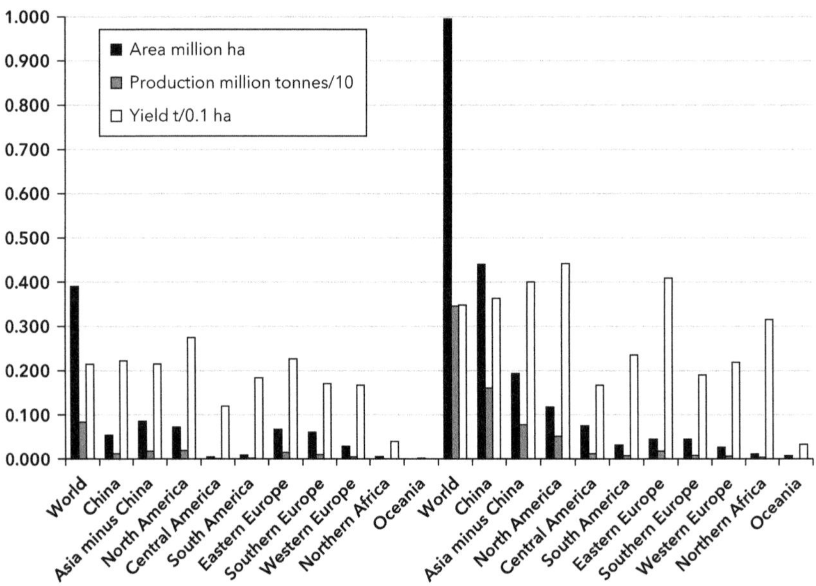


**Figure 1** FAO (2017) regional production, area and yield of walnuts in 1985 at the start of the major expansionary phase (left side) and most recent data from 2014 (right side).

of global production. European and West Asian orchards are shifting from traditional production methods and local varieties to intensive agronomy and newer regional or global varieties (http://www.frutas-hortalizas.com/Fruits/About-Walnut.html). For example FAS (2017) indicates 25% of Ukraine production has shifted from declining traditional small farms to commercial plantations in recent times which accounts for a production increase of 20% between 2012/13 and 2015/16. Major growth has occurred in Asian, particularly Chinese, production. Here production has largely depended on expansion using local high-quality but often still lower-yielding varieties (see nut photos in Zhi et al., 2009) into a large range of environments. For example over 33 000 ha has been planted in wet, subtropical mountain areas in Guangxi province, China (McNeil and Zeng, 2013). There, trees are often on rocky (often limestone) slopes, in small village plots inter-planted with companion crops (e.g. mulberry) using local and Yunnan bred *J. regia* × *J. sigillata* hybrids with local government extension agents providing fertilizer, trees and information. However, Chinese production systems, research and agronomy is substantially improving (e.g. Sun et al., 2017) meaning China is producing higher-quality walnuts in significant volumes with higher yields. Figure 2 shows a new corporate orchard just after planting in Qinghai. It covered 30 ha, was irrigated and intensively managed using best available grafted varieties and is not unusual for China. In the Northeast and West of China plantations are often larger with widely used extensively tested Chinese varieties on flatter, more fertile, loess soils. In other Asian regions (e.g. Kyrgyz Republic; McNeil, 2013b) production is based on disaggregated collection from native walnut forests (increasingly with organic certification) or from predominantly small farm woodlots. However, throughout the area traditional methods are being

**Figure 2** Walnut: protection of young trees from winter frost damage in Guide County, Qinghai province, China. Courtesy: Hilary Ma.

replaced by larger (often communal) operations and new varieties with more advanced agronomy (Singh et al., 2012).

Increased demand has arisen from population growth plus higher per capita consumption. The higher consumption arises in part from expanded wealth of traditional consumers who see the crop as an enhancement of their diet associated with increased wealth (particularly in China). In addition, particularly in the west, increased demand is associated with the recognition of the proven health benefits (Guasch-Ferré et al., 2017; Bamberger et al., 2017) associated with consumption of walnuts. Consumption has been made easier as well by the consequent expansion in recipes and methods of consumption (McNeil, 2014a). Figure 3 shows that from 1961 till about 2000 global consumption per capita of walnuts remained steady between 0.16 and 0.20 kg/capita.year. Thus, expanded demand over this period was principally through global population growth. From 2000 till 2013 global demand expanded from 0.2 to 0.5 kg/capita.year. This trend was observed across a broad range of countries and included developed, less developed, European, Asian, walnut producers and non-producers (Table 2). Over the same period US farm gate prices rose from 0.62$/lb in 2000 to 1.85$/lb in 2013 (falling back in 2015 to 0.81$/lb as US production rose following reduced effects of the Californian drought) (USDA NASS figures quoted in http://www.capitalpress.com/California/20161020/californias-multi-billion-dollar-nut-boom-keeps-going). However, prices subsequently recouped very quickly back to previous peak prices (https://data.ers.usda.gov/reports.aspx?programArea=fruit&stat_year=2009&top=5&H

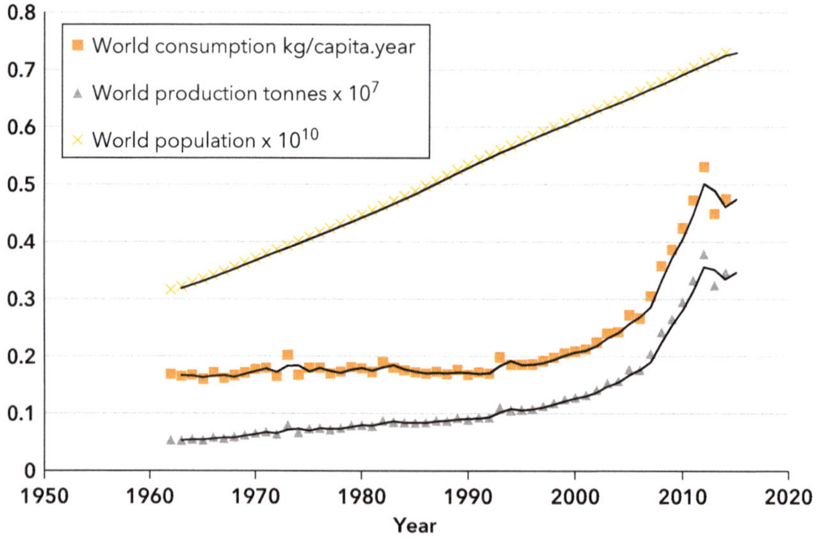


**Figure 3** FAO (2017) global walnut production, consumption and population by year.

**Table 2** Growth in walnut consumption from 1985 base for a variety of countries across the world including major producers and import-only countries

| | | Per capita consumption relative to 1985 base | |
|---|---|---|---|
| Country | Major producer | 1999 | 2013 |
| Japan | No | 3.95 | 5.16 |
| United Kingdom | No | 0.75 | 1.15 |
| Canada | No | 0.64 | 1.02 |
| Saudi Arabia | No | NA | 1.08 |
| China | Yes | 2.29 | 12.43 |
| Turkey | Yes | 0.91 | 1.44 |

NA: Not applicable.
*Source*: FAO (2017).

ardCopy=True&RowsPerPage=25&groupName=Tree%20nuts&commodityName=Walnuts&ID=17851#Pbaae46c90e2d4c6dba739eb13126c478_2_477). Rising prices associated with rising production and consumption thus indicate a demand-driven desire for walnuts based on increased desirability.

The result of this expansion has been the creation of a diverse group of production regions with often highly specific issues requiring research and development. Some regions have considerably expanded production areas into new walnut habitat (e.g. Guangxi; McNeil, 2014a) or have introduced new production systems in old habitats (e.g. Ukraine). Others have greatly increased productivity of existing systems (e.g. the United States), or completely new areas have started producing (e.g. Chile, New Zealand, Australia). These new production methods and areas are often associated with limited experience in agronomy and production methods and little specifically relevant published material to go by. In other areas, emergence of new pests (e.g. longicorn beetle in South and Southwest China), diseases and physiological issues (e.g. increased sunburn arising from global warming in the United States; FAS, 2017) not previously encountered in production require attention. There is considerable need for supply chains to require greater emphasis on environmental concerns, quality, safety and marketing as greater amounts of often more variable product are being produced and supplied to more demanding consumers with growers seeking to maintain or improve prices. From present suppliers this means highest quality is needed to maintain markets. From new, less sophisticated, growers higher quality is needed for market access. For Southern Hemisphere growers the ability to ship through the wet humid tropics to Northern Hemisphere markets is essential. From high-yielding intensive orchards there is a need to use ever more sophisticated methods to increase yields (e.g. molecular breeding; Neale, 2017). For less sophisticated producers there is a need to improve harvesting, drying, storage and distribution methods to achieve high, consistent quality and higher prices.

The following sections will look at the situation and advances occurring or needed with respect to these major issues both globally and regionally.

## 2 Key issues for a diverse industry

Walnut (genus *Juglans*) consist of a large number of species with a number of diverse uses. Their wood has been used for centuries in China and Europe (Avanzato, 2014). Their nuts have been used for consumption, decoration and playthings (Wang and He, 2006). Walnut tissues (e.g. leaves, bark) are rich in bioactives and the nuts may be used for carbonization, fuel and other uses. This review will primarily concentrate on the use of *Juglans regia* (and some hybrids) for producing edible kernels. However, extraction of the maximum value from waste such as shells (used in California for burning and heat generation) and land (e.g. intercropping in China) can considerably alter the economics of production.

Walnut fruit production is a highly diverse global industry with many issues specific to regions or modes of production and other issues in common across wide parts of the globe. This chapter will not cover all aspects of walnut production but rather look particularly at issues which fit into one or more of the following categories: (1) widespread, (2) new and emerging, (3) locally critical, (4) require novel solutions, (5) are at the scientific forefront, (6) harmonize the global market from producer to consumer. Breeding, consumer health and safety, pest and disease control are more extensively covered elsewhere but this chapter will briefly look at parts of these issues where they interact with other physiological, agronomical and production issues.

### *2.1 Bacterial blight control*

Walnut bacterial blight is caused by the bacterium *Xanthomonas arboricola* pv. Juglandis. It is the main disease affecting walnut production worldwide and has been for a considerable time (Miller, 1934; Frutos, 2010). Miller (1951) wrote '*Walnut blight is the most widespread and destructive disease of the Persian (English) walnut in the Pacific Northwest. It causes greater annual crop loss than all of the other diseases of walnuts combined*'. This remains true today of any area with a wet spring where the disease is not well controlled (Lang and Evans, 2010) where losses can exceed 80% (Miller, 1934; Bandi et al., 2015). Occasionally other nut and leaf diseases are mistaken for blight, chiefly walnut anthracnose (*Gnomonia leptostyla*) which can cause brown spots in the spring and lead to premature leaf drop in the autumn. The causative agent for walnut blight has been sequenced including genes associated with virulence, Cu and antibiotic resistance (Retamales et al., 2017). There is considerable interaction between blight infection, water and fertilizer management and nut quality.

Management of the disease has been based on (1) escape, either by avoiding areas with wet springs (e.g. Wang et al., 2014; Xinjiang, China, Duan et al., 2017), or by using late leafing cultivars (e.g. Solar, 2006) and (2) combining detailed understanding of the disease epidemiology with targeted use of predominantly copper, bactericidal and antibiotic sprays (Ninot et al., 2002; Jenkins et al., 2010) with weather/phenology-based decision models (Lang and Evans, 2010) reducing spray dosage. However, in old orchards there is still considerable build-up of bacterial copper and antibiotic (Retamales et al., 2017) resistance as well as copper residues in the soil and on the nuts (Epstein and Bassein, 2001). Blight-resistant genotypes are sought (Frutos, 2010) to allow expansion of walnut production areas to new environments such as has happened in Guangxi. However, outside of China few claims of highly resistant useful cultivars are made. Screens for tolerance of 25 000 seedling trees in Turkey (Akca et al., 2015) and 11 selected genotypes in Hungary (Bandi et al., 2015) found only moderate resistance often not holding up in other regions or conditions (Arrieta and Diaz, 2007). Many productive and high-quality walnuts including new releases do not have high levels of resistance (e.g. UC Davis releases; Ivanhoe, a 2010 release is susceptible and Durham (2016) and Solarno (2013) have only low susceptibility, http://ccfruitandnuts.ucanr.edu/files/239180.pdf).

There is thus a need to develop alternative control methods.

### 2.1.1 Hybrids

A new possibility for walnut blight control is the use of walnut hybrids. The disease can affect a wide range of *Juglans* species and hybrids (Frutos, 2010). McNeil (2016) showed for *J. sigillata* × *J. regia* hybrids grown in Australia and Guangxi province, blight incidence was much less than predicted by a disease incidence model (Lang and Evans, 2010) that worked well for *J. regia* and *J. sigillata* in the same locations. Thus some *J. sigillata* × *J. regia* hybrids may contain useful resistance to walnut blight. Xi et al. (2007) have also suggested resistance in these hybrids. *J. sigillata* should be included into advanced breeding work for management of blight and opening up new production environments.

### 2.1.2 Bacteriophages

McNeil et al. (2001) and Romero-Suarez (2006) were the first to use bacteriophages (viruses that infect and can kill specific bacteria) as a means to control blight in walnut. Their efforts were based on two observations: (1) bacteriophages which attack *X. arboricola* pv. Juglandis existed in large numbers in orchard soil and canopies globally and (2) the normal collapse of bacterial numbers in the buds during the walnut growing season was accompanied by a massive increase in phage numbers in the canopy. They thus characterized phages and tried to use them as biocontrol agents.

Field trials found survivability on the leaves was inhibited by sunlight with the source and type of phage and protectants being important for getting phages to survive and function. Soil phages (generally lytic) were less successful in the field with poorer survivability than ones isolated from the canopy (generally lysogenic) consistent with the review of Jones et al. (2012). Romero-Suarez et al. (2012) further characterized phages from New Zealand while Dömötör et al. (2016) characterized *X. arboricola* pv. Juglandis phages from Hungary for further developing the biocontrol option. Kocharunchitt et al. (2009) found that several phages were needed to cover all strains with biocontrol of *Salmonella* and temporary acquired resistance could occur that limited phage utility. There may thus be significant potential in the use of bacteriophages to control walnut blight. However, considerably more effort will be needed to commercialize a viable product.

### 2.1.3 Conclusions

Bacterial blight is one of the most critical limitations for walnut production worldwide. It reduces yields and adversely affects quality of nuts and destroys new growth. Physiological/weather-dependent spraying with active copper and antibiotics can control the disease in situations where the disease pressure is not extreme. Appropriate phenology can also lessen disease pressure. However, ultimately control must be through breeding for reduced susceptibility. The most successful method at present is to use *J. regia* × *J. sigillata* crosses. Additionally, evidence exists for natural control via bacteriophages and development of this method holds considerable opportunity.

## 2.2 Use of molecular and biotechnological methods

McGranahan and Leslie (2012) have reviewed walnut breeding methods and objectives. Selection and crossing of walnuts has been, and still is, the key means for improvement of walnut. Significant natural stands for selection of *J. regia* still exist in Europe and Asia. For example Akca et al. (2015) selected among 25 000 seedlings in Turkey for improved disease resistance, tree type, quality and yield. McNeil and Savage (2001) report trials comparing local and global selections in New Zealand and UC Davis reports on the performance of latest walnut releases in California (http://fruitandnuteducation.ucdavis.edu/fruitnutproduction/Walnut/Walnut_Cultivar_Table/). Martínez-García et al. (2017) have determined breeding values for 15 000 trees in the UC Davis breeding programme. However, in addition, use of molecular and biotechnological methods is now well embedded into walnut development programmes globally. These methods include molecular characterization of the whole genome and particular genes, marker development, characterization of pests' genomes, comparison of DNA-based relationships and creation of genetic modifications as well as creating in vitro protocols to support these methods and for direct use in plant development. Associated methods used in transcriptome, proteome and metabolome analysis are also being developed. Zhou et al. (2017) have published a full transcriptome analysis of both *J. regia* 'Yuanfeng', an elite walnut cultivar in China, and *J. cathayensis*, a close relative from China, which possesses some superior traits.

To date there has been little work on proteomics in walnuts. Of particular significance are the allergen proteins which have been characterized both before and after roasting, Downs et al. (2016) reported that '*individual walnut allergens respond differently to thermal processing, and the detection of these proteins by LC-MS/MS is dependent on the protein in question, its susceptibility to proteolytic digestion, the degree of thermal processing, and the sample preparation methodology*'. Most enzyme (e.g. Araji et al., 2014; Zhu et al., 2015b) work relates more to individual enzymes than the large-scale overall proteomic analysis.

Metabolomics research primarily deals with the metabolome during fruit maturation (Rao et al., 2016; Araji et al., 2014) and the metabolites (e.g. phenolics, quinones; and fatty acids; tocopherols) that may affect the health profile (Savage et al., 1999, 1998; McNeil and Savage, 2000).

### 2.2.1 Molecular breeding

An excellent example of a walnut molecular breeding programme is provided by Neale (2017) for the UC Davis Walnut Improvement Program 2013–18 (WIP). They sequenced the Chandler genome (Martínez-García et al., 2016) then 29 other genomes including five additional species. These data are now being used to investigate the molecular basis of various pathways such as water use efficiency, non-structural polyphenols and shell/kernel characteristics to enable use of molecular breeding technologies for breeding improved varieties. To enable this, a high-throughput sequencing method for genotyping thousands of trees in the Walnut Improvement Program (WIP) was developed. The genotype data and WIP trait data are then brought together using statistical and computing methods into a genetic analysis to identify for breeding molecular markers associated with the 28 yield, nut, phenology and tree architecture traits routinely measured in the WIP thus uniting the quantitative genetics analysis of the same traits (Martínez-García et al., 2017) into a global breeding collaboration. Other molecular programmes also exist, for example sequencing *Juglans* spp. chloroplasts (Hu et al., 2017a) and determining the (slow) rate of genetic change (Luo et al., 2015) of walnuts.

Genetic modification (GM) and improvement of walnuts is possible provided genes are identified for use and in vitro propagation can be easily carried out. Payghamzadeh and Kazemitabar (2011) reviewed the methods available indicating micro-propagation is reasonably easy for walnut trees. Protocols have considerably improved since then (e.g. Licea-Moreno et al., 2015; Asayesh et al., 2017a,b). Leslie et al. (2015) have provided a protocol for *Agrobacterium*-mediated transformation of somatic embryos with selectable markers that has taken genetically modified trees to the field. McGranahan and Leslie (2012) cite examples of using RNAi gene silencing to protect against crown gall disease and Bti insertion for insect protection. Goharrizi, et al. (2016) have used bacterial fld genes to increase salinity tolerance in walnut somatic embryos, Yang et al. (2017) have identified abscisic acid-dependent stress response genes in walnuts and Xu et al. (2017) have identified a vacuolar ATPase gene involved in mannitol-induced osmotic stress tolerance. These and others are all targets for future GM but may have to await greater public acceptance before commercial varieties are developed. Potentially, acceptance may be greater for rootstock genes; however, Liu et al. (2017) found there was evidence for translocation of transformed mRNA products from rootstocks to the scion.

### 2.2.2 Marker use

Less problematically than GM, a wide variety of molecular markers have been developed for walnuts. These include isozymes, RFLP, RAPD, ISSR and SSR (McGranahan and Leslie, 2012). Marker-assisted backcross breeding also has been used to develop blackline-resistant scions for more than 20 years. More recently microsatellite markers have been used to select for early maturity (Ebrahimi et al., 2017). SCAR markers for locating specific genes directly (Cao and Niu, 2015) and SLAF-seq markers have located anthracnose resistance QTLs (Zhu et al., 2015b). Transcriptome analysis has been used to characterize the red-leaf characteristic of walnuts and identify a large suite of genes involved with pigment expression (Chen et al., 2017). Once identified, genes of interest can be sequenced and compared to alternative sources of the genes, for example *PIP2* and *GAI* genes (Mohseniazar et al., 2017) to study biochemistry, for early selection of variants and carriers in traditional breeding or direct GM. Sequencing as well as EST-SSRs, SSRs and microsatellites have also been used to construct *Juglans* spp. relationship trees (Hu et al., 2017b; Han et al., 2016; Wang et al., 2015) and demonstrate natural hybrids of *J. regia* and *J. cathayensis* in the wild (Shu et al., 2016).

### 2.2.3 Walnut pest molecular genetics

Walnut pest and disease organisms' germplasm has also been molecularly characterized (Section 2.1 for blight) to better understand epidemiology and potentially control the pest or disease more appropriately. Wang and Tang (2017) have produced the complete mitochondrial sequence of *Paleosepharia posticata*, an important walnut leaf pest to determine the relationship of the pest to related species and confirm how many races are involved in different regions. Zerillo et al. (2014), using SNPs, examined the population structure of *Geosmithia morbida,* the causal agent of thousand cankers disease to infer population structure and the recent evolution of the pest in California. Chakraborty et al. (2016) used deep RNA-Seq profiling to characterize the genus of microbes associated with walnut trees in California.

### 2.2.4 Conclusions

Molecular methods have become mainstream in walnut breeding, stress breeding, pest management and population understanding. However, ultimately these methods must result in acceptable improvements in varieties, product quality or agronomic management of crops to prove their worth as they can be expensive, particularly if there is not a clear goal associated with their use. This is now starting to occur and should accelerate in the future. What has not yet occurred for walnuts has been a substantial amount of work in proteomics which is needed in such areas as allergens and enzyme production.

## 2.3 Nutraceuticals and extractives

McNeil (2013b) and Hayes et al. (2016) give reviews of walnut composition and human health aspects. Nuts for life (https://www.nutsforlife.com.au/latest-research/) gives a continuously updated list of global research on all aspects of nut consumption (including walnuts) and health issues. McNeil (2013b) gives detailed analyses of walnut kernels for macronutrients, minerals, vitamins and lipids. They are rich sources of energy (27.4 KJ/g), protein (~15%) and fats (~65%). They are low in sugars (<3%) and rich in fibre (~7%). 100 g provides greater than the RDI for tocopherols, Mn and Cu. They are also rich in Fe, Mg and P. Their fat composition can vary considerably among locations and varieties (Zwarts et al., 1999) but is dominated by polyunsaturated (~0.47 g/(g kernel), mostly 18:2) and monounsaturated (~0.09 g/(g kernel), mostly 18:1) oils with low levels of saturated oils (~0.06 g/(g kernel), mostly 16:0). This composition contributes to their reduced storability compared with other nuts. They are rich in arginine (14.4%), glutamic acid (17.9%) and aspartic acid (11.6%). They are also rich in most of the essential amino acids. They contain biologically active allelopathic compounds (and dyes) such as juglones (Soderquist, 1973). They are used in traditional medicine for their health giving properties attributed to their appearance, taste and richness in energy and proteins (Avanzato, 2014). Recently there has been considerable interest in walnuts for their scientifically validated claims of health-giving properties (e.g. blood cholesterol reduction; Savage et al., 2001) for three loose groups of compounds: (1) fatty acids, (2) anti-oxidants and (3) others.

### 2.3.1 Nutritional benefits

McNeil (2013b) summarized the scientifically demonstrated health benefits of walnuts including lowered cholesterol and blood lipids, better weight management, reduced inflammation and improved heart health across unselected and special populations. Considerable additional research has been completed on nuts since then which has been reviewed by Mayhew et al. (2015). A significant amount of research has also been done specifically on walnuts since then which is summarized in Table 3 and confirms the health benefits of walnuts.

### 2.3.2 Bioactives

Walnut tissues have been explored for, and found to be rich in, biologically active materials. Their involvement in human health through consumption of kernels (particularly polyphenols and allergens) has been discussed elsewhere including a review of polyphenols (Sánchez-González et al., 2017). Male inflorescences (consumed in China) and their bioactive components and nutritional profiles have also been investigated (Zhang et al., 2017a). Outside the human body Juglone, plumbagin (naphthoquinones) and other phenolics are present in a range of walnut tissues including green husks and bark. They are bactericidal, have considerable activity against phenoloxidase and potential value as insecticidal compounds (Hu et al., 2017c) and have anti-cancer activities and potential (Sreelatha et al., 2014; Spaccarotella et al., 2008; Schlörmann et al., 2017). In excess of 100 bioactive compounds have been found and isolated in significant amounts from various *Juglans* species and tissues (Grace et al., 2014; Wang et al., 2017). What is now needed is to take these chemicals from the experimental to the commercial stage.

### 2.3.3 Safety

The safety of walnuts falls into three main categories: (1) physical and chemical contamination, (2) allergenic issues and (3) microbiological contamination. All of these have recently been reviewed by McNeil (2013b). Even with their high PUFA value and low smoke point (Bockisch, 1998), walnuts degrade slowly in

**Table 3** List of recent trials specifically demonstrating human health benefits of walnuts

| Health benefit | Sub-area | Publication | Conclusion with respect to walnuts |
|---|---|---|---|
| Brain health | Cognitive capability | Tian et al. (2016) | Improved learning and memory function in college students |
| Brain health | Repair of function (rat) | Haider et al. (2017) | Impairments caused by oxidative stress in rats significantly restored by pre-administration of walnut |
| Brain health | Trial design | Rajaram et al. (2017) | Trial results will provide information on benefits of walnuts on cognition |
| Brain health | Review | Poulose et al. (2014) | Reviews evidence for the beneficial effects of consuming a walnut-rich diet on brain health |
| Brain health | Review | Poulose et al. (2014) | Polyphenolic compounds in walnuts reduce the oxidant and inflammatory load on brain cells and improve interneuronal signalling |
| Cancer benefits | Prostate biomarkers | Spaccarotella et al. (2008) | Improves biomarkers of prostate and vascular status |
| Cancer benefits | Gene expression | Schlörmann et al. (2017) | Chemopreventive effects on risk for colon cancer by inducing detoxification genes (CAT, GSTT2), growth inhibition and apoptosis in colon adenoma |
| Diabetes benefits | Diet consequences | Njike et al. (2016) | Walnut inclusion in the diets of adults at risk for diabetes led to an increase in intake of other healthful foods |
| Diabetes benefits | Blood pressure | Katz et al. (2012) | Improves endothelial function in overweight adults with visceral adiposity |
| Diabetes benefits | Blood sugar | Zibaeenezhad et al. (2016) | Consumption of walnut oil improved blood glucose level |
| Eye health | Trial design | Rajaram et al. (2017) | Trial results will provide information on benefits of walnuts on age related eye health |
| Gut health | Biota changes | Byerley et al. (2017) | Enriched the microbiota for probiotic-type bacteria |
| Overall health | Review | Hayes et al. (2016) | Overall review and confirmation of the benefits of walnut consumption |

*(Continued)*

**Table 3** (*Continued*)

| Health benefit | Sub-area | Publication | Conclusion with respect to walnuts |
|---|---|---|---|
| Cardiovascular health | Walnut component effects | Berryman et al. (2013) | Oil favourably affected endothelial function, whole walnuts increased cholesterol efflux. Explains in part the cardiovascular benefits of walnuts |
| Cardiovascular health | Review | Guasch-Ferre et al. (2017) | Nut consumption including walnuts improves cardiovascular health across a range of assessment factors |
| Oxidative stress benefits | Blood composition | Haddad et al. (2014) | Walnut meal acutely increased postprandial γ-tocopherol and catechins and attenuated some measures of oxidative stress |
| Oxidative stress benefits | Review | Sánchez-González et al. (2017) | Assess potential effect of walnut polyphenols in health maintenance and disease |
| Plasma lipids | Review | del Gobbo et al. (2015) | Reduces cholesterol, total and LDL, ApoB and triglycerides. Dose rather than nut type most important |
| Plasma lipids | Blood composition | Bamberger et al. (2017) | Improved the lipid profile independent of the recommended macronutrient replacement |
| Weight management | Available energy | Baer et al. (2016) | Provide less energy than expected |
| Weight management | Blood composition | Rock et al. (2017) | Attenuated the postprandial insulin response, but gave similar satiety to cottage cheese |
| Weight management | Weight loss | Smith et al. (2015) | Switching walnut for CHO caused weight loss, other foods for example meat and cheese caused weight gain |
| Weight management | Metabolic syndrome | Selma et al. (2016) | Consuming walnuts reduces markers for metabolic syndrome |
| Weight management | Stability | Katz et al. (2012) | Addition of walnuts to the diet does not lead to weight gain |

shell (Osterberg et al., 2001) but break down more rapidly out of shell (Savage at al., 1998, 1999, 2001) or when processed (Stark et al., 2000; Vanhanen and Savage, 2006). They thus can contain rancid oils and breakdown products that can be detrimental to health. However, they generally constitute a small part of the diet (mean global per capita consumption is 0.4% of our diet, derived from Fig. 3), thus they have relatively little adverse cumulative effect on human health even if rancid or slightly contaminated with agrichemicals or mycotoxins.

Walnut processing typically consists of three levels (McNeil, 2013b): (1) basic hand harvested, sun-dried in the open with hand shelling giving opportunities for environmental contamination (typical of developing areas e.g. Kyrgyz Republic), (2) machine harvested and processed, forced air dried (e.g. the United States). Here there are opportunities for metal and glass contamination but these are minimized by extensive specialist equipment (air separators, gravity tables, colour and sound sorters), (3) somewhere between 1 and 2 (e.g. much of Chinese production). The most common physical issues are shell pieces particularly with mechanical shelling that can lead to tooth damage and choking hazards. Shells can be removed by sound sorters and other cleaning equipment (McNeil and Evans, 2014). Thus walnuts often have warnings (McNeil, 2013a,b) for shell (62% of tested packs for the United Kingdom, 0% France and Hong Kong) and choking (33% UK, 33% France, 13% Hong Kong).

Chemical contamination in walnuts generally is low due to relatively limited chemical use. However, there is considerable organic production (both wild forests and horticultural orchards) and sale of organic walnuts (McNeil, 2013b) generally attracts price premiums (http://www.crackernut.co.nz/kerake l.html). Copper contamination (from blight control Section 2.1) can be an issue through environmental build up with time as it has generally washed off the fruits before harvest.

McNeil (2013b) has detailed the issues surrounding allergens in walnuts. They are one of the more common causes of anaphylaxis (reviewed by McWilliam et al., 2015) which can be fatal (Matti, 2017). Traditionally allergies to peanuts and all tree nuts have been considered as strongly correlated. However, Couch et al. (2017) have indicated that co-allergy between a variety of tree nuts and peanuts may be less common than previously believed. There is thus interest in the use of immunotherapy to suppress allergen sensitivity both for the particular nut tested and others. The 'Walnut Oral Immunotherapy for Tree Nut Allergy (WOIT)' trial is determining if walnut immunotherapy (http s://clinicaltrials.gov/ct2/show/NCT01546753) alters the immune system and reduces tree nut allergy. While relatively rare, recalls for undeclared walnuts in food products still occur (https://www.fda.gov/Safety/Recalls/ucm575054 .htm) requiring research into detecting walnut allergens using real-time PCR of foods (Linacero et al., 2016), reducing their effects in susceptible individuals

(WOIT trial) and eliminating them from walnuts. This is important for the safety of consumers as well as to enhance the reputation of the industry.

Microbiological contamination falls into two main categories: (1) toxins (principally aflatoxin) and (2) disease-causing microbes (principally, *E. coli, Salmonella* and *Listeria*). The regulatory environment frequently requires testing for these, especially for materials arriving from less advanced production environments. Less frequent than allergen recalls for walnuts, McNeil (2013b) lists a number of recalls. See recent US recalls at: https://www.fda.gov/Safety/Recalls/ArchiveRecalls/default.htm.

Simulated storage for 8 days and drying will reduce *Salmonella, E. coli* and *Listeria* populations by 3–5 log units (Frelka et al., 2016). In spite of this high reduction in bacterial numbers and studies indicating walnuts are not good hosts for bacteria (McNeil, 2013b), concern over microbiological contamination persists. Thus removal of grazing animals from orchards well before harvest is recommended though wild fauna are also capable of causing faecal contamination. Walnut samples (~4000) from California handlers across four harvests (2010–13) were tested for *Salmonella* and *E. coli* giving 4, low level, positive samples for *Salmonella* and no positive *E. coli* samples (Davidson et al., 2015). Zhang et al. (2017b) found low levels of *Salmonella* in 8/628 retail samples of shelled walnuts collected throughout the United States. This was similar to other raw nuts and similar foods at retail outlets but much higher than when leaving the handlers, suggesting issues during transfer to retail outlets. However, the situation for walnuts grown elsewhere may not be as good. Riyaz-Ul-Hassan et al. (2003) found high bacterial loads (30% > $10^6$ CFU/g) in Indian walnuts including *E. coli* and *Salmonella*.

Aflatoxin contamination can be an important issue for walnuts particularly with more now coming from developing regions. Walnuts usually come into contact with the ground during harvest and they may be contaminated there with *Aspergillus*. McNeil (2013b) indicated many samples taken from India and Turkey containing high levels of aflatoxins though surveys in Australia and the United States (principally locally grown nuts) showed very low levels. Yilmaz (2017) found 64% of retail walnut kernels in Sakarya City centre were contaminated with aflatoxin. One specimen of walnut (10.26–5.06 μg/kg) exhibited the maximum contamination levels of AFT and AFB1, respectively. This exceeded the recommended AFB1 (5 μg/kg) limit defined by the Turkish Food Codex. In Iran, Habibipour et al. (2016) gave the rather weak conclusion *'the content of aflatoxin in* ***most*** *of the walnut samples did not reach to maximum tolerable limit for aflatoxin B1 in EU'.* Thus considerable effort needs to be expended on monitoring and developing new and improving systems for *Aspergillus* control (e.g. chemosensitization; Kim et al., 2015) to ensure a safe supply of walnuts worldwide. In particular, nuts produced using less developed systems need to be routinely tested and their sources trained in aflatoxin reduction methods.

### 2.3.4 Conclusions

Walnuts, in addition to the excellent taste of fresh nuts, also have been shown to have great health benefits. They may also have specific compounds from various tissues with direct pharmacological uses. Ongoing research needs to continue to work with these and continue to demonstrate the value of walnut consumption. However, safety is taken as a basic requirement and there are issues with walnuts in both allergic effects and potential to carry harmful bacteria and aflatoxins. While these potential contaminants are well scrutinized in developed region systems, other areas of the world still have issues that can tarnish the name of walnuts. Thus, research to generate rapid screening tests and chains that minimize likelihood of contamination and contact with soil and faecal material need to continue to be developed to maintain public confidence in the product.

## 2.4 Value chains

Walnuts, as with other nuts, proceed through a number of steps as they enter into commerce. They are grown, harvested, processed (shelled, in shell, oil, flour etc.), sorted (into mixes, or grades), packaged, distributed, displayed, sold and utilized. Throughout this process there may be periods of storage and or transport and ownership changes. There may be changes in light level, atmosphere, humidity, temperature and biological action. Section 2.3.3 has indicated there is potential for biological contamination during these processes. Additionally, the chains coming from wild harvest or minimally managed orchards in remote less-developed regions of the world may differ substantially from those originating in California where farms are large, controlled storage is available, processes are highly mechanized, harvest is rapid and timely, rapid drying to a stable moisture content reduces mould and there are food safety management systems (e.g. HACCP, ISO 22000 etc.). In reality potential maximum quality will be set in the orchard by the growing and harvesting conditions used (McNeil and Evans, 2014) and then the rest of the chain will be involved in maintaining that quality and adding value by making the products of greater desirability to consumers (e.g. through; packaging type, size and contents; guaranteed freshness; mixes; delivery channels).

### 2.4.1 Quality management

Walnuts, unlike most nuts (hazelnuts, almonds, pine nuts, macadamias and peanuts) are split on shelling and thus have direct exposure of the flesh to the air. This can particularly be the case with mechanical shelling which can also damage the pellicle. Additionally, walnuts are high (Zwarts et al., 1999) in polyunsaturated fatty acids (PUFA) which have a tendency to degrade to free fatty acids and go rancid. This process is relatively slow in shell (Osterberg et al., 2001) but faster for kernels (Savage at al., 1998, 1999, 2001) and flours or oils (Stark et al., 2000; Vanhanen and Savage, 2006).

Targett (2014) has extensively reviewed and researched quality components of walnut kernels and their measurement and causes for loss of quality. He showed that walnuts darkened considerably when exposed to light for periods of 15–45 days at 1083 lux (Fig. 4). Within local supermarkets lux values varied from 500 to 1900 lux at the time of the trial. Many walnut packages that are used globally expose large amounts of walnuts to such levels of light exposure (McNeil, 2013a). Quality parameters, both hedonistic (taste, texture, rancidity, aftertaste) and instrument measurable (colour, free fatty acids, peroxide value) have been found to be strongly correlated. Targett (2014) found for hedonistic scores correlations ranged from 0.46 to 0.94 all with $P < 0.001$. Colour and palatability scores had an $R^2$ of 0.84, and colour and peroxide values had an $R^2$ of 0.82. Evans et al. (2016) and McNeil et al. (2014) have developed a scoring system which they have used with growers in Australia to assess quality. However, changes in colour within a variety by itself gives a good overall assessment of alterations in quality associated with treatment effects on walnut kernels. Colour alone may not be an indicator among varieties as there are some excellent tasting red, grey and brown varieties in China and Europe. The more detailed scoring method of McNeil et al. (2014) has also been used to assess quality in three separate global markets (France, the United Kingdom and Hong Kong; McNeil, 2013a) suggesting different concerns of various markets among price and quality parameters that have been quantified by McNeil and

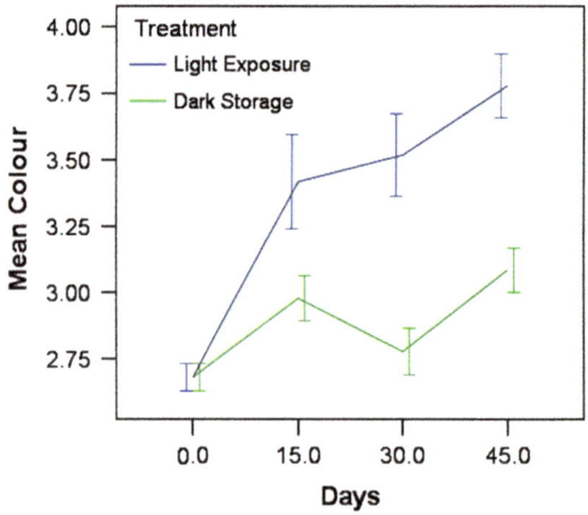


**Figure 4** Effect of storage of Lara walnuts in the light (1083 lux) at 25°C on mean colour score (higher numbers = darker and lower quality; see McNeil and Evans, 2014). Error bars indicate ±1 standard error (Targett, 2014).

Felgate (2014) and McNeil (2014c). These show considerable quality/price trade-offs in UK purchasing behaviour that lead to some nuts being 2 years old and rancid in store. Fuentealba et al. (2017) have now given some physiological explanation for the different colour and quality characteristics of walnuts linking the improved quality of cooler growing regions to key metabolism precursors. They suggest that the increases in arbutin levels found in cooler growing areas which produced lighter walnuts could help develop a method to manage walnut colour.

Table 4 shows that 45 days of exposure of walnut kernels even at normal $O_2$ levels and temperatures leads to increased colour scores (0.31 and 0.59 for Lara and Chandler respectively) and thus reductions in quality. However, both increased $O_2$ and temperature lead to much more rapid deterioration with colour score increasing by 2.8. Lack of $O_2$ ($N_2$ atmosphere) and low temperatures both maintained walnut quality over the period investigated indicating suitable storage conditions are relatively easily achievable. Peroxide values for these nuts were also assessed and these showed similar responses

**Table 4** The effect of $O_2$ level, temperature and variety on the colour score of walnut kernels stored in the dark in sealed plastic bags

| Variety | $O_2$ level (%) | Temperature (°C) | Change in colour score | SE |
|---|---|---|---|---|
| Lara | 0 | 4 | **0.00** | 0.06 |
| | | 25 | −0.19 | 0.01 |
| | | 37 | −0.25 | 0.09 |
| | 20 | 4 | 0.13 | 0.08 |
| | | 25 | **0.31** | 0.08 |
| | | 37 | 1.60 | 0.11 |
| | 100 | 4 | 0.36 | 0.10 |
| | | 25 | 0.84 | 0.09 |
| | | 37 | **2.81** | 0.01 |
| Chandler | 0 | 4 | **−0.05** | 0.14 |
| | | 25 | −0.36 | 0.05 |
| | | 37 | 0.07 | 0.14 |
| | 20 | 4 | 0.04 | 0.12 |
| | | 25 | **0.59** | 0.06 |
| | | 37 | 1.62 | 0.06 |
| | 100 | 4 | 0.32 | 0.08 |
| | | 25 | 1.13 | 0.13 |
| | | 37 | **2.80** | 0.12 |

*Source*: Targett (2014).
SE = Standard Error.
Increase in colour score indicates loss of quality.
Bolded values indicate the best, central and worst treatment combinations.

to colour confirming the use of colour change as a quality score. Targett (2014) also found that keeping the nuts in shell with nitrogen atmosphere prevented any deterioration over 45 days at 25°C but that even in-shell nuts exposed to air deteriorated over this time. Chandler at 30% of the rate of kernels with Lara at 75% the rate of kernels. This indicated there can be varietal response in quality to differences in storage conditions.

For more rapid and quantitatively consistent assessment of walnut kernel characteristics there needs to be automation of the assessment methods for which Targett (2014) developed a basis for future refinement. Colour score saturation ($R^2 = 0.72$), intensity ($R^2 = 0.76$), red ($R^2 = 0.73$), green ($R^2 = 0.85$) and blue ($R^2 = 0.81$) all showed good agreement between measured and visual colour scores. Given a larger sample it is expected a combination of individual scores will give a good colour quality assessment. NIR spectral data were also tested against measured data for colour scores ($R^2 = 0.82$), free fatty acids ($R^2 = 0.98$) and peroxide values ($R^2 = 0.96$). Validation sets for free fatty acids and peroxides gave reduced but still highly significant $R^2$ values of 0.87 and 0.56, respectively. There thus appears to be potential in using NIR measures to test kernel quality.

### 2.4.2 Value management

Given this wealth of knowledge on management of quality for walnuts, how well do supply chains operate? In developed regions such as Australia and the United States producers and processors manage well. Microbiological contamination is low, walnuts (with shells) are stored in cooled bunkers till ready for shelling and distribution. Walnuts shipped from Australia to Europe across the equator are sent in refrigerated (4°C) humidity controlled (60% RH) bulk containers. This is not necessarily the case from other regions. McNeil (2013a) showed for the samples he collected in Europe the closest (France) and furthest suppliers (Chile) had the highest quality while Chinese walnuts were of significantly lower quality. In the final distribution phase often the chain failed badly. Retail walnut kernel packs were surveyed for four simple packaging improvements (>50% light exclusion, resealable, identified origin and inert atmosphere). In Hong Kong, France and the United Kingdom the packs, average scores were 1.6, 2.9 and 2.2 respectively. Similarly, samples labelled as 'halves' had only 10% halves, UK walnut kernels had an average age at purchase of 14 months with 7 months till use by date. Associated with this age was a strong drop in quality (McNeil, 2013a) and a reduction in retail price. Indeed of the 81 samples 45% had quality scores assessed as being suitable for cooking only with 5% unsuitable for any use.

There does exist a market for higher priced, fresh, high-quality walnuts. Waitzmann (1998) and Waitzmann and McNeil (2000) demonstrated this with respect to consumer preferences in the New Zealand market. It is within this space that the entire New Zealand walnut production exists and the market is supplied by prepared to order processor 'A Cracker of a Nut' (http://www.crackernut.co.nz/). They market conventional, varietal, wild and organic nuts throughout New Zealand to a loyal customer base of chefs, individuals, local and specialist stores. Developing the larger bulk walnut retail market more on these lines will not only return premium prices but also expand demand provided the whole chain actually produces the additional value.

### 2.4.3 Conclusions

Information on how to assess and maintain walnut quality is readily available (McNeil and Evans, 2014; McNeil et al., 2014; McNeil, 2013b; Targett, 2014), and used extensively through most of the industrialized world growing regions. However, these methods are only slowly taking hold in many less developed production areas even though their importance has often been recognized (McNeil, 2013b). Of even greater importance is the failure of the retail and near-retail end of the market to recognize the greater perishability of walnuts than other nuts and the value of purchasing quality walnuts rather than cheap walnuts (McNeil and Felgate, 2014; McNeil, 2014c; McNeil and Felgate, 2012). Consequently old, rancid, poorly packaged and stored nuts are the retail norm in many regions of the world. A common statement the author has encountered in many countries is 'I do not like walnuts'. What the consumer really means is 'I do not like the rancid walnuts as presented to me in the supermarket or health store'. When these people were presented with fresh, in-shell walnuts in taste panels convened by the author in the United Kingdom, Australia and New Zealand their entire attitude changes. It is thus imperative that walnut producers generate the quality required and pass on information down the chain to consumers to maintain the growing demand needed as new walnut plantings come on stream.

## 2.5 Walnut reproduction

There still exist large areas of natural and planted seedling walnut forests and old low productivity varieties in Asia and Eastern Europe such as China, Iran, Hungary and the Kyrgyz Republic (McNeil, 2013b; Avanzato, 2014). Increasingly, old walnut stands are being successfully top worked (Rezaee et al., 2014) to new varieties which the author has personally observed in Sichuan, Shanxi, Shaanxi and Qinghai provinces in China. Generally the new bud or scion graft needs to be warm (≥25°C), protected from sunburn, wrapped to prevent drying out and high sap pressure avoided by cutting phloem below the union. The growth of a new bud on a large tree can be extremely rapid, weak at the bud junction and have spindly growth requiring support. Thus other branches are left intact for competition and support of the new bud branches during the first years. Most new plantings globally are clonal scions budded onto either clonal or seedling rootstock or are reproduced by micro-propagation of rootstock, or own rooted as trees. Development of these methods has been carried out comprehensively at a number of locations including the walnut breeding programme at UC Davis (2017) and Shanxi Academy of Forestry (2011) with considerable international germplasm exchange (Ciceoi et al., 2012). A major issue within many of the less developed and rapidly expanding walnut areas is the quality and certification of planting materials. Ren et al. (2014) have indicated critical issues for Fenzhou walnut development which include '*certification of scion origin, finding a better orchard system, improving nursery management*'. While most orchards in China are uniformly of high quality using local varieties, the author has observed walnut orchards with highly variable phenology and nut quality that were all supposed to be the same grafted variety. This attests to the accuracy of Ren et al.'s (2014) comments. Basically the goals have been to create scion/rootstock combinations with early and high yields, high nut quality and disease resistance.

### 2.5.1 Micro-propagation

Initially all clonal work on walnuts required seedling rootstocks. Nurseries need to be aware that walnut seeds require stratification in wet, cold (4°C for 8 weeks) conditions with the seed also presenting a mechanical barrier (Vahdati et al., 2012). Stratification requirements are less in the more subtropical *J. sigillata* and *J. sigillata* × *J. regia* hybrids (McNeil, *pers. comm.*). Walnut budding and grafting can be difficult and various means can enhance success such as hot callusing which is still extensively used in cooler growing areas such as New Zealand and Bulgaria (Gandev, 2017). Recently however, summer field patch budding of same season seedlings has been carried out (>80% success rate observed by McNeil) in blocks containing 100 000s of trees. This method works well in hotter growing areas such as China (Shanxi Academy of Forestry, 2011), Australia and California (UC Davis, 2017). Epicotyl grafting has also been successfully demonstrated as a method for rapid propagation with success rates of up to 70% (Qureshi et al., 2016). Micro-propagation has been used for reproduction of hybrids (Meier-Dinkela and Wenzlitschke, 2017), somatic embryo culture (Leslie et al., 2015) and for production of plants for direct field growing of selected varieties of rootstock and own rooted trees (Tuan et al., 2017). Micro-propagation can be difficult because of high callusing temperatures (Gandev, 2017), phenolics and inhibitory compounds in the explants (Cheniany et al., 2013), endophytic bacterial populations which can be positive or negative for growth (Pham et al., 2017), varietal differences, plus disease and desiccation issues (Tuan et al., 2017). Connell et al. (2010) have suggested yield efficiency of own rooted Chandler walnuts may be lower than for grafted trees. Hasey et al. (2014) have shown that own rooted walnuts or walnuts on selected clonal rootstocks can have reduced *Agrobacterium* infections and blackline disease compared to seedling rootstocks. In Guangxi typically *J. sigillata* rootstock is used because of its lower root disease frequency in the wet subtropical climate (McNeil and Zeng, 2013) and Browne et al. (2015) have identified clonal hybrid rootstocks with good resistance to *Phytophthora cinnamomi* and *P. citricola*.

Ex vitro acclimation of micro-propagated walnuts is a major issue that has required understanding of the physiological response of the walnut plants. Thin leaves maintain numerous, wide open stomata even when transferred to the drier ex vitro environment (Maleki Asayesh et al., 2017a) in spite of high betaine and proline levels in the walnut leaves which may have been expected to reduce the desiccation effect. However, K and Na ions were low due to low transpiration in the high humidity in vitro environment. Management therefore requires reducing populations of, or shutting, stomata. Maleki Asayesh et al. (2017b) showed reducing humidity in the culture vessels over the last couple of weeks using increased osmotic solutions can achieve this outcome. Another alternative was to increase the $CO_2$ concentration of the headspace which increased osmotic potential of the in vitro propagules and decreased stomatal aperture leading to improved survival (Vahdati et al., 2017).

### 2.5.2 Conclusions

In spite of physiological and disease issues walnut propagation is predominantly through use of selected seedlings with patch budding in the field. Using clonal rootstock with disease resistance and high performance is becoming more popular but ongoing difficulty in micro-propagation means the method cannot yet produce the millions of very cheap trees per year presently being planted in China.

## 2.6 New issues with pests, diseases, physiology and the environment

Integrated pest management (IPM) of walnuts is well covered in Strand (2003) with updates at http://ipm.ucanr.edu/PMG/selectnewpest.walnuts.html as well as by Wiman et al. (2017) for a cooler wetter region and Sharma et al. (2012) for an Indian perspective. Thus the next paragraphs take an overview of new and emerging non-biological (and to a lesser degree biological) threats that are being observed globally as area of walnuts expands rapidly into new regions, local conditions change in response to a warming earth and new issues arise in older regions potentially coming from outside areas and responding to increased trade and movement of people and walnut equipment, plants and nuts.

### 2.6.1 Non-biological threats

Rapidly expanding plantings can create significant problems associated with lack of regional experience and infrastructure. During a number of visits to China I have observed the enormous increase in trees being planted which has led to considerable variability in the quality, reliability, labelling accuracy and pest- and disease-free status of planting material. As a result, many orchards have had to top graft large old trees to healthier, new or correct varieties. There has also been extreme variability in the growth of new stands due to poor land preparation or suitability, variable quality planting material and poor management and the need to replace locally unsuited varieties or abandon some areas. For example, in Guangxi province the varieties needed are generally blight resistant Yunnan bred *J. regia* × *J. sigillata* crosses on *Phytophthora*-tolerant rootstocks. Where external varieties have been used they have not fared well. In Sichuan, plantings show extreme variability due to waterlogging and root diseases in some old rice soils and compaction in others. Similarly, harvesting, storing, drying and processing equipment and infrastructure lag behind planting and production. There is no quick fix for the difficulties of rapid expansion but extension efforts, planning and staged infrastructure construction are needed to minimize problems.

### 2.6.2 Climate change

Global warming can be a major issue in a number of walnut growing areas. Increasing temperatures can be an issue both directly and less directly through reduced water availability as experienced by California (2010–17) and increased solar radiation exposure. Fulton and Buchner (2015) have produced a management strategy for irrigation management in California to minimize the adverse drought effects. As an example of the reality of growing crops in a warming earth; in spring 2015 the daily maximum temperatures for Leeton (near the large Australian walnut orchards in the MIA) averaged 4.75°C above those experienced in the same period in 2016 which was around the normal climate expectation (http://www.bom.gov.au/climate/data/stations/). These high spring temperatures (possibly also associated with reduced winter chill as walnuts generally have a high chill requirement) damaged flowering and reduced walnut yields from 12 000 to 6000 tonnes (https://static1.squaresp ace.com/static/5770b8d4b8a79bc1087f2cfc/t/5a0e600dec212d62c91ee24 2/1510891602830/Webster_AR_LR-links.pdf). Winter chill needs may be less in Guangxi, China, where local varieties are adapted to low chill. In addition, high summer temperatures with intense sunlight led to major issues with walnut sunburn across the region. Conversely, increased temperatures may be of benefit in cooler walnut growing regions such as New Zealand's South Island and Tasmania provided they are not associated with increased spring frosts, reduced irrigation availability or in-migration of pests with the changing climate. Thus there is a need to conduct research to better understand how to manage global warming via pest management strategies, varietal development and agronomy.

A great challenge going forward will be to maintain the productivity of walnut plantations under the potential climate change scenarios (IPCC, 2015) with increased $CO_2$ level, raised sea levels, increased temperature, increased severity and variability of weather events, increased evapotranspiration and shifting climate patterns with respect to location, latitude and altitude. Therefore models of regional suitability are needed for long-term plantation crops such as walnuts to determine appropriate future locations of plantations of different varieties. Su and Wang (2017) looked at spatio-temporal changes in hardwood, including *Juglans,* trees in a region of China which is a pre-requisite to understanding what shifts may occur with climate change. In their 2011 review of ecophysiology of walnut in response to environmental stresses and potential acclimation to climate change Gauthier and Jacobs (2011) indicated there was little available information to understand the likely impacts of climate change on walnut production. Since then a number of authors have modelled responses to help clarify the situation. Phenological models are now available particularly for US varieties (Luedeling et al., 2009). Luedeling (2012) has consequently been able to review the effects of climate change on winter chill for fruit and nut trees and provide recommendations. Barrueto et al. (2017) have modelled present and future walnut distribution in Nepal and concluded, '*climatic requirements for the cultivation of Juglans are fulfilled across a large part of Nepal at present and in the future*' but suitable areas will shrink marginally under reasonable climate predictions. '*However, suitable areas shift considerably in spatial and altitudinal terms, meaning that some currently productive areas will become unproductive in the future, while currently unproductive ones will become productive. We conclude that the consideration of macro- and microclimatic changes in agricultural planning is essential to long-term agricultural success in Nepal.*' More recently an analysis of 16 Australian fruit and nut growing districts for Chandler walnuts indicates considerable risk going forward (Darbyshire et al., 2016), greater than for the other fruit and nut trees reviewed. Site and time period as well as cultivar and species all interact with the chill models meaning there is no consistent broad-scale pattern. A consequence of this is that flowering by main crop and pollinizer varieties may become uncoupled leading to an additional level of complexity. As a result of these and other model runs it can be concluded that in addition to the stresses of high temperatures (e.g. sunburn) and low temperatures (frost) plus potential shifts in pest and disease ranges which have already been discussed there is potential for winter chill requirements to be unmet in some growing areas. Charrier et al. (2011) indicate both environmental and genetical controls exist for walnut phenology while Guardia et al. (2016) also indicate consequences of direct stresses such as frosts are at least partly also under genetical control making breeding possible. This is particularly the case with walnuts where near relatives (e.g. *J. sigillata*; McNeil, 2016) with different ecological ranges can bring in specific characteristics. Earlier sections have indicated low winter chill, high-quality varieties have been developed in China with *J. sigillata* crosses (McNeil and Zeng, 2013) and large tree top-working is possible (Rezaee et al., 2014). Thus the phenological climate requirements of walnut trees may need to be managed by breeding and variety change to compensate for climate change. However, this will require effort and preparation ahead of time in a long-term crop like walnuts. There are also limits to what adaptation is available particularly for extreme temperatures and possible loss of irrigation water availability.

As indicated not all climate change will be negative for all regions. Some new areas will be opened up at higher altitudes with increasing temperatures (Barrueto et al., 2017) though this will not be a net benefit. However, increased $CO_2$ is likely to result in increased productivity given all other factors remaining the same. This will either be a direct effect on photosynthesis or a reduction in transpiration via reduced need to evaporate water through stomatal restriction to attain the same level of $CO_2$ uptake. Short-term trials on young walnut trees by Delaire et al. (2005) gave a linear relationship between $CO_2$ level and carbon uptake. Based on their data 3% of the increased productivity of walnut trees between 1985 and 2014 could be attributed to rising $CO_2$. While walnut trees have not been included in any present Free Air $CO_2$ Enrichment (FACE) trial a number of other trees have. These studies indicated that a 41% increase in $CO_2$ improved growth and water use in large deciduous forest trees (Bader et al., 2010) and coffee trees (Damatta et al., 2016) of the order of 40–50% if other nutrient requirements are met. Where water is strongly limiting forest trees do not respond as well to elevated $CO_2$ (Duursma et al., 2016). Irrigated (e.g. California and Australia) and well-watered (e.g. Guangxi) walnut orchards are expected to behave more like the first two examples. With a 15.5% increase in global $CO_2$ between 1985 and 2014 (IPCC, 2015) and assuming similar responses to other trees at least a portion, of the order of 3–16%, of the 63% increase in yield between these dates (Fig. 1, Table 1) is likely to be directly attributed to rising $CO_2$ rather than improved management and varieties. However, it is equally realistic to blame climate change and consequent drought (2011–17) in California for at least some of the restriction to expansion and yields of the US walnut industry (Table 1). This again emphasizes the mixed effects of climate change.

### 2.6.3 Biological threats

For walnuts as with other crops there are always threats of new pests and diseases affecting production. A good example is thousand canker disease (TCD) which is primarily a disease of *Juglans nigra* (black walnut) that appears to have originated in the western United States (http://www.thousandcankers.com/). The causal agent is the fungus *Geosmithia morbida*, vectored by the walnut twig beetle, *Pityophthorus juglandis*. It has also been associated with other *Juglans* species. Specifically with *J. regia* in Italy (Montecchio et al., 2014) and *J. regia* and paradox rootstock in California (Yaghmour et al., 2014). It seems to be associated with *Fusarium solani* in *Juglans regia* in both early and late stages in both Italy and California (Montecchio et al., 2015). These three components together develop the full potential of TCD. Leslie et al. (2010) indicate there is great risk for the walnut industry in California both in terms of production and distribution of materials from and to the Californian breeding programme. The origins of the disease remain unclear. However, considerable effort is being expended on basic epidemiological data gathering (Hefty et al., 2017; Zerillo et al., 2014). At present control is based on quarantine to prevent spread (http://www.thousandcankers.com/) into new US and European (Montecchio et al., 2016) locations. No chemical control measures are at present available for the disease and while suggestions have been made that drought may increase risk there is no good evidence at present to support this contention (http://ipm.ucanr.edu/PMG/r881101211.html). Two IPM approaches are being tested. These are the use of commercial strains of entomopathogenic fungi *Beauveria bassiana* and *Metarhizium brunneum* (Castrillo et al., 2017) and trapping or confusion with pheromones (Seybold et al., 2012). TCD is therefore an emerging and spreading threat of unknown final potential requiring considerable global effort to bring under control.

Additionally, many existing regional issues have not been dealt with adequately in existing growing areas, for example stem borers in China. In Guangxi and Sichuan, longicorn (longhorn) beetles (*Batocera lineolate, Batocera horsfieldi*) are a major pest without a good method of control. They have also been a pest in India for a considerable period (Rahman and Khan, 1942). Control generally centres around killing the beetles or larvae individually and maintaining good tree health. Li et al. (2013) have had some success with the parasitoid *Dastarcus helophoroides* and this and other biological control agents are in use in China. Pheromone or plant attractants (Yang et al., 2013) development would seem to be possible methods of control but little research has been done to date. Longicorns are not the only local borer attacking walnuts in China. In Shanxi a stem boring beetle, associated with a disease complex that caused weeping cankers, has been observed causing considerable damage to a hill orchard (McNeil, *pers. comm.*).

In addition to these new diseases, and unmanaged old diseases Table 5 shows there has been a wide distribution of first reports of a range of diseases of walnuts around the world over the last 5 years. Similarly insect pests have spread to new regions: walnut husk fly (*Rhagoletis completa*) arriving in Europe (Switzerland 1988) and spreading so far to at least seven other countries (Verheggen et al., 2017). While it can achieve up to 80% loss if left uncontrolled good phytosanitary and other control mechanisms (e.g. ethephon to advance harvest and thus avoid damage from late flights, UCIPM, 2017) that have developed can keep losses below 10%. However, the new requirements for husk fly control may put in jeopardy other parts of the system. Good cover crops make good phytosanitary orchard clean up more difficult (Grant et al., 2006). Thus there is an ongoing need to continue to find control measures relevant to and adaptable to the new areas for new diseases and pests as well as maintain either quarantine or good cleanliness in individual farms to reduce the risk of new disease outbreaks as long as possible.

### 2.6.4 Conclusions

Both biological and non-biological threats have arisen with the expansion of walnut production and the passage of time. Many of these problems are amenable to resolution or at least reduction through research into improved management in general, targeted breeding, matching of genotype and changing environments plus improved IPM methods including new biological control methods such as attractants and pathogens. These researches need to be combined with molecular approaches, field management and extension to ensure that new threats do not seriously affect the growing industry.

## 2.7 Production issues

### 2.7.1 Temperature

Walnuts are being planted more extensively with irrigation in drier continental climates (e.g. North and West China; Central NSW, Australia) where lack of maritime influence and clear skies rapidly lead to large temperature fluctuations throughout the year. At the same time global temperatures and climate variability are increasing due to $CO_2$-induced global warming (IPCC, 2015). This is creating more issues with low temperature and freeze injuries (autumn to spring), heat and sunburn damage (spring to autumn) and combined freeze/thaw/sunburn damage (autumn to spring) of walnut trees, flowers and fruits. High spring temperatures (Section 2.6.1) can lead to flower

**Table 5** First reports of walnut diseases over the last 5 years

| Location | Pathogen | Result | Host | Citation |
|---|---|---|---|---|
| China | *Diplodia seriata* | Twig die back | *J. regia* | Zhang et al. (2017c) |
| China | *Fusarium solani* | Wilt | *J. regia* | Sun et al. (2016) |
| Italy | *Geosmithia morbida, Pityophthorus juglandis* | Thousand canker disease | *J. regia* | Montecchio et al. (2014) |
| California | *Geosmithia morbida, Pityophthorus juglandis* | Thousand canker disease | *J. regia + paradox* | Yaghmour et al. (2014) |
| Italy | *Fusarium solani* | Wilt | *J. regia* | Montecchio et al. (2015) |
| Korea | *Phyllactinia juglandis* | Powdery mildew | *J. mandshurica* | Lee et al. (2016) |
| China | *Fusarium solani* | Wilt | *J. sigillata* | Zheng et al. (2015) |
| China | *Colletotrichum fioriniae* | Leaf spot | *J. regia* | Zhu et al. (2015)a |
| Italy | *Phytophthora taxon walnut* | Dieback | *J. regia* | Ginetti et al. (2014) |
| Hungary | *Brenneria nigrifluens* | Bacterial canker | *J. regia* | Végh et al. (2014) |
| Serbia | *Brenneria nigrifluens* | Shallow bark canker | *J. regia* | Popović et al. (2013) |
| Korea | *Neofusicoccum parvum* | Branch die back | *J. regia* | Cheon et al. (2013) |
| California | *Lasiodiplodia citricola, Neoscytalidium dimidiatum* | Graft union death | *J. regia* | Chen et al. (2013) |

loss (as can spring freezes) then tree and fruit sunburn. Winter freeze injuries and winter and spring sunburn can lead to direct tree damage and leaf loss and also act as entry points for walnut diseases, in particular being one factor that can increase blight damage (Section 2.1). Ultimately therefore either, or both, genetic and management methods need to be developed for reducing effects of temperature stress. Guardia et al. (2013) suggest *J. regia* is one of the more sensitive *Juglans* species to low autumn temperature damage. Charrier et al. (2013) have developed models for linking physiology and frost hardiness as a prelude to understanding how to manage and describe tolerance. Temperature-induced damage can be measured by electrolyte leakage, proline accumulation and specific enzymatic activities among others (Zhai et al., 2016).

Winter freeze followed by sunburn and subsequent blight infection has been extensively observed on the equatorial side of young trees by the author in Shaanxi, Shanxi and Qinghai provinces, China. Two major methods of control are to either paint the tree trunks white (acrylic paint) or to surround them with protective insulation which can be plastic, snow, dirt or straw depending on availability (Figure 2). The goal is to prevent ice nucleation in the stems which causes high water potential gradients and redistribution of water to the ice front causing dehydration of and damage to the bark cells. Cavitation then occurs with embolism formation followed by cell death and potential for disease entry when thawing followed. This is particularly likely on the equatorial side (Charra-Vaskou et al., 2016) where there may be repeated rapid freeze/thaw events. Sunburn may then exacerbate the damage as the bark areas have poor water connections, and already damaged cells. The use of protective insulation may also have the added benefit of slowing any freeze/thaw processes that do occur as slower cooling has been shown to induce better cold tolerance in buds (Choudhary et al., 2015).

Snyder (2000) gives a comprehensive description of causes of frosts (in particular spring frosts which kill buds and flowers and are globally particularly damaging to walnuts for kernel production) and how to manage them. Generally sprinkler application of water (to protect via latent heat of freezing release) is not available for walnut orchards to protect from frosts due to tree size and water normally being applied below the canopy level (though it may still have some effect at that level). Damage through ice nucleation by bacteria on buds decreases if ice nucleating bacteria are reduced. Theoretically, but not so far practically, this can be accomplished using pesticides (such as copper compounds (Kocide; Jenkins et al., 2010) or by applying competitive non-ice nucleating bacteria. Passive protection via site selection (away from cold air drainage sites, used extensively in the hill areas of China), ensuring high soil heat content (keeping top 30 cm moist, heat absorptive mulch covering) can reduce local advective frost damage. With inversion layer frosts wind machines and helicopters are used extensively in wealthier production areas where the cost can be spread over a large orchard (e.g. Tasmania).

Breeding for frost tolerance or avoidance is possible using late leafing out and late flowering varieties. However, this reduces season length for the crop and thus potentially growth rates and yields. Guardia et al. (2016) found both autumn and winter frost hardiness correlated inversely with secondary annual growth measured at breast height. Autumn acclimation to frost has been observed in walnuts (particularly wood types, Charrier et al., 2011) suggesting solutions may be available via manipulation of autumn management (e.g. late N/water availability). GM breeding for walnuts is now possible and one potential method for increasing high and low temperature stress tolerance is to overexpress the *JrsHSP17.3* gene for a small heat shock protein (sHSP) from *J. regia* (Zhai et al., 2016). Transformed lines displayed less cell damage, lower electrolyte leakage rate, malondialdehyde and $H_2O_2$ content, and higher activities of catalase, glutathione transferase, superoxide dismutase and peroxidase as well as more accumulation of proline than controls.

High temperatures can adversely affect walnut through winter sunburn on stems and summer nut sunburn. In particular sunburn increases susceptibility to a range of diseases and pest infestations as well as leading to darkening of affected kernels and quality loss. A good, well-watered canopy can reduce excess nut light exposure and reduce temperatures via evaporative cooling. However, excess canopy growth can reduce nut quality (Lampinen et al., 2010). Nut sunburn occurs above 50°C which is not usual for air temperatures but occurs with additional direct sun exposure. Active protection is possible. In China high value, playing walnuts are protected individually by bundling leaves around the nuts. Spray on sunburn protectants based on Kaolin have been trialled but generally with no benefit and only a small reduction in leaf temperature (Rosati et al., 2006).

Heat/cold damage causes can be ameliorated to some degree by keeping trees well-watered, protected from winter extremes and selecting suitable varieties and growing locations. There is potential to select varieties suited to a particular environment as well as transform trees with specific genes and select more suitable genes using molecular markers.

### 2.7.2 Nutrient management

Ramos (1997) in his management manual covers nutrient deficiency in walnuts. A comprehensive walnut fertilization guide for NPK in California is provided by the CDFA (2017). In particular it gives soil and leaf analysis methods and recommendations. Anderson (2006) gives an efficient nitrogen budgeting method for walnuts which has been used in the Californian BIOS programme (Grant et al., 2003) and can be expanded to other nutrients. Sambeek and Reid

(2009) give leaf analysis levels for pecan and black walnut which should provide some value for *J. regia*. McNeil and Evans (2014) summarize information on critical values, adverse effects of excess N on vegetative growth (excessive nut shading) and potential post-harvest quality effects of poor fertilizer regimes not based on critical values and nutrient budgets. Much of the rest of the world does not have such detailed information. The author visited and reviewed numerous walnut programmes in China between 2010 and 2017 among other things observing tree nutritional status and fertilizer practices. All were applying both organic and bag fertilizer and had fertilizer recommendations based on experiment and local knowledge. However, at the time, none provided specific local guidelines based on locally validated soil and leaf analyses. However, Hu et al. (2015, 2016) have looked at spectral leaf analyses as a faster alternative to chemical analysis as a way to rapidly measure leaf N and K and produce these types of fertilizer recommendations in China. Catch up with fertilizer management and recommendations of many of the less developed regions with California is essential. Different values may be needed to Californian recommendations considering the high rainfall and stony limestone soil base for example in much of the growing region of Guangxi, the growth of walnuts on the highly eroded Loess Plateau in Shaanxi and the use of old rice paddy soils in Sichuan. In Guangxi for example leaf fall was observed to be early due to leached nitrogen which may have reduced autumn photosynthesis, overwinter carbohydrate storage and ultimately yield potential. High water throughput may also deplete K in light soils. The author has observed this in sandy/rocky New Zealand soils where excess irrigation has been applied. Similarly Bravo et al. (2017) found in Chile that late summer nitrogen was of benefit there for improving spring growth. Visual inspection of orchards in China indicates micronutrient deficiencies are common and not always adequately overcome. Keshavarz et al. (2011) have indicated in Iran that foliar application of Zn and B can benefit both vegetative and reproductive growth as well as nut quality. Excess micronutrients can also be an issue (e.g. NaCl, B, Mn, Cu). Chatzistathis et al. (2015) showed liming was able to reduce leaf concentrations of Mn and Cu and overcome excess levels. Thus the optimal fertilizer management system involves use of soil and leaf analyses to compare levels to optimal values. Then soil application and/or leaf application of foliar nutrients should be made to overcome the measured limitation. Soil amendment to improve or reduce availability (e.g. through application of lime to change soil pH) can be of benefit as an alternative to direct application for some nutrients. However, most systems rely on application of the nutrient in an available form (e.g. Fe chelates, or foliar if bound by soil).

As well as the direct effects of nutrient applications on walnuts there is the potential for biological interactions with fertility to affect productivity. Of interest is the potential interactions among walnut nutrient status, pests and biological control agents. Mace and Mills (2017) looked at this without any really clear conclusions for a walnut-aphid-parasitoid system. In addition, with much of the walnut expansion occurring in remote less developed areas there is considerable interest in the use of plant beneficial rhizospheric microorganism for better use of existing nutrients or to fix nitrogen. The area has been reviewed by Meena et al. (2017) who conclude there is great potential. What little work that has been done with walnuts does sometimes show advantages but not consistently in a field situation (Yu et al., 2012). Their trials attempted with marginal success to use microbes to increase both N and P availability. Intercropping with other species (particularly N fixers) and animals is a promising method for improving under tree fertility which will be looked at in the next section. Grant et al. (2003) have tried to bring all of these aspects together (intercropping, reduced chemical and increased organic fertilizer use, leaf analysis and nutrient budgeting, IPM, habitat enhancement) in the Californian BIOS programme. Their results indicate such programmes can be a success with their overall summary being '*We found no significant differences between BIOS and conventional blocks in nut quality or yields*' but with significantly reduced chemical input.

### 2.7.3 Intercropping of walnuts

While not carried out to a great extent except for the production of pasture, ground cover or N fixing mulches in most developed world walnut systems intensive intercropping with walnuts is commonplace in much of the developing world. Wolz and DeLucia (2018) reviewed research on alley cropping globally and suggest walnut and poplars are the dominant trees in the temperate zone. Vegetables, cereals, pulses and other trees (e.g. mulberry) are intercropped with walnuts in China (McNeil, *pers. comm.*). In addition animals (chickens, pigs, sheep) are frequently grazed among walnut trees in China and sheep in Australia and New Zealand. There are also mixed stands of walnuts in forestry systems for example N fixing *Robinia pseudoacacia*. Intercropping can be an important source of income during the early years. Malhotra and McNeil (2010) estimated an annual net return of ~1300 Rs/ha from intercropping in an Indian situation over the non-bearing years and beyond. This was after basic management costs of the trees had been covered and was sufficient to cover about 30% of the land hire and depreciation costs. Figure 2 shows a new walnut planting in Qinghai, China, which had alfalfa planted between the rows for its N fixation, soil protection and feed values. Querné et al. (2017) have found that % nitrogen derived from the air by alfalfa increases when grown under walnuts as does light use efficiency. Thus even in more mature trees intercropping with alfalfa increased physical productivity. The issue in intensive horticultural systems is whether the increase in complexity (and hence management costs) and possible reduction of early walnut yields of the system are adequately compensated for. Part of the increased complexity is also the need to only use agri-chemicals compatible with both crops which can be an issue particularly in highly mechanized systems. The goal of intercropping is to gain an increase in land equivalency ratio (LER). Specifically that is to produce more than each crop alone. Or more accurately from a business sense to produce more profit per unit input. With walnuts, production in the early years is zero so it is not difficult with additional inter-row planting to increase returns provided the intercrop does not substantially delay or reduce walnut production. Duan et al. (2017) found that walnuts (1–2 year old) intercropped with wheat had an LER of 1.62. Walnut yields were reduced by 27% and 11% in years 1 and 2 with wheat yields reduced by 10% and 27% in years 1 and 2. The reduction could clearly be related to intense root competition with significant reductions in root parameters for both walnut and wheat. In this instance when walnut yields are low the value of the wheat greatly exceeds (~8x) the loss of walnuts. Zhang et al. (2016) intercropped the nitrogen fixing crop *Vigna radiata* with walnuts and found (1) under N deficiency the *V. radiata* fixed additional N and improved walnut root growth and (2) under N adequacy the *V. radiata* competed with the walnut and reduced walnut root growth. This exemplifies the difference between less developed (usually case 1) and more intensive developed plantations (usually case 2) helping to explain the difference in popularity of intercropping between regions.

Finally, when looking at walnut intercropping systems it is important to look at the effects of intercropping on soil health and properties. Lu et al. (2015) demonstrated on the Loess Plateau of China that walnut orchards without intercropping (but not with intercropping) lead to reductions in soil organic carbon and soil nitrogen. Thus as well as capturing value from the intercrop there is value for the walnut trees in the long term through improved soils. It is for these ecological reasons that cover crops are usually grown in the United States, Australian, New Zealand and other intensive walnut orchards. Grant et al. (2006) have produced a comprehensive manual for cover crops in these situations including a description of advantages and disadvantages. Cover crops improve soil health by adding C and N, helping water penetration, salvaging nutrients for release again on mulching, encourage beneficial insects and protect the soil surface. Usually they are not used for grazing for fear of *Salmonella* contamination and damage by animals to irrigation systems, though grazing is used in Australia and New Zealand. Generally the cover crop is either mowed and left or incorporated. In both Europe (Cardinael et al., 2015) and China (Duan et al., 2017) intercropping with cover crops can lead to deeper walnut root systems in agroforestry systems. However, cover crops also have their challenges, increasing water demand, requiring additional irrigation management and infrastructure, increasing frost risk, acting as host for nematodes, costing money to establish and mow, reducing pesticide choices and making sanitation more difficult. In these systems cover crops are basically a cost valued for the environmental benefits rather than a source of additional revenue as is common for intercrops in less developed regions with cheaper labour.

### 2.7.4 Water management

Water availability (both high and low) can affect yield and quality of walnuts both directly and indirectly. Critical issues related to water use are nut size, colour, insect pressure, kernel fill, mould (and aflatoxin) as well as yield. Generally supplying as much water as the crop needs so that it is able to extract water from the soil as required is the goal of any water management system. To this end complex irrigation models often with remote sensing have been developed for specific crops or more generally to maximize production from limited amounts of irrigation water (e.g. Shin and Jung, 2014). UC Davis provides irrigation management systems for California growers based on local research (http://ucmanagedrought.ucdavis.edu/PDF/Lampinen%20et%20al%202004.pdf). A water-deficit management system as described in Fulton and Buchner (2015) is essential to accurately monitor water use in walnuts. These can be soil based, water budget based or leaf or stem potential based. Direct soil water monitoring using implanted blocks (Hanson et al., 2000) is one of the most common direct measurement methods for irrigation scheduling. Shackel et al. (2010) have used midday stem water potential in almonds and walnuts to achieve water use reductions of 10–50% without yield reduction, depending largely on soil type. Detailed methods for use of stem water potential are given in Fulton et al. (2014). Heavier soils with greater water-holding capacity generally give greater reductions in water use without yield loss with controlled deficit irrigation. Controlled deficit irrigation (CDI) can improve nut quality without reducing yield and can reduce excess vegetative growth in older trees and thereby minimize pruning and shading. It may also improve yield per unit of applied water. However, there are often complex interactions and generally yield reductions (Buchner et al., 2008). The management of vegetative growth through CDI may be more important in view of the increased planting populations now being used along with early yielding varieties to reduce the lag phase between planting and significant and maximum yields. Of more importance for improved irrigation water use efficiency may be methods that reduce evaporation such as mulching (Grant et al., 2006) and subsurface drip irrigation which is increasingly being used within California as a means to manage water use (Ayars et al., 2015). Similarly, better parameterization of the evapotranspiration models can lead to more accurate knowledge on when water needs to be applied (Anderson et al., 2017).

### 2.7.5 Population density, pruning and harvest management

Population density is a major issue with walnut growers. High densities with precocious varieties initially lead to good early returns. In year 2 in Xinjiang, China, the cultivar Wen 185 achieved 240 kg/ha of walnut fruits (Duan et al., 2017) at a population of 417 stems/ha. However, at these high populations orchards are expensive to plant (helped in China by low tree prices and government inputs), can make equipment access difficult and in the absence of dwarfing rootstocks may lead to major problems with tree pruning management later on. Low planting populations lead to long lag times before significant production causing delays in getting a return on investment. Individual tree deaths are major issues and top-working to new cultivars is difficult and expensive. Ren et al. (2014) thus indicate *'planting at a reasonable density'* is needed for creating a profitable Fenzhou walnut industry in Shanxi, China. Van Kralingen and Adem (2010) have evaluated walnut population in an Australian context and determined that higher density orchards make economic sense. Californian recommendations are for 160 stems per ha in Sacramento with micro-sprinklers (Hasey et al., 2015). Their analysis suggests yields of $625/ acre in year 4/5 equating to 415 kg/ha. Timing and returns are similar to those quoted by Malhotra and McNeil (2010) for India.

Pruning of walnuts is necessary for two purposes: (1) management of tree form and (2) removing excess vegetative mass that can lead to nut shading. Adem (2012) gives an excellent video presentation of pruning walnut trees for shape. Reduction of excess vegetation in older trees can be partly managed by CDI (Section 2.7.4). However, ultimately some pruning is required as excess shade can reduce nut quality (Lampinen et al., 2010). Most experiments in orchards showed that excess early pruning reduced yields if not quality though significant mature stand pruning was needed to maintain quality even if at the cost of some yield in year of pruning (McNeil and Evans, 2014; Lampinen et al., 2015). Ultimately therefore minimal early pruning beyond initial shape formation is recommended with later pruning needed to maintain light access to mature orchards.

Harvest management can also be an important aspect of walnut management with ethephon applied in many growing areas (California, Australia, Chile, Mexico) to promote uniform or staggered harvest. UCIPM (2017) suggests that ethephon can *'speed up hull dehiscence, accelerate maturity, and promote fruit abscission'*, *'use of ethephon on the earlier-maturing varieties avoids the late season walnut husk fly and navel orangeworm flights'*. With navel orangeworm (NOW) it is critical for hull split and egg laying not to coincide so there is the potential for altered maturity to affect damage in either direction depending on timing of these two events as well as a need to relate control measures to tree physiology. Financial analysis indicates its use can increase prices by producing lighter coloured nuts as well as making savings at harvest and increasing yield. When green walnuts were mechanically dehulled 4 days after an ethephon spray (0.3%), throughput capacity, dehulling efficiency and percentage were found significantly higher (Hussain et al., 2016). Shaking efficiency can also be increased by ethephon treatment (Yalçin et al., 2014). More recently interest has been growing in fresh green walnuts in China which decay rapidly even with cool storage. Wang et al. (2016) showed ethephon use (500 mg/L) can '*extend storage life and maintain kernel quality of fresh in-husk walnuts through the manipulation of phenolic metabolism*'.

### *2.7.6 Conclusions*

Achieving acceptable yields in walnuts requires establishing a system that is suited to the location and economic status of the region. For example the value of intercrops differ considerably among regions. However, several activities do show through as common problems. Tree canopy needs to be developed quickly to achieve early yields. To do this, optimal amounts of water, nutrients, protection from stresses, soil management, IPM pest and disease management all need to be followed. After yield, nut quality becomes paramount and managing the orchard and harvest for quality is the greatest single contributor to profitability in intensive orchard systems (McNeil and Evans, 2014). For most of these issues there are recommendations based on experimental validation which can greatly increase profitability in whichever system is in use.

## 3 Case study

McNeil (2013b, 2014a,b), McNeil and Zheng (2013) and Scheuring (1997) have characterized very different walnut industries in Australia, New Zealand, California, Guangxi and Kyrgyz Republic. The goal has been to understand better the different trajectories of expansion and create a more general understanding of successful industry development and the research, social, economic and environmental conditions that can assist or hinder the industry and, if needed, the changes that need to be made.

### *3.1 Industry expansion*

The case studies presented in McNeil (2014b) are summarized using the new industry analytical tool he developed and are presented in Table 6. This tool can be used to determine where efforts should be directed in trying to develop a new industry. Some of the issues may be amenable to scientific research (to be discussed in more detail in the next section), others require financial or social management. For example the low size of the New Zealand

**Table 6** Evaluation of the fit between the walnut industries in three regions and McNeil's new industry evaluation tool

| | Measure | New Zealand | Australia | Guangxi |
|---|---|---|---|---|
| 1 Description | | | | |
| 1.1 Size | ha | 500 | 3000 | 30 000 |
| 1.2 Expansion rate | ~ha/yr over 10 years | 30 | 300 | 3000[a] |
| 1.3 Capital value retention | Proportion sound investments to date | Moderate/high | Moderate[b] | High |
| 1.4 Profitability | ROI | High | High | High |
| 1.5 Entrepreneurial businesses | Existence | Yes | Yes | Yes |
| 1.6 High-quality entrepreneurs and innovation | Number and diversity of products, processes, vertical and horizontal relationships | Highest | High | Least |
| 1.7 Potential size | Tonnes | Large (export) Small (local) | Large (export) Moderate (local) | Large (export and local) |
| 2 Risk management | | | | |
| 2.1 Production risk | Prices/environmental/ agronomy viable opportunity | Moderate[c] | Moderate | Moderate |
| 2.2 Financial risk | Operations losing capital | Moderate/ low | Moderate/high[b] | Low |
| 2.3 Market risk | Local production/ consumption | Low | Low | Low |
| 2.4 Societal risk | Licence to operate issues | Low | Low | Low |
| 2.5 Government input | Level | Low | Moderate/low | High |

| | Measure | New Zealand | Australia | Guangxi |
|---|---|---|---|---|
| 3 Entrepreneur access/control | | | | |
| 3.1 Land available | Existing owners | Low | Low/moderate | High |
| | New owners | High | High | Low[d] |
| | *Overall* | High | High | High |
| 3.2 Market access | Opportunities | High | High | High/moderate[e] |
| 3.3 Finance available | Risk capital available | Low | Moderate | High[f] |
| 3.4 Society views | *Issues with entrepreneurs* | Low | Low | Low |
| | *Availability of entrepreneurs* | High | High | High |

[a] Approximate based on observations and personal communications.
[b] Losses by early entrants (Vecon, Gunns) but assets remain, high government input in Guangxi, low $ input in New Zealand.
[c] Production packages available in all locations but climate, pest and disease issues in Australia, New Zealand and Guangxi and varietal yield issues in New Zealand and Guangxi.
[d] In other more developed Chinese provinces/regions this can be high.
[e] Enhanced by major rural road development projects.
[f] Both many landholders with small inputs and large government direct input.
Black backgrounds indicate major issues and grey backgrounds possible lesser issues.
*Source*: updated from McNeil (2014b).

industry (1.1 in Table 6) means there is little in the way of economies of scale for that industry. Two solutions are available to get around this: (1) that used in New Zealand has been to concurrently use two options: (a) concentrate on a local high-value market and (b) reduce cost of investment by having growers with other objectives (e.g. rural lifestyle) invest in walnut ventures. Option (a) is amenable to research development (Waitzmann, 1998; Waitzmann and McNeil, 2000) while option (b) is a social issue rather than a research issue. (2) The second solution is to use financial implements to achieve expansion as was done through the taxation system which increased apparent value of Managed Investment Schemes in Australia. These were used for the major expansion of walnuts into the MIA area of Australia (https://www.agwltd.com.au/). In Guangxi this small size problem was eliminated by massive injection of state resources.

Similarly, other issues can be identified. The high input of government funding for research, development and extension in Guangxi has fuelled expansion there (2.5 in Table 6). Considerably lesser levels of government RD&E investment in other regions (New Zealand and Australia) are hampering expansion. Government investment can support appropriate R&D so is important to scientific research but does not necessarily directly involve scientific research. All three regions have moderate production risks (2.1 in Table 6). For example Australia is presently dominated by exports for its walnut production and is thus subject to global price risk. Guangxi and New Zealand are presently dominated by local sales and are not therefore so subject to global price risk. All three regions have environmental risks and need research to ameliorate them. These risks include low temperatures causing damage and slow growth in New Zealand, high temperatures and unstable water availability in the MIA area potentially causing flowering loss and yield reduction in Australia and high rainfall and minimal processing capability in Guangxi significantly affecting the ability to achieve high-quality nuts with good storage. Similarly, production risks also include a range of specific pest and disease issues such as; longicorn beetle and nutrient deficiencies in Guangxi, frost damage and blight in New Zealand and the Tasmanian growing area of Australia.

## 3.2 Conclusions

Analyses have been carried out using the evaluation tool of McNeil (2014b) of a range of regional walnut industries expanding at different rates. This tool has identified potential impediments to expansion and allowed evaluation of financial, production and social issues. All of these need to be examined and their relative importance and potential solutions determined before investments should be made in developing new industries. This tool can therefore be used by other areas to prioritize their development needs for walnut industry expansion.

## 4 Future trends

Future trends in research fall under several categories. (1) New research at the forefront of science which is novel and often blue sky, plus the expansion of this research into more practical applications as well as confirmation of the extent to which it may be used. (2) Targeted research that is directed at highly specific issues, amenable to scientific solutions, which are preventing the development of expanded industries in new regions, or are threatening existing production areas. (3) Ongoing stepwise research creating and refining general improvements in systems (e.g. new slightly improved varieties) and improving management of internal threats/issues (e.g. TCD) or external threats/issues (e.g. water availability) that are needed by existing areas to maintain their profitability. Specific research technologies and targets will spread across all three areas for example traditional and molecular breeding will be needed for new forefront developments as well as creating adapted varieties and dealing with new threats. Similarly increased environmental concerns will threaten old production systems and will need novel science which will operate in widely divergent areas as well as having a more restricted tool kit to deal with other new threats. Typically, more responsive systems (e.g. real-time water monitoring) using detailed knowledge of interactions (e.g. pest/disease epidemiology) will be needed to reduce inputs and make their outcomes more targeted, certain and beneficial.

### *4.1 Novel science*

One of the greatest scientific areas for walnut research going forward is the expansion of molecular methods including advanced genomics, transcriptome, proteome and metabolome research. These areas can greatly improve our understanding in terms of evolutionary relationships, genes, genetic/environment interactions, physiology, health effects, bioactives and population structures. The great challenge within this area is to both conduct the novel research and to tie it into practical benefits for growers and consumers. Thus clearer understanding of gene function in management of stresses (e.g. frost, salinity, heat stress), pest and disease control, phenology and tree structure and walnut quality characteristics as well as in terms of effects on human health need to be developed. Then a means needs to be determined to incorporate the improved genes into production. This could be done by searching among relatives, better understanding complementary crosses or ultimately by direct GM.

### *4.2 New region science*

Figure 5 shows a stylized breakdown of the production/supply/use chain for a typical new industry compared to an old established industry. In this framework 'new' can have a variety of meanings: new to industrial agriculture, new to a region, new to a particular use. Globally 'new' for walnuts means new to a region for example Australia, Chile, New Zealand and Guangxi. While the previous case study (Table 6) has shown not all issues in a new region are amenable to scientific investigation it has indicated at least in these three case studies that there are issues needing scientific research. Typically the research needs in this situation are therefore two fold. The first is to solve the greatest threat to production in the area, (a) in Fig. 5. The second is to maximize the area of competitive advantage, which is why there is any interest in growing the new crop in the region, (b) in Fig. 5. In this instance the research will need to be highly targeted but may also be novel. Examples of the research suggested by (a) in Fig. 5 include pheromone control of longicorn beetles in China, establishment of leaf critical concentrations in new areas, bacteriophage control of walnut blight, new hybrids for expanded production zones (e.g. Guangxi), precocious fruiting for intensive orchards. Examples of the research suggested by (b) in Fig. 5 include demonstration of greater health benefits and quality benefits to increase local consumption in New Zealand, improved marketing of out-of-season walnuts from the Southern Hemisphere and growing of disease-resistant, precocious trees in the mountainous, humid subtropics (Guangxi).

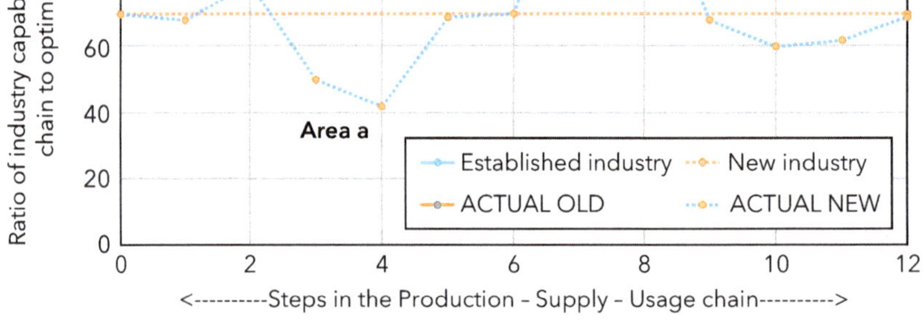


**Figure 5** Stylized diagram of production chains for a well-established and newly establishing industry. New industries generally are not as efficient as the well-established industry but can exist because of specific competitive advantage in some areas (Area b). However, some areas are worse than others and need immediate improvement (Area a) then, over time, improvement continues on all fronts.

### 4.3 New threat science

The changing climate with additional droughts in California and sunburn in Australia as well as the spread of TCD to walnuts in the last decade exemplifies the need for maintaining ongoing research capacity to deal with new potentially destructive issues as they arise. Whether these threats are the spread of existing problems from one area to another, spread to walnut from other species or the development of entirely new concerns through climate change, all will need to be dealt with. Many will be regional and in some situations may make use of data from elsewhere. Others may be regional or even global and need whole new approaches using biological control or breeding of new lines. It is difficult to foretell what will be needed. The critical issue is to have capability at hand to deal with the issue, to be prepared to take new approaches where existing ones are problematic and to switch resources from lower priority research when needed.

## 5 Conclusion

In terms of walnut research the general conclusions are that there needs to be a base available for dealing with issues as they arise, that new methods need to be included but that there always needs to be a clear understanding and basis for the research that is being done. In this system the shifting high-priority work will be the research that is carried out.

## 6 Where to look for further information

The US Foreign Agricultural Service provides data on global nut supply including walnuts (e.g. https://apps.fas.usda.gov/psdonline/circulars/TreeNuts.pdf). The California walnut Commission (https://walnuts.org/) provides popular information as well as information on marketing (https://walnuts.org/report/inventory-report-august-31-2017/) carryover stocks and recent US production. University of California Davis supplies comprehensive information on all aspects of walnuts in the United States including research, germplasm, extension, marketing, health effects and recipes (http://fruitsandnuts.ucdavis.edu/datastore/?ds=391&reportnumber=612&catcol=2806&categorysearch=Walnut). The UC ANR database (http://ucanr.edu/sites/cawalnut/CategoriesListing/) provides considerable information on all aspects of walnut production. Approximately 1300 walnut recipes are available from http://allrecipes.com/recipes/. The International Society of Horticultural Science (ISHS) through its *Acta Horticulturae* series (http://www.ishs.org/acta-horticulturae) has published the proceedings of the International Walnut Symposia (1-7 published, 2017 meeting in Chile available soon) as well as other nut symposia (e.g. from the

IHC, 2014). Through its *Scriptae Horticulturae* volume 17 (Avanzato, 2014) ISHS has published a comprehensive treatise on walnuts country by country globally. A Chinese manual on walnut production is also available (Zhang Zhi et al., 2009). The Food and Agricultural Organisation of the UN provides a database of global production, trade and consumption of walnuts from 1961 to the present (www.fao.org/faostat/). Similarly detailed trade data for many of the walnut producing countries are available through their own statistical websites for example UK trade data can be downloaded from https://www.uktradeinfo.com/Pages/Home.aspx.

For searching of the scientific literature, use Google Scholar (https://scholar.google.com/), which is freely available and has 20000 articles with walnut or *juglans* in the title. Web of Science has 3900 articles with walnut or *juglans* in the title (https://webofknowledge.com/) and Scopus has 3560 (https://www.scopus.com/) and also allows access to patents. Both also provide excellent resources for scientific published papers on walnut but require login credentials. All three databases are fully searchable.

## 7 References

Adem, H. 2009. Best practice management for establishing a walnut orchard. Australian Walnut Industry Development report NT06001. Victorian Department of Primary Industries.

Adem, H. 2010. *Australian Walnut Industry Development*. Horticulture Australia, Sydney, Australia. ISBN: 0734124481.

Adem, H. H. 2012. Training walnut trees. https://www.youtube.com/watch?v=y9G3MZuB4U8 (accessed 3 December 2017).

Akca, Y., Bilgen, Y. and Ercisli, S. 2015. Selection of superior Persian walnut (*Juglans regia* L.) from seedling origin in Turkey. *Acta Scientiarum Polonorum Hortorum Cultus* 14(3): 103–14.

Anderson, K. K. 2006. Guide to efficient nitrogen fertilizer use in walnut orchards. UCANR Publication number 21623. UCANR, Oakland, CA, 19pp.

Anderson, R. G., Ferreira, J. F. S., Jenkins, D. L., da Silva Dias, N. and Suarez, D. L. 2017. Incorporating field wind data to improve crop evapotranspiration parameterization in heterogeneous regions. *Irrigation Science* 35(6): 533–47.

Araji, S., Grammer, T. A., Gertzen, R., Anderson, S. D., Mikulic-Petkovsek, M., Veberic, R., Phu, M. L., Solar, A., Leslie, C. A. and Dandekar, A. M. 2014. Novel roles for the polyphenol oxidase enzyme in secondary metabolism and the regulation of cell death in walnut. *Plant Physiology* 164(3): 1191-203. doi:10.1104/pp.113.228593.

Arrieta, A. and Díaz, B. 2007. Native genotypes of the walnut tree (*Juglans regia* L) potentially tolerant to *Xanthomonas campestris pv juglandis*: prospection and selection. Cost 873, WG3/WG4 Joint Meeting, Murcia 2007. http://www.cost873.ch/_uploads/_files/m_Atanasio_murcia.pdf (accessed 20 November 2017).

Asayesh, Z. M., Vahdati, K. and Aliniaeifard, S. 2017a. Investigation of physiological components involved in low water conservation capacity of in vitro walnut plants. *Scientia Horticulturae* 224: 1-7.

Asayesh, Z. M., Vahdati, K., Aliniaeifard, S. and Askari, N. 2017b. Enhancement of *ex vitro* acclimation of walnut plantlets through modification of stomatal characteristics *in vitro*. *Scientia Horticulturae* 220: 114–21.

Avanzato, D. 2014. Following walnut footprints (*Juglans regia* L.): Cultivation and culture, folklore and history, traditional and uses. *Scripta Horticulturae* 17: 448.

Ayars, J. E., Fulton, A. and Taylor, B. 2015. Subsurface drip irrigation in California-here to stay? *Agricultural Water Management* 157: 39–47.

Bader, M. K.-F., Siegwolf, R. and Körner, C. 2010. Sustained enhancement of photosynthesis in mature deciduous forest trees after 8 years of free air $CO_2$ enrichment. *Planta* 232(5): 1115–25.

Baer, D. J., Gebauer S. K. and Novotny J. A. 2016. Walnuts consumed by healthy adults provide less available energy than predicted by the Atwater Factors. *Journal of Nutrition* 146(1): 9–13.

Bamberger, C., Rossmeier, A., Lechner, K., Wu, L., Waldmann, E., Stark, R. G., Altenhofer, J., Henze, K. and Parhofer, K. G. 2017. A walnut-enriched diet reduces lipids in healthy Caucasian subjects, independent of recommended macronutrient replacement and time point of consumption: A prospective, randomized, controlled trial. *Nutrients* 9(10): article 1097, 14pp.

Bandi, A., Hevesi, M., Szani, Z. and Tóth, M. 2015. Assessment of bacterial blight tolerance of Persian walnut based on immature nut test. *Notulae Scientia Biologicae* 7: 253–7. doi:10.15835/nsb.7.2.9620.

Barrueto, A. K., Merz, J., Hodel, E. and Eckert, S. 2017. The suitability of *Macadamia* and *Juglans* for cultivation in Nepal: An assessment based on spatial probability modelling using climate scenarios and *in situ* data. *Regional Environmental Change* 18: 859–71.

Berryman, C. E., Grieger, J. A., West, S. G., Chen, C. Y., Blumberg, J. B., Rothblat, G. H., Sankaranarayanan, S. and Kris-Etherton, P. M. 2013. Acute consumption of walnuts and walnut components differentially affect postprandial lipemia, endothelial function, oxidative stress, and cholesterol efflux in humans with mild hypercholesterolemia. *Journal of Nutrition* 143(6): 788–94.

Bockisch, M. 1998. *Fats and Oils Handbook*. American Oil Chemists Society Press, Champaign, IL, p. 838.

Bravo, K., Marcolini, G., Sorrenti, G., Baldi, E., Quartieri, M. and Toselli, M. 2017. Effect of time of application on nitrogen uptake, partitioning, and remobilization in walnut trees. *Journal of Plant Nutrition* 40(5): 719–25.

Browne, G. T., Leslie, C. A., Grant, J. A., Bhat, R. G., Schmidt, L. S., Hackett, W. P., Kluepfel, D. A., Robinson, R. and McGranahan, G. H. 2015. Resistance to species of phytophthora identified among clones of *Juglans microcarpa* × *J. Regia*. *HortScience* 50(8): 1136–42.

Buchner, R. P., Fulton, A. E., Gilles, C. K., Prichard, T. L., Lampinen, B. D., Shackel, K. A., Metcalf, S. G., Little, C. C. and Schwankl, L. J. 2008. Effects of regulated deficit irrigation on walnut grafted on 'Northern California Black' or 'Paradox' rootstock. *Acta Horticulturae* 795: 141–6.

Byerley, L. O., Samuelson, D., Blanchard, E., Luo, M., Lorenzen, B. N., Banks, S., Ponder, M. A., Welsh, D. A. and Taylor, C. M. 2017. Changes in the gut microbial communities following addition of walnuts to the diet. *Journal of Nutritional Biochemistry* 48: 94–102.

Cao, F. J. and Niu, J. X. 2015. Cloning and chromosome localisation of the linked SCAR marker (P1-P2 (762)) gene in early walnut (Dode). *Bangladesh Journal of Botany* 44(5): 719–25.

Cardinael, R., Mao, Z., Prieto, I., Stokes, A., Dupraz, C., Kim, J. H. and Jourdan, C. 2015. Competition with winter crops induces deeper rooting of walnut trees in a Mediterranean alley cropping agroforestry system. *Plant and Soil* 391(1–2): 219–35.

Castrillo, L. A., Mayfield, A. E., Griggs, M. H., Camp, R., Mudder, B., Taylord, A. and Vandenberg, J. D. 2017. Mortality and reduced brood production in walnut twig beetles, *Pityophthorus juglandis* (Coleoptera: Curculionidae), following exposure to commercial strains of entomopathogenic fungi *Beauveria bassiana* and *Metarhizium brunneum*. *Biological Control* 114: 79–86.

CDFA. 2017. California fertilization guidelines: Walnut. https://apps1.cdfa.ca.gov/FertilizerResearch/docs/Walnut.html (accessed 4 December 2017).

Chakraborty, S., Britton, M., Martínez-García, P. J. and Dandekar, A. M. 2016. Deep RNA-Seq profile reveals biodiversity, plant-microbe interactions and a large family of NBS-LRR resistance genes in walnut (*Juglans regia*) tissues. *AMB Express* 6(1): Article number 12, pp. 1–13.

Charra-Vaskou, K., Badel, E., Charrier, G., Ponomarenko, A., Bonhomme, M., Foucat, L., Mayr, S. and Ameglio, T. 2016. Cavitation and water fluxes driven by ice water potential in *Juglans regia* during freeze-thaw cycles. *Journal of Experimental Botany* 67(3): 739–50. doi:10.1093/jxb/erv486.

Charrier, G., Bonhomme, M., Lacointe, A. and Améglio. T. 2011. Are budburst dates, dormancy and cold acclimation in walnut trees (*Juglans regia* L.) under mainly genotypic or environmental control? *International Journal of Biometeorology* 55(6): 763–74.

Charrier, G., Poirier, M., Bonhomme, M., Lacointe, A. and Ameglio, T. 2013. Frost hardiness in walnut trees (*Juglans regia* L.): How to link physiology and modelling? *Tree Physiology* 33(11): 1229–41. doi:10.1093/treephys/tpt090.

Chatzistathis, T., Alifragis, D. and Papaioannou, A. 2015. The influence of liming on soil chemical properties and on the alleviation of manganese and copper toxicity in *Juglans regia, Robinia pseudoacacia, Eucalyptus* sp. and *Populus* sp. plantations. *Journal of Environmental Management* 150: 149–56. doi:10.1016/j.jenvman.2014.11.020.

Chen, S. F., Fichtner, E., Morgan, D. P. and Michailides, T. J. 2013. First report of *Lasiodiplodia citricola* and *Neoscytalidium dimidiatum* causing death of graft union of English walnut in California. *Plant Disease* 97(7): 993.

Chen, X., Xu, L., Wei, H. R., Zhu, D. Z., Tan, Y., Fu, M. R. and Liu, Q. Z. 2017. Transcriptome analysis, marker discovery and pigment biosynthesis of red-leaf *Juglans regia*. *Journal of Applied Biotechnology* 5(2): 20–45.

Cheniany, M., Ebrahimzadeh, H., Vahdati, K., Preece, J. E., Masoudinejad, A. and Mirmasoumi, M. 2013. Content of different groups of phenolic compounds in microshoots of *Juglans regia* cultivars and studies on antioxidant activity. *Acta Physiologiae Plantarum* 35(2): 443–50.

Cheon, W., Kim, Y. S., Lee, S. G., Jeon, Y. H. and Chun, I.-J. 2013. First report of branch dieback of walnut caused by *Neofusicoccum parvum* in Korea. *Plant Disease* 97(8): 1114.

Choudhary, R., Malik, S. K. and Chaudhury, R. 2015. A study of freezing behaviour for cryopreservation in walnut winter buds. *Research Journal of Biotechnology* 10(4): 17–20.

Ciceoi, R., Stanica, F., Wang, G., Chang, Y. M. and Wu, J. 2012. Sino-Romanian collaboration for walnut breeding. *Acta Horticulturae* 940: 225–32.

Couch, C., Franxman, T. and Greenhawt, M. 2017. Characteristics of tree nut challenges in tree nut allergic and tree nut sensitized individuals. *Annals of Allergy, Asthma & Immunology* 118(5): 591–6.e3.

Connell, J. H., Olson, W. H., Limberg, J. and Metcalf, S. G. 2010. Effects of various roots on 'Chandler' walnut catkin and pistillate bloom, tree growth, yield, and nut quality. *Acta Horticulturae* 861: 237–44.

Damatta, F. M., Godoy, A. G., Menezes-Silva, P. E., Martins, S. C. V., Sanglard, L. M. V. P., Morais, L. E., Torre-Neto, A. and Ghini, R. 2016. Sustained enhancement of photosynthesis in coffee trees grown under free-air $CO_2$ enrichment conditions: Disentangling the contributions of stomatal, mesophyll, and biochemical limitations. *Journal of Experimental Botany* 67(1): 341–52.

Darbyshire, R., Measham, P. and Goodwin, I. 2016. A crop and cultivar-specific approach to assess future winter chill risk for fruit and nut trees. *Climatic Change* 137(3–4): 541–56.

Davidson, G. R., Frelka, J. C., Yang, M., Jones, T. M. and Harris, L. J. 2015. Prevalence of *Escherichia coli* O157:H7 and *Salmonella* on in-shell California walnuts. *Journal of Food Protection* 78(8): 1547–53. doi:10.4315/0362-028X.JFP-15-001.

Del Gobbo, L. C., Falk, M. C., Feldman, R., Lewis, K. and Mozaffarian, D. 2015. Effects of tree nuts on blood lipids, apolipoproteins, and blood pressure: Systematic review, meta-analysis, and dose-response of 61 controlled intervention trials. *American Journal of Clinical Nutrition* 102(6): 1347–56.

Delaire, M., Frak, E., Sigogne, M., Adam, B., Beaujard, F. and Le Roux, X. 2005. Sudden increase in atmospheric $CO_2$ concentration reveals strong coupling between shoot carbon uptake and root nutrient uptake in young walnut trees. *Tree Physiology* 25(2): 229–35.

Dömötör, D., Frank, T., Rákhely, G., Doffkay, Z., Schneider, G. and Kovács, T. 2016. Comparative analysis of two bacteriophages of *Xanthomonas arboricola pv. juglandis*. *Infection, Genetics and Evolution* 43: 371–7. doi:10.1016/j.meegid.2016.06.011.

Downs, M. L., Baumert, J. L., Taylor, S. L. and Mills, E. N. C. 2016. Mass spectrometric analysis of allergens in roasted walnuts. *Journal of Proteomics* 142: 62–9. doi:10.1016/j.jprot.2016.04.045.

Duan, Z. P., Gan, Y. W., Wang, B. J., Hao, X. D., Xu, W. L., Zhang, W. and Li, L. H. 2017. Interspecific interaction alters root morphology in young walnut/wheat agroforestry systems in northwest China. *Agroforestry Systems* (in press). doi:10.1007/s10457-017-0133-2.

Duursma, R. A., Gimeno, T. E., Boer, M. M., Crous, K. Y., Tjoelker, M. G. and Ellsworth, D. S. 2016. Canopy leaf area of a mature evergreen Eucalyptus woodland does not respond to elevated atmospheric [$CO_2$] but tracks water availability. *Global Change Biology* 22(4): 1666–76.

Ebrahimi, A., Zarei, A., Fardadonbech, M. Z. and Lawson, S. 2017. Evaluation of genetic variability among 'Early Mature' *Juglans regia* using microsatellite markers and morphological traits. *PEER Journal* 5: e3834. doi:10.7717/peerj.3834.

Epstein, L. and Bassein, S. 2001. Pesticide applications of copper on perennial crops in California, 1993 to 1998. *Journal of Environmental Quality* 30: 1844–7.

Evans, K. J., McNeil, D. L., Jack, C. and Lang, M. D. 2016. Establishing quality parameters for Australian walnuts. *Acta Horticulturae* 1109: 23–8. doi:10.17660/ActaHortic.2016.1109.4.

FAO. 2017. www.fao.org/faostat/ (accessed 15 November 2017).

Foreign Agricultural Service. 2017. Tree nuts: World markets and trade. https://apps.fas.usda.gov/psdonline/circulars/TreeNuts.pdf (accessed 15 November 2017).

Frelka, J. C., Davidson, G. R. and Harris, L. J. 2016. Changes in aerobic plate and Escherichia coli-coliform counts and in populations of inoculated foodborne pathogens on in-shell walnuts during storage. *Journal of Food Protection* 79(7): 1143–53.

Frutos, D. 2010. Mini-review: Bacterial diseases of walnut and hazelnut and genetic resources. *Journal of Plant Pathology* 92: S1.79–S1.85.

Fuentealba, C., Hernández, I., Saa, S., Toledo, L., Burdiles, P., Chirinos, R., Campos, D., Brown, P. and Pedreschi, R. 2017. Colour and in vitro quality attributes of walnuts from different growing conditions correlate with key precursors of primary and secondary metabolism. *Food Chemistry* 232: 664–72.

Fulton, A. and Buchner, R. 2015. Drought strategies for California walnut production. ANR Publication 8533. University of California. https://anrcatalog.ucanr.edu/pdf/8533.pdf.

Fulton, A., Grant, J., Buchner, R. and Connell, J. 2014. Using the pressure chamber for irrigation management in walnut, almonds and prunes. ANR publication 8503, May 2014, 27p.

Gandev, S. 2017. Walnut propagation using a hot water installation and growing the obtained plants in containers. *Bulgarian Journal of Agricultural Science* 23(1): 83–5.

Gauthier, M.-M. and Jacobs, D. F. 2011. Walnut (*Juglans* spp.) ecophysiology in response to environmental stresses and potential acclimation to climate change. *Annals of Forest Science* 68(8): 1277–90.

Ginetti, B., Ragazzi, A. and Moricca, S. 2014. First report of *Phytophthora* taxon walnut in Lombardy, North Italy. *Plant Disease* 98(3): 424.

Goharrizi, M. A. S. B., Dejahang, A., Tohidfar, M., Darbandi, A. I., Carillo, N., Hajirezaei, M. R. and Vahdati, K. 2016. Agrobacterium mediated transformation of somatic embryos of Persian walnut using *fld* gene for osmotic stress tolerance. *Journal of Agricultural Science and Technology* 18(2): 423–35.

Grace, M. H., Warlick, C. W., Neff, S. A. and Lila, M. A. 2014. Efficient preparative isolation and identification of walnut bioactive components using high-speed counter-current chromatography and LC-ESI-IT-TOF-MS. *Food Chemistry* 158: 229–38.

Grant, J., Bentley, W., Pickel, C. and Groh-Lowrimore, J. 2003. BIOS approach tested for controlling walnut pests in San Joaquin Valley. *Californian Agriculture* 57(3): 86–92. doi:10.3733/ca.v057n03p86.

Grant, J., Anderson, K. K., Pritchard, T., Hasey, J., Bugg, R. L., Thomas, F. and Johnson, T. 2006. Cover crops for walnut orchards. UCANR publication 21627, Oakland, CA, 20p. ISBN-13: 978-1-601-397-6.

Guardia, M., Diaz, R., Save, R. and Aleta, N. 2013. Autumn frost resistance on several walnut species: Methods comparison and impact of leaf fall. *Forest Science* 59(5): 559–65. doi:10.5849/forsci.12-094.

Guardia, M., Charrier, G., Vilanova, A., Save, R., Ameglio, T. and Aleta, N. 2016. Genetics of frost hardiness in *Juglans regia* L. and relationship with growth and phenology. *Tree Genetics and Genomes* 12(5): article 83. doi:10.1007/s11295-016-1038-3.

Guasch-Ferré, M., Liu, X., Malik, S. V., Sun, Q., Willett, W. C., Manson, J. E., Rexrode, K. M., Li, Y., Hu, F. B. and Bhupathiraju, S. N. 2017. Nut consumption and risk of cardiovascular disease. *Journal of the American College of Cardiology* 70(20): 2519-32. doi:10.1016/j.jacc.2017.09.035.

Habibipour, R., Tamandegani, P. R. and Farmany, A. 2016. Monitoring of aflatoxin G1, B1, G2, and B2 occurrence in some samples of walnut. *Environmental Monitoring and Assessment* 188(12): article 669.

Haddad, E. H., Gaban-Chong, N., Oda, K. and Sabaté, J. 2014. Effect of a walnut meal on postprandial oxidative stress and antioxidants in healthy individuals. *Nutrition Journal* 13(1): 4. doi:10.1186/1475-2891-13-4.

Haider, S., Batool, Z., Ahmad, S., Siddiqui, R. A. and Haleem, D. J. 2017. Walnut supplementation reverses the scopolamine-induced memory impairment by restoration of cholinergic function via mitigating oxidative stress in rats: A potential therapeutic intervention for age related neurodegenerative disorders. *Metabolic Brain Disease* 33: 39-51.

Han, H., Woeste, K. E., Hu, Y., Dang, M., Zhang, T., Gao, X.-X., Zhou, H., Feng, X., Zhao, G. and Zhao, P. 2016. Genetic diversity and population structure of common walnut (*Juglans regia*) in China based on EST-SSRs and the nuclear gene phenylalanine ammonia-lyase (PAL). *Tree Genetics and Genomes* 12(6): Article 111.

Hanson, B., Orloff, S. and Peters, D. 2000. Monitoring soil moisture helps refine irrigation management. *Californian Agriculture* 54(3): 38–42. doi:10.3733/ca.v054n03p38.

Hasey, J. K., Kluepfel, D. A. and Anderson, K. K. 2014. Crown gall incidence and severity: Seedling walnut rootstock versus clonally propagated rootstock. *Acta Horticulturae* 1050: 305–8.

Hasey, J. K., Buchner, R. P., Klonsky, K., Sumner, D., Anderson, N. and Stewart, D. 2015. Sample costs to establish and produce English walnuts. UC Cooperative extension Service, CA, 20p. https://coststudyfiles.ucdavis.edu/uploads/cs_public/78/a7/78a775e3-6488-49d4-b234-b3eaf9171a55/15walnutssacvalleyfinaldraftjan4.pdf.

Hayes, D., Angove, M. J., Tucci, J. and Dennis, C. 2016. Walnuts (*Juglans regia*) chemical composition and research in human health. *Critical Reviews in Food Science and Nutrition* 56(8): 1231–41.

Hefty, A. R., Seybold, S. J., Aukema, B. H. and Venette, R. C. 2017. Cold tolerance of *Pityophthorus juglandis* (Coleoptera: Scolytidae) from northern California. *Environmental Entomology* 46(4): 967–77.

Hu, Z., Pan, C., Xiao, B. and Pan, X. 2015. Spectral characteristic parameter-based models for foliar nitrogen concentration estimation of *Juglans regia*. *Nongye Gongcheng Xuebao/Transactions of the Chinese Society of Agricultural Engineering* 31(9): 180–6.

Hu, Z.-Z., Pan, C.-D., Xiao, B. and Pan, X. 2016. Spectral characteristic parameters estimation models for potassium content of *Juglans regia* 'Wen185' leaves. *Chinese Journal of Applied Ecology* 27(5): 1393–400.

Hu, Y. H., Dang, M., Feng, X. J., Woeste, K. and Zhao, P. 2017a. Genetic diversity and population structure in the narrow endemic Chinese walnut *Juglans hopeiensis* Hu: Implications for conservation. *Tree Genetics and Genomes* 13(4): article 91. doi:10.1007/s11295-017-1172-6.

Hu, Y., Woeste, K. E. and Zhao, P. 2017b. Completion of the chloroplast genomes of five Chinese *Juglans* and their contribution to chloroplast phylogeny. *Frontiers in Plant Science* 7: 1955. URL=https://www.frontiersin.org/article/10.3389/fpls.2016.01955.

Hu, W., Du, W., Bai, S., Lv, S. and Chen, G. 2017c. Phenoloxidase, an effective bioactivity target for botanical insecticide screening from green walnut husks. *Natural Product Research*. doi:10.1080/14786419.2017.1380015.

Hussain, S. Z., Ahad, T., Rather, A. H. and Naik, H. R. 2016. Development of walnut dehulling machine and assessment of its performance using Ethephon and Tween-80 as pre-treatments for hull loosening. *Journal of Food Science and Technology* 53(6): 2835–43.

IPCC. 2015. Climate change 2014 synthesis report. ISBN: 978-92-9169-143-2. https://www.ipcc.ch/pdf/assessment-report/ar5/syr/SYR_AR5_FINAL_full_wcover.pdf (accessed 4 December 2017).

Jenkins, T. A., Marsh, C., Lang, M. D., Vanestre, J., Walter, M. and Obanor, F. 2010. Walnut blight sustainable management research in New Zealand. *Acta Horticulturae* 861: 479–88.

Jones, J. B., Vallad, G. E., Iriarte, F. B., Obradović, A., Wernsing, M. H., Jackson, L. E., Balogh, B., Hong, J. C. and Momol, M. T. (2012). Considerations for using bacteriophages for plant disease control. *Bacteriophage* 2(4): 208–14. doi:10.4161/bact.23857.

Katz, D. L., Davidhi, A., Ma, Y., Kavak, Y., Bifulco, L. and Njike, V. Y. 2012. Effects of walnuts on endothelial function in overweight adults with visceral obesity: A randomized, controlled, crossover trial. *Journal of the American College of Nutrition* 31(6): 415–23.

Keshavarz, K., Vahdati, K., Samar, M., Azadegan, B. and Brown. P. H. 2011. Foliar application of zinc and boron improves walnut vegetative and reproductive growth. *HortTechnology* 21(2): 181–6.

Kim, J. H., Chan, K. L. and McConnell, L. L. 2015. Augmenting the efficacy of fungal and mycotoxin control via chemosensitization. *Outlooks on Pest Management* 26(4): 171–5.

Kocharunchitt, C., Ross, T. and McNeil, D. L. 2009. Use of bacteriophages as biocontrol agents to control *Salmonella* associated with seed sprouts. *International Journal of Food Microbiology* 128(3): 453–9. doi:10.1016/j.ijfoodmicro.2008.10.014.

Lampinen, B. D., Grant, J. A., Metcalf, S. G., Negron, C., Martinelli, A. and Rutto, L. K. 2010. Walnut production and quality as influenced by orchard and within tree canopy environment. *Acta Horticulturae* 861: 204–14.

Lampinen, B. D., Edstrom, J. P., Metcalf, S. G., Stewart, W. L., Negron, C. M. and Contador, M. L. 2015. Howard walnut trees can be brought into bearing without annual pruning. *California Agriculture* 69(2): 123–8.

Lang, M. D. and Evans, K. J. 2010. Mini-review: Epidemiology and status of walnut blight in Australia. *Journal of Plant Pathology* 92(Suppl. 1): S1.49–S1.55.

Lee, S. H., Lee, C. K., Cho, S. E., Park, J. H. and Shin, H. D. 2016. First report of powdery mildew caused by *Phyllactinia juglandis* on *Juglans mandshurica* in Korea. *Plant Disease* 100(1): 214.

Leslie, C. A., Seybold, S. J., Graves, A. D., Cranshaw, W. and Tisserat, N. 2010. Potential impacts of thousand cankers disease on commercial walnut production and walnut germplasm conservation. *Acta Horticulturae* 861: 431–4.

Leslie, C. A., Walawage, S. L., Uratsu, S. L., McGranahan, G. and Dandekar, A. M. 2015. Walnuts (*Juglans*). *Methods in Molecular Biology* 1224: 229–41. doi:10.1007/978-1-4939-1658-0_19.

Li, J., Yang, Z., Mei Z., Feng, B., Wang. P. and Wang, X. 2013. Control effects on *Batocera lineolate* (Coleoptera:Cerambycidae) attacking walnut trees by releasing parasitoid *Dastarcus helophoroides* (Coleoptera:Bothrideridae). *Chinese Journal of Biological Control*. http://en.cnki.com.cn/Article_en/CJFDTotal-ZSWF201302004.htm (accessed 25 November 2017).

Licea-Moreno, R. J., Contreras, A., Morales, A. V., Urban, I., Daquinta, M. and Gomez, L. 2015. Improved walnut mass micro-propagation through the combined use of phloroglucinol and FeEDDHA. *Plant Cell, Tissue and Organ Culture* 123(1): 143–54.

Linacero, R., Ballesteros, I., Sanchiz, A., Prieto, N., Iniesto, E., Martinez, Y., Pedrosa, M. M., Muzquiz, M., Cabanillas, B., Rovira, M., Burbano, C. and Cuadrado, C. 2016. Detection by real time PCR of walnut allergen coding sequences in processed foods. *Food Chemistry* 202: 334–40.

Liu, X., Walawage, S. L., Leslie, C. A., Dandekar, A. M., Tricoli, D. M., Hu, H., Huang, Y., Zhang, J., Xv, C., Huang, J. and Zhang, Q. 2017. *In vitro* gene expression and mRNA translocation from transformed walnut (*Juglans regia*) rootstocks expressing DsRED fluorescent protein to wild-type scions. *Plant Cell Reports* 36(6): 877–85.

Lu, S., Meng, P., Zhang, J., Yin, C. and Sun, S. 2015. Changes in soil organic carbon and total nitrogen in croplands converted to walnut-based agroforestry systems and orchards in southeastern Loess Plateau of China. *Environmental Monitoring and Assessment* 187(11): article 688, 9p.

Luedeling, E. 2012. Climate change impacts on winter chill for temperate fruit and nut production: A review. *Scientia Horticulturae* 144: 218–29. doi:10.1016/j.scienta.2012.07.011.

Luedeling, E., Zhang, M., McGranahan, G. and Leslie, C. 2009. Validation of winter chill models using historic records of walnut phenology. *Agricultural and Forest Meteorology* 149(11): 1854–64. doi:10.1016/j.agrformet.2009.06.013.

Luo, M.-C., You, F. M., Li, P., Wang, J.-R., Zhu, T., Dandekar, A. M., Leslie, C. A., Aradhya, M., McGuire, P. E. and Dvorak, J. 2015. Synteny analysis in Rosids with a walnut physical map reveals slow genome evolution in long-lived woody perennials. *BMC Genomics* 16(1): 707. doi:10.1186/s12864-015-1906-5.

Mace, K. C. and Mills, N. J. 2017. Nitrogen-mediated interaction: A walnut-aphid-parasitoid system. *Environmental Entomology* 45(4): 891–6.

Malhotra, S. P. and McNeil, D. L. 2010. Economics of walnut production in India. *Acta Horticulturae* 861: 229–36.

Martínez-García, P. J., Crepeau, M. W., Puiu, D., Gonzalez-Ibeas, D., Whalen, J., Stevens, K. A., Paul, R., Butterfield, T. S., Britton, M. T., Reagan, R. L., Chakraborty, S., Walawage, S. L., Vasquez-Gross, H. A., Cardeno, C., Famula, R. A., Pratt, K., Kuruganti, S., Aradhya, M. K., Leslie, C. A., Dandekar, A. M., Salzberg, S. L., Wegrzyn, J. L., Langley, C. H. and Neale, D. B. 2016. The walnut (*Juglans regia*) genome sequence reveals diversity in genes coding for the biosynthesis of non-structural polyphenols. *The Plant Journal* 87: 507–32. doi:10.1111/tpj.13207.

Martínez-García, P. J., Famula, R. A., Leslie, C., McGranahan, G. H., Famula, T. R. and Neale, D. B. 2017. Predicting breeding values and genetic components using generalized linear mixed models for categorical and continuous traits in walnut (*Juglans regia*). *Tree Genetics and Genomes* 13(5): article 109. doi:10.1007/s11295-017-1187-z.

Matti, M. 2017. Man dies of anaphylaxis following exposure to walnuts used in sandblasting. *Allergic Living*, Fall issue. https://www.allergicliving.com/2017/10/23/man-dies-of-anaphylaxis-following-exposure-to-walnuts-used-in-sandblasting/ (accessed 22 November 2017).

Mayhew, A. J, De Souza, R. J, Meyre, D., Anand, S. S. and Mente, A. 2015. A systematic review and meta-analysis of nut consumption and incident risk of CVD and all-cause mortality. *British Journal of Nutrition* 115(2): 212-25.

McGranahan, G. and Leslie, C. 2012. Walnut. In Badenes, M. L. and Byrne, D. H. (Eds), *Fruit Breeding. Handbook of Plant Breeding*, Volume 8. Springer, Boston, MA, pp. 827–46. doi:10.1007/978-1-4419-0763-9_22.

McNeil, D. L. 2010. A global perspective of walnut production. *Acta Horticulturae* 861: 25–30.

McNeil, D. L. 2013a. Retail walnut quality - an analysis of walnut quality and value in three major global markets. *Australian Nut Grower Journal* 27(1): 37–40.

McNeil, D. L. 2013b. Improving the quality and safety of walnuts. In: Harris, L. (Ed.), *Improving the Safety and Quality of Nuts*. Woodhead Publishing, UK, pp. 245–73. ISBN:978-0857092663.

McNeil, D. L. 2014a. Following walnut footprints in Australia. In: *Following Walnut Footprints (Juglans regia L.): Cultivation and Culture, Folk and History, Traditions and Use*. International Society for Horticultural Science, Leuven, Belgium. *Scripta Horticulturae* 17: 49–59. ISBN:978-94-6261-003-3.

McNeil, D. L. 2014b. Case studies in walnut industry development; Guangxi, Australia and New Zealand. *Proceedings of the 12th Rural Entrepreneurship Conference*, Harper Adams University. https://cdn.harper-adams.ac.uk/document/page/153_David-McNeil-Studies-in-the-walnut-industry.pdf (accessed 4 December 2017).

McNeil, D. L. 2014c. Market chain insights created by empirical modelling of inputs to the UK nut market. *British Food Journal* 116(12): 1960–75.

McNeil, D. L. 2016. Walnut blight disease development on Guangxi walnuts. *Acta Horticulturae* 1109: 261–4.

McNeil, D. L. and Evans, K. J. 2014. *Improved Management of Walnut Quality Factors Under Grower Control*. University of Tasmania, Hobart, Australia, p. 72. ISBN:978-1-63175-123-5.

McNeil, D. L. and Felgate, M. K. 2012. Sales and promotion in the United Kingdom nut and walnut retail market. *Australian Nut Grower Journal* 26(4): 32–4.

McNeil, D. L. and Felgate, M. K. 2014. Analysis of UK walnut market import purchase behaviour'. *British Food Journal* 116(3): 405–18. doi:10.1108/BFJ-03-2012-0071 2014.

McNeil, D. L. and Savage, G. P. 2000. Some New Zealand sources of healthy oils-olive and nut oils. *Proceedings of the Nutrition Society of New Zealand* 25: 10–18.

McNeil, D. L. and Savage, G. P. 2001. New Zealand walnut variety trials 1985–1999. *Acta Horticulturae* 544: 157–64.

McNeil, D. L. and Zeng, L. 2013. Walnuts in Guangxi Province China. *Australian Nut Grower Journal* 27(4): 28–30.

McNeil, D. L., Romero, S., Kandula, J., Stark, C., Stewart, A. and Larsen, S. 2001. Bacteriophages: A potential biocontrol agent against walnut blight (*Xanthomonas campestris* pv. *juglandis*). *New Zealand Plant Protection* 54: 220–4.

McNeil, D. L., Evans, K. J. and Adem, H. H. 2014. Creating, maintaining and assessing walnut quality from grower to market. *Acta Horticulturae* 1050: 299–303.

McWilliam, V., Koplin, J., Lodge, C., Tang, M., Dharmage, S. and Allen, K. 2015. The prevalence of tree nut allergy: A systematic review. *Current Allergy and Asthma Reports* 15(9): Article 54, 13p.

Meena, V. S., Meena, S. K., Verma, J. P., Kumar, A., Aeron, A., Mishra, P. K., Bisht, J. K., Pattanayak, A., Naveed, M. and Dotaniya, M. L. 2017. Plant beneficial rhizospheric microorganism (PBRM) strategies to improve nutrients use efficiency: A review. *Ecological Engineering* 107: 8–32.

Meier-Dinkela, A. and Wenzlitschke, I. 2017. Micropropagation of mature *Juglans* hybrids. *Acta Horticulturae* 1155: 85–92.

Miller, P. W. 1934. Walnut blight and its control in the Pacific Northwest USA. Circulation no. 333. US Department of Agriculture, Washington, DC, p. 13.

Miller, P. W. 1951. The control of walnut blight in the Pacific Northwest. Extension Circular 488. Federal Cooperative Extension Service. Oregon State College, Corvallis, OR.

Mohseniazar, M., Tohidfar, M. and Vahdati, K. 2017. In silico coding sequence analysis of walnut *GAI* and *PIP2* genes and comparison with different plant species. *Majallah-i 'Ulum-i Bāghbānī* 30(3): 480–90.

Montecchio, L., Fanchin, G., Simonato, M. and Faccoli, M. 2014. First record of Thousand Cankers Disease fungal pathogen *Geosmithia morbida* and walnut twig beetle *Pityophthorus juglandis* on *Juglans regia* in Europe. *Plant Disease* 98(10): 1445. doi:10.1094/PDIS-07-14-0719-PDN.

Montecchio, L., Faccoli, M., Short, D. P. G., Fanchin, G., Geiser, D. M. and Kasson, M. T. 2015. First report of *Fusarium solani* phylogenetic species 25 associated with early stages of thousand cankers disease on *Juglans nigra* and *Juglans regia* in Italy. *Plant Disease* 99(8): 1183.

Montecchio, L., Vettorazzo, M. and Faccoli, M. 2016. Thousand cankers disease in Europe: An overview. *EPPO Bulletin* 46(2): 335–40.

Neale, D. B. 2017. Development of Molecular Breeding in the Walnut Improvement Program. https://walnuts.org/news/development-molecular-breeding-walnut-improvement-program-david-b-neale-dept-plant-sciences-uc-davis/ (accessed 15 November 2017).

Ninot, A., Aletà, N., Moragrega, C. and Montesinos, E. 2002. Evaluation of a reduced copper spraying program to control bacterial blight of walnut. *Plant Disease* 86(6): 583–7.

Njike, V. Y., Yarandi, N., Petraro, P., Ayettey, R. G., Treu, J. A. and Katz, D. L. 2016. Inclusion of walnut in the diets of adults at risk for type 2 diabetes and their dietary pattern changes: A randomized, controlled, cross-over trial. *BMJ Open Diabetes Research & Care* 4(1): e000293.

Osterberg, K., Savage, G. P. and McNeil, D. L. 2001. Oxidative stability of walnuts during long term storage in shell. *Acta Horticulturae* 544: 591-7.

Payghamzadeh, K. and Kazemitabar, S. K. 2011. In vitro propagation of walnut - A review. *African Journal of Biotechnology* 10(3): 290–311.

Pham, N. T., Meier-Dinkel, A., Höltken, A. M., Quambusch, M., Mahnkopp, F. and Winkelmann, T. 2017. Endophytic bacterial communities in in vitro shoot cultures derived from embryonic tissue of hybrid walnut (*Juglans × intermedia*). *Plant Cell, Tissue and Organ Culture* 130(1): 153–65.

Popović, T., Ivanović, Z., Živković, S., Trkulja, N. and Ignjatov, M. 2013. First report of *Brenneria nigrifluens* as the causal agent of shallow-bark canker on walnut trees (*Juglans Regia*) in Serbia. *Plant Disease* 97(11): 1504.

Poulose, S., Miller, M. and Shukitt-Hale, B. 2014. Role of walnuts in maintaining brain health with age. *Journal of Nutrition* 144(4): 561S–566S.
Querné, A., Battie-laclau, P., Dufour, L., Wery, J. and Dupraz, C. 2017. Effects of walnut trees on biological nitrogen fixation and yield of intercropped alfalfa in a Mediterranean agroforestry system. *European Journal of Agronomy* 84: 35–46.
Qureshi, S. N., Javaid, K., Wani, I. A., Muzafar, M., Mehraj, S., Mushtaq, R., Basu, Y. A. and Padder, B. A. 2016. Effect of rootstock age and type of scion wood on epicotyl grafting in walnut (*Juglans regia* L.) under Kashmir conditions. *Ecology, Environment and Conservation* 22(3): 1435–56.
Rahman, K. A. and Khan, A. W. 1942. A study of the life-history and control of *Batocera horsfieldi* hope (lamiidæ: Coleoptera) A borer pest of walnut tree in the Punjab. *Proceedings of the Indian Academy of Sciences – Section B* 15(4): 201–5.
Rajaram, S., Valls-Pedret, C., Cofán, M., Sabaté, J., Serra-Mir, M., Pérez-Heras, A. M., Arechiga, A., Casaroli-Marano, R. P., Alforja, S., Sala-Vila, A., Doménech, M., Roth, I., Freitas-Simoes, T. M., Calvo, C., López-Illamola, A., Haddad, E., Bitok, E., Kazzi, N., Huey, L., Fan, J. and Ros, E. 2017. The Walnuts and Healthy Aging Study (WAHA): Protocol for a nutritional intervention trial with walnuts on brain aging. *Frontiers in Aging Neuroscience* 8(January): Article 333.
Ramos, D. E. 1997. *Walnut Production Manual*. UCANR Publications, Technology & Engineering, p. 320. ISBN:13 9781879906334.
Rao, G. D., Sui, J. K. and Zhang, J. G., 2016. Metabolomics reveals significant variations in metabolites and correlations regarding the maturation of walnuts (*Juglans regia* L.). *Biology Open* 5(6): 829–36. doi:10.1242/bio.017863.
Ren, C., Wei, C. and Bo, R. 2014. Strategies for the comprehensive development of the Fenzhou walnut industry. *Acta Horticulturae* 1050: 361–8.
Retamales, J., Segovia, C., Alvarado, R., Nuñez, P. and Santander, J. 2017. Draft genome sequence of *Xanthomonas arboricola pv. juglandis* J303, isolated from infected walnut trees in southern Chile. *Genome Announc.* 5: e01085-17. doi:10.1128/genomeA.01085-17.
Rezaee, R., Hassani, D. and Vahdati, K. 2014. Long term trials on top-working of walnut trees in Iran. *Acta Horticulturae* 1050: 197–202.
Riyaz-Ul-Hassan, S., Verma, V., Malik, A. and Qazi, G. N. 2003. Microbiological quality of walnut kernels and apple juice concentrate. *World Journal of Microbiology & Biotechnology* 19: 845–50.
Rock, C. L., Flatt, S. W., Barkai, H.-S., Pakiz, B. and Heath, D. D. 2017. A walnut-containing meal had similar effects on early satiety, CCK, and PYY, but attenuated the postprandial GLP-1 and insulin response compared to a nut-free control meal. *Appetite* 117: 51–7.
Romero-Suarez, S. 2006. Ecology of *Xanthomonas campestris pv. juglandis* bacteriophages isolated from walnut orchards in New Zealand. PhD thesis, Lincoln University. https://researcharchive.lincoln.ac.nz/handle/10182/1650?show=full (accessed 15 November 2017).
Romero-Suarez, S., Jordan, B. and Heinemann, J. A. 2012. Isolation and characterization of bacteriophages infecting *Xanthomonas arboricola pv. juglandis*, the causal agent of walnut blight disease. *World Journal of Microbiology and Biotechnology* 28(5): 1917–27. doi:10.1007/s11274-011-0992-z.
Rosati, A., Metcalf, S. G., Buchner, R. P., Fulton, A. and Lampinen, B. D. 2006. Physiological effects of kaolin applications in well-irrigated and water-stressed walnut and almond trees. *Annals of Botany* 98: 267–75. doi:10.1093/aob/mcl100.

Sambeek, J. W. and Reid, W. 2009. Foliar analyses for improved nut production: Step 2. http://nutgrowing.org/feature1b.htm (accessed 4 December 2017).
Sánchez-González, C., Ciudad, C. J., Noé, V. and Izquierdo-Pulido, M. 2017. Health benefits of walnut polyphenols: An exploration beyond their lipid profile. *Critical Reviews in Food Science and Nutrition* 57(16): 3373–83.
Savage, G. P., McNeil, D. L. and Dutta, P. C. 1998. Vitamin E content and oxidative stability of fatty acids in walnut oils. *Proceedings the Nutrition Society of New Zealand* 23: 64–9.
Savage, G. P., Dutta, P. C. and McNeil, D. L. 1999. Fatty acid and tocopherol contents and oxidative stability of walnut oils. *Journal of the American Oil Chemists' Society* 76: 1059–63.
Savage, G. P., McNeil, D. L. and Dutta, P. C. 2001. Some nutritional advantages of walnuts. *Acta Horticulturae* 544: 557–63.
Scheuring, A. F. 1997. *Eighty Years of Excellence: A History of Diamond Walnut Growers*. Diamond Walnut Growers Inc., Stockton, CA, 114p. ISBN:0-9659264-0-0.
Schlörmann, W., Lamberty, J., Ludwig, D., Lorkowski, S. and Glei, M. 2017. *In vitro*-fermented raw and roasted walnuts induce expression of *CAT* and *GSTT2* genes, growth inhibition, and apoptosis in LT97 colon adenoma cells. *Nutrition Research* 47: 72–80.
Selma, M. V., González-Sarrías, A., Salas-Salvadó, J., Andrés-Lacueva, C., Alasalvar, C., Örem, A., Tomás-Barberán, F. A. and Espín, J. C. 2016. The gut microbiota metabolism of pomegranate or walnut ellagitannins yields two urolithin-metabotypes that correlate with cardiometabolic risk biomarkers: Comparison between normo-weight, overweight-obesity and metabolic syndrome. *Clinical Nutrition* 37(3): 897–905. doi:doi:10.1016/j.clnu.2017.03.012.
Seybold, S. J., King, J. A., Harris, D. R., Nelson, L. J., Hamud, S. M. and Chen, Y. 2012. Diurnal flight response of the walnut twig beetle, *Pityophthorus juglandis* Blackman (Coleoptera: Scolytidae), to pheromone-baited traps in two northern California walnut habitats. *Pan-Pacific Entomologist* 88(2): 231–47.
Shackel, K., Buchner, R., Connell, J., Edstrom, J., Fulton, A., Holtz, B., Lampinen, B., Reil, W., Stewart, W. and Viveros, M. 2010. Midday stem water potential as a basis for irrigation scheduling. *ASABE – 5th National Decennial Irrigation Conference Proceedings*, 5–8 December 2010, Phoenix Convention Center, Phoenix, Arizona USA IRR10-9993. (doi:10.13031/2013.35841).
Shanxi Academy of Forestry. 2011. Report on achievements of Shanxi Academy of Forestry walnut program, 1985–2010, 37p (Chinese and English).
Sharma, R. M., Pandey, M. K. and Shankar, U. 2012. Pest management in walnut: An overview. In: Abrol, D. P. and Uma Shankar (Eds), *Ecologically Based Integrated Pest Management*, . New India Publishing Agency, New Delhi, India, , pp. 765–85. www.nipabooks.com. Available from: https://www.researchgate.net/publication/303485932_Pest_Management_in_Walnut_An_Overview (accessed 25 November 2017).
Shin, Y. and Jung, Y. 2014. Development of irrigation water management model for reducing drought severity using remotely sensed soil moisture footprints. *Journal of Irrigation and Drainage Engineering* 140(7): article 04014021.
Shu, Z., Zhang, X., Yu, D., Xue, S. and Wang, H. 2016. Natural hybridization between Persian walnut and Chinese walnut revealed by simple sequence repeat markers. *Journal of the American Society for Horticultural Science* 141(2): 146–50.

Singh, S. R., Ahmed, N., Mir, J. I. and Abid Mir, A. 2012. Present status and future strategies for walnut production in India. https://www.biotecharticles.com/PDF_Articles/Recent_advances_in_walnut_production_BA_4098.pdf (accessed 15 November 2017).

Smith, J. D., Hou, T., Ludwig, D. S., Rimm, E. B., Willett, W., Hu, F. B. and Mozaffarian, D. 2015. Changes in intake of protein foods, carbohydrate amount and quality, and long-term weight change: Results from 3 prospective cohorts. *American Journal of Clinical Nutrition* 101(6): 1216–24.

Snyder, R. L. 2000. Principles of frost protection. University of California Davis, CA, 26p. http://biomet.ucdavis.edu/frostprotection/Principles%20of%20Frost%20Protection/PFPlong/PFPlong.pdf (accessed 4 December 2017).

Soderquist, C. J. 1973. Juglone and allelopathy. *Journal of Chemical Education* 50(11): 782–3. doi:10.1021/ed050p782.

Solar, A. 2006. Evaluation of some perspective walnut genotypes in Slovenia. *Acta Horticulturae* 705: 131–6.

Spaccarotella, K. J., Kris-Etherton, P. M., Stone, W. L., Bagshaw, D. M., Fishell, V. K., West, S. G., Lawrence, F. R. and Hartman, T. J. 2008. The effect of walnut intake on factors related to prostate and vascular health in older men. *Nutrition Journal* 7: 13.

Sreelatha, T., Kandhasamy, S., Dinesh, R., Shruthy, S., Shweta, S., Mukesh, D., Karunagaran, D., Balaji, R., Mathivanan, N. and Perumal, P. T. 2014. Synthesis and SAR study of novel anticancer and antimicrobial naphthoquinone amide derivatives. *Bioorganic and Medicinal Chemistry Letters* 24(15): 3647–51.

Stark, C., McNeil, D. L. and Savage, G. P. 2000. The effect of storage conditions on the stability of peroxide values of New Zealand grown walnuts. *Proceedings of the Nutrition Society of New Zealand* 25: 43–54.

Strand, L. L. 2003. *Integrated Pest Management for Walnuts*, 3rd edition. UC ANR Communication Services, USA, 136p. ISBN-13: 978-1-879906-62-4.

Su, J. J. and Wang, X. C. 2017. Spatio-temporal variations in climate-growth relationships of three hardwood tree species across the north Zhangguangcai mountains, Northeast China. *Shengtai Xuebao/Acta Ecologica Sinica* 37(5): 1484–95.

Sun, Z. X., Hsiang, T., Li, J. K., Zhou, Y., Deng, J. X. and Wang, F. 2016. First report of Fusarium wilt caused by *Fusarium solani* on walnut in central China. *Journal of Plant Pathology* 98(3): 2.

Sun, Q., Wang, X., Liu Z., Dun, Y., Huang, J., Hu, H. and An, F. 2017. High yield cultivation techniques of two fine walnut varieties, Qingxiang and Liaohe 1. *Agricultural Science & Technology* 18(6): 1051–7.

Targett, P. 2014. Improving walnut quality assessment and storage practices. Honours thesis, University of Tasmania, 114p.

Tian, S., Feng, D., Yao, K., Zhang, S., Xia, J., Zhou, M., Gao, Q. and Ma, Y. 2016. A human feeding trial of walnut milk on memory improvement. *Journal of Chinese Institute of Food Science and Technology* 16(3): 36–41.

Tuan, P. N., Meier-Dinkel, A., Höltken, A. M., Wenzlitschke, I. and Winkelmann, T. 2017. Factors affecting shoot multiplication and rooting of walnut (*Juglans regia* L.) in vitro. *Acta Horticulturae* 1155: 525–30.

UC Davis. 2017. Fruit and nut research information. http://fruitandnuteducation.ucdavis.edu/fruitnutproduction/Walnut/Walnut_Cultivar_Table/ (accessed 30 November 2017).

UCIPM. 2017. Walnut using Ethephon. http://ipm.ucanr.edu/PMG/r881900711.html (accessed 3 December 2017).

Vahdati, K., Aslani Aslamarz, A., Rahemi, M., Hassani, D. and Leslie, C. 2012. Mechanism of seed dormancy and its relationship to bud dormancy in Persian walnut. *Environmental and Experimental Botany* 75: 74–82.

Vahdati, K., Asayesh, Z. M., Aliniaeifard, S. and Leslie, C. 2017. Improvement of *Ex Vitro* desiccation through elevation of $CO_2$ concentration in the atmosphere of culture vessels during in vitro growth. *HortScience* 52(7): 1006–12.

van Kralingen, H. and Adem, H. H. 2010. The effect of tree establishment cost and planting densities on the economic performance of walnut production. *Acta Horticulturae (ISHS)* 861: 221–8. http://www.actahort.org/books/861/861_29.htm

Vanhanen, L. P. and Savage, G. P. 2006. The use of peroxide value as a measure of quality for walnut flour stored at five different temperatures using three different types of packaging. *Food Chemistry* 99: 64–9.

Végh, A., Tóth, A., Zámbó, Á. Borsos, G., Palkovics, L. 2014. First report of bacterial bark canker of walnut caused by *Brenneria nigrifluens* in Hungary. *Plant Disease* 98(7): 988.

Verheggen, F., Verhaeghe, A., Giordanengo, P., Tassus, X. and Escobar-Gutiérrez, A. 2017. Walnut husk fly, *Rhagoletis completa* (Diptera: Tephritidae), invades Europe: invasion potential and control strategies. *Applied Entomology and Zoology* 52(1): 1–7.

Waitzmann, D. 1998. The New Zealand walnut market: An investigation into customer buying behaviour. Thesis for Diplom-Ingenieur fur Gartenbau at Hochschule fur Technik und Wirtshaft (FH) Dresden, 166p.

Waitzmann, D. and McNeil, D. L. 2000. Overview of the New Zealand walnut market. *The Tree Cropper, Official Journal of the NZTCA* 22: 10–17.

Wang, Y. and He, Y. 2006. *Walnut Play and Appreciation* (核桃把玩与鉴赏 (Chinese)). Beijing mei shu she ying chu ban she, Beijing, China, 143p. ISBN:9787805013404.

Wang, Q. and Tang, G. 2017. Genomic and phylogenetic analysis of the complete mitochondrial DNA sequence of walnut leaf pest *Paleosepharia posticata* (Coleoptera: Chrysomeloidea). *Journal of Asia-Pacific Entomology* 20(3): 840–53.

Wang, G. A., Shang, X. Y., Zhang, Q., Abduxukur, Y. and Huang, M. M. 2014. The walnut in Xinjiang. *Acta Horticulturae*. 1050: 29–34. doi:10.17660/ActaHortic.2014.1050.1.

Wang, H., Wu, W., Pan, G. and Pei, D. 2015. Analysis of genetic diversity and relationships among 86 Persian walnuts (*Juglans regia* l.) genotypes in Tibet using morphological traits and SSR markers. *Journal of Horticultural Science and Biotechnology* 90(5): 563–70.

Wang, J., Liang, S., Ma, H., Zhang, P. and Shi, W. 2016. Effects of ethephon on fresh in-husk walnut preservation and its possible relationship with phenol metabolism. *Journal of Food Science* 81(8): C1921–7.

Wang, T.-M., Liu, J., Yi, T., Zhai, Y.-J., Zhang, H., Chen, H.-B., Cai, S.-Q., Kang, T.-G. and Zhao, Z.-Z. 2017. Multi-constituent identification in root, branch, and leaf extracts of *Juglans mandshurica* using ultra high performance liquid chromatography with quadrupole time-of-flight mass spectrometry. *Journal of Separation Science* 40(17): 3440–52.

Wiman, N. G., Pscheidt, J. W. and Peachey, E. 2017. Walnut: 2017 Pest Management Guide for the Willamette Valley. EM 8421. Oregon State University Extension Service, 9p. Revised April 2017. https://catalog.extension.oregonstate.edu/sites/catalog/files/project/pdf/em8421_0.pdf.

Wolz, K. J. and DeLucia, E. H. 2018. Alley cropping: Global patterns of species composition and function. *Agriculture, Ecosystems and Environment* 252: 61–8.

Xi, X. L., Fang, W. L., Fan, Z. Y., Zhang, X. X., Huang, Q. and Yang, Z. B. 2007. Breeding of new precocious walnut cultivar 'Yunxin Gaoyuan'. *China Fruits* 2: 7–9.

Xu, Z. G., Zhao, Y. L., Ge, Y., Peng, J., Dong, M. and Yang, G. Y. 2017. Characterization of a vacuolar H+-ATPase G subunit gene from *Juglans regia* (JrVHAG1) involved in mannitol-induced osmotic stress tolerance. *Plant Cell Reports* 36(3): 407–18. doi:10.1007/s00299-016-2090-z.

Yaghmour, M. A., Nguyen, T. L., Roubtsova, T. V., Hasey, J. K., Fichtner, E. J., DeBuse, C., Seybold, S. J. and Bostock, R. M. 2014. First report of *Geosmithia morbida* on English walnut and its paradox rootstock in California. *Plant Disease* 98(10): 1441. doi:10.1094/PDIS-06-14-0569-PDN.

Yalçin, M., Acican, T., Alibaş, K., Ertürk, U., Akça, Y. and Polat, R. 2014. Effects of ethephon application on different harvesting methods and hulling of walnut husk. *Acta Horticulturae* 1050: 323–30.

Yang, H., Yang, W., Yang, C.-P., Zhu, T.-H., Huang, Q., Han, S. and Xiao, J.-J. 2013. Electrophysiological and behavioural responses of the white striped longhorned beetle, *Batocera lineolata,* to the diurnal rhythm of host plant volatiles of holly, *Viburnum awabuki*. *Journal of Insect Science* 13: 85. doi:10.1673/031.013.8501.

Yang, G. Y., Zhang, W. H., Sun, Y. D., Zhang, T. T., Hu, D. and Zhai, M. Z. 2017. Two novel WRKY genes from *Juglans regia*, *JrWRKY6* and *JrWRKY53*, are involved in abscisic acid-dependent stress responses. *Biologia Plantarum* 61(4): 611–21. doi:10.1007/s10535-017-0723-x.

Yilmaz, S. O. 2017. The contamination rate of aflatoxins in ground red peppers, dried figs, walnuts without shell and seedless black raisins commercialized in Sakarya City Center, Turkey. *Italian Journal of Food Science* 29(4): 591–8.

Yu, X., Liu, X., Zhu, T.-H., Liu, G.-H. and Mao, C. 2012. Co-inoculation with phosphate-solubilzing and nitrogen-fixing bacteria on solubilization of rock phosphate and their effect on growth promotion and nutrient uptake by walnut. *European Journal of Soil Biology* 50: 112–17.

Zerillo, M. M., Caballero, J. I., Woeste, K., Graves, A. D., Hartel, C., Pscheidt, J. W., Tonos, J., Broders, K., Cranshaw, W., Seybold, S. J. and Tisserat, N. 2014. Population structure of *Geosmithia morbida*, the causal agent of thousand cankers disease of walnut trees in the United States. *PLoS ONE* 9(11): article e112847.

Zhai, M. Z., Sun, Y. D., Jia, C. X, Peng, S. B., Liu, Z. X. and Yang, G. Y. 2016. Over-expression of *JrsHSP17.3* gene from *Juglans regia* confer the tolerance to abnormal temperature and NaCl stresses. *Journal of Plant Pathology* 50(5): 549–58. DOI: 10.1007/s12374-015-0507-9.

Zhang, C.-P., Meng, P., Zhang, J.-S. and Wan, X.-C. 2016. Effects of intercropped *Vigna radiata* on root hydraulic conductance and photosynthetic characteristics of *Juglans regia* seedlings. *Forest Research* 29(1): 110–16.

Zhang, W.-E., Wang, C.-L., Shi, B.-B. and Pan, X.-J. 2017a. Effect of storage temperature and time on the nutritional quality of walnut male inflorescences. *Journal of Food and Drug Analysis* 25(2): 374–84.

Zhang, G., Hu, L., Melka, D., Wang, H., Laasri, A., Brown, E. W., Strain, E., Allard, M., Bunning, V. K., Musser, S. M., Johnson, R., Farakos, S. M. S., Scott, V. N., Pouillot, R., Van Doren, J. M. and Hammack, T. S. 2017b. Prevalence of salmonella in cashews, hazelnuts, macadamia nuts, pecans, pine nuts, and walnuts in the United States. *Journal of Food Protection* 80(3): 459–66.

Zhang, M., Zhang, Y. K., Geng, Y. H., Zang, R. and Wu, H. Y. 2017c. First report of *Diplodia seriata* causing twig dieback of English walnut in China. Plant Disease. 101(6): 1036.

Zheng, L., Peng, Y., Zhang, J., Ma, W. J., Li, S. J. and Zhu, T. H. 2015. First report of *Fusarium solani* causing root rot of *Juglans sigallata* Dode in China. *Plant Disease*. 99 (1): 159.

Zhi, Z., Wang, H., Xia, W. H., Gang, Z. S. 2009. *Walnut Production Support High Quality and High Technology* (Chinese Edition). China Agriculture Press, p. 180. ISBN 10: 7109134296, ISBN 13: 9787109134294.

Zhou, Z., Han, M., Hou, M., Deng, X., Tian, R., Min, S., Zhang, J. 2017. Comparative study of the leaf transcriptomes and ionoms of *Juglans regia* and its wild relative species *Juglans cathayensis*. *Acta Physiologiae Plantarum* 39(10): article 224.

Zhu, Y. Z., Liao, W. J., Zou, D. X., Wu, Y. J. and Zhou, Y. 2015a. First report of leaf spot disease on walnut caused by *Colletotrichum fioriniae* in China. *Plant Disease* 99(2): 289.

Zhu, Y. F., Yin, Y. F., Yang, K. Q., Li, J. H., Sang, Y. L., Huang, L., Fan, S. 2015b. Construction of a high-density genetic map using specific length amplified fragment markers and identification of a quantitative trait locus for anthracnose resistance in walnut (*Juglans regia* L.). *BMC Genomics* 16: article: 614. doi:10.1186/s12864-015-1822-8.

Zibaeenezhad, M., Aghasadeghi, K., Hakimi, H., Yarmohammadi, H. and Nikaein, F. 2016. The effect of walnut oil consumption on blood sugar in patients with diabetes mellitus type 2. *International Journal of Endocrinology and Metabolism* 14(3): article e34889.

Zwarts, L., Savage, G. P. and McNeil, D. L. 1999. Fatty acid content of New Zealand grown walnuts (*Juglans regia L.*). *International Journal of Food Sciences and Nutrition* 50: 189–94.

# Chapter 11

## Advances in cultivation of almonds: effects of genotypes, environment and cultural techniques

*Ettore Barone and Francesco Sottile, Università di Palermo, Italy*



## 1 Introduction

Almond (*Prunus dulcis* (Mill) D.A. Webb syn. *P. amygdalus* Batsch) was one of the first fruit tree species to be domesticated (Abbo et al., 2015) and is the most important temperate nut crop in the world. Almonds are deciduous and monoecious, typically self-sterile, native to central Asia from where they moved westward, perhaps together with other wild almond species, following the trade routes of emerging civilizations (Gradziel, 2011) as early as 2000 BC. Almonds arrived in Greece prior to 300 BC and gradually spread throughout the entire Mediterranean basin (Janick, 2005) via Sicily. Greek colonists and Phoenicians facilitated their distribution (Egea and Garcia, 1975). In the eighth century, the tree was already cultivated in southern France, arriving in England and Germany in the ninth century (Avanzato and Vassallo, 2006). Almonds were introduced to North America by the Spanish colonists in the sixteenth century and spread from there to the Southern hemisphere countries with Mediterranean climates including Australia, Chile, Argentina and South Africa.

In the Mediterranean regions almond cultivation proliferated in those areas characterized by mild winters and hot dry summers and on hillsides to escape frost. Due to the almond's ancient reputation of being adaptable to

http://dx.doi.org/10.19103/AS.2018.0042.20

poor and dry soils, orchards traditionally were grown on marginal soils and with little care and irrigation. This traditional Mediterranean pattern is still typical of most almond production in Italy, Greece, Spain and other European and Asian countries. However, in the newer almond-producing countries, following a 'transition period' (Kester and Ross, 1996), a significant cultural shift occurred, which now characterizes the modern almond industry.

### *1.1 Almond-growing models*

There are now three different almond models, with different locations and degrees of diffusion at the regional level, sometimes all three within a single producing country (Sottile et al., 2014).

The first is the old, traditional almond system. This is based upon old, local, less-efficient cultivars grafted onto drought-tolerant almond seedling rootstocks, and orchard systems in poor, marginal soils, without irrigation, with unsatisfactory yields and characterized by a low level of specialization and a high degree of structural variability (in terms of tree age, density etc.) that is often impossible or difficult to convert and modernize. This still represents the most widespread system in the Mediterranean region.

The second employs farming practices including mechanization, pruning, irrigation, fertilization, pest and orchard-floor management (modern almond system) on fertile, deep, well-drained soils, with a restricted number of 'modern' selected cultivars of the paper-shell type and appropriate pollinizers grafted on rootstocks other than almond seedlings. This model is used in California and Australia. In California, due to new soft-shell cultivars, nematode-resistant rootstocks and reliable irrigation, almond acreage has more than tripled from 1965 to 1995 and again from 1996 to 2017 (Almond Board of California, 2017a). Similar growing trends are reported for Australia (Almond Board of Australia, 2017). Through this model California production per ha has increased to more than 4 t $ha^{-1}$ (Gradziel and Martínez-Gómez, 2013).

The last is the new high-density plantations (HDP). These are high-input, very high-return cropping systems, with tightly spaced hedgerow systems, dwarfing rootstocks, spur[1]-type cultivars and intensive orchard management. These HDPs are projected to have accelerated production, improved yields, full mechanization and, therefore, cost reductions. In spite of initial underwhelming results (Monastra et al., 1998) they are still considered promising and thus under investigation in Europe as well as in other producing countries (Duncan, 2012; Kumar et al., 2012). However, other studies observed that while these high-density systems might be more productive in years 3–4 or so, after that

1 Short lateral shoots that are the main flowering and fruit bearing units in mature almond trees.

they lag well behind traditional plantings since the narrow row spacings devote more space to drive rows and ultimately intercept less light (Lampinen, *pers. comm.*).

This chapter will survey the recent advances in almond tree eco-physiology, and how it can be used to improve cultivation with special emphasis on the key factors determining almond tree yield potential as affected by the specific orchard system adopted.

### 1.2 Almond yield components

As with all the other temperate fruit tree species, almonds have a 2-year bearing cycle involving in the first year the establishment of bearing potential (vegetative growth and flower bud development) and in the second year both that of reproductive potential (flowering) and that of weight potential (nut growth) (Kester et al., 1996). As a whole, the final crop tonnage is, in turn, the end result of the total number of marketable nuts and their average kernel weight, the number of kernels being the most important factor (Reidel et al., 2001). All these phases of vegetative and reproductive cycles are in most cases dependent on photosynthetic activity, even if factors other than photosynthesis must be considered (Faust, 1989). In almond, yield has been shown to increase with increasing midday canopy PAR interception in the PAR range from about 20% to 90% (Lampinen et al., 2012) so that, reportedly, with every 1% increase in light interception an increase of +56 kg ha$^{-1}$ higher yield potential has to be expected. A complex interaction among three factors (i.e. genotype–environment–cultural management), including carry-over effects of the previous year, is responsible for the productive result and the fructification regularity over the years. This is true both in terms of nut quantity and quality, so that optimal cultivar(s)/environment/orchard design and management combinations must be pursued on a case-by-case basis.

## 2 Genotype

A proper choice of the scion/rootstock combination, well suited for a specific orchard system, is an essential pre-requisite for the success of an almond orchard. Recent research aimed at defining the limits of almond productivity showed that, among the tree characteristics under genetic control, bearing habit is a crucial determinant for tree yield (Gradziel and Lampinen, 2011).

### 2.1 Almond bearing habits

Almond cultivars exhibit three different bearing habits (Gülcan, 1985): (1) most production on 1-year-old shoot (e.g. cv. Ai and the Italian (Apulian) genotypes); (2) most production on spurs (e.g. cv. Tuono) and (3) mixed, that is on both spurs and long shoots (as in 'Mission' and 'Nonpareil'). From an energetic point of view, spurs are more efficient than shoots in balancing between reproductive and vegetative activity costs. A combination on the same tree of both bearing habits is considered advantageous (Gradziel and Martínez-Gómez, 2013). From these studies on almond bearing habits emerged the importance to final productivity of factors such as tree architecture and tree size for light interception, the total accumulated carbohydrate reserves, total number of flowers on a tree rather than relative fruit set (Tombesi et al., 2017), number of nuts per tree (in a range between 20000 and 30000) and kernel size (optimum about 1 g), as well as the importance of spur leaf area for spur flowering and survival in the next year (Heerema et al., 2008). In other words, these last findings indicate that individual spurs tend to alternate bear with only a small percentage of spurs flowering the year after bearing (Lampinen et al., 2011). Relative fruit set in almond appears to be negatively associated with current-year spur leaf area (Tombesi et al., 2015). Fruiting spurs had 50% less leaf area than non-fruiting spurs because they had smaller and fewer leaves (Valdebenito et al., 2017). Lampinen et al. (2011) concluded that survival of fruiting spurs with a leaf area <28 $cm^2$ was significantly reduced.

## 2.2 Other genetic factors affecting almond fertility

Another key factor for almond productivity, also under genetic control, is represented by self-sterility (gametophytic self-incompatibility), which affects the majority of commercial almond cultivars. These can be pooled in groups of reciprocal compatibility (Asai et al., 1996; Ortega et al., 2014; Wirthensohn et al., 2011). As a consequence, proper pollinizer cultivar choice and arrangement is of paramount importance for sound, commercial yields whereas insufficient cross-pollination is the most frequent cause of unsatisfactory productive results (Asai et al., 1996). Thus, growing self-compatible almonds in solid orchards of single cultivar would be desirable, as this would result in lower costs for crop management and potentially higher yields (Godini and Palasciano, 1998; Ortega and Dicenta, 2006) of quality kernels (Kodad and Socías I Company, 2008) without negative effect on the fruit's agronomic traits (Martínez-García et al., 2011). Consequently most almond breeding programmes have promoted the development of new self-fertile cultivars (Socías I Company, 2002; Socías I Company et al., 2010) and the exploitation of the available already known self-fertile cultivars (Godini, 2002). A comprehensive review of self-(in)compatibility almond genotypes can be found in López et al. (2006).

### 2.3 Genotype effects on almond eco-physiological aspects and nut quality

Genotype is responsible for differences observed in several eco-physiological aspects. Eichi et al. (2014) demonstrated that water relations, water use efficiency (WUE) and leaf anatomy in almond trees differed among genotypes. In particular, leaf anatomy, namely a compact mesophyll tissue, can limit the amount of water loss during the hot and dry summers of Mediterranean climates. These results clearly indicated that it is possible to decrease water use without reducing yield by selecting new varieties with improved WUE. Differences in the degree of drought tolerance by different scion genotypes have been reported by several authors (Herralde et al., 2001; Gomes-Laranjo et al., 2005; Palasciano et al., 2014). The influence of genotype on kernel, shell and hull quality characteristics has been recently reviewed by Socías I Company et al. (2008). A thorough review of almond breeding can be found in Gradziel and Martínez-Gómez (2013).

### 2.4 Rootstock genotype

Rootstocks for almond have been less studied than varieties and limited comparative test data are available. 'Nemaguard' and 'Lovell' peach stocks are preferred by almond growers in California, whereas hybrids of peach and almond, particularly GF677, have come into relatively wide usage more recently and have become popular in the Mediterranean areas as substitutes for traditional almond seedlings which in turn have been dominant for centuries in rain-fed almond orchards (Godini and Palasciano, 1998; Socías I Company et al., 2010). Recently released new rootstocks include hybrid genotypes selected for their ease of propagation, graft compatibility, vigour, drought resistance, specific unfavourable soil adaptation, and pest and disease resistance (Grasselly and Olivier, 1989; Sedgley and Collins, 2002; Felipe, 2009; Pinochet, 2009; Bielsa et al., 2016a,b). In order to enhance environmental sustainability, rootstock selections are under evaluation as non-fumigant approaches to managing *Prunus* replant disease (PRD), a poorly understood soil-borne disease complex that suppresses early growth and cumulative yield in replanted almond orchards (Browne et al., 2013). Further perspectives of other rootstock breeding programmes, especially for nematode resistance, are summarized in Duval (2016). A recent comprehensive review on almond rootstocks is given by Rubio-Cabetas (2016).

## 3 Cultural environment

The cultural environment, that is the pedo-climatic prevailing conditions at the orchard site location, can strongly affect many eco-physiological aspects of almond trees, and thus, directly or indirectly, influence the proper cultivar/rootstock combination and, as a whole, the productive result of the almond orchard (Barbera et al., 1994).

### 3.1 Temperature

One of the main aspects to be considered is the seasonal variation of the temperature. Seasonal changes both in leaf development and stomatal characteristics have been reported as a consequence of the seasonal increasing temperature regime resulting in improved stress responses (Camposeo et al., 2011). Almond kernel quality traits such as tocopherol concentration have been reported to depend on the genotype and the growing management of the orchard and the climatic conditions, with increased levels at higher temperatures and in water-deficit conditions (Kodad et al., 2018).

### 3.2 Flower bud dormancy

Another fundamental eco-physiological factor is represented by the thermal requirements to overcome dormancy. In order to regularly overcome winter flower bud endodormancy the tree must be exposed to a specific, genetically controlled, amount of chilling, generally from late October through mid-December in the northern hemisphere. Even if almond is considered a low chilling species with respect to other temperate fruit species (Baldini, 1957), an inadequate level of low ambient temperatures, common in warmer areas, can be detrimental for the regularity of both vegetative and reproductive cycles (Kester et al., 1996). After the release from endodormancy the buds are now entering a new dormancy phase, the so-called 'ecodormancy' (Lang, 1987) and, prior to bloom, begin to respond positively to warm temperatures so that, as a whole, time of bloom is considered dependent on the winter progression of temperatures affecting the different stages of dormancy (Bernad and Socías I Company, 1995). It has been calculated that most almond cultivars have chilling requirements between 220 and 600 chill units, whereas the span of heat requirements to overcome ecodormancy ranges from 5500 to 9300 growing degree hours (Rattigan and Hill, 1986; Alonso et al., 2005), that is favourable growth temperatures.

### 3.3 Scion genotype adaptation to the cultural environment and the goals of breeding

As a consequence of genetically driven chill and heat requirements, almond varieties that fulfil both their thermal requirements early are also early blooming, hence highly susceptible to frost damage, depending upon the environment.

Low temperatures as well as other unfavourable weather conditions (rain, wind) that are frequent during bloom can also reduce yields by interfering with pollinating insects, mainly honeybees (Bernad and Socías I Company, 1995). Thus, obtaining late (March) or extra-late (early April) flowering cultivars continues to be one of the main objectives of almond breeding, although with much more attention in the Mediterranean region or in Australia than in California (Wirthensohn and Sedgley, 2002; Dicenta et al., 2011; Socías I Company et al., 2012, 2015). On the contrary, at low latitudes, in warm and arid regions, where low-chill accumulation is common, early blooming, followed by early nut maturation, remains a useful agronomic trait for escaping or minimizing the exposure to the extreme summer climatic conditions of heat and drought (Alonso Segura et al., 2017). Furthermore, it should be considered that flower buds in late-flowering cultivars are likely subjected to a more intense competition with sprouting vegetative buds for water and nutrients (Martínez-Gómez et al., 2017). However, no indication for the date of bloom and the kernel size has been detected (Socías I Company et al., 2014).

### 3.4 Flowering time determinants and variety classification

It has been reported that chill, rather than the subsequent heat requirement, is the most important determinant of flowering time (Egea et al., 2003; Martínez-Gómez et al., 2017). However, this is true only for warmer areas whereas, for cold environments, the opposite may be true (Alonso Segura et al., 2017) where delayed/prolonged bloom is generally expected. In any case the flowering time must be interpreted as the resultant combined effect of the relative duration of chill and heat requirements completion driven by the environment (Caruso et al., 1992). For further understanding of the flowering process new advancements and opportunities are needed from the integration of recent studies including phenotypic, genetic, genomic, transcriptomic and metabolomic studies from the beginning of dormancy until flowering (Sánchez-Pérez et al., 2014; Hosseinpour et al., 2018). Almond cultivars can be classified for their blooming time in five groups (Martínez-Gómez et al., 2017): extra-early (e.g. 'Desmayo Largueta'), early ('Marcona'), late ('Nonpareil', 'Tuono'), extra-late ('Mardía') and ultra-late ('Tardona').

### 3.5 Fruiting time and duration

Flowering time is not the only critical characteristic for successfully adapting almond or peach to different environments (Marra et al., 2008; Martínez-Gómez et al., 2017). Fruiting time and duration also are important. Four combination traits can be recognized among almond cultivars: early flowering and early ripening cultivars (e.g. 'Ramillete'); early flowering and late ripening cultivar

('Desmayo Largueta'); late flowering and early ripening ('Antoñeta'); and late flowering and late ripening ('Wawona') (Martínez-Gómez et al., 2006). In order to enhance fruit set and productivity, besides ensuring adequate pollination, foliar boron application protocols immediately post-harvest (0.8–2.1 kg $ha^{-1}$ in September) can be adopted to positively affect fruit set and yield (Nyomora et al., 1999). Tombesi et al. (2010) reported that temperatures in the first 90 DAFB are the primary factor influencing the time of nut maturity in almond cultivars. Additionally, they suggested that the rates of fruit development that occur early in the season can have a strong effect on 'programming' development for the rest of the season.

### *3.6 Soil*

Almonds are adapted to a wide range of soil and water conditions. The most favourable soils being loamy soils rather than sandy soils. Nevertheless, soils once considered too shallow and heavy have been successfully planted using low volume irrigation methods and adapted rootstocks (Micke and Kester, 1998). However, in soils with high clay content and poor drainage, flooding damage will likely occur, and, in response to this, rootstock characteristics may offer a different tolerance to root anoxia. In a recent study (Ziegler et al., 2017) Myrobalan plum was confirmed to have the greatest tolerance to soil flooding, whereas Nemared (peach) rootstock was the most sensitive, with Monegro (peach × almond hybrid) having an intermediate behaviour. The first responses were decreases in photosynthesis, stomatal conductance and water potential. Peach × almond hybrid rootstocks have been found to be more tolerant to sodium and chloride than peach rootstocks (Doll et al., 2014). The evaluation of optimal soils for almond growing was thoroughly examined by Fulton et al. (1996).

## 4 Cultural management

Recent advances in almond orchard design and management include soil and tree management practices, that is orchard configuration, soil-plant-water relationships, weed control, irrigation, nutrition, pollination, training, and pruning and pest management. Not all these aspects will be covered in this chapter. All the aforementioned aspects are strongly interrelated to specific environments and scion/rootstock combinations, and thus should be interpreted following a comprehensive holistic approach. Additionally, most of the available experimental results reflect the agronomic context in which they were obtained and, therefore, must be generally regarded on a case-by-case basis. An interesting example showing the interactions between pollination and post-pollination processes is offered by Klein et al. (2015). In this mentioned study, Klein et al. clearly indicated that the influences of pollination on fruit tree yield interact with the plant availability of nutrients and water and that excess pollination can reduce fruit quality and the production of leaves for photosynthesis so that high fruit yield may come at the expense of the tree's vegetative performance, such as quantity and photosynthetic efficiency of foliage, with inherent long-term consequences for the tree.

## 4.1 Irrigation

Irrigation can be considered as the main limiting factor for almond tree cultivation in the Mediterranean regions (Hutmacher et al., 1994). Advances in irrigation management have been achieved with the use of plant-based eco-physiological approaches to estimate acceptable levels of plant water stress (midday stem water potential - SWP (Shackel et al., 1997; Shackel, 2011)) rather than on soil- or environmental-based measurements. A significant correlation has been found between leaf water potential and canopy temperature, measured with infrared thermometry or other remote infrared sensors, suggesting that the latter could be an advantageous and less time-consuming technique to monitor water stress than others (García-Tejero et al., 2015). The negative effects of water stress on yield, both in terms of kernel size (Girona et al., 1993) and fruit load (Goldhamer and Viveros, 2000), are well documented (Esparza et al., 2001), as well as the negative effect on leaf function and persistence (Klein et al., 2001). Nevertheless in California the application of a correct regulated deficit of irrigation (RDI) strategy (i.e. a mild-to-moderate stress level (−14 to −18 bars of measured SWP) imposed during hull split period) allowed to obtain consistent water savings of more than 12.7 cm, or about 13%, of applied water without negative effects either on short- or medium-term orchard productivity (Stewart et al., 2011) or on kernel nutrition quality (Zhu et al., 2014). The application of proper irrigation during the growing season as well as post-harvest has proven to be helpful to minimize spur death and to positively influence subsequent bloom and fruit set in almond (Esparza et al., 2001; Goldhamer and Smith, 1995). Previous studies in almonds also supported the positive effects of RDI, including reduced hull rot, improved hull split and control of excessive vegetative growth (Goldhamer et al., 2006; Teviotdale et al., 2001; Shackel, 2003), also under subsurface drip irrigation conditions (Romero et al., 2004). Additionally, as an effect of reduced vegetative growth, maintaining more compact canopies with RDI rather than pruning would also lessen the tree management expenses and the amount of wood requiring disposal, thereby moderating environmental impact and air quality degradation resulting from burning (Goldhamer et al., 2006). A recent study (Goldhamer and Fereres, 2017) combining both eco-physiological (midday shaded leaf water potential - SLWP) and environmental-based (reference evapotranspiration: $ET_0$) measurements has been carried out to identify an optimal irrigation level for maximizing yield. The obtained results suggested that in the high evaporative demand, southern San Joaquin Valley of California, maximum marginal water productivity (WPM) was 0.30 kg $m^{-3}$ at 1080 mm of applied water, falling to zero at 1290 mm, that is about 10% lower irrigation requirements than full $ET_c$ (Goldhamer and Fereres, 2017). The long-term financial feasibility of cropping almond under different deficit irrigation (DI) strategies in a water scarcity context has been recently analysed by Alcon et al. (2013) and led the authors to conclude that in such a context, DI is a financially feasible alternative to FI (full irrigation). Additionally it should be mentioned that the application of a correct and reliable strategy of RDI may also benefit the environment by reducing water and nitrogen fertilizer leaching below the root zone (Shackel, 2011).

## 4.2 Drought resistance as affected by scion/rootstock combination

In researches concerning soil-plant-water relationships the role of rootstock and of scion/rootstock interaction is probably underestimated. Espadafor et al. (2017) reported that contrasting values of tree water status developed in different RDI studies with similar climatic characteristics suggest that, in addition to soil differences, there may be significant cultivar/rootstock differences in response to water. In a warm Mediterranean cultural environment and for super-high-density plantations, a recent preliminary comparative trial of different scion/rootstock combinations has identified that the new Rootpac40 rootstock exhibited better agronomic performance (higher SWP values) for its medium capacity to control tree vigour when compared to the more (Garnem) and the less (Rootpac20) vigorous rootstocks (Ben Yahmed et al., 2016). In general terms, the less vigorous combinations are not recommended for rain-fed or low irrigated orchards, as trees take longer to reach the commercial size and, therefore, their maximum production (Dicenta et al., 2016). In a recent study designed to assess three *Prunus* rootstocks's response to drought stress by measuring genotype differences in different physiological parameters, the two hybrid genotypes 'P.2175' × 'Garnem'-3 hybrid and OP-'P.2175' (*P. cerasifera*) showed a better adaptive response to drought than the 'Garnem' genotype (*P. amygdalus* Batsch × *P. persica* (L.) Batsch), confirming the profitability of rootstock breeding programmes which combine abiotic tolerances in a new generation of complex interspecific hybrids (Dirlewanger et al., 2004; Amador et al., 2012; Bielsa et al., 2016a,b).

## 4.3 Mineral nutrition: nitrogen and potassium

Maintaining adequate soil moisture is essential to plant nutrition. The choice of rootstock can affect the ability of a tree to take up nutrients, whereas the scion can create differences in sensitivity to low nutrient availability (Brown and Uriu, 1996). Nitrogen is important for tree growth and productivity. An increase in flower number is stimulated by application of high levels of nitrogen, which is generally associated with increased vigour and delayed flowering (Sedgley, 1990). On the contrary, lower leaf N content is correlated with spur mortality supporting the hypothesis of a strong competition for N between fruit and spur (Saa and Brown, 2014). Fruit growth and embryo growth (kernel fill) are both characterized by a high N demand (Nortes et al., 2009), the former accounting for 60% of the total annual fruit N demand (Muhammad et al., 2015). During the growing season, once the leaves are fully expanded, the N and K requirements for vegetative growth are largely satisfied. Fruits continue to accumulate N and K until harvest (Brown and Uriu, 1996). The presence of fruit on a spur is responsible for a significant reduction in leaf concentrations of N, P, K, S, Zn and Cu when compared to non-fruiting spurs (Saa et al., 2017), thus suggesting diverse nutritional and mineral requirements by fruiting and non-fruiting trees. In a complex study on light interception undertaken by Zarate-Valdez et al. (2015), for leaf nitrogen and yield prediction on a commercial almond orchard located in the southern San Joaquin Valley of California on sandy loam soils with the Nonpareil and Monterey (pollinator) varieties, it was found that N fertilization had a significant effect on individual tree fractional PAR intercepted by the canopy, LAI, leaf nitrogen content and nut yield, in the tested N fertilization rate range (150–400 kg $ha^{-1}$). Further, leaf nitrogen content, more than other nutrients, explained up to 76% of the yield variation over a period of two or three years. Saa et al. (2016) have reported that increased N application, in excess of tree demand, significantly increased the incidence of hull rot in fruiting spurs probably due to an increase in susceptibility to infection during certain stages of hull split. In fact, N application rate altered the pattern of fruit development and delayed the transition from the 'closed hulls' stage to the 'beginning of hull split' stage. In another experiment the application of soil and foliar N treatments effectively increased spur leaf area, fruit and leaf nitrogen concentration, but no effect was found as far as spur survival and/or return boom of any spur type are concerned, independent of the N rate (Saa et al., 2017). The antagonistic effect of fruit load on return bloom has been explained both in terms of negative hormonal effect exerted by the growing embryos and of competition for photosynthates between kernels and the concurrent process of flower bud initiation (Tombesi et al., 2011). A significant increase in mortality of spurs that have already fruited has been recognized by Reidel et al. (2001) as a major factor in the lower return bloom and reduced yields of low-K trees. In the same research carried out in California on the effects of K applications on leaf K concentrations and yields with 'Nonpareil', maximum yields have been correlated with leaf K values equal to or greater than 1.4%. Potassium deficiency has been proven to negatively affect the photosynthetic capacity of almond tree canopies both as a result of reductions in light interception and decreases in the leaf photosynthetic rate (accelerated premature leaf senescence and premature leaf abscission) (Basile et al., 2003). Effects of K-deficiency are suggested to be cumulative, and especially pronounced following years of heavy cropping (Reidel et al., 2004). Yield-based determination of nutrient critical values in almonds – a concept associated with optimal yield levels – to support nutrient management and fertilization decisions has been reviewed by Saa et al. (2013). The results published so far clearly support the criterion, as reported by Muhammad et al. (2018), that nitrogen and potassium fertilization rates should be based on expected yield and tree N and K status, with no further benefits to be expected when leaf content exceeds the threshold of 2.5% and 1%, respectively.

## 4.4 Sustainable orchard soil management practices

In a recent meta-analysis study of soil carbon sequestration rates in Mediterranean woody crops (Vicente-Vicente et al., 2016) it has been shown that implementation of recommended management practices (RMPs), which include plant cover between rows, minimum or no tillage, and off- and on-farm organic matter amendments, has increased C sequestration compared to conventionally managed cropping systems. However, the results suggest that the adoption of RMPs may be a sustainable and efficient way to mitigate climate change. Other recent advances in the field of sustainable orchard soil management practices (Almagro et al., 2016) have indicated that reduction and suppression of tillage under almonds led to an increase in soil water content in semi-arid agroecosystems. Reduced tillage and reduced tillage combined with green manure treatments showed lower soil erosion rates and higher crop yields of almonds when compared to conventional tillage treatment (ploughed 3–5 times per year). This research highlights the central role of vegetation cover under reduced tillage treatments in restoring or sustaining soil physical properties and quality without increasing soil $CO_2$ emissions to the atmosphere, enhancing carbon sequestration and crop yields, while controlling soil erosion. Further refinements of this research concluded and summarized that improved soil management can be beneficial for (1) maintenance of soil fertility and nutrient cycling, (2) prevention of soil erosion and water contamination, (3) increased soil water infiltration and storage capacity, and (4) making agroecosystems more resilient against the impacts of climate change. Finally, it has been confirmed that incorporating C inputs from plant or crop residues rather than reducing tillage frequency contribute to mitigation of $CO_2$ emissions from agroecosystems (Almagro et al., 2017). Additionally, the maintenance of a living ground cover within commercial almond orchards could provide better habitat and resources for potential wild pollinators, particularly native bees, than orchards without ground cover (Saunders et al., 2013).

## 5 Conclusion

Advanced genotype(s) both for scion varieties and rootstocks in combination with proper environment and cultural management are together powerful tools to ensure further development of the global almond industry. The availability of well-suited new fruit tree genotypes of good quality is generally recognized as one of the main factors in the successful establishment of new almond orchards and expansion to new areas (Sottile et al., 2014). However, increasing yields is largely a result of improved orchard management, including water, agrochemicals and labour, all of which are becoming more restricted (Gradziel and Kester, 1996). While most almond farmers have adopted best management practices for orchard-floor management, irrigation and crop nutrition, many of these cultural practices are not environmentally sustainable, hence the focus on 'sustainable intensification' (Garnett et al., 2013). This will require more research including developing appropriate measurement systems and understanding of trade-offs between yields and environmental impacts (Expo, 2015). The research needs are diverse but have much in common with other fruit crops. As it was already concluded years ago by Tilman et al. (2001) there is still a need to enhance efficiency of crop N, P and water use by breeding and biotechnology. Advances in precision agriculture that decrease N and P inputs are needed, as are methods that manage soil organic matter and microbial communities to reduce nutrient leaching and to optimize soil fertility. In order to encourage almond farmers to adopt more sustainable management practices and to compensate them for eventual major costs and/or potential decreases in income, it may be helpful to provide direct payments to growers for services offered to the ecosystem (Galati et al., 2015), as well as the investments in research supported by growers' organizations. A positive example of this new approach is represented by that adopted in California. There, almond farmers and processors created and adopted a formal definition of sustainability specific to almonds based on the three pillars of sustainability: a system that is socially equitable, economically viable and ecologically sound, including all aspects of the almond life cycle and, overall, based on research into irrigation scheduling and nitrogen budgeting (Almond Board of California, 2017b).

## 6 Acknowledgement

We are indebted to Louise Ferguson for the thorough review of most of the manuscript.

## 7 Where to look for further information

A completely new fundamental text is *Almonds: Botany, Production and Uses*, 2017 by Socías I Company, R. and Gradziel, T.M. (Eds), ISBN 9781780643540. Up-to-date detailed information about the almond industry is available at the website of the Almond Board of California (www.almonds.com) and Australia (www.australianalmonds.com.au). A wealth of more than 40 years' research information on almond is represented by proceedings of the GREMPA meetings supported by the FAO_CIHEAM Network on nuts (see Vargas, 2014), mostly published on *Options Méditerranéennes - Séries A* or synthesized on the FAO-CIHEAM - *Nucis-Newsletter*, or by the ISHS International Symposium on Almond and Pistachio Proceedings of which the first was held in Agrigento (Italy) in 1993 and the last in Adelaide (Australia) in 2017, published in *Acta Horticulturae* (see n° 373, 470, 591, 726, 912, 1028) and available at www.actahort.org. A fundamental handbook for growers and researchers is provided by the *Almond Production Manual*, 1996 by Micke, W.C. (Ed.). Division of Agriculture and Natural Resources Publication 3364. University of California, Oakland, CA.

Reviews on specific aspects of almond cultivation and biology can be retrieved in selected volumes of *Horticultural Reviews*, some of them are listed in the present bibliography.

For almond nutritional and human health aspects, access the following sources:

US Department of Agriculture, Agricultural Research Service, Nutrient Data Laboratory. USDA National Nutrient Database for Standard Reference, Release 28. Version Current: May 2016. http://www.ars.usda.gov/ba/bhnrc/ndl.

Grosso, G. and Estruch, R. (2016), 'Nut consumption and age-related disease', *Maturitas*, 84, pp. 11–16.

For further information on sustainability and sustainable development goals consult also:

Rudel, T. K., Schneider, L., Uriarte, M., Turner II, B. L., DeFries, R., Lawrence, D., Geoghegan, J., Hecht, S., Ickowitz, A., Lambin, E.F., Birkenholtz, T., Baptista, S. and Grau, R. (2009), 'Agricultural intensification and changes in cultivated areas, 1970–2005', *Proceedings of the National Academy of Sciences*, 106(49), pp. 20675–80. doi:10.1073/pnas.0812540106.

FAO (2012), *Coping with Water Scarcity: An Action Framework for Agriculture and Food Security*, FAO, Rome, Italy.

FAO (2014), *Building a Common Vision for Sustainable Food and Agriculture – Principles and Approaches, American Journal of Evaluation*, FAO, Rome, Italy. doi:10.1177/109821408300400423.

Tilman, D., Balzer, C., Hill, J. and Befort, B.L. (2011), 'Global food demand and the sustainable intensification of agriculture', *Proceedings of the National Academy of Sciences*, 108(50), pp. 20260–4. doi:10.1073/pnas.1116437108.

Borlaug, N. (2007), 'Feeding a hungry world', *Science*, 318(5849), p. 359. doi:10.1126/science.1151062.

Rockström, J., Williams, J., Daily, G., Noble, A., Matthews, N., Gordon, L., Wetterstrand, H., DeClerck, F., Shah, M., Steduto, P., de Fraiture, C., Hatibu, N., Unver, O., Bird, J., Sibanda, L. and Smith, J. (2017), 'Sustainable intensification of agriculture for human prosperity and global sustainability', *Ambio*, 46(1), pp. 4–17. doi:10.1007/s13280-016-0793-6.

The Royal Society (2009), Reaping the benefits: science and the sustainable intensification of global agriculture. RS Policy document 11/09 Issued: October 2009 RS1608, ISBN: 978-0-85403-784-1. http://royalsociety.org/reapingthebenefits.

WRI (2013), 'Creating a sustainable food future', World Resources Report. Available at: http://www.worldresourcesreport.org.

World Bank (2008), 'Agriculture for development', world development report 2008. World Bank, Washington DC. doi:10.1596/978-0-8213-7233-3.

## 8 References

Abbo, S., Gopher, A. and Lev-Yadun, S. (2015), 'Fruit domestication in the near east', in Janick, J. (Ed.), *Plant Breeding Reviews: Volume 39*, John Wiley & Sons, Inc. doi:10.1002/9781119107743.ch07.

Alcon, F., Egea, G. and Nortes, P. A. (2013), 'Financial feasibility of implementing regulated and sustained deficit irrigation in almond orchards', *Irrigation Science*, 31(5), pp. 931–41. doi:10.1007/s00271-012-0369-6.

Almagro, M., de Vente, J., Boix-Fayos, C., García-Franco, N., Melgares de Aguilar, J., González, D., Solé-Benet, A. and Martínez-Mena, M. (2016), 'Sustainable land management practices as providers of several ecosystem services under rainfed Mediterranean agroecosystems', *Mitigation and Adaptation Strategies for Global Change*, 21(7), pp. 1029–43. doi:10.1007/s11027-013-9535-2.

Almagro, M., Garcia-Franco, N. and Martínez-Mena, M. (2017), 'The potential of reducing tillage frequency and incorporating plant residues as a strategy for climate change mitigation in semiarid Mediterranean agroecosystems', *Agriculture, Ecosystems and Environment*, 246, pp. 210–20. doi:10.1016/j.agee.2017.05.016.

Almond Board of Australia (2017), 'Australian almond insights 2016–17', Almond Board of Australia. Available at: growing.australianalmonds.com.au.

Almond Board of California (2017a), 'Almond Almanac 2017', Annual report, Almond Board of California.

Almond Board of California (2017b), 'Growing good: Almond sustainability 2017', Document #2017CN0267.

Alonso, J. M., Ansón, J. M., Espiau, M. T. and Socías I Company R. (2005), 'Determination of endodormancy break in almond flower buds by a correlation model using the average temperature of different day intervals and its application to the estimation of chill and heat requirements and blooming date', *Journal of the American Society for Horticultural Science*, 130(3), pp. 308–18.

Alonso Segura, J. M., Socías I Company, R. and Kodad, O. (2017), 'Late-blooming in almond: A controversial objective', *Scientia Horticulturae*, 224, pp. 61–7. doi:10.1016/j.scienta.2017.05.036.

Amador, M. L., Sancho, S., Bielsa, B., Gomez-Aparisi, J. and Rubio-Cabetas, M. J. (2012), 'Physiological and biochemical parameters controlling waterlogging stress tolerance in *Prunus* before and after drainage', *Physiologia Plantarum*, 144(4). doi:10.1111/j.1399-3054.2012.01568.x.

Asai, W. K., Micke, W. C., Kester, D. E. and Rough, D. (1996), 'The evaluation and selection of current varieties', in Micke, W. C. (Ed.), *Almond Production Manual*, University of California, Division of Agriculture and Natural Resources, Oakland, CA, pp. 52–60.

Avanzato, D. and Vassallo, I. (2006), 'Following almond footprints – Sulle orme del mandorlo', *Scripta Horticulturae*, 4.

Baldini, E. (1957), 'La difesa degli alberi da frutto dalle gelate primaverili', in *Atti Accademia dei Georgofili*, Firenze, Italy.

Barbera, G., La Mantia, T., Monastra, F., De Palma, L. and Schirra, M. (1994), 'Response of Ferragnes and Tuono almond cultivars to different environmental conditions in southern Italy', *Acta Horticulturae*, 373, pp. 125–8. International Society for Horticultural Science (ISHS), Leuven, Belgium. doi:10.17660/ActaHortic.1994.373.16.

Basile, B., Reidel, E. J., Weinbaum, S. A. and DeJong, T. M. (2003), 'Leaf potassium concentration, $CO_2$ exchange and light interception in almond trees (*Prunus dulcis* (Mill) D.A. Webb)', *Scientia Horticulturae*, 98(2), pp. 185–94. doi:10.1016/S0304-4238(02)00214-5.

Ben Yahmed, J., Ghrab, M. and Ben Mimoun, M. (2016), 'Eco-physiological evaluation of different scion-rootstock combinations of almond grown in Mediterranean conditions', *Fruits*, 71(3), pp. 185–93. doi:10.1051/fruits/2016003.

Bernad, D. and Socías I Company, R. (1995), 'Characterization of some self-compatible almonds. II. Flower phenology and morphology', *HortScience*, 30(2), pp. 321–4.

Bielsa, B., Rubio-Casetas, M. J., Felipe, A. J., Gómezaparisi, J. and Socías I Company, R. (2016a), 'Rootstock trial of eight GxN interespecific hybrids in almond', *Options Méditerranéennes Serie A: Seminaires Méditerranéennes*, 119, pp. 183–6. XVI GREMPA Meeting on almonds and pistachios: Meknes – Royaume du Maroc, 12–14 mai 2015.

Bielsa, B., Leida, C. and Rubio-Cabetas, M. J. (2016b), 'Physiological characterization of drought stress response and expression of two transcription factors and two LEA genes in three *Prunus* genotypes', *Scientia Horticulturae*, 213. doi:10.1016/j.scienta.2016.11.006.

Brown, P. and Uriu, K. (1996), 'Nutrition deficiencies and toxicities: Diagnosing and correcting imbalances', in *Almond Production Manual*, UC ANR Publication, University of California, Division of Agriculture and Natural Resources, Oakland, CA.

Browne, G. T., Lampinen, B. D., Holtz, B. A., Doll, D. A., Upadhyaya, S. K., Schmidt, L. S., Bhat, R. G., Udompetaikul, V., Coates, R. W., Hanson, B. D., Klonsky, K. M., Gao, S., Wang, D., Gillis, M., Gerik, J. S. and Scott Johnson, R. (2013), 'Managing the almond and stone fruit replant disease complex with less soil fumigant', *California Agriculture*, 67(3). doi:10.3733/ca.v067n03p128.

Camposeo, S., Palasciano, M., Vivaldi, G. A. and Godini, A. (2011), 'Effect of increasing climatic water deficit on some leaf and stomatal parameters of wild and cultivated almonds under Mediterranean conditions', *Scientia Horticulturae*, 127(3), pp. 234–41. doi:10.1016/j.scienta.2010.09.022.

Caruso, T., Motisi, A., Marra, F. P. and Barone, E. (1992), 'The use of phenoclimatic models to characterize environments for chilling and heat requirements of deciduous fruit trees: Methodological approaches and initial results', *Advances in Horticultural Science*, 6(2), pp. 65–73. Available at: https://www.jstor.org/stable/42881628#%0Ahttps://www.researchgate.net/publication/237021495_The_use_of_phenoclimatic_models_to_characterize_environments_for_chilling_and_heat_requirements_of_deciduous_fruit_trees_methodological_approaches_and_initial_resu.

Dicenta, F., Ortega, E., Martínez-Gómez, P., Sánchez-Pérez, R., Martínez-García, P. J., Cremades, T., Rubio, M., García-Gusano, M. and Egea, J. (2011), 'Breeding late-flowering almonds in the CEBAS-CSIC, Murcia, Spain', *Acta Horticulturae*, 912, pp. 385–9. doi:10.17660/ActaHortic.2011.912.56.

Dicenta, F., Egea, J., Martínez-Gómez, P. and Sánchez-Pérez, R. (2016), 'Almond breeding: Important issues and challenges for research', *Options Méditerranéennes Serie A: Seminaires mediterraneens*, 119, pp. 23–8.

Dirlewanger, E., Cosson, P., Howad, W., Capdeville, G., Bosselut, N., Claverie, M., Voisin, R., Poizat, C., Lafargue, B., Baron, O., Laigret, F., Kleinhentz, M., Arús, P. and Esmenjaud, D. (2004), 'Microsatellite genetic linkage maps of myrobalan plum and an almond-peach hybrid – Location of root-knot nematode resistance genes', *Theoretical and Applied Genetics*, 109(4), pp. 827–38. doi:10.1007/s00122-004-1694-9.

Doll, D. A., Duncan, R. A., Hendricks, L. and Micke, W. (2014), 'Performance of six almond rootstocks with long-term exposure to sodium and chloride', *Acta Horticulturae*, 1028, pp. 273–5.

Duncan, R. (2012), 'The effect of high density planting and minimal pruning on the productive life of an almond orchard', University of California. Available at: cestanislaus.ucdavis.edu.

Duval, H. (2016), 'Genetic diversity of almond rootstocks. The INRA *Prunus* rootstock breeding program', in Kodad, O., López-Francos, A., Rovira, M. and Socías I Company, R. (Eds), *XVI GREMPA Meeting on Almonds and Pistachios*. CIHEAM, Zaragoza, pp. 163–5 (Options Méditerranéennes: Série A. Séminaires Méditerranéens; n. 119).

Egea, L. and Garcia, J. E. (1975), 'Lignes de recherche actuelle sur l'amandier au CEBAS, Murcie. Objectifs et recherches observations', *2nd Coll. GREMPA, CIHEAM*, Montpellier, France.

Egea, J., Ortega, E., Martínez-Gómez, P. and Dicenta, F. (2003), 'Chilling and heat requirements of almond cultivars for flowering', *Environmental and Experimental Botany*, 50(1). doi:10.1016/S0098-8472(03)00002-9.

Eichi, V. R., Tyerman, S. D. and Wirthensohn, M. G. (2014), 'Water relations and mesophyll anatomy in almond leaves', *Journal of the American Society for Horticultural Science*, 139(6), pp. 627–33.

Espadafor, M., Orgaz, F., Testi, L., Lorite, I. J., González-Dugo, V. and Fereres, E. (2017), 'Responses of transpiration and transpiration efficiency of almond trees to moderate water deficits', *Scientia Horticulturae*, 225, pp. 6–14. doi:10.1016/j.scienta.2017.06.028.

Esparza, G., DeJong, T. M., Weinbaum, S. A. and Klein, I. (2001), 'Effects of irrigation deprivation during the harvest period on yield determinants in mature almond trees', *Tree Physiology*, 21(14), pp. 1073–9.

Expo (2015), EU Scientific Steering Committee: The role of research in global food and nutrition security. Discussion paper.

Faust, M. (1989), *Physiology of Temperate Zone Fruit Trees*, Wiley, New York, NY.

Felipe, A. J. (2009), '"Felinem", "Garnem", and "Monegro" almond × peach hybrid rootstocks', *HortScience*, 44(1), pp. 196–7.

Fulton, A. E., Wildman, W. E., Begg, E. L. and Huntington, G. L. (1996), 'The evaluation and modification of physical soil problems', in Micke, W. C. (Ed.), *Almond Production Manual*. Publication 3364, University of California, Division of Agriculture and Natural Resources, Oakland, CA, pp. 20–8.

Galati, A., Gristina, L., Crescimanno, M., Barone, E. and Novara, A. (2015), 'Towards more efficient incentives for agri-environment measures in degraded and eroded vineyards', *Land Degradation and Development*, 26(6). doi:10.1002/ldr.2389.

García-Tejero, I. F., Hernández, A., Rodríguez, V. M., Ponce, J. R., Ramos, V., Muriel, J. L. and Durán-Zuazo, V. H. (2015), 'Estimating almond crop coefficients and physiological response to water stress in semiarid environments (SW Spain)', *Journal of Agricultural Science and Technology*, 17(5), pp. 1255–66.

Garnett, T., Appleby, M. C., Balmford, A., Bateman, I. J., Benton, T. G., Bloomer, P., Burlingame, B., Dawkins, M., Dolan, L., Fraser, D., Herrero, M., Hoffmann, I., Smith, P., Thornton, P. K., Toulmin, C., Vermeulen, S. J. and Godfray, H. C. J. (2013), 'Sustainable intensification in agriculture: Premises and policies', *Science Magazine*, 341, pp. 33–4. doi:10.1126/science.1234485.

Girona, J., Marsal, J., Cohen, M., Mata, M. and Miravete, C. (1993), 'Physiological, growth and yield responses of almond (*Prunus dulcis* L.) to different irrigation regimes', *Acta Horticulturae*, 335, pp. 389–98. International Society for Horticultural Science (ISHS), Leuven, Belgium. doi:10.17660/ActaHortic.1993.335.47.

Godini, A. (2002), 'Almond fruitfulness and role of self-fertility', *Acta Horticulturae*, pp. 191–203.

Godini, A. and Palasciano, M. (1998), 'Growth and yield of four self-fruitful and four self-unfruitful almonds on three rootstocks: A thirteen-year study', *Acta Horticulturae*, 470, pp. 200–7. doi:10.17660/ActaHortic.1998.470.27.

Goldhamer, D. A. and Fereres, E. (2017), 'Establishing an almond water production function for California using long-term yield response to variable irrigation', *Irrigation Science*, 35(3), pp. 169–79. doi:10.1007/s00271-016-0528-2.

Goldhamer, D. A. and Smith, T. (1995), 'Single-season drought irrigation strategies influence almond production', *California Agriculture*, 49(1), pp. 19–22. doi:10.3733/ca.v049n01p19.

Goldhamer, D. A. and Viveros, M. (2000), 'Effects of preharvest irrigation cutoff durations and postharvest water deprivation on almond tree performance', *Irrigation Science*, 19(3), pp. 125–31.

Goldhamer, D. A., Viveros, M. and Salinas, M. (2006), 'Regulated deficit irrigation in almonds: Effects of variations in applied water and stress timing on yield and yield components', *Irrigation Science*, 24(2), pp. 101–14. doi:10.1007/s00271-005-0014-8.

Gomes-Laranjo, J., Coutinho, J. P., Galhano, V. and Cordeiro, V. (2005), 'Characterization of almond varieties in irrigated and non-irrigated conditions', *Options Méditerranéennes Serie A: Seminaires mediterraneens*, 63, pp. 311–19.

Gradziel, T. M. (2011), 'Origin and dissemination of almond', in Janick, J. (Ed.), *Horticultural Reviews*, John Wiley & Sons, Inc., Hoboken, NJ, pp. 23-81. doi:10.1002/9780470872376.ch2.

Gradziel, T. M. and Kester, D. E. (1996), 'Genetic improvements', in Micke, W. C. (Ed.), *Almond Production Manual*, UC ANR Publication, Division of Agriculture and Natural Resources, University of California, Oakland, CA, pp. 70–5.

Gradziel, T. M. and Lampinen, B. L. (2011), Defining the limits of almond productivity to facilitate marker assisted selection and orchard management, *Acta Horticulturae*, 912, 47–52.

Gradziel, T. M. and Martínez-Gómez, P. (2013), 'Almond breeding', in Janick, J. (Ed.), *Plant Breeding Reviews*, John Wiley & Sons, Inc. doi:10.1002/9781118497869.ch4.

Grasselly, C. and Olivier, G. (1989), 'Possibilités d'amélioration des porte-greffes de l'amandier par hybridation interspécifique', *Options Méditerranéennes Serie A: Seminaires mediterraneens*, 5, pp. 47–9.

Gülcan, R. (Ed.). (1985), *Almond Descriptors (revised)*, IBPGR Secretariat Rome.

Heerema, R. J., Weinbaum, S. A., Pernice, F. and Dejong, T. M. (2008), 'Spur survival and return bloom in almond [*Prunus dulcis* (Mill.) D.A. Webb] varied with spur fruit load, specific leaf weight, and leaf area', *Journal of Horticultural Science and Biotechnology*, 83(2), pp. 274–81.

Herralde, F. D. E., Save, R., Biel, C., Batlle, I. and Vargas, F. J. (2001), 'Differences in drought tolerance in two almond cultivars: "Lauranne" and "Masbovera"', *Cahiers Options Méditerranéennes*, 56, pp. 149–54.

Hosseinpour, B., Sepahvand, S., Kamali Aliabad, K., Bakhtiarizadeh, M., Imani, A., Assareh, R. and Salami, S. A. (2018), 'Transcriptome profiling of fully open flowers in a frost-tolerant almond genotype in response to freezing stress', *Molecular Genetics and Genomics*, 293, pp. 151–63. doi:10.1007/s00438-017-1371-8.

Hutmacher, R. B., Nightingale, H. I., Rolston, D. E., Biggar, J. W., Dale, F., Vail, S. S. and Peters, D. (1994), 'Growth and yield responses of almond (*Prunus amygdalus*) to trickle irrigation', *Irrigation Science*, 14(3), pp. 117–26. doi:10.1007/BF00193133.

Janick, J. (2005), 'The origin of fruits, fruit growing and fruit breeding', *Plant Breeding Reviews*, 25, pp. 255–320. doi:10.1002/9780470650301.ch8.

Kester, D. E. and Ross, N. W. (1996), 'History', in Micke, W. C. (Ed.), *Almond Production Manual*, University of California, Division of Agriculture and Natural Resources, Oakland, CA, pp. 1–2.

Kester, D. E., Martin, G. C. and Labavitch, J. M. (1996), 'Growth and development', in Micke, W. (Ed.), *Almond Production Manual*, University of California, Division of Agriculture and Natural Resources, Oakland, CA, pp. 90–7.

Klein, I., Esparza, G., Weinbaum, S. A. and DeJong, T. M. (2001), 'Effects of irrigation deprivation during the harvest period on leaf persistence and function in mature almond trees', *Tree Physiology*, 21(14), pp. 1063–72.

Klein, A.-M., Hendrix, S. D., Clough, Y., Scofield, A. and Kremen, C. (2015), 'Interacting effects of pollination, water and nutrients on fruit tree performance', *Plant Biology*, 17(1), pp. 201–8. doi:10.1111/plb.12180.

Kodad, O. and Socías I Company, R. (2008), 'Fruit quality in almond as related to the type of pollination in self-compatible genotypes', *Journal of the American Society for Horticultural Science*, 133(3), pp. 320–6.

Kodad, O., Socías I Company, R. and Alonso, J. (2018), 'Genotypic and environmental effects on tocopherol content in almond', *Antioxidants*, 7(1), p. 6. doi:10.3390/antiox7010006.

Kumar, D., Ahmed, N. and Verma, M. K. (2012), 'Studies on high density planting in almond in Kashmir valley', *Indian Journal of Horticulture*, 69(3), pp. 328–32.

Lampinen, B. D., Tombesi, S., Metcalf, S. G. and DeJong, T. M. (2011), 'Spur behaviour in almond trees: Relationships between previous year spur leaf area, fruit bearing and mortality', *Tree Physiology*, 31(7), pp. 700–6. doi:10.1093/treephys/tpr069.

Lampinen, B. D., Browne, G. T., Udompetaikul, V., Upadhyaya, S. K., Slaughter, D., Holtz, B., ..., Klonsky, K. (2012), 'A mobile platform for measuring canopy photosynthetically active radiation interception in tree crops', *HortTechnology*, 22(2), pp. 237–44.

Lang, G. A. (1987), 'Dormancy: A universal terminology', *HortScience*, 22, pp. 817–920.

López, M., Vargas, F. J. and Batlle, I. (2006), 'Self-(in)compatibility almond genotypes: A review', *Euphytica*, 150(1-2). doi:10.1007/s10681-005-9009-z.

Marra, F. P., Campisi, G., La Mantia, M., Barone, E. and Caruso, T. (2008), 'Esigenze bio-termiche e stima del periodo di sviluppo del frutto in cultivar di pesco a differente epoca di maturazione', in Vaio, C. Di, Damiano, C. And Fideghelli, C. (Eds), *VI Convegno Nazionale Sulla Peschicoltura Meridionale*, Caserta, 6–7 marzo.

Martínez-García, P. J., Ortega, E. and Dicenta, F. (2011), 'Self-pollination does not affect fruit set or fruit characteristics in almond (*Prunus dulcis*)', *Plant Breeding*, 130(3). doi:10.1111/j.1439-0523.2010.01816.x.

Martínez-Gómez, P., Sánchez-Pérez, R. and Dicenta, F. (2006), 'Fruit development in almond', *Acta Horticulturae*, 726, pp. 241-6.

Martínez-Gómez, P., Prudencio, A. S., Gradziel, T. M. and Dicenta, F. (2017), 'The delay of flowering time in almond: A review of the combined effect of adaptation, mutation and breeding', *Euphytica*, 213(8). doi:10.1007/s10681-017-1974-5.

Micke, W. C. and Kester, D. E. (1998), 'Almond growing in California', *Acta Horticulturae*, 470, pp. 21-8.

Monastra, F., Martelli, G. and Chiarotti, A. (1998), 'Economic advantages of high-density planting of almond orchards', *Options Méditerranéennes Serie A: Seminaires mediterraneens*, 85, pp. 81–5.

Muhammad, S., Sanden, B. L., Lampinen, B. D., Saa, S., Siddiqui, M. I., Smart, D. R., Olivos, A., Shackel, K. A. and DeJong, T. and Brown, P. H. (2015), 'Seasonal changes in nutrient content and concentrations in a mature deciduous tree species: Studies in almond (*Prunus dulcis* (Mill.) D. A. Webb)', *European Journal of Agronomy*, 65, pp. 52–68. doi:10.1016/j.eja.2015.01.004.

Muhammad, S., Sanden, B. L., Saa, S., Lampinen, B. D., Smart, D. R., Shackel, K. A., DeJong, T. M. and Brown, P. H. (2018), 'Optimization of nitrogen and potassium nutrition to improve yield and yield parameters of irrigated almond (*Prunus dulcis* (Mill.) D. A. webb)', *Scientia Horticulturae*, 228. doi:10.1016/j.scienta.2017.10.024.

Nortes, P. A., Gonzalez-Real, M. M., Egea, G. and Baille, A. (2009), 'Seasonal effects of deficit irrigation on leaf photosynthetic traits of fruiting and non-fruiting shoots in almond trees', *Tree Physiology*, 29(3), pp. 375–88. doi:10.1093/treephys/tpn032.

Nyomora, A. M. S., Brown, P. H. and Krueger, B. (1999), 'Rate and time of boron application increase almond productivity and tissue boron concentration', *HortScience*, 34(2), pp. 242–5.

Ortega, E. and Dicenta, F. (2006), 'Self-fertilization in homozygous and heterozygous self-compatible almonds', *Scientia Horticulturae*, 109(3), pp. 288–92. doi:10.1016/j.scienta.2006.04.017.

Ortega, E., Gómez, E. M. and Dicenta, F. (2014), 'Almond self-incompatibility genetics: Recent progress and future prospects', *Acta Horticulturae*, 1028, pp. 99–102.

Palasciano, M., Logoluso, V. and Lipari, E. (2014), 'Differences in drought tolerance in almond cultivars grown in Apulia region (Southeast Italy)', *Acta Horticulturae*, 1028, pp. 319–24.

Pinochet, J. (2009), '"Greenpac", A new peach hybrid rootstock adapted to mediterranean conditions', *HortScience*, 44(5), pp. 1456–7.

Rattigan, K. and Hill, S. J. (1986), 'Relationship between temperature and flowering in almond', *Australian Journal of Experimental Agriculture*, 26, pp. 399–404.

Reidel, E. J., Brown, P. H., Duncan, R. A. and Weinbaum, S. A. (2001), 'Almond productivity as related to tissue potassium', *Better Crops*, 85(3), pp. 21-3.

Reidel, E. J., Brown, P. H., Duncan, R. A., Heerema, R. J. and Weinbaum, S. A. (2004), 'Sensitivity of yield determinants to potassium deficiency in "Nonpareil" almond (*Prunus dulcis* (Mill.) D.A. Webb)', *Journal of Horticultural Science and Biotechnology*, 79(6), pp. 906–10.

Romero, P., Botia, P. and Garcia, F. (2004), 'Effects of regulated deficit irrigation under subsurface drip irrigation conditions on vegetative development and yield of mature almond trees', *Plant and Soil*, 260(1–2), pp. 169-81. doi:10.1023/B:PLSO.0000030193.23588.99.

Rubio-Cabetas, M. J. (2016), 'Almond rootstocks: Overview', *Options Méditerranéennes Serie A: Seminaires mediterraneens*, 119, pp. 133–43.

Saa, S. and Brown, P. H. (2014), 'Fruit presence negatively affects photosynthesis by reducing leaf nitrogen in almond', *Functional Plant Biology*, 41(8), pp. 884–91. doi:10.1071/FP13343.

Saa, S., Muhammad, S. and Brown, P. H. (2013), 'Development of leaf sampling and interpretation methods and nutrient budget approach to nutrient management in almond (*Prunus dulcis* (Mill.) D.A.Webb)', *Acta Horticulturae*, 984, pp. 291-6.

Saa, S., Peach-Fine, E., Brown, P., Michailides, T., Castro, S., Bostock, R. and Laca, E. (2016), 'Nitrogen increases hull rot and interferes with the hull split phenology in almond (*Prunus dulcis*)', *Scientia Horticulturae*, 199. doi:10.1016/j.scienta.2015.12.027.

Saa, S., Fernández, E., Muhammad, S., Río, A. O.-D., DeJong, T. M., Laca, E. and Brown, P. (2017), 'Increases in leaf nitrogen concentration and leaf area did not enhance spur survival and return bloom in almonds (*Prunus dulcis* [Mill.] DA Webb)', *Acta Physiologiae Plantarum*, 39(4). doi:10.1007/s11738-017-2401-1.

Sánchez-Pérez, R., Del Cueto, J., Dicenta, F. and Martínez-Gómez, P. (2014), 'Recent advancements to study flowering time in almond and other *Prunus* species', *Frontiers in Plant Science*, 5, p. 334. doi:10.3389/fpls.2014.00334.

Saunders, M. E., Luck, G. W. and Mayfield, M. M. (2013), 'Almond orchards with living ground cover host more wild insect pollinators', *Journal of Insect Conservation*, 17(5), pp. 1011–25. doi:10.1007/s10841-013-9584-6.

Sedgley, M. (1990), 'Flowering of deciduous perennial fruit crops', in Janick, J. (Ed.), *Horticultural Reviews*, Timber Press, pp. 223–64. doi:10.1002/9781118060858.ch5.

Sedgley, M. and Collins, G. (2002), 'Almond improvement in Australia. Amélioration de l'amandier en Australie', *Fruits*, 57(2), pp. 129–34. doi: 10.1051/fruits:2002012.

Shackel, K. (2003), 'Deficit irrigation management during hull-split', in *31st Almond Research Conference, Modesto, CA, The Almond Board of California.*

Shackel, K. A. (2011), 'A plant-based approach to deficit irrigation in trees and vines', *HortScience*, 46(2), pp. 173–7.

Shackel, K. A., Ahmadi, H., Biasi, W., Buchner, R., Goldhamer, D., Gurusinghe, S., Hasey, J., Kester, D., Krueger, B., Lampinen, B., McGourty, G., Micke, W., Mitcham, E., Olson, B., Pelletrau, K., Philips, H., Ramos, D., Schwankl, L., Sibbett, S., Snyder, R., Southwick, S., Stevenson, M., Thorpe, M., Weinbaum, S. and Yeager, J. (1997), 'Plant water status as an index of irrigation need in deciduous fruit trees', *HortTechnology*, 7(1), pp. 23–9.

Socías I Company, R. (2002), 'Latest advances in almond self-compatibility', *Acta Horticulturae*, 591, 205–12.

Socías I Company, R., Kodad, O., Alonso, J. M. and Gradziel, T. M. (2008), 'Almond quality: A breeding perspective', in Janick, E. (Ed.), *Horticultural Reviews*, John Wiley & Sons, Inc., pp. 197–238. doi:10.1002/9780470380147.ch3.

Socías I Company, R., Rubio-Cabetas, J. M., Alonso, J. M., Kodad, O. and Gómez Aparisi, J. (2010), 'An overview of almond cultivars and rootstocks: Challenges and perspectives', *Options Méditerranéennes Serie A: Seminaires mediterraneens*, 214(94), pp. 205–14.

Socías I Company, R., Alonso, J. M., Kodad, O. and Gradziel, T. M. (2012), 'Almond', in Badenes, M. L. and Byrne, D. H. (Eds), *Fruit Breeding*, Springer Science & Business Media. doi:10.1007/978-1-4419-0763-9_18.

Socías I Company, R., Alonso Segura, J. M., Kodad, O., Espada, J. L. and Andreu, J. (2014), 'Nut size and blooming time in almond', *Nucis-Newsletter* (May 2014), pp. 14–16.

Socías I Company, R., Kodad, O., Ansón, J. M. and Alonso, J. M. (2015), '"Vialfas" almond', *HortScience*, 50(11), pp. 1726–8.

Sottile, F., Barone, E., Barbera, G. and Palasciano, M. (2014), 'The Italian almond industry: New perspectives and ancient tradition', *Acta Horticulturae*, 1028, 401–7.

Stewart, W. L., Fulton, A. E., Krueger, W. H., Lampinen, B. D. and Shackel, K. A. (2011), 'Regulated deficit irrigation reduces water use of almonds without affecting yield', *California Agriculture*, 65(2). doi:10.3733/ca.v065n02p90.

Teviotdale, B. L., Goldhamer, D. A. and Viveros, M. (2001), 'Effects of deficit irrigation on hull rot disease of almond trees caused by *Monilinia fructicola* and *Rhizopus stolonifer*', *Plant Disease*, 85(4), pp. 399–403.

Tilman, D., Fargione, J., Wolff, B., Antonio, C. D., Dobson, A., Howarth, R., Schindler, D., Schlesinger, W.H., Simberloff, D. and Swackhamer, D. (2001), 'Forecasting agriculturally driven environmental change', *Science*, 292(5515), pp. 281–4. doi:10.1126/science.1057544.

Tombesi, S., Scalia, R., Connell, J., Lampinen, B. and Dejong, T. M. (2010), 'Fruit development in almond is influenced by early Spring temperatures in California', *Journal of Horticultural Science and Biotechnology*, 85(4), pp. 317–22. doi:10.1080/14620316.2010.11512674.

Tombesi, S., Lampinen, B. D., Metcalf, S. and Dejong, T. M. (2011), 'Relationships between spur- and orchard-level fruit bearing in almond (*Prunus dulcis*)', *Tree Physiology*, 31(12), pp. 1413–21. doi:10.1093/treephys/tpr119.

Tombesi, S., Lampinen, B. D., Metcalf, S. and Dejong, T. M. (2015), 'Spur fruit set is negatively related with current-year spur leaf area in almond', *HortScience*, 50(2), pp. 322–5.

Tombesi, S., Lampinen, B. D., Metcalf, S. and DeJong, T. M. (2017), 'Yield in almond is related more to the abundance of flowers than the relative number of flowers that set fruit', *California Agriculture*, 71(2), pp. 68–74. doi:10.3733/ca.2016a0024.

Valdebenito, D., Tombesi, S., Tixier, A., Lampinen, B., Brown, P. and Saa, S. (2017), 'Spur behavior in almond trees (*Prunus dulcis* [Mill.] DAWebb): Effects of flowers, fruit, and "June drop" on leaf area, leaf nitrogen, spur survival and return bloom', *Scientia Horticulturae*, 215. doi:10.1016/j.scienta.2016.11.050.

Vargas, F. J. (2014), 'Influence of research on modernization of almond and pistachio culture in mediterranean area during the last forty years and some challenges for the future', *Acta Horticulturae*, 1028, pp. 19–26.

Vicente-Vicente, J. L., García-Ruiz, R., Francaviglia, R., Aguilera, E. and Smith, P. (2016), 'Soil carbon sequestration rates under Mediterranean woody crops using recommended management practices: A meta-analysis', *Agriculture, Ecosystems and Environment*, 235. doi:10.1016/j.agee.2016.10.024.

Wirthensohn, M. G. and Sedgley, M. (2002), 'Almond breeding in Australia', *Acta Horticulturae*, 591, pp. 245–8.

Wirthensohn, M., Rahemi, M. and Fernández I Martí, A. (2011), 'Identification of self-incompatibility genotypes and DNA fingerprinting of some Australian almond cultivars', *Acta Horticulturae*, 912, pp. 561–6.

Zarate-Valdez, J. L., Muhammad, S., Saa, S., Lampinen, B. D. and Brown, P. H. (2015), 'Light interception, leaf nitrogen and yield prediction in almonds: A case study', *European Journal of Agronomy*, 66. doi:10.1016/j.eja.2015.02.004.

Zhu, Y., Taylor, C., Sommer, K., Wilkinson, K. and Wirthensohn, M. (2014), 'Effect of deficit irrigation on almond kernel constituents', 1028, *Acta Horticulturae*, pp. 221–3.

Ziegler, V. H., Ploschuk, E., Weibel, A. and Insausti, P. (2017), 'Short-term responses to flooding stress of three *Prunus* rootstocks', *Scientia Horticulturae*, 224, pp. 135–41. doi:10.1016/j.scienta.2017.06.009.

# Chapter 12

## Advances in breeding of chestnuts

*Teresa Barreneche, BFP, INRA, Université de Bordeaux, France; Roberto Botta, Università di Torino, Italy; and Cécile Robin, BIOGECO, INRA, Université de Bordeaux, France*



### 1 Introduction

The chestnut tree belongs to the genus *Castanea* (order Fagales, family *Fagaceae*). *Castanea* species (of which the number varies depending on the authors, Mellano et al. 2012; Nelson et al. 2014) are naturally distributed in three geographic domains in the Northern hemisphere. Firstly, a large Mediterranean Southern European domain (Iberian Peninsula, Italy, Balkans, Mediterranean Islands) which extends to Northwest Europe (England, Belgium) and to Western Asia (Northeast Turkey, Armenia, Georgia, Azerbaijan and Syria). Secondly, an American domain (eastern coast of Northern America) and thirdly, an Asian domain (China, Japan, Korea, Taiwan). Four *Castanea* species are grown and valued for their nuts and/or wood: *Castanea sativa Mill.* in Europe and Minor Asia, *C. dentata* (Marsh.) Borkh. in Northern America, *C. mollissima* (Blume) and *C. crenata* (Sieb & Zucc.) in Asia (Table 1). *C. sativa* and *C. dentata* are the tallest tree species and are also used for timber production.

Chestnut species are monoecious species, and cross-pollination is necessary to obtain seeds. The chestnut species has the feature of a mixed pollination syndrome, involving wind and insects. However, to date, the relative importance of each mode of pollination remains poorly understood

http://dx.doi.org/10.19103/AS.2018.0042.16

**Table 1** Main characteristics of chestnut species (Solignat and Chapa, 1975; Pereira-Lorenzo et al., 2012; http://www.acf.org/resources/identification/chinese-american-chestnuts/)

| | *Castanea dentata* (Marshall) Borkh. American Chestnut | *Castanea crenata* Siebold & Zucc. Japanese Chestnut | *Castanea mollissima* Blume. Chinese Chestnut | *Castanea sativa* Mill. European Chestnut |
|---|---|---|---|---|
| Tree | Grows to 30–40 m high and 1–1.5 m diameter | Small trees not higher than 10 m and sometimes shrub like | Medium size, trees of average height 12 m and 0.70–0.80 m diameter | Robust trees up to 30 m and more than 1 m diameter. May exceed 500 years of age |
| Stem | Stems smooth and hairless. Stem colour: reddish brown to dark green. Sharp stipules. Small, numerous lenticels on stem | Stems smooth and hairless, not ribbed. Stem colour: brown to yellow. Small lenticels on the stem | Hairy stems and hairy leaf veins. Stem colour: tan to pea green. Large lenticels on stem | Stem is robust, ribbed (five thicker than those of the other chestnut species). Hairless with numerous lenticels |
| Leaves | Elongated leaf. Large, prominent teeth on edge; bristle on teeth curves inward. Blade tapers sharply to meet stem at base of leaf blade. Light green underside on leaves exposed to the sun | Very small teeth on leaf edges. Leaf elongated, with almost parallel sides | Leaf is oval shaped. Teeth are smaller. Base of leaf blade is rounded. Leaf is thick and waxy feeling | Prominent, triangular teeth on leaf edges. Leaf elongated, with rounded base. Long petiole |
| Buds | Pointed buds that angle away from the stem | Rounded buds | Rounded buds that hug the stem. Stipules cover the buds | Triangular buds larger than those of the other chestnut species |
| Burs | A dense mass of long, slender spines. Spines are 2–3 cm long, 0.5 mm thick | A mass of spines that interlock in a thatched pattern | A sparse mass of short, thick spines. Spines are 1–2 cm long, 1 mm thick | A dense mass of long, stiff spines. Spines are 2–3 cm long, 0.75 mm thick |
| Nuts | 2–3 nuts in each bur. Nuts are relatively small, 1/2 to 1 inch in diameter. Nuts are hairy over 1/3 to 2/3 of length from pointed end. Tips are pointed. Vascular bundles in a sunburst pattern on hilum end. Very sweet, non-astringent, easy to peel | 2–3 nuts in each bur. Nuts are relatively large, 1–2 inches in diameter. Very little surface hair; only at tip. Tips are pointed. Vascular bundles in a diffuse pattern on hilum end. Not sweet, astringent. Adherent pellicle | Up to three nuts per bur. Nuts are relatively large, 3/4 to 2 inches in diameter. Only the tips of the nuts are hairy. Tips of chestnuts are rounded. Vascular bundles in a diffuse pattern on hilum end. Sweet nuts with easily removable pellicle. Two crops/year | 2–3 nuts in each bur. Nuts are relatively large, 1–2 inches in diameter. Surface hair only at tip. Tips are pointed. Vascular bundles in a large star pattern on hilum end. Sweet nuts. Adherent pellicle to peel easily |

(De Oliveira et al. 2001; Giovanetti and Aronne 2011; Hasegawa et al. 2015). Chestnut trees form two types of inflorescences: unisexual catkins with only male flowers, and bisexual catkins in which female flowers appear in clusters of three, at the base of the central stem on which the male flowers are arranged. Moreover, four different types of male flowers can be distinguished according to their morphology: astaminate, brachystaminate, meso staminate and longistaminate. Because astaminate flowers do not produce pollen, and brachystaminate and mesostaminate flowers produce a very low to low quantity of pollen, only catkins with longistaminate flowers are functionally bisexual (Solignat and Chapa 1975; Mert and Soylu 2006). All *Castanea* species are diploid species, with 2n = 2x = 24 chromosomes (Nelson et al. 2014) and can cross-pollinate.

In all these regions chestnuts have long been a significant food source. Some evidence of chestnut cultivation, found in archaeological sites and palynological data, suggested that it was a component of an advanced form of agriculture. In China, the study of archeaobiological samples revealed that chestnut cultivation started about 4500 years BP (Li et al. 2007). In Japan, chestnut has been associated with the hunter-gatherer culture of the Jomon period (14000–300 BP; Kotobuki 2002). In Europe, *C. sativa* has been cultivated for its fruits, together with other trees such as olive tree and walnut in Europe since about 3700 BP (Conedera et al. 2004). In America, the spread of chestnut from about 2500 BP in New England was probably due to agriculture intensification and assisted migration by Native Americans, who used and traded chestnut as a food and wood resource (Russell 1987). Because of this early domestication, many chestnut cultivars have been selected and bred for hundreds of years. Nowadays, in these chestnut-growing regions, nut production is still partly based on traditional orchards and locally adapted cultivars. In Europe, Pereira-Lorenzo et al. (2012) described more than 1000 varieties, which were vegetatively propagated by grafting on seedlings in Italy, France, Spain, Portugal and Turkey.

In the last two centuries, diseases and pests from Asia have been introduced in chestnut plantations in both Northern America and Southern Europe (Anagnostakis 1987). These have jeopardized indigenous populations and some of the local selected chestnut varieties. Ink disease is the first pandemic chestnut disease and the most destructive one, and has been reported as early as the nineteenth century in Europe as well as in Southeastern US (Crandall et al. 1945; Grente 1961). The causal agents, identified by Petri in 1917, are fungi-like organisms (*Phytophthora cambivora* and *Phytophthora cinnamomi* belonging to the Oomycetes). They live in the soil and infect roots and collars of plants. Typical symptoms include root rot resulting in the wilting and death of the trees, and collar and trunk necrotic lesions with black exudates. Devastating epidemics that occurred in the nineteenth and twentieth centuries have strongly contributed to the decline of chestnut cultivation in these regions. Nowadays, *P. cinnamomi* and *P. cambivora* are widely spread in Europe and cause significant losses in chestnut orchards (Vannini and Vettraino 2001; Vettraino et al. 2005; Choupina et al. 2014; Costa et al. 2011). No control method other than the use of resistant varieties proved to be efficient to reduce losses caused by ink disease in chestnut plantations. Breeding for resistance to ink disease has therefore been a key issue for European chestnut breeders.

The chestnut blight disease was first described in 1904 in a New York zoo. *Cryphonectria parasitica,* its causal agent, was introduced in the United States on imported Japanese trees. This pathogenic fungus has had a dramatic impact on American native chestnut–oak forests, causing the quasi extinction of *C. dentata* and significant changes in this ecosystem (Anagnostakis 1987). The fungus is air-dispersed through spores and infects chestnut tree trunks and branches using natural or artificial wounds as entry holes. It colonizes and kills the bark and wood tissues, which results in typical canker symptoms, severe hydraulic transport failure, and in the death of the upper part of the tree. Infected trees may stay alive but nut production is highly impacted. More often, the chestnut stumps die after several years of repeated infections. *C. parasitica* originates from China and Japan, where it causes very little damage to the different chestnut species. *C. dentata* appeared to be highly susceptible to this invasive pathogenic species. Thus *C. parasitica* spread within 50 years throughout all American oak-chestnut forests where *C. dentata* trees dominated before the arrival of the disease. It was then reported in Europe in 1938 in Italy. Nowadays, it is present in all European countries where *C. sativa* grows and is cultivated as a forest or fruit species (Rigling and Prospero 2018). The Japanese origin of the American populations was confirmed by recent genetic studies. By contrast, in Europe multiple introductions occurred, involving both the original centre (Japan and China) and the first introduction area (North America, Dutech et al. 2012).

Although *C. sativa* trees are almost as susceptible as *C. dentata* (Anagnostakis 1992), in Europe, disease severity is lower than in the United States. Indeed, in an increasing number of regions and countries, *C. sativa* trees are recovering from *C. parasitica* infections. The healing of cortical cankers results in superficial damage to the tree. This recovery is due to the spread in European *C. parasitica* populations of a mycovirus (CryphonectriaHypoVirus 1) which decreases the virulence of the pathogenic fungus, a phenomenon known as hypovirulence (Nuss 1992; Heiniger and Rigling 1994). Although it is naturally spreading in Europe, this virus is used as a biological control agent in orchards, and the fungal disease is no longer considered a problematic disease in these managed orchards. By contrast, the natural regulation of chestnut blight was not achieved by *C. parasitica* mycoviruses in Eastern America (Milgroom and Cortesi 2004). American breeding programs were aimed at selecting trees resistant to blight with the goal of restoring the chestnut ecosystem.

Among pests, the Asian gall wasp (*Dryocosmus kuriphilus* Yasumatsu, Hymenoptera: Cynipidae) is a major threat to chestnut. In fact its accidental introduction in areas where the wasp has no natural enemies has caused severe yield losses (Kato and Hijii 1997; Sartor et al. 2015). This is due to its exponential reproductive ability (all adults are females and can lay up to 100–150 eggs), and it spread quickly via nursery plant materials and passive transportation. Native to China, the pest was found in Japan (1941), Korea (1963), the United States (1974), Nepal (1999), Canada (2012) and in Europe (Cho and Lee 1963; Payne et al. 1975; Brussino et al. 2002; Moriya et al. 2003; Abe et al. 2007; Huber and Read 2012). The chestnut gall wasp is univoltine and it reproduces by thelytokous parthenogenesis. In mid-June to mid-August, adults lay eggs into the newly formed buds; in the following year, the larvae induce the formation of galls that develop at the time of budburst, on leaves, shoots and at the base of male catkins (Moriya et al. 1990; Kato and Hijii 1997). The presence of the galls reduces vegetative growth and can cause severe yield losses, reaching up to 65–85% in highly susceptible genotypes. Control strategies are based on the development of resistant cultivars, and on biological control using the parasitoid *Torymus sinensis* Kamijo (Hymenoptera: Torymidae), the latter emerging as the most effective method (Rieske et al. 2007; Quacchia et al. 2008; Ferracini et al. 2018), while chemical treatments are rarely effective due to the protection ensured to larvae by the gall tissue.

The first chestnut breeding programs sought to exploit the sources of resistance to these exotic diseases, but differed by the selection criteria (mainly ink disease in Europe and chestnut blight in the United States) and by the main production objective (mainly nuts in Europe and wood in the United States). Therefore, the need to select for improved fruit production and quality has emerged in Europe and in Japan. In Europe after the Second World War, *C. sativa* cultivation and industry extended to areas where modern chestnut orchards could be established (Bounous 2002). When European populations migrated to southern hemisphere countries (Australia, New Zealand, Chile and Argentina), they brought with them their traditional chestnut culture. For example, in Australia, although chestnuts were first introduced during the 1850s gold rush, the area planted with chestnuts considerably increased with the increasing European immigration in the 1950s (Klinac et al. 1995). In New Zealand, old trees reminiscent of first introductions can still be observed. Although *C. crenata* trees were introduced as well as *C. sativa* trees, few hybrids had been characterized in the southern and northern parts of the country (Oraguzie et al. 1999). The societal demand for chestnut in these countries has considerably increased. Development of all these new plantations raised the necessity to breed high-yielding varieties and rootstocks adapted to a more intensive management of orchards. This became an additional key issue of more recent breeding programs that focus on pollination, and on qualities such as nut size, taste and peel-ability. Therefore, several breeding programs were initiated.

## 2 Breeding for resistance to diseases threatening chestnut orchards

Epidemics in Europe and Northern America have prompted the use of Asian species and interspecies hybrids. Previously, investigations were carried out to assess the resistance of Asian species to these diseases introduced from Asia and the feasibility of restoration of chestnut orchards and forests through replantation. Prior to the nineteenth century, landowners and botanists made the first introductions of *C. mollissima* in Northern America (Anagnostakis 2012; McCleary et al. 2013). Crandall et al. (1945) reported field tests of Asian chestnuts for ink disease resistance. In Europe, *C. crenata* was introduced in France as early as the beginning of the twentieth century (1904), to test its resistance to ink disease (Prunet 1915). In Spain, chestnuts from Korea and from Japan were introduced from 1917 to 1940 in the Basque country and after 1940 in Galicia (Fernández-López 2011). In Italy, Japanese and Chinese seedlings were introduced in 1920 as starting material for further programs aimed at improving resistance, nut production and wood (Bagnaressi 1956).

Due to these early works, resistance to ink disease and chestnut blight was confirmed in *C. molissima* and *C. crenata* (Anagnostakis 2012; Claper 1954; Dufrenoy 1930; Schad et al. 1952). However, the introduced genotypes were not adapted to European or American environmental conditions and did not match the grower's and consumer's demands. Moreover, they cannot be valued as rootstocks for European fruit varieties due to graft incompatibility. The use and the massive importation of Asian germplasm were limited by quarantine regulations, linked to the risk of introducing exotic chestnut pathogens along with the plants. Therefore, the Plant Quarantine Act was enacted in the United States in 1912. Chestnut importations from Asia into France were forbidden in 1922, in order to prevent any introduction of *C. parasitica* which was spreading in the United States at this time but was not yet described in Europe (Mangin 1922). Therefore, a new screening of a larger set of *C. crenata* or *C. mollissima* genotypes in the United States or in Europe was not possible, and resistance sources had to be found in those already introduced.

In spite of differences in flowering time and pollen incompatibilities between species, natural hybridization occurred within *Castanea* genus. The first controlled crosses between different chestnut species were done in the United States by Van Fleet in 1907 using *C. crenata* as a source of resistance to blight disease, and then by Clapper with *C. mollissima* (Anagnostakis 2012). Thanks to these pioneering works, cross-combination breeding appeared as a good strategy to improve the genetic resistance of chestnut trees to *C. parasitica*, with which they have not coevolved. Indeed Chinese × American, and Japanese × American hybrids, exhibited resistance to blight disease. In 1981 Burnham initiated a new breeding program with the goal to replace American chestnut with hybrids that are indistinguishable from American chestnut but resistant to blight. His idea was to back-cross American chestnuts with first-generation hybrids in order to introgress as many genes as possible from *C. dentata* (Burnham 1981). Twenty-five years and three back-cross generations later, the selected genotypes proved to be resistant with similar morphological characteristics to that of *C. dentata* (Diskin et al. 2006).

In France, progenies generated by open pollination of Asian chestnut trees were evaluated in the field, and they proved to be more resistant than European chestnut to ink disease. Couderc in France (1920) and Gallastegui in Spain (1926) carried out the first cross-controlled hybridization. By 1940, several breeding programs had been initiated in Europe: in France (Schad et al. 1952), and in Spain and Portugal (Fernandez-Lopez 2011). The French program, which resulted in the creation of the most used hybrid chestnut rootstocks, is detailed below.

## 3 Breeding for ink-resistant rootstocks: the French program

In 1946, a breeding program started in France to select hybrids resistant to ink disease, to be used for reforestation or as rootstocks for nut production (Schad et al. 1952). Such a program addressed important and related questions such as screening and vegetative propagation methodology, but also genetics of ink disease resistance traits (heritability and determinism) and chestnut reproduction biology, which are still being investigated nowadays.

Several methods can be used to screen chestnut plants for their resistance to *P. cinnamomi* and *P. cambivora*. Assessment of resistance after natural infection in infested fields was the first method used (Schad et al. 1952). Resistance of open progenies from *C. crenata* plants was assessed with this method, which allowed testing of a high number of individuals in real conditions of infections. Lower mortality due to ink disease was thus observed in hybrid progenies compared to pure *C. sativa* ones. However, such a screening test was not optimal: most of the tested plants, particularly those with an intermediate level of resistance, were lost; quantification of resistance level was not possible; the pathogen source could not be controlled quantitatively and qualitatively; and other mortality causes could interact with ink disease. To overcome these problems, controlled inoculations of excised organs or whole plants with *Phytophthora* spp. isolates were systematically used after 1950 (Schad et al. 1952). Development of different screening tests and their comparison have been the subject of several studies at INRA (French National Institute for Agricultural Research, Salesses et al. 1993; Ramos-Guedes-Lafargue and

Salesses 1999; Robin et al. 2006), and in Europe (Fernandez-Lopez et al. 2001; Miranda-Fontaiña et al. 2007). At the same time, variability in aggressiveness among *P. cinnamomi* isolates was explored (Robin and Desprez-Loustau 1998) and the effects of root infections by *P. cinnamomi* on chestnut physiology was quantified (Maurel et al. 2001, 2004). It is now well documented that although the main symptom of ink disease, the root infection, results in water dysfunction (decrease in stomatal conductance, carbon assimilation, predawn water potential), these changes in chestnut physiology are dependent on root infection level. This vulnerability of root tissues is well correlated with the susceptibility of stem tissues, and thus can be assessed by stem inoculation. Chestnut individuals, selected as rootstock candidates for their resistance to ink disease, were tested for their grafting compatibility and their adaptation to different environments (Breisch 1995).

Defining the genetic determinism and diversity of the main defence mechanisms in *Castanea* spp. to *Phytophthora* spp. is a major scientific challenge. Early studies (Claper 1954; Veitez et al. 1982) and recent results obtained converge to show that the resistance to ink disease in *Castanea* spp. is partial and quantitative. The analysis of *C. mollissima* × *C. dentata* families allowed the identification of one major quantitative trait loci (QTL) for *P. cinnamomi* resistance (Zhebentyayeva et al. 2014). More recently, a second QTL located in another chromosome has been located in a family derived from a *C. sativa* × *C. crenata* cross (Santos et al. 2017a). Nevertheless, these two QTLs only explain a small part of the trait phenotypic variance (respectively 9% and 13%), which suggests that other regions of the genome may be involved in the determinism of this character and need to be investigated. Functional studies have identified several candidate genes involved in the ink resistance mechanisms in chestnut. These genes are coding for cystatin, beta 1,3 glucanase and thaumatin-like protein (Serrazina 2004). Differential gene expression analysis following *Phytophthora* inoculation suggests that *C. crenata* implements a larger defence gene cluster than *C. sativa* (Serrazina et al. 2015). Among these genes, eight candidate genes, including the Cast_Gnk2-like gene, have been recently studied (Santos et al. 2017b).

The first results of the French chestnut breeding program were *C. crenata* × *C. sativa* varieties: 'Marigoule', 'Ferosacre', 'Marsol', 'Maraval' and 'Précoce Migoule' (Breisch 1995; Breisch and Hennion 2004). Selected first as ink disease-resistant varieties for timber production, they were used very quickly after their release as rootstocks and became very popular as such in several countries. These hybrids were open progenies collected on *C. crenata* mother trees, imported several years before in Southwestern France by INRA and which proved to have adapted to the local ecological conditions. The genetic basis was very narrow due to the limited number of these Japanese parents. Schad et al. (1952) reported chestnut clonal propagation by stool layering as the best technique to multiply these hybrid genotypes. Although variation was observed between and within species, the large-scale propagation of rootstocks was first done by this method in France. In the 1980s a new chestnut breeding program was initiated as well as studies on *in vitro* propagation techniques (Chevre et al. 1983; Chevre and Salesses 1984). The breeding strategy consisted of enlarging the parent genetic sources, by including parents from the three species (*C. crenata*, *C. mollissima* and *C. sativa*) and by increasing not only their number but their origin too. For rootstocks the first objective was, in addition to ink disease resistance, vigour, adaptation to a larger set of ecological conditions, and graft compatibility (Chapa and Bernhard 1984). This led initially to the selection of the 'Maridonne' and 'Marlhac' rootstocks. Between 1987 and 1996, 6000 hybrids, obtained by controlled pollination, were selected and evaluated. This evaluation made it possible to retain 100 rootstocks. Forty-one rootstock candidates showing the desired traits were tested for their layering propagation ability. Among them, five were retained. Their graft compatibility with a wide range of French *C. sativa* cultivars is currently being assessed (N. Pasquet, Invenio, pers. comm.). Pending the release of new rootstocks, eight hybrid clones, all showing a satisfactory level of ink disease resistance, can be used and three of them are particularly recommended ('Maraval', 'Marsol', 'CHR 151', Table 2).

These programs addressed ink disease problems in several French areas consisting of intensive and new chestnut plantations with high-yielding fruit varieties. In mountainous southern areas, edaphic and climatic conditions strongly limit the use of hybrid clones. Maintaining chestnut tree cultivation in these areas urgently requires new rootstocks adapted to these specific ecological conditions and to the indigenous traditional varieties. Considering the large amount of variation in ink disease resistance and the existence of resistant genotypes in *C. sativa* (Schad et al. 1952; Robin et al.

**Table 2** Rootstocks resistant to ink disease (interspecific hybrids) currently commercialized in Europe

| Name and code | Frost susceptibility induction | Graft compatibility | Vigour induction | References |
|---|---|---|---|---|
| *Marsol CA 07* | Low | Broad | High | Breisch and Hennion (2004) |
| *Maraval CA 74* | Low | Broad | Low | |
| Ferosacre CA90 | Very high | Broad | High | |
| Marlhac CA118 | High | Moderate | Moderate | |
| *'CHR-151' (HS)* | Low | Broad | High | Pereira-Lorenzo et al. (2012) |
| 'CHR-161' (100) | Low | Broad | NA | |
| 'CHR-162' (7521) | Low | Broad | High | |
| 'CHR-168' (110) | Low | Broad | High | |

Italic and underlined: the rootstocks widely used at present.

2006) a new program is being developed by INRA together with different stakeholders who participate in the selection of candidate genotypes (Barreneche and Robin, unpublished). The first step in this program consists of selecting trees that are locally adapted and resistant to ink disease. To achieve this goal, basal shoots were removed from adult chestnut trees surviving from ink disease foci in areas where commercialized rootstocks could not be deployed. These shoots were used to confirm their resistance to *P. cinnamomi* level under controlled inoculation conditions. Seed orchards established with the selected genotypes will allow assessment as to whether their progenies could be used as rootstocks. Because natural hybridization is occurring in relatively high proportion in countries where Asian chestnuts have been introduced, identification of the genealogy of these candidate trees will be performed with molecular markers (Pereira-Lorenzo et al. 2010; Fernandez-Lopez 2011).

## 4 Breeding for resistance to gall wasps threatening chestnut orchards

The first breeding program aimed at improving resistance to *Dryokosmus kuriphilus* was carried out in Japan from 1947 to 1962. In 1952 only a few cultivars of *C. crenata* showed some degree of resistance ('Akanaka', 'Shikatsume', 'Kishine' and 'Ginyose'). At the time, different types of plant responses were observed and it was initially thought that the resistance may be due to mechanical barriers that partially or totally impeded oviposition. On the contrary, it was found that the gall wasp was laying eggs in 'Ginyose' buds. In this case the resistance was due to a reaction mechanism of the bud tissue surrounding the larval cell that, in early spring, was found to be necrotic. As a result the larva cannot feed in the gall and dies. Therefore, the presence of two mechanisms of resistance or reduced sensitivity was hypothesized, a biochemical and a physical mechanism.

Starting from the material found to be resistant, the Horticultural Research Station of Tsukuba developed new varieties through traditional breeding, combining resistance to adequate yield and quality. Cultivars 'Tanzawa', 'Ibuki' and 'Tsukuba' were bred in 1959, followed by 'Ishizuchi' in 1968. These were planted extensively throughout Japan allowing recovery of the chestnut industry (Pereira Lorenzo et al. 2012).

The early studies on the genetic basis of resistance showed that through crossbreeding of 'Ginyose' with susceptible plants, fully resistant progenies are obtained. This suggested that the gene involved in the resistance of 'Ginyose' was in homozygosity (Kajiura 1955). Later, Shimura (1973) reported lower rates of success from crossings between resistant and susceptible cultivars and between resistant cultivars. In the same period, however, a new biotype of the insect able to overcome the resistance was found in developed galls of the resistant cultivars (Shimura, 1972). This forced the resumption of the breeding programs and the search for new control strategies. Two cultivars ('Kunimi' and 'Shiho') with improved resistance to gall wasp were released, together with the establishment of the biological control by the parasitoid *Torymus sinensis*, introduced from China (Saito 2009; Moriya et al. 2003).

The use of cultivars bearing resistance, or low susceptibility to the pest, combined with the use of biological control by natural parasitoid enemies has helped to overcome the emergency of the infestation and yield losses in Japan. Breeding is now oriented towards the improvement of nut quality by means of interspecific hybridization between Chinese and Japanese chestnut.

Other breeding strategies adopted in Korea and China were based on the selection of cultivars resistant to the cynipid by focusing on some varieties of *C. mollissima* and hybrids *C. crenata* × *C. mollissima*, with late bud formation. In this case, at the time of adult emergence, the cynipid does not find buds of appropriate size for oviposition (Norton et al. 1987).

In the United States, most of the breeding efforts have been devoted to obtaining chestnut trees resistant to canker blight. More recently, the genetic improvement activity of nut cultivars has tried to introduce resistance factors to *D. kuriphilus* using *C. pumila* and *C. ozarkensis* (American and Chinese chinquapins), and *C. henryi* (Anagnostakis 2012). Their resistance has not been thoroughly investigated, but individuals of these species and their hybrids have been observed to be free of infestation in plantings with other chestnut species that were highly infested. Following these observations, renewed effort has been put into breeding with chinquapins for resistance to gall wasp and produce trees for the chestnut industry and for the forest plantings that will not be damaged by the pest (Anagnostakis 2014).

Following the introduction of the cynipid in Italy in 2002, a study aimed at evaluating the susceptibility of local and international *C. sativa* and hybrid cultivars under controlled conditions was carried out starting in 2004 until 2013. The observations showed both the presence of quantitative variation of the trait with large differences in terms of infestation among cultivars and the occurrence of resistance (no gall development in spring) in *C. sativa* and Eurojapanese hybrids, including 'Bouche de Bétizac', 'Marlhac' and 'Maridonne' (all *C. sativa* × *C. crenata*) (Sartor et al. 2015). The factors involved in quantitative susceptibility are still unknown, but they may include bud size and texture, and the presence of volatile substances in the bark (Huang et al. 1990). In the case of hybrid cultivars, results confirmed the genetic origin of the mechanism of inhibition of larva and gall development from the *C. crenata* parent. The resistance trait was investigated in 'Bouche de Bétizac' where histological examinations revealed the presence in dormant buds of eggs and larvae at the first instar. Larvae failed to reach the second instar in spring due to a hypersensitive reaction by the plant (Dini et al. 2012), a mechanism that appears similar to the one found in 'Ginyose'. Further studies on a progeny 'Bouche de Bétizac' × 'Madonna' (*C. sativa,* a very susceptible cultivar) revealed that the trait has a simple Mendelian inheritance and were able to locate the trait on a recently developed saturated map using single nuclear polymorphism (Torello Marinoni et al. 2017) and single nuclear polymorphism markers (Torello Marinoni et al. unpublished data).

Currently some 40 cultivars resistant to the cynipid are reported (Sartor and Botta 2011; Sartor et al. 2015) and, with the exception a cultivar of *C. mollissima* and two of *C. sativa*, the others are all *C. crenata* varieties or hybrids involving this species. Studies on susceptibility in Italy have also shown the presence of resistance in wild *C. sativa* individuals, currently under investigation with the two *C. sativa* cultivars found resistant by Sartor et al. (2015), in order to understand if the genetic bases of the trait are the same as in 'Ginyose' and 'Bouche de Bétizac'.

The biological control based on *Torymus sinensis* has been successful in different parts of the world and, in most cases, is able to reduce the infestation to levels compatible with the chestnut cultivation for nut production. However, the studies on host-pathogen interaction and the search for strategies based on quantitative and qualitative resistance are still of relevant interest for future breeding programs.

## 5 Breeding for fruit quality

The methods used in chestnut for breeding, especially for fruit quality, are clonal selection (selective breeding) and cross-combination breeding. Depending on the final purpose, fresh or processed fruits, evaluation criteria for chestnut fruit quality breeding varies (Pereira-Lorenzo et al. 2012). In chestnut as in many crops, in particular in fruit trees, before the genetic improvement itself, germplasm screening of genotypes showing the desired traits, especially fruit quality, was carried out. Resulting from centuries of natural and human selection, a multitude of sweet chestnut landraces are available, although they are increasingly threatened. In Europe and Asia, these local cultivars were characterized, initially using morphological and phenological criteria (Chapa 1987; Breisch 1995; Botu et al. 1999; Liu 1992; Pereira-Lorenzo et al. 2006; Serdar et al. 2010; Mellano et al. 2012; Grygorieva et al. 2017). Often-studied traits are those recommended by the International Union for the Protection of New Varieties of Plants (UPOV), which have been revised (Bolvanský and Mendel 2001). These may be completed by specific criteria defined by the germplasm curators or by the breeders, since Biodiversity International (https://www.bioversityinternational.org) has not published any specific descriptors for chestnut germplasm evaluation. Isozymes, and more recently, molecular markers, have allowed for improvement to the genetic diversity picture of the cultivated chestnut, highlighting the great richness of local chestnut germplasm (Buck et al. 2003; Pereira-Lorenzo et al. 2006, 2010, 2011, 2017; Solar et al. 2005; Nishio et al. 2011; Mellano et al. 2012; Torello Marinoni et al. 2013). In addition to the screening of local cultivars, surveys were also performed in different regions, often located at or close to the origin centres of the different chestnut species. The aim was to find wild or 'naturalized' chestnut types with superior nut and yield characteristics (Botu et al. 1999; Serdar 1999; Podjavorsek et al. 1999; Pandit et al. 2013; Prgomet et al. 2014; Ormeci et al. 2016). Phenotyping existing genetic resources is a crucial step before starting breeding, and therefore several research programs aimed at improving identification, in situ and ex situ evaluation, collection, propagation and regeneration of chestnut germplasm are currently underway in several countries (e.g. Botu et al. 2009; Torello Marinoni et al. 2013; Martin et al. 2017). The objective is to conserve chestnut diversity and to use it for breeding and selection of new varieties. To reach this goal, it is necessary to characterize chestnut germplasm not only with morphological traits, but also with new genetic markers that are currently being developed.

In natural populations or within local cultivars, the most successful genotypes were retained for further evaluation in experimental conditions similar to those of production orchards. This clonal selection implemented in different countries has allowed the development of some of the top chestnut fruit varieties (European and Asian ones) commercialized today (Bergougnoux et al. 1978; Breisch 1995; Bounous et al. 1997; Solar and Štampar 2010; Serdar et al. 2011). On the other hand, the same method was used to identify candidate progenitors carrying interesting traits and those suitable for cross-combination breeding.

Hybridization is the second method used for nut quality breeding (Bounous et al. 1997; Pereira-Lorenzo et al. 2016). In France, progenies generated by open and controlled pollination of Asian chestnut trees were exploited for selection of superior nut quality trait genotypes. Among the genotypes already selected for their resistance to ink disease, some ('Marigoule', 'Bournette', 'Maridonne', 'Bouche de Bétizac') also exhibited desirable fruit quality traits (Breisch 1995). In addition, following the introduction of gall wasp in Europe 'Bouche de Bétizac' was found to be resistant to this pest (Sartor et al. 2009). The fruit breeding program designed in the 1980s targeted fresh and processed fruits. Thousands of first-generation interspecific hybrids were obtained by controlled crosses, 6000 seedlings were evaluated, and from those, ten nut cultivars were retained. Currently three of them have entered a release process: one promising cultivar for the nut industry 'Bellefer®' will soon be available for producers (N. Pasquet, Invenio, pers. comm., Table 3). Hybridizations of European and Japanese chestnuts was also the method

**Table 3** List of new interesting chestnut cultivars in different countries

| Country | Name or code of cultivars | Status | References |
|---|---|---|---|
| France | 'BelleFer' | Interspecific hybrid (*C. sativa* × *C. crenata*) CPVO patented | N. Pasquet, Invenio, pers. comm. |
| Greece | EPR.4<br>EPR.5<br>EPR 8<br>EPR 25 | Selections from the wild (*C. sativa*) | Manthos and Botu (2013) |
| Japan | 'Mikuri' | Intraspecific hybrid (*C. crenata*) released in 2009 | Saito et al. (2015) |
| | 'Porotan' | Intraspecific hybrid (*C. crenata*) released in 2006 | Saito et al. (2009) |
| Korea | 'Jahong' | Selection from indigenous old chestnut individuals (*C. crenata*) | Kim et al. (2014) |
| Romania | 'Tismana T.1'<br>'Horezu M.1' | Selections from the wild (*C. sativa*) | Manthos and Botu (2013) |
| Turkey | 'Ersinop'<br>'Erfelek'<br>'Eryayla'<br>'Serdar'<br>'Ünal' | Selections from the wild (*C. sativa*), registered in 2009 and 2010 | Serdar and Macit (2010) and Serdar et al. (2013) |
| USA | 'Emalyn's Purple'<br>'Hong Kong'<br>'Jenny'<br>'Kyoung'<br>'Nanjing Special'<br>'Ness'<br>'Patterson'<br>'Payne'<br>'YGF'<br>'Yixian Large'<br>'Boitano'<br>'Yolo Grande'<br>'Bergantz'<br>'Gillet'<br>'Szego' | Selections from pure *C. mollissima*<br>*Castanea sativa* selection<br>Interspecific hybrid (*C. sativa* × *C. mollissima*)<br>Interspecific hybrids (*C. sativa* × *C. crenata*)<br>Complex hybrid | Nave (2017) and www.chestnutgrowers.org/MikeNave-NewCultivars.pdf |

exploited in Italy in order to select cultivars with superior quality nuts. Two Japanese sources were used ('Tsukuba' and 'Tanzawa') allowing the release of cultivars 'Lusenta' and 'Primato' which showed high yield and good quality nuts, and were also tolerant to chestnut blight (Bounous et al. 1997). In addition, the latter has been proved to be resistant to gall wasp (Sartor et al. 2007). In Japan, recent studies based on two Japanese chestnut intraspecific mapping families state that some nut traits, such as nut weight and pericarp splitting, are quantitative. For each of these traits at least one significant QTL was found explaining more than 20% of the phenotypic variance (Nishio et al. 2018). These results are promising for further implementation of marker-assisted selection (MAS). Both the fresh market and industry require easy peeling fruits, since removing the kernel pellicle is a laborious and costly process. The Japanese chestnut showed a thick pellicle extensively invaginated into the nuts and very difficult to remove from the kernel. It has been suggested that the pellicle adhesiveness in Japanese chestnut is related to phenolic compounds accumulated in the pellicle (Tanaka and Kotobuki 1998). Therefore, another main goal of the Japanese breeding program was the release of easy peel cultivars. This program is detailed below.

## 6 Breeding for peel-ability: the Japanese experience

In 1947, a breeding program started in Japan at the National Institute of Fruit Tree Science, focused on the selection of cultivars with quality nuts and easy peeling (Pereira-Lorenzo et al. 2012). The latter being crucial since for a long time, all Japanese chestnuts were supposed to be difficult to peel (Miller et al. 1996). On the contrary, Chinese chestnuts were known for their pellicle, which is easily removed from the kernel (Bounous et al. 1997). The Japanese breeding program for easy peel-ability is addressing issues such as screening methodology and inheritance, and determinism of easy peel pellicle.

To screen genotypes for easy peeling, nuts were first roasted and the pellicle was removed with a knife. This method was expensive and time consuming (Tanaka and Kotobuki 1992). More recently, the high oil peeling (HOP) method has been developed. It consists of deep frying the nuts in oil and allows more rapid evaluation of pellicle peel-ability (Shoda et al. 2006).

The breeding of this trait was addressed first by hybridization of Japanese chestnut with Chinese chestnut (Bounous et al. 1997). Investigating interspecific hybrids and their progenies, the easy peeling of the kernel pellicle appeared to be influenced by the pollen parent. In general, nuts from crosses with Chinese pollen were easy to peel while those coming from crosses with Japanese pollen presented a pellicle difficult to remove from the kernel. In fact, it was demonstrated that a wide range of responses occurred depending on the Chinese parent genotype, suggesting a large variability for this trait in the Chinese chestnuts. Nevertheless, promising sources of easy peel-ability were identified in Chinese chestnut cultivars such as 'Houji-360' and 'Houiji-480' (Tanaka and Kotobuki 1992). Easy peel pellicle genetic determinism in Chinese chestnut seemed to be quantitative, controlled by QTLs (Nishio et al. 2013). This research line was not pursued because the progenies of the interspecific hybrids carried undesirable characteristics (i.e. small nuts), together with the easy peel pellicle trait (Takada et al. 2017).

Another direction of the Japanese easy peel breeding program consisted of screening Japanese selections and cultivars. Thanks to the HOP method, one selection 'Porotan' showing a very easy peeling pellicle associated with good nut quality was found and released in 2006 (Saito et al. 2009). The genetic control of the easy peeling trait of 'Porotan' was assessed using a modified half diallel cross. A single major gene (P) controlled this trait, 'Porotan' carrying both recessive alleles (*p/p*) (Takada et al. 2012). Genetic mapping of the gene was undertaken using two *C. crenata* intraspecific families resulting from crosses between Japanese chestnuts ['Tanzawa' (*P/p*) × 'Porotan' (*p/p*) and 550–40 (*P/p*) × 'Tanzawa' (*P/p*)]. Single sequence repeat markers closely linked to the locus responsible for easy peel pellicle, especially two of them PRD52 and PEB62, were identified making possible the implementation of MAS of this trait (Nishio et al. 2013). Recently, variability for peel pellicle trait in Japanese chestnuts was investigated, 34 genotypes consisting of 18 cultivars, 15 selections and one wild clone ('Shibaguri 37') of *C. crenata* chestnuts were evaluated. Genetic differences in pellicle peel-ability were assessed and the variation of peel-ability appeared continuous, suggesting quantitative gene effects. In addition, the mean broad sense heritability estimated over the three years of study was high (0.67). These promising results offer new directions for easy peel pellicle Japanese breeding, and represent an alternative to the use of the 'Porotan' gene which often results in inbreeding (Takada et al. 2017).

## 7 Research in breeding as a contribution to enhanced and sustainable crop production

In 1999 Breisch listed the main problems that chestnut growers have to face to improve nut yields (Breisch 1999). Twenty years later, some of these problems (including difficulties in establishing new orchards, labour cost for harvesting etc) have been or are being solved by advances in propagation and harvesting techniques, and others (like pruning, irrigation, weeding and fertilization) by experimentation with cultivation practices.

Research in chestnut breeding has resulted in a higher number of registered rootstocks and fruit varieties. The release of rootstocks resistant to ink disease has allowed the yield to be maintained or increased in several regions, in both the ancient and new world. Selecting chestnuts, seeded or clonally propagated, resistant to diseases and ecologically adapted to several environmental conditions, is an urgent issue which is currently being addressed.

Breeding efforts focused on *Castanea* spp. for many decades were combined with research on chestnut vegetative propagation techniques. As with many other large seeded hardwood species, chestnut trees are difficult to multiply due to their high tannin content. Most of the research regarding conventional chestnut propagation techniques has been using *C. sativa* (Nelson et al. 2014). Although chestnut trees have been considered for a long time as difficult to multiply by cutting, methods are currently available to obtain copies from adult trees by using sprouting stems from adventitious buds and rooting tunnels where optimal conditions (about 24°C and 90–100% of humidity) can be maintained (Fernández-López et al. 1992; Lopez-Villamor et al. 2017). This clonal propagation method is currently used for breeding programs in Galicia, because it is simple, easy to deploy and gives results in a short time (Fernández-López et al. 1992). Candidate genes involved in rooting ability have been identified (Lopez-Villamor et al. 2017). Understanding how these genes are regulated could help to improve propagation techniques and therefore facilitate breeding programs. Moreover, these advances when transferred to the nursery industry could meet the growing demand for registered clonal rootstocks.

*In vitro* propagation methods comprise micropropagation (axillary shoot multiplication), organogenesis (adventitious shoot production) and somatic embryogenesis (Nelson et al. 2014). In France, methods of *in vitro* vegetative multiplication of ink disease rootstocks have been transferred to private companies (such as Invenio) to increase the availability of grafted chestnut seedlings. The first two *in vitro* culture techniques were mainly developed in Europe on *C. sativa*, whereas the somatic embryogenesis required for gene transfer in *C. dentata* was developed in the United States (Nelson et al. 2014; Corredoira et al. 2017). It now constitutes an effective propagation system, essential for *C. dentata* restoration.

The development of biological control methods for chestnut blight and gall wasp did not mitigate these diseases, and losses in chestnut production are still attributable to these pests. Breeding efforts are being made in order to select resistant fruit varieties, which could be used in order to achieve a more complete disease control. A better knowledge of susceptibility to chestnut weevil (*Curculio elephas*), carpophagous insects (*Cydia* spp.), nut rots caused by several fungi (*Gnomoniopsis*, *Phomopsis*, *Sclerotinia* spp.) or by physiological disease, is also required to improve crop quality. For example, internal kernel breakdown, a physiological problem observed in Michigan chestnut orchards, was associated with pollination by *C. mollissima* trees (Fulbright et al. 2014).

Progress made by American researchers in understanding mechanisms of resistance to chestnut blight has been outstanding (Worthen et al. 2010; Nelson et al. 2014). The main objective of this work was the restoration of the American chestnut forest. However, these results could be used for the selection of fruit varieties with reduced susceptibility or increased tolerance to chestnut blight. Initially, genetic analysis of hybrid *C. dentata* × *C. mollissima* progenies suggested that three genes conferred resistance to chestnut blight, enabling the design of an American back-cross strategy (Steiner et al. 2017). Genetic modification was also carried out as early as 1956. The first approach was the so-called 'mutational breeding,' which used an ionizing radiation treatment of the nuts to induce random mutations (Nelson et al. 2014). Trees obtained from the irradiated nuts (M1 generation) were planted and gave M2 and then M3 generations in which chestnut blight resistance was evaluated (Dietz 1978, https://www.acf.org/md/orchards/sugarloaf/). A second strategy is currently in progress in the United States, consisting of the transfer of an extraspecific gene for resistance to *C. parasitica* (Newhouse et al. 2014; Steiner et al. 2017). These transgenic chestnut trees express a wheat gene coding for an enzyme (an oxalate oxidase) which detoxifies the acid oxalic. This fungal toxic compound being responsible for the pathogenicity of the fungus, its degradation in infected tissues highly decreases tree susceptibility. The Food and Drug Administration authorized the production of 10 000 transgenic chestnut trees for distribution to the public, and restoration of the chestnut grove in Tully County (New York). However, even in Asian species resistance to *C. parasitica* is not complete, which suggests that chestnut blight will still be problematic for resistant trees, whatever their pedigree, if they are threatened by multiple stress factors (biotic or abiotic).

## 8 Future trends and conclusion

Recent developments in the field of tree genome mapping and sequencing have opened up new perspectives for breeders, with different applications that rely on an in depth knowledge of the genome (Nelson et al. 2014). One way to improve chestnut breeding is the development of MAS for the creation of either fruit varieties (using original roots or grafting) or rootstocks, producing high-quality fruits and resistance to abiotic and biotic stress. Nowadays used for breeding resistance to chestnut blight in the United States (Nelson et al. 2014), MAS has been developed for easy peel pellicle in Japan with limited results (Takada et al. 2017). Nevertheless, it is currently implemented for other traits such as harvesting data and splitting pericarp (Nishio et al. 2018). The recent development of multiple genomic resources (such as markers and maps), currently available (see Section 9), should help to enhance the use of this potential selection method. Studies of the genetic determinism of other important adaptive traits (including resistance to abiotic stress, growth, productivity etc) are needed to reinforce our understanding of their molecular basis.

Obtaining new genetic maps of chestnut trees, which are under development in Europe, will make it possible to enrich the genetic maps already available. These involve *C. sativa* (Casasoli et al. 2001), *C. mollissima* (Kubisiak et al. 2013), *C. sativa* × *C. crenata* (Santos et al. 2017a), (*C. sativa* × *C. crenata*) × *C. sativa* (Torello Marinoni et al. 2017) and *C. crenata* (Nishio et al. 2018). Using these enhanced maps could allow new QTLs to be identified. At the same time, the comparison of all the maps might confirm the QTL already assessed, allowing the identification of sound candidate genomic regions across pedigrees for the traits of interest. Due to the efficient characterization of the main traits of local cultivars extensively researched in chestnut, and to the multitude of markers now available (including single sequence repeats and single nuclear polymorphisms), genome-wide association studies might be envisaged. In addition, following transcriptomic investigations and abiotic and biotic gene expression studies, many candidate genes useful for association genetics studies for adaptive traits are also now available. Besides the multiple genomic resources available in chestnut, the ones developed in oak *Quercus robur* might be used too. Indeed, the chestnut is phylogenetically close to the oak (Manos et al. 2008) with which it has a very strong macrosynteny and collinearity (Barreneche et al. 2004; Casasoli et al. 2006; Durand et al. 2010). Therefore, the transfer of the genetic and molecular information from one genus to another becomes possible.

Ultra-high-density linkage maps and availability of new markers become a reality, opening up large research perspectives, as for instance, the study of the genetic basis of incompatibility and pollen limitation, which are two features often associated (Hasegawa et al. 2009; Burd 1994; Larson and Barrett 2000). Male sterility is observed very often in European chestnuts particularly in the marron type (Klinac and Lelieveld 1995; Pereira-Lorenzo et al. 2016), but also in interspecific hybrid cultivars released in Europe and other countries in the last decades. These varieties largely planted in modern plantations face difficult pollination that needs to be solved by the selection of good pollinator varieties or a relevant orchard management system. Research programs are currently under development in several countries with the aim to improving identification, *in situ* and *ex situ* evaluation, collection, propagation and regeneration of chestnut germplasm (e.g. Botu et al. 2009; Torello Marinoni et al. 2013; Martin et al. 2017; Lusini et al. 2014). The objective is to conserve chestnut diversity and to use it for breeding and selection of new varieties. To reach this goal, it is necessary to characterize chestnut germplasm not only with morphological traits but also with new genetic markers. Several European countries have created *ex situ* germplasm collections, but for now no European core collection, representing European chestnut diversity, is currently available. In Italy, an *ex situ* collection is located at the Chestnut Regional Repository in Chiusa di Pesio, Cuneo Province (Northwestern Italy), on the border with France (Mellano et al. 2012). In France, the INRA *ex situ* chestnut collection is maintained at Villenave d'Ornon near Bordeaux. This is currently being duplicated to be set up in the two main traditional chestnut production regions (Southwestern and Southeastern France) thus ensuring its durability. In Spain, the CIF Lourizán at Pontevedra (Galicia) has a collection of grafted Galician traditional varieties of *Castanea sativa* [Furones-Perez and Fernández-López, 2009 and https://lourizan.xunta.gal/en/botanical-garden-collections/collections]. Moreover, in France many NGO and amateur associations of several regions preserved chestnut landraces, as in Spain and in Portugal where *in situ* and on-farm conservation is frequent. The Romanian national chestnut collection (*ex situ* and *in situ* conservation) is preserved by the University of Craiova in the Fruit Growing Research and Extension Station (UCv-SCDP) in Vâlcea (Botu et al. 2017). In Turkey, the Aegean Agricultural Research Institute coordinates the protection and preservation of genetic diversity, with other institutes such as the Ataturk Horticultural Central Research Institute in Yalova which maintains chestnut cultivars (Karagôz, 2003).

In the future one of the main challenges that chestnut will have to face is climate change. In temperate regions, climate change scenarios forecast an increase of mean temperature which will impact substantially both the phenology and the water use efficiency of the trees. Frequently, phenotypic variability of these adaptive traits among chestnut from contrasting climatic environments has been assessed (Fernandez-Lopez et al. 2005; Lauteri et al. 1997; Pliura and Eriksson 2002; Tchatchoua and Aravanopoulos 2010a,b). Furthermore, QTLs underlying phenology and water use efficiency have been located in *C. sativa* map (Casasoli et al. 2001), and their regulation appears to be due to many genetic factors of low and moderate effect (Casasoli et al. 2004). In addition, comparative QTL mapping has allowed the identification of common chromosomal regions affecting the same quantitative trait in chestnut and oak (Casasoli et al. 2006). Abiotic gene expression studies on oak have identified candidate genes for water use efficiency and bud burst (Porth et al. 2005; Derory et al. 2006; Kremer et al. 2012). This knowledge and all the genomic resources available, open up the possibility of selecting new cultivars adapted to climate change. In particular, trait selection via the implementation of innovative breeding methods such as MAS appear to be possible in the near future.

## 9 Where to look for further information

Several websites provide information on chestnut for different topics:

Chestnut production and industry (http://www.fao.org/faostat/en/#home):

China, Turkey and the Republic of Korea are the top three countries for chestnut production (FAOSTAT, 2016); together these countries produce more than 90% of the world's chestnuts. Other major chestnut producing countries include Italy, Greece, Portugal, Japan, Spain and the Democratic People's Republic of Korea.

Chestnut restoration in the United States:

- The American Chestnut Foundation is a non-profit corporation. Relying on citizen science, its mission is to save the American chestnut tree: https://theconversation.com/new-genetically-engineered-american-chestnut-will-help-restore-the-decimated-iconic-tree-52191; https://www.acf.org/

Chestnut production in the United States:

- List of chestnut cultivars and their characteristics, prepared by S. L. Anagnostakis (January 2013): http://www.ct.gov/caes/lib/caes/documents/publications/fact_sheets/plant_pathology_and_ecology/cultivars_of_chestnut_a-z.pdf
- Information on new interesting chestnut cultivars and where they can be found in the United States: www.chestnutgrowers.org/MikeNave-NewCultivars.pdf
- Information on chestnut orchards: horticulture, pest management, news, experts. . .: http://www.canr.msu.edu/chestnuts/establishing_orchards/

Chestnut production in Europe:

- Information concerning European chestnut producers: www.areflh.org
- Information concerning European Fruit research institutes: www.eufrin.org

Chestnut germplasm:

- UPOV Guidelines: http://www.upov.int/edocs/mdocs/upov/en/twf_46/tg_124_4_proj_2.pdf
- INRA Genetic resources data base: https://urgi.versailles.inra.fr/siregal/common/taxon/form.do

Genomic resources:

- Transcriptome: https://www.hardwoodgenomics.org/Transcriptome-assembly/1963054
- Chestnut genome: https://www.hardwoodgenomics.org/; https://www.ncbi.nlm.nih.gov/genome/?term=Castanea

## 10 References

Abe, Y., Melika, G. and Stone, G. N. (2007), 'The diversity and phylogeography of cynipid gallwasps (Hymenoptera, Cynipidae) of the Eastern Palearctic and their associated communities', *Oriental Insect* 41, 169–212.

Anagnostakis, S. L. (1987), 'Chestnut blight: The classical problem of an introduced pathogen', *Mycologia* 79(1), 23–37.

Anagnostakis, S. L. (1992), 'Measuring resistance of chestnut trees to chestnut blight', *Canadian Journal of Forest Research* 22, 568–71.

Anagnostakis, S. L. (2012), 'Chestnut breeding in the United States for disease and insect resistance', *Plant Disease* 96, 1392–403.

Anagnostakis, S. L. (2014), 'A preliminary report on Asian chestnut gall wasp on species and hybrids of chestnut in Connecticut', *Acta Horticulturae* 1019, 21–2.

Bagnaressi, U. (1956), 'Osservazioni morfo-biologiche sulle provenienze di castagno giapponese cultivate in Italia', *Centro di Studio sul Castagno Stazione Sperimentale di Selvicoltura di Firenze* 3, 7–48.

Barreneche, T., Casasoli, M., Russell, K., Akkak, A., Meddour, H., Plomion, C., Villani, F. and Kremer, A. (2004), 'Comparative mapping between *Quercus* and *Castanea* using simple-sequence repeats (SSRs)', *Theoretical Applied Genetics* 108, 558–66.

Bergounoux, F., Verlhac, A., Breisch, H. and Chapa, J. (1978), *Le Châtaignier Production et Culture*, Comité National Interprofessionnel de la Châtaigne et du Marron, Nimes, France.
Bolvanský M. and Mendel L. (2001), 'Revised descriptor list for the evaluation of genetic resources of European chestnut (*Castanea sativa* Mill.)', *Forest Snow and Landscape Research* 76, 439–44.
Botu, M., Achim, G. and Turcu, E. (1999), 'Evaluation of some chestnut selections from the population formed into the ecological conditions from the North-East of Oltenia', *Acta Horticulturae* 494, 77–83.
Botu, M., Botu, I., Neagoe, A. and Papachatzis, A. (2009), 'Evaluation of sweet chestnut cultivars and selections into the Valcea area', *Acta Horticulturae* 844, 311–18.
Botu, M., Botu, I., Achim, G., Preda, S., Scutelnicu, A. and Giura, S. (2017), 'Conservation of fruit tree genetic resources and their use in the breeding process', *The Annals of 'Valahia' University of Targoviste'*, DOI:10.1515/agr-2017-0013.
Bounous, G. (2002), *Il Castagno. Coltura, ambiente ed utilizzazione in italia e nel mondo*, Edagricole- Calderini, Bologna, 311pp.
Bounous, G., Botta, R., Bellini, E. and Giannini, R. (1997), 'Il Miglioramento Genetico del Castagno (*Castanea* spp.)', *Il Simposio Internazionale su stato dell'arte e prospective del miglioramiento genetico dei fruttiferi- castagno*, 10 October 1997, Faenza, Italy.
Breisch, H. (1995), *Châtaignes et Marrons*, CTIFL, Paris, France.
Breisch, H. (1999), 'Highlights of chestnut industry in France', *Acta Horticulturae* 464, 23–6.
Breisch, H. and Hennion B. (2004), 'Les porte-greffe du châtaignier, vers plus de résistance', *Infos CTIFL janvier-février* 2004: 24-08.
Brussino, G., Bosio, G., Baudino, M., Giordano, R., Ramello, F. and Melika, G. (2002), 'A dangerous exotic insect threatening European chestnut', *Informatore Agrario* 58(7), 59–61
Buck, E. J., Hadonou, M., James, J., Blakesley, D. and Russel, K. (2003), 'Isolation and characterization of polymorphic microsatellites in European chestnut (*Castanea sativa* Mill.)', *Molecular Ecology Notes* 3, 239–41.
Burd, M. (1994), 'Bateman's principle and plant reproduction: The role of pollen limitation in fruit and seed set', *The Botanical Review* 60(1), 83–139.
Burnham, C. R. (1981), 'Blight resistant American chestnut: There's hope', *Plant Disease* 65, 459–60.
Casasoli, M., Mattioni, C., Cherubini, M. and Villani, F. (2001), 'A genetic linkage map of European chestnut (*Castanea sativa* Mill.) based on RAPD, ISSR and isozyme markers', *Theoretical Applied Genetics* 102(8), 1190–9.
Casasoli, M., Pot, D., Plomion, C., Monteverdi, M. C., Barreneche, T., Lauteri, M. and Villani, F. (2004), 'Identification of QTLs affecting adaptive traits in *Castanea sativa* Mill.', *Plant Cell Environment* 27, 1088–101.
Casasoli, M., Derory, J., Morera-Dutrey, C., Brendel, O., Porth, I., Guelh, J. M., Villani, F. and Kremer, A. (2006), 'Comparison of quantitative trait loci for adaptive traits between oak and chestnut based on an expressed sequence tag consensus map', *Genetics* 172(1), 533–46
Chapa, J. (1987), 'Chataignes et marrons, variétés inscrites au catalogue officiel', *Arboriculture Fruitière* 399, 21–30.
Chapa J. and Bernhard, R. (1984), 'Etude d'un programme de création de nouvelles variétés de chataigniers', *Congreso internacional sobre el castaño*, Lourizan-Pontevedra, pp. 253–64.

Chevre, A. M., Gill, S. S., Mouras, A. and Salesses, G. (1983), '*In vitro* vegetative multiplication of chestnut', *Journal of Horticultural Science* 58, 23–9.
Chevre, A. M. and Salesses, G. (1984), 'Choice of explants for chestnut micropropagation', *Acta Horticulturae* 212, 517–23
Cho, D. Y. and Lee, S. O. (1963), 'Ecological studies on the chestnut gall wasp, *Dryocosmus kuriphilus* Yasumatsu, and observation of the damages of the chestnut trees by its insect', *Korean Journal of Plant Protection* 2, 47–54.
Choupina, A. B., Estevinho, L. and Martins, I. M. (2014), 'Scientifically advanced solutions for chestnut ink disease', *Applied Microbiology and Biotechnology* 98, 3905–9.
Claper, R. B. (1954), 'Chestnut breeding techniques and results', *Journal of Heredity* 45, 106–14.
Conedera, M., Krebs, P., Tinner, W., Pradella, M. and Torriani, D. (2004), 'The cultivation of *Castanea sativa* (Mill.) in Europe, from its origin to its diffusion on a continental scale', *Vegetation History and Archaeobotany* 13(3), 161–79.
Corredoira, E., Martínez, M. T., Cernadas, M. J. and San José, M. C. (2017), 'Application of biotechnology in the conservation of the genus *Castanea*', *Forests* 8, 394.
Costa, R., Santos, C., Tavares, F., Machado, H., Gomes-Laranjo, J., Kubisiak, T. and Nelson, C. D. (2011), 'Mapping and transcriptomic approaches implemented for understanding disease resistance to *Phytophthora cinnamomi* in *Castanea sp.*', *BMC Proceedings* 5(Suppl. 7), O18.
Crandall, B. S., Gravatt, G. F. and Ryan, M. M. (1945), 'Root disease of *Castanea* species and some coniferous and broadleaf nursery stocks, caused by *Phytophthora cinnamomi*', *Phytopathology* 35, 162–80.
Derory, J., Léger, P., Garcia, V., Schaeffer, J., Hauser, M. T., Salin, F., Luschnig, C., Plomion, C., Glössl, J. and Kremer, A. (2006), 'Transcriptome analysis of bud burst in sessile oak (*Quercus patraea*)', *New Phytologist* 170, 723–38.
Dietz, A. (1978), 'The use of ionizing radiation to develop a blight resistant American chestnut, *Castanea dentata*, through induced mutations', *in* MacDonald, W. L., Cech, F. C., Luchok, J. and Smith, H. C. (Eds), *Proceedings of the First international Chestnut Conference*, West Virginia University Press, Morgantown, WV, pp. 17–20.
Dini, F., Sartor, C. and Botta, R. (2012), 'Detection of a hypersensitive reaction in the chestnut hybrid 'Bouche de Bétizac' infested by *Dryocosmus kuriphilus* Yasumatsu', *Plant Physiology and Biochemistry* 60, 67–73.
Diskin, M., Steiner, K. and Hebard, F. V. (2006), 'Recovery of American chestnut characteristics following hybridization and backcross breeding to restore blight-ravaged *Castanea dentata*', *Forest Ecology and Management* 223, 439–47.
De Oliveira, D., Gomes, A., Ilharco, F A., Manteigas, A M. and Ramalho, J. (2001), 'Importance of insect pollinators for the production in the chestnut, *Castanea sativa*', *Acta Horticulturae* 561, 269–73.
Dufrenoy, J. (1930), 'La lutte contre la maladie des châtaigniers', *Annales des Epiphyties* 1, 3–49.
Durand, J., Bodénès, C., Chancerel, E., Frigerio, J. M., Vendramin, G., Sebastiani, F., Buonamici, A., Gailing, O., Koelewijn, H. P., Villani, F., Mattioni, C., Cherubini, M., Goicoechea, P. G., Herrán, A., Ikaran, Z., Cabané, C., Ueno, S., Alberto, F., Dumoulin, P. Y., Guichoux, E., de Daruvar, A., Kremer, A. and Plomion, C. (2010), 'A fast and cost-effective approach to develop and map EST-SSR markers: Oak as a case of study', *BMC Genomics* 11, 570.
Dutech C, Barres B, Bridier J, Robin, C., Milgroom, M. G. and Ravigné, V. (2012), 'The chestnut blight fungus world tour: Successive introduction events from diverse origins in an invasive plant fungal pathogen', *Molecular Ecology* 21, 3931–46.

FAOSTAT (2016), http://www.fao.org/faostat/en/#data/QC

Fernandez-Lopez, J. (2011), 'Identification of the genealogy of interspecific hybrids between *Castanea sativa*, *Castanea crenata* and *Castanea mollissima*', *Journal of Forest Systems* 20(1), 65–80.

Fernandez-Lopez, J., Pereira, S. and Miranda-Fontaina, E. (1992), 'Fog and substrate conditions for chestnut propagation by leafy cuttings', *Symposium Proceedings Mass Production Technology for Genetically Improved Fast Growing Species, AFOCEL/IUFRO*, I, pp. 379–83.

Fernandez-Lopez, J., Vazquez-Ruiz-De-Ocenda, R. A., Díaz-Vazquez, R. and Pereira-Lorenzo, S. (2001), 'Evaluation of resistance of *Castanea* sp. clones to *Phytophthora sp.* using excised chestnut shoots', *Forest Snow and Landscape Research* 76(3), 451–4.

Fernandez-Lopez, J., Zas, R., Diaz, R., Villani, F., Cherubini, M., Aravanopoulos, F. A., Alizoti, P. G., Eriksson, G., Botta, R. and Mellano, M. G (2005), 'Geographic variability among extreme European wild chestnut populations', *Proceedings of the 3rd Intern. Chestnut Congress*, 20–23 October 2004, Chaves, Portugal, *Acta Horticulturae*, 693, 181–6.

Ferracini, C., Ferrari, E., Pontini, M., Saladini, M. A. and Alma, A. (2018), 'Effectiveness of *Torymus sinensis*: A successful long-term control of the Asian chestnut gall wasp in Italy', *Journal of Pest Science* (in press), 1–7.

Fulbright, D. W., Stadt, S., Medina-Mora, C., Mandujano, M., Donis-González, I. R. and Serdar, U. (2014), 'Kernel breakdown appears when hybrid *Castanea* cultivars are pollinized by *Castanea mollissima*', *Acta Horticulturae* 1019, 91–8.

Furones-Perez, P. and Fernández-López, J. (2009), 'Morphological and phenological description of 38 sweet chestnut cultivars (*Castanea sativa* Miller) in a contemporary collection', *Spanish Journal of Agricultural Research* 7(4), 829–43

Giovanetti, M. and Aronne, G. (2011), 'Honey bee interest in flowers with anemophilous characteristics: First notes on handling time and routine on *Fraxinus ornus* and *Castanea sativa*', *Bulletin of Insectology* 64(1), 77–82.

Grente, J. (1961), 'La maladie de l'encre du châtaignier', *Annales des Epiphyties* 12, 5–59.

Grygorieva, O., Klymenko, S., Brindza, J., Schubertová, Z., Nikolaieva, N. and Šimková, J. (2017), 'Morphometric characteristics of sweet chestnut (*Castanea sativa* Mill.) fruits', *Potravinarstvo Slovak Journal of Food Sciences* 11(1), 288–95.

Hasegawa, Y., Suyama, Y. and Seiwa, K. (2009), 'Pollen donor composition during the early phases of reproduction revealed by DNA genotyping of pollen grains and seeds of *Castanea crenata*', *New Phytologist* 182(4), 994–1002.

Hasegawa, Y., Suyama, Y. and Seiwa, K. (2015), 'Variation in pollen-donor composition among pollinators in an entomophilous tree species, *Castanea crenata*, revealed by single-pollen genotyping', *PLoS One* 10(3), e0120393.

Heiniger, U. and Rigling, D. (1994), 'Biological control of chestnut blight in Europe', *Annual Review of Phytopathology* 32, 581–99.

Huang, H. W., Norton, J. D., Smith, D. A. and Slayden, P. W. (1990), 'Characteristics of chestnut gall wasp resistant Chinese chestnut', *Annual report - Northern Nut Growers' Association* 81, 29–32.

Huber, J. T. and Read, J. (2012), 'First record of the oriental chestnut gall wasp, *Dryocosmus kuriphilus* Yasumatsu (Hymenoptera: Cynipidae), in Canada', *Journal of the Entomological Society of Ontario* 143, 125–8.

Kajiura, M. (1955), 'The breeding of new varieties of Japanese chestnut resistant to chestnut gall wasp (*Dryocosmus kuriphilus* "Yasumatsu")', *Report of the 14th*

*International Horticultural Congress* held at The Hague-Scheveningen, Netherlands, 29 August-6, 1955. 1-2, pp. 1243–49.
Karagôz, A. (2003), 'Plant genetic resources conservation in Turkey', *Acta Horticulturae* 598, 17–25.
Kato, K. and Hijii, N. (1997), 'Effects of gall formation by *Dryocosmus kuriphilis* Yasumatsu (Hym., Cynipidae) on the growth of chestnut trees', *Journal of Applied Entomology* 121, 9–15.
Kim, M-J., Lee, U. and Park, Y. (2014), '"Jahong": A new indigenous Korean chestnut cultivar with reddish burrs', *HortScience* 49(12), 1588–9.
Klinac, D. J., Lelieveld, J. F. and Knowles R. D. (1995), 'Time of flowering and pollination in some New Zealand chestnut selections, and introduced cultivars', *New Zealand Journal of Crop and Horticultural Science* 23, 331–40.
Kotobuki, K. (2002), 'Situazione mondiale, Giappone', *Il Castagno. Coltura, ambiente ed utilizzazioni in Italia e nel mondo*, pp. 245–9.
Kremer, A., Abbott, A. G., Carlson, J. E., Manos, P. S., Plomion, C., Sisco, P., Staton. M. E., Ueno, S. and Vendramin, G. G. (2012), 'Genomics of *Fagaceae*', *Tree Genetics & Genomes* 8, 583–610.
Kubisiak, T. L., Nelson, C. D., Staton M. E., Zhebentyayeva, T., Smith, C., Olukolu B. A., Fang G. C., Hebard, F. V, Anagnostakis, S., Wheeler, N., Sisco, P. H., Abbott, A. G. and Sederoff, R. R. (2013), 'A transcriptome-based genetic map of Chinese chestnut (*Castanea mollissima*) and identification of regions of segmental homology with peach (*Prunus persica*)', *Tree Genetics & Genomes* 9, 557-71
Larson, B. M. H. and Barrett, S. C. H. (2000), 'A comparative analysis of pollen limitation in flowering plants', *Biological Journal of the Linnean Society* 69(4), 503–20.
Lauteri, M., Scartazza, A., Guido, M. C. and Brugnoli, E. (1997), 'Genetic variation in photosynthetic capacity, carbon isotope discrimination and mesophyll conductance in provenances of *Castanea sativa* adapted to different environments', *Functional Ecology* 11, 675–83.
Li, X., Zhou, X., Zhou, J., Dodson, J., Zhang, H. and Shang, X. (2007), 'The earliest archaeobiological evidence of the broadening agriculture in China recorded at Xishanping site in Gansu Province', *Science in China - Series D, Earth Sciences* 50(11), 1707–14.
Liu, L. (1992), 'A study on germplasm diversity of chestnut and the proposals for its conservation and utilization', *Journal of Plant Resources and Environment* 1, 18–22.
Lopez-Villamor, A., Miguez-Soto, B. and Fernandez-Lopez, J. (2017), 'Adventitious root formation in *Castanea* sp semi-hard cuttings is under moderate genetic control caused mainly by non-additive genetic variance', *Canadian Journal of Forest Research* 47, 946–56.
Lusini, I., Velichkov, I., Pollegioni, P., Chiocchini, F., Hinkov, G., Zlatanov, T. Cherubini, M. and Mattioni, C. (2014), 'Estimating the genetic diversity and spatial structure of Bulgarian *Castanea sativa* populations by SSRs: Implications for conservation', *Conservation Genetics* 15, 283–93.
Mangin, L. (1922), 'Sur la reconstitution des châtaigneraies détruites par l'encre', *Compte rendus de l'Académie d'Agriculture de France* 8, 144–51.
Manos, P. S., Cannon, C. H. and Oh, S.-H. (2008), 'Phylogenetic relationships and taxonomic status of the paleoendemic *Fagaceae* of Western North America: Recognition of a new genus *Notholithocarpus*', *Madroño* 55, 181–90.

Manthos, I. and Botu, M. (2013), 'Valuable sweet chestnut selections (*Castanea sativa* Mill.) from Epirus-Greece and Northern Oltenia -Romania', *South Western Journal of Horticulture, Biology and Environment* 4(2), 91–101.

Martín, M. A., Mattioni, C., Cherubini, M., Villani, F. and Martín, L. M. (2017), 'A comparative study of European chestnut varieties in relation to adaptive markers', *Agroforestry Systems* 91, 97–109.

Maurel, M., Robin, C., Capdevielle, X., Loustau, D. and Desprez-Loustau M. L. (2001), 'Effects of variable root damage caused by *Phytophthora cinnamomi* on water relations of chestnuts saplings', *Annals of Forest Sciences* 58, 639–51.

Maurel, M., Robin, C., Simoneau, T., Loustau, D., Dreyern E. and Desprez-Loustau, M. L. (2004), 'Stomatal conductance and root-to-shoot signalling in chestnut saplings exposed to *Phytophthora cinnamomi* or partial soil drying', *Functional Plant Biology* 31, 41–51.

McCleary, T., McAllister, M., Coggeshall, M. and Romero-Severson, J. (2013), 'EST-SSR markers reveal synonymies, homonymies and relationships inconsistent with putative pedigrees in chestnut cultivars', *Genetic Resources and Crop Evolution* 60, 1209–22.

Mellano, M. G., Beccaro, G. L., Donno, D., Torello Marinoni, D., Boccacci, P., Cerutti, A. K. and Bounous, G. (2012), '*Castanea* spp. biodiversity conservation: Collection and characterization of the genetic diversity of an endangered species', *Genetic Resources and Crop Evolution* 59, 1727–41.

Mert, C. and Soylu, A. (2006), 'Flower and stamen structures of male-fertile and male-sterile chestnut (*Castanea sativa* Mill.) cultivars', *Journal of the American Society for Horticultural Science* 131(6), 752–9

Milgroom, M. G. and Cortesi, P. (2004), 'Biological control of chestnut blight with hypovirulence: A critical analysis', *Annual Review of Phytopathology* 42, 311–38.

Miller, G., Miller, D. D. and Jaynes R. A. (1996), 'Chestnuts', *in* Janick, J. and Moore, J. N. (Eds), *Fruit Breeding, Volume III, Nuts*, Wiley Inc, New York, NY, pp. 99–123.

Miranda-Fontaiña, M. E., Fernandez-Lopez, J., Vettraino, A. M. and Vannini, A. (2007), 'Resistance of *Castanea* clones to *Phytophthora cinnamomi*: Testing and genetic control', *Silvae Genetica* 56(1), 11–21.

Moriya, S., Inoue, K. and Mabuchi, M. (1990), 'The use of *Torymus sinensis* (Hymenoptera, Torymidae) for controlling the chestnut gall wasp, Dryocosmus kuriphilus Hymenoptera, Cynipidae), Japan', *FFTC-NARC International Seminaron 'The Use of Parasitoids and Predators to Control Agricultural Pests'*, Tsukuba Science City, Ibaraki-ken, 21pp.

Moriya, S., Shiga, M. and Adachi I. (2003), 'Classical biological control of the chestnut gall wasp in Japan', *Proceedings of the 1st International Symposium on biological control of arthropods*, Honolulu, Hawaii, 14–18 January 2002, United States Department of Agriculture, Forest Service, Washington DC, pp. 407–15

Nave, M. (2017), 'Interesting new Chinese chestnut cultivars for 2017', *The Chestnut Grower* 18(4), 8–10.

Nelson, C. A., Powell, W. A., Merkle, S. A., Carlson, J. E., Hebard, F. V., Islam-Faridi, N., Staton, M. E. and Georgi, L. (2014). 'Biology and Biotechnology: Section 1-Biotechnology of Trees: Chestnut', *in* Ramawat, K. G., Merillon, J. M. and Ahuja, M. R. (Eds), *Tree Biotechnology*, CRC Press, Boca Raton (USA), London (UK) and New York (USA), pp. 3–35.

Newhouse, A. E., McGuigan, L. D., Baier, K. A., Valletta, K. E., Rottmann, W. H., Tschaplinski, T. J., Maynard, C. A. and Powell, W. A. (2014), 'Transgenic American chestnuts show enhanced blight resistance and transmit the trait to T1 progeny', *Plant Science* 228, 88–97.

Nishio, S., Yamamoto, T., Terakami, S., Sawamura, Y., Takada, N. and Saito, T. (2011), 'Genetic diversity of Japanese chestnut cultivars assessed by SSR markers', *Breeding Science* 61, 109-20.

Nishio, S., Takada, N., Yamamoto, T., Terakami, S., Hayashi, T., Sawamura, Y. and Saito, T. (2013), 'Mapping and pedigree analysis of the gene that controls the easy peel pellicle trait in Japanese chestnut (*Castanea crenata* Sieb. Et Zucc.)', *Tree Genetics & Genomes* 9, 723–30.

Nishio, S., Terakami, S., Matsumoto, T., Yamamoto, T., Takada, N., Kato, H., Katayose, Y. and Saito, T. (2018), 'Identification of QTLs for agronomic traits in the Japanese chestnut (*Castanea crenata* Sieb. Et Zuuc.) breeding', *The Horticultural Journal*. DOI:10.2503/hortj.OKD-093.

Norton, J. D., Harris, H., and Conaty, T. E. (1987), 'Resistance to chestnut gall wasp in Chinese chestnut', *in* Burnett, M. S. and Wallace, R. D. (Eds), *Chestnuts and Creating a Commercial Chestnut Industry: Proceedings of the Second Pacific Northwest Chestnut Congress*, Chestnut Growers Exchange, Portland, OR, pp. 30–2.

Nuss, D. L. (1992), 'Biological control of chestnut blight: An example of virus-mediated attenuation of fungal pathogenesis', *Microbiological Reviews* 56, 561–76.

Oraguzie, N. C., Paterson, A. M. and McNeil, D. L. (1999), 'Origin and relationships of New Zealand chestnut (*Castanea* sp. *Fagaceae*) selections reflect patterns of graft failure', *Plant Systematics and Evolution* 218, 193–204.

Ormeci, Y., Akca, Y. and Ercisli, S. (2016), 'Selection of promising chestnuts (*Castanea sativa*) among wild growing trees from southern Mediterranean region forests of Turkey', *Journal of Forestry Research* 27(2), 349–55.

Pandit, A. H., Mir, M. A., Kour, A. and Bhat, K. M. (2013), 'Variability and selection of chestnut (*Castanea sativa* Mill.) genotypes in Srinagar district of Kashmir Valley', *Pakistan Journal of Agricultural Sciences* 50(2), 205–9.

Payne, J. A.; Menke, A. S. and Schroeder, P. M. (1975), '*Dryocosmus kuriphilus* Yasumatsu (Hymenoptera: Cynipidae), an oriental chestnut gall wasp in North America', *USDA CEIR* 25, 903–5.

Pereira-Lorenzo, S., Diaz-Hernandez, M. B. and Ramos-Cabrer, A. M. (2006), 'Use of highly discriminating morphological characters and isozymes in the study of Spanish chestnut cultivars', *Journal of the American Society for Horticultural Science* 131(6), 770–9.

Pereira-Lorenzo, S., Lourenço Costa, R. M., Ramos-Cabrer, A. M., Marques Ribeiro, C. A., Serra da Silva, M. F., Manzano, G. and Barreneche, T. (2010), 'Variation in grafted European chestnut and hybrids by microsatellites reveals two main origins in the Iberian Peninsula', *Tree Genetics & Genomes* 6(5), 701–15.

Pereira-Lorenzo, S., Costa, R. M., Ramos-Cabrer, A. M., Ciordia-Ara, M., Marques Ribeiro, C. A., Borges, O. and Barreneche, T. (2011), 'Chestnut cultivar diversification process in the Iberian Peninsula, Canary Islands and Azores', *Genome* 54, 301–15.

Pereira-Lorenzo, S., Ballester, A., Corredoira, E., Vieitez, A. M., Agnonastakis, S., Costa, R., Bounous, G., Botta, R., Beccaro, G. L., Kubisiak, T. L., Conedera, M., Krebs, P., Yamamoto, T., Sawamura, Y., Takada, N., Gómez-Laranjo, J. and Ramos-Cabrer, A.

M. (2012), 'Chapter 19: Chestnut', *in* Badenes, M. L. and Byrne, D. H. (Eds), *Fruit Breeding*, Springer, pp. 729–70.

Pereira-Lorenzo, S., Costa, R., Anagnostakis, S., Serdar, U., Yamamoto, T., Saito, T., Ramos-Cabrer, A. M., Ling, Q., Barreneche, T., Robin, C., Botta, R., Contessa, C., Conedera, M., Martín, L. M., Martín, A., Gomes-Laranjo, J., Villani, F. and Carlson, J. E. (2016), 'Chapter 15: Interspecific hybridization of chestnut', *in Polyploidy and Interspecific Hybridization for Crop Improvement*, CRC Press, pp. 379–408.

Pereira-Lorenzo, S., Ramos-Cabrer, A., Barreneche, T., Mattioni, C., Villani, F., Díaz-Hernández, M. B, Martín, L. M. and Martín, A. (2017), 'Database of European chestnut cultivars and definition of a core collection using simple sequence repeats', *Tree Genetics & Genomes* 13, 114–19.

Pliura A. and Eriksson, G. (2002), 'Genetic variation in juvenile height and biomass of open-pollinated families of six *Castanea sativa* Mill. populations in a 2 × 2 factorial temperature × watering experiment', *Silvae Genetica* 51, 152–60.

Podjavorsek, A., Štampar, F., Solar, A. and Batic, F. (1999), 'Morphological variation in chestnut fruits (*Castanea sativa* Mill.) in Slovenia', *Acta Horticulturae* 494, 129–32.

Porth, I., Scotti-Saintagne, C., Barreneche, T., Kremer, A. and Burg, K. (2005), 'Linkage mapping of osmotic induced genes of oak', *Tree Genetics & Genomes* 1, 31–40.

Prgomet, I., Marinoni Torello, D., Donno, D., Prgomet, Z., Mujić, I., Beccaro, G. L. and Bounous, G. (2014), 'Genetic diversity of European chestnut (*Castanea sativa* Mill) in Istria and Primorsko-Goranska County (Croatia)', *Acta Horticulturae* 1043, 87–90.

Prunet, A. (1915), 'La maladie des châtaigniers et la reconstitution des châtaigneraies', *Annales des Epiphyties* 67–100.

Quacchia, A., Moriya, S., Bosio, G., Scapin, I. and Alma, A. (2008), 'Rearing, release and settlement prospect in Italy of *Torymus sinensis* the biological control agent of the chestnut gall wasp *Dryocosmus kuriphilus*', *Biological Control* 53, 829– 839.

Ramos Guedes-Lafargue, M. and Salesses, G. (1999), 'Ink disease resistance: Some preliminary elements from the study of different crosses', *Acta Horticulturae* 464, 355–61.

Rieske, L. K. (2007), 'Success of an exotic gall maker, *Dryocosmus kuriphilus*, on chestnut in the USA: A historical account', *EPPO Bulletin* 37, 172–4.

Rigling, D. and Prospero, S. (2018), '*Cryphonectria parasitica*, the causal agent of chestnut blight: Invasion history, population biology and disease control', *Molecular Plant Pathology* 19(1), 7-20

Robin, C. and Desprez-Loustau, M. L. (1998), 'Testing variability in pathogenicity of *Phytophthora cinnamomi*', *European Journal of Plant Pathology* 104, 465–75.

Robin, C., Morel, O., Vettraino, A. M., Perlerou, C., Diamandis, S. and Vannini, A. (2006), 'Genetic variation in susceptibility to *Phytophthora cambivora* in European chestnut (*Castanea sativa*)', *Forest Ecology and Management* 226, 199–207.

Russell, E. W. B. (1987), 'Pre-blight distribution of *Castanea dentata* (Marsh.) Borkh', *Bulletin of the Torrey Botanical Club* 114(2), 183–90.

Saito, T. (2009), 'Japan. Book chapter in following chestnut footprints (*Castanea* spp.) - cultivation and culture, folklore and history, tradition and uses', *Scripta Horticulturae* 9, 96–9.

Saito, T., Kotobuki, K., Sawamura, Y., Abe, K., Terai, O., Shoda, M., Takada, N., Sato, Y., Hirabayashi, T., Sato, A., Nishibata, T., Kashimura, Y., Kozono, T., Fukuda, H., Kihara, K., Suzuki, K. and Uchida, M. (2009), 'New Japanese chestnut cultivar "Porotan"', *Bulletin of the National Institute of Fruit Tree Science* 9, 1–9.

Saito, T., Kotobuki, K., Sawamura, Y., Takada, N., Hirabayashi, T., Sato, A., Shoda, M., Terai, O., Nishibata, T., Kashimura, Y., Abe, K., Nishio, S., Kihara, T., Suzuki K. and Uchida, M. (2015), 'New Japanese chestnut cultivar 'Mikuri' (*Castanea crenata* Sieb. et Zucc)', *Bulletin of the National Institute of Fruit Tree Science* 16, 1-9.

Salesses, G., Ronco, L., Chauvin, J. E. and Chapa J. (1993), 'Amélioration génétique du châtaignier mise au point de tests d'évaluation du comportement vis-à-vis de la maladie de l'encre', *L'arboriculture fruitière* 458, 23-31.

Santos, C., Nelson, D. C., Zhebentyayeva, T., Machado, H., Gomes-Laranjo, J. A. and Costa, R. L. (2017a), 'First interspecific genetic linkage map for *Castanea sativa* × *Castanea crenata* revealed QTLs for resistance to *Phytophthora cinnamomi*', *PLoS One* 12(9), e0184381.

Santos, C., Duarte, S., Tedesco, S., Fevereiro, P. and Costa R. L. (2017b), 'Expression profiling of *Castanea* genes during resistant and susceptible interactions with the oomycete pathogen *Phytophthora cinnamomi* reveal possible mechanisms of immunity', *Frontiers in Plant Science* 8, 1-14.

Sartor, C. and Botta, R. (2011), 'Dossier Cinipide del Castagno. Effetti dell'infestazione e ricerca di resistenze', *Sherwood* 177, 17-19.

Sartor, C., Mellano, M. G., Quacchia, A., Alma, A. and Botta, R. (2007), 'Cinipide galligeno del Castagno: Prospettive di impiego di strategie da affiancare alla lotta biologica', *Italus Hortus* 14, 130-3.

Sartor, C., Botta, R., Mellano, M. G., Beccaro, G. L., Bounous, G., Torello Marinoni, D., Quacchia, A. and Alma, A. (2009), 'Evaluation of susceptibility to *Dryocosmus kuriphilus* Yasumatsu (Hymenoptera: Cynipidae) in *Castanea sativa* Miller and in hybrid cultivars', *Acta Horticulturae* 815, 289-97.

Sartor, C., Dini, F., Torello Marinoni, D., Mellano, M. G., Beccaro, G. L., Alma, A., Quacchia, A. and Botta, R. (2015), 'Impact of the Asian wasp *Dryocosmus kuriphilus* (Yasumatsu) on cultivated chestnut: Yield loss and cultivar susceptibility', *Scientia Horticulturae* 197, 454-60.

Schad, C., Solignat, G., Grente, J. and Venot, P. (1952), 'Recherches sur le châtaignier à la Station de Brive', *Annales de l'Amélioration des plantes* 3, 369-458.

Serdar, U. (1999), 'Selection of chestnuts (*C. sativa* Mill.) in Sinop vicinity', *Acta Horticulturae* 494, 327-32.

Serdar, U. and Macit, I. (2010), 'New advances in chestnut growing in the Black Sea region in Turkey', *Acta Horticulturae* 866, 303-8.

Serdar, U., Beyhan, N., Demirsoy, L. and Demirsoy, H. (2010), 'Some phenological properties of chestnut genotypes in the Black Sea region, Turkey', *Acta Horticulturae* 866, 135-42

Serdar, U., Demirsoy, H. and Demirsoy, L. (2011), 'Morphological and phenological chacterisitics of "Ersinop" and "Eryayla" chestnut cultivars', *American-Eurasian Journal of Agriculture and Environmental Sciences* 10(4), 684-91

Serdar, U., Dermisoy, H. and Dermisoy, L. (2013), 'Two sweet chestnut cultivars from the Anatolian region: "Unal" and "Erfelek"', *Journal of the American Pomological Society* 67(3), 162-80

Serrazina, S. (2004), 'Isolamento e Caracterização de Genes de Resistência à Doença da Tinta em *Castanea sativa* Mill.', PhD Dissertation, Universidade de Lisboa, 301pp.

Serrazina, S., Santos, C., Machado, H., Pesquita, C., Vicentini, R., Pais, M. S., Sebastiana, M. and Costa R. L. (2015), '*Castanea* root transcriptome in response to *Phytophthora cinnamomi* challenge', *Tree Genetics & Genomes* 11, 1-19.

Shimura, I. (1972), 'Breeding of chestnut varieties resistant to chestnut gall wasp *Dryocosmus kuriphilus* Yasumatsu', *Japan Agricultural Research Quarterly* 6(4), 224–30.

Shimura, I. (1973), 'Studies on the breeding behaviors of several characters in Chestnut, *Castanea* spp. III. Damaged Variation in the chestnut cultivars by the chestnut gall wasp, *Dryocosmus kuriphilus* Yasumatsu', *Japanese Journal of Breeding* 13(6), 311–19.

Shoda, M. N., Tanaka, K., Tanaka, N., Sawamura, Y. and Kotobuki, K. (2006), 'A method for quickly removing pellicles from chestnuts by deep frying cooking oil', *Bulletin of the National Institute of Fruit Tree Science* 5, 21–7

Solar A. and Štampar, F. (2010), 'Promising chestnut genotypes from Slovenia', *Acta Horticulturae* 866, 259–64.

Solar, A., Podjavorsek, A. and Štampar, F. (2005), 'Phenotypic and genotypic diversity of European chestnut (*Castanea sativa* Mill.) in Slovenia - opportunity for genetic improvement', *Genetic Resources and Evolution* 52(4), 381–94.

Solignat, G. and Chapa, J. (1975), 'Principales espèces de châtaignier', *Châtaignes et Marrons* INVUFLEC, Chap II, pp. 15–28.

Steiner, K. C., Westbrook, J. W., Hebard, F. V., Goergi, L. L., Powell, W. A. and Fitzsimmons, S. F. (2017), 'Rescue of American chestnut with extraspecific genes following its destruction by a naturalized pathogen', *New Forests* 48(2), 317–36

Takada, N., Nishio, S., Yamada, M., Sawamura, Y., Sato, A., Hirabayashi, T. and Saito, T. (2012), 'Inheritance of easy-peeling pellicle trait of Japanese chestnut cultivar "Porotan"', *Horticultural Science* 47(7), 845–7.

Takada, N., Yamada, M., Nishio, S., Sawamura, Y., Sato, A., Onoue, N. and Saito, T. (2017), 'Existence of genetic differences in pellicle peelability in Japanese chestnut (*Castanea crenata* Sieb. et Zucc.) cultivars and selections with difficult-peeling pellicles', *The Horticultural Journal* 86(4). 10.2503/hortj.OKD-030

Tanaka K. and Kotobuki, K. (1992), 'Comparative ease of pellicle removal among Japanese chestnut (*Castanea crenata* Sieb. et Zucc.) and Chinese chestnut (*C. mollissima* Blume) and their hybrids', *Journal of Japanese Society for Horticultural Science* 60(4), 811–19.

Tanaka K. and Kotobuki, K. (1998), 'The role of phenolic substances in the adhesion between the pellicle and kernel of Japanese chestnut under experimental conditions', *Japanese Society for Horticultural Science* 67, 1–8.

Tchatchoua D. T. and Aravanopoulos, F. A. (2010a), 'Evaluation of selected European chestnut (*Castanea sativa*) provenances-I: Inter-provenance genetic variation', *Acta Horticulturae* 866, 203–14.

Tchatchoua D. T. and Aravanopoulos, F. A. (2010b), 'Evaluation of selected European chestnut (*Castanea sativa*) provenances-II: Intra-provenance family variation', *Acta Horticulturae* 866, 215–24.

Torello Marinoni, D., Akkak, A., Guaraldo, P., Boccacci, P., Ebone, A., Viotto, E., Bounous, G. Ferrara, A. M. and Botta, R. (2013), 'Genetic and morphological characterization of chestnut (*Castanea sativa* Mill.) germplasm in Piedmont (North-western Italy)', *Tree Genetics & Genomes* 9(4), 1017–30.

Torello Marinoni, D., Sogo, N., Portis, E., Valentini, N., Sartor, C., Dini, F., Ruffa, P., Oglietti, S., Martino, G., Akkak, A. and Botta R. (2017), 'Development of a genetic linkage map for molecular breeding of chestnut', *Proceedings VI International Symposium on Chestnut*, Samsun (Turkey), 9–13 October 2017. *Acta Horticulturae* (in press).

Vannini, A. and Vettraino, A. M. (2001), 'Ink disease in chestnuts: Impact on the European chestnut', *Forest Snow Landscape Research* 350, 345–50.

Vettraino, A. M., Morel, O., Perlerou, C., Robin, C., Diamandis, S. and Vannini, A. (2005), 'Occurrence and distribution of *Phytophthora* species in European chestnut stands, and their association with ink disease and crown decline', *European Journal of Plant Pathology* 111, 169–80.

Vieitez, E., Vieitez, M. L. and Vieitez, A. M. (1982), 'El castaño y su resistancia a *Phytophthora cinnamomi* y *P. cambivora*', *Anales de Edafologia y Agrobiologia* 41(9–10), 1991–8.

Worthen, L. M., Woeste, K. E. and Michler, C. H. (2010), 'Breeding American chestnuts for blight resistance', *Plant Breeding Reviews* 33, 305–35.

Zhebentyayeva, T., Chandra, A., Abbott, A. G., Staton, M. E., Olukolu, B. A., Hebard, F. V., Georgi, L. L., Jeffers, S. N., Sisco, P. H., James, J. B. and Nelson, C. D. (2014), 'Genetic and genomic resources for mapping resistance to *Phytophthora cinnamomi* in chestnut', *Acta Horticulturae* 1019, 263–70.

# Chapter 13

## Advances in cultivation of chestnuts

*Ümit Serdar, Ondokuz Mayıs University, Turkey; Toshihiro Saito, Institute of Fruit Tree and Tea Science, Japan; Beatriz Cuenca, Empresa de Transformacion Agraria S.A (TRAGSA) Maceda, Spain; Burak Akyüz, Ondokuz Mayıs University, Turkey; José Gomes Laranjo, University of Tras-Os-Montes (UTAD) CITAB, Portugal; Gabriele Beccaro and Giancarlo Bounous, University of Turin, Italy; Rita Lourenço Costa and Patrícia Fernandes, Instituto Nacional de Investigação Agrária e Veterinária, Portugal; Maria Gabriella Mellano, University of Turin, Italy; and Pereira Lorenzo Santiago, Universidad de Santiago de Compostela, Spain*



## 1 Introduction

According to paleobotanic studies, the genus chestnut may have existed in the earth's Northern Hemisphere since the Cenozoic Era. The southwest region of China was the plant's region of origin, while *Castanea mollissima* was the "mother" specie. Chestnut was dispersed in two directions from its region of origin, via North America (*C. dentata*) and Europe (*C. sativa*). A further species derived from *C. mollissima* has also been observed in Japan (*C. crenata*). In Europe, Anatolia is known as the oldest home of the chestnut, from where it dispersed west to European countries.

Chestnut (*Castanea*) is part of the family Fagaceae and is a multipurpose tree that produces nuts, timber and honey, as well as holds an important place in the landscape and culture of many countries. Chestnut belongs to the genus *Castanea* and grows between 37° (Canary Islands) and 51° latitude (northern France) in the Northern Hemisphere (Fernandéz-López and Alía,

http://dx.doi.org/10.19103/AS.2018.0042.22

2003). According to 2017 figures, the world's annual chestnut production was 2,327,495 t. China is the leading producer with a production of 1 939 719 t. Turkey follows China, producing 62 904 t. The Republic of Korea is in the third place with 52 764 t, while Italy is in the fourth place with 52 356 t. Other significant producer countries are Greece (36 000 t), Portugal (29 875 t), Japan (18 700 t) and Spain (15 623 t) (FAOSTAT, 2019). According to FAOSTAT figures, Bolivia is the second-largest producer with a production of 85 047 t. However, the plant often called 'chestnut' in Bolivia is not actually *Castanea* (Bounous, Personal Communications). Chestnut has recently been introduced to Australia, Argentina, Chile and New Zealand in response to increasing demand. The world's total chestnut plantation area covers 603 076 ha. More than half of this area is located in China (335 904 ha) (FAOSTAT, 2019). In China, chestnut is considered an important food for the rural population, with cultivation strongly promoted by the government. In Europe, cultivation of chestnut is in decline and it is often grown in low-income mountainous regions with a low population density. Climate change is producing new problems and making it necessary to adapt existing cultivation methods or to develop new methods of chestnut production.

The aim of this chapter is to describe the current state of global cultivation of the genus *Castanea*, in terms of its ecological characterization, propagation methods and multifunctional uses. The chapter will examine chestnut's economic profitability, and the importance of wood and nut production for promoting the resilience and sustainability of chestnut cultivation. The chapter describes specific regional characteristics of chestnut cultivation around the world.

## 2 Ecological conditions affecting chestnut growth

Chestnut is known for its climate plasticity (Villani et al., 2001), growing in the Northern Hemisphere between 27ºN and 53ºN. It can be grown in both deciduous and subtropical-tropical climates at an altitude between 0 and 1 300 m. Long-term mean annual air temperature varies from 7 to 16°C in the Northern Hemisphere to 20°C in Minas Gerais (Brazil) in the Southern Hemisphere. The annual precipitation requirement for chestnut varies between 700 and 1 500 mm; nevertheless, around 200 mm is recommended during the period between flowering and one month before harvesting. However, chestnut can also be grown under more extreme rainfall conditions, between 550 mm (in Slovakia, southern Portugal and Spain) and 2 300 mm of precipitation (the eastern Black Sea Region of Turkey and northern Spain). During the winter in Michigan, USA, when vegetative activity largely stops, chestnut can tolerate temperatures from about −8 to −6 C°, and sometimes as low as −25 to −30°C. However, growth may occur too early when the winter is mild and plants risk being damaged by spring frosts. Freezing injury occurs frequently in young trees and has become one of the biggest problems for production in Japan.

Freezing injury occurs mainly in 2- to 4-year-old trees, especially in the early winter or between the end of the dormant stage and the beginning of the next growing season (Horimoto and Araki, 1999), because the freezing tolerance of trees does not sufficiently increase during this period. The incidence of freezing injury is the highest closest to the ground, decreasing further up the tree (Sakai, 1967). It has been observed that the fewest freezing injuries are seen in orchards located >150 m above sea level, while many more injuries have been seen in orchards located at <150 m, while the incidence was lower in orchards with slopes of >5 degrees than in orchards growing on flat land (Mizuta and Oribe, 2015). Severe freezing injuries have occurred in orchards cultivated on land that had been converted from paddy fields using grey lowland soil (Mizuta and Oribe, 2015). Very large amounts of rainfall during the winter increase the occurrence of freezing injury (Kamio and Mizuta, 2013). The advance of cold hardening in autumn and the delay of cold dehardening in spring are induced in low soil moisture conditions, while high soil moisture results in an increase in the water content of the current season's shoots and a rapid decline in the freezing tolerance during the cold dehardening period (Sakamoto et al., 2015). In addition, soil in orchards with no incidence of freezing injury has both a low pH and a high permeability coefficient (Inoue et al., 2014). These results suggest that high water content in soil tends to induce freezing injury by increasing the water content inside the tree and lowering its freezing tolerance.

European chestnut needs about 2700–3000° growing degree-days (GDDs) between January and the end of nut fall in October, which is calculated by the sum of air temperature values in degrees Celsius above $T_0$ = 6°C) (Gomes-Laranjo et al., 2018). The optimal growth temperature in *C. sativa* is around 28°C ($T_{100}$); growth is reduced by half at a temperature of 32°C.

Chestnut orchard cultivation requires deep and light soils, 2–4% organic matter and a pH between 5 and 6.5. Chestnuts do not favour alkaline soils. However, as the potassium content of the soil increases, the plant can tolerate alkaline soils. Chestnut does not favour wet and cold conditions, especially excessively wet forest soils, such as the upper parts of river streams, marshes and flooded natural sites, but it also suffers in locations that are too dry and warm, though it tolerates these conditions (e.g. Southeast Spain) better than wet conditions.

## 3 Chestnut management

Generally speaking, chestnut cultivation is managed in two different ways: (1) coppice stands for timber production and (2) nut production based on grafted trees in orchards or coppice populations.

### 3.1 Coppice for timber production

Chestnut wood is remarkably durable and is therefore favoured for exterior uses such as telephone poles, balcony supports, railings of all kinds, fence posts and supporting poles for vineyards. Chestnut wood can be used for high-quality flooring, window frames, doors and furniture.

Chestnut has the highest regeneration capability of all deciduous trees. The European chestnut, *C. sativa*, is suitable for management in coppice forests in short rotations. Coppices may be cut back every 6 years to produce the thinnest shoots to be used in basket making. The shoots (or suckers) may be allowed to grow into large poles over 10–15 years and are used for basket making or as poles for vines to trail around. The best-quality timber may be obtained after 30–40 years.

According to standard coppice forestry practice, clear cutting is done during the winter and new shoots in the form of thick clumps begin to grow in spring. During the following years, these shoots are in competition with one another. After 7 years, 5–8 shoots per stump succeed in dominating, while the other ones remain suppressed and gradually die back. After the first thinning and through to the 14th year of growth, fierce competition continues among the coppice shoots, which leads to the formation of dense, closed stands with straight and tall trees. The lack of light inside the stands promotes natural pruning. At this stage, the second thinning is performed, which removes 30–35% of the volume. The stand is maintained without any further application until the end of the rotation cycle, which is 20–22 years, depending on the quality of the site and the development of the stand. The trees reach a height of 17–22 m and a diameter of 18–25 cm during this period. A final thinning is carried out during the 25th year in which the best trees are selected and the nearest competing trees are removed. This rotation time is suitable for the production of timber for fence posts, telephone poles, mines supports and other uses. However, there is a trend for increasing the rotation to 30 or even 40 years in order to produce more valuable construction timber (Diamandis, 2009).

### 3.2 Chestnut cultivation in orchards

There are three methods of chestnut cultivation: grafting of natural coppice populations, planting and grafting seedlings, and planting grafted rootstocks.

#### 3.2.1 Grafting of natural coppice populations

Individual trees or coppice stands may be grafted, but it is difficult to manage natural populations because they grow in irregular patterns, typically in less accessible areas such as mountain slopes.

### 3.2.2 Planting and grafting seedlings

To begin with, rootstocks of 1- or 2-years old are planted in orchards at distances of 10 m (100 trees/ha) to 12 m (69 trees/ha) apart. When hybrid rootstocks are used, distances can be shortened to 7–8 m (156–204 trees/ha). One or two years after being planted, rootstocks are grafted using suitable grafting methods in March–May or in September.

### 3.2.3 Planting grafted rootstocks

This is the best method of chestnut cultivation because it allows popular cultivars to be grafted on approved rootstocks. Grafting can save one year of growth. Chestnut growers are encouraged to obtain grafted plants from reliable nurseries and plant them with the recommended spacing. Modern orchards equipped with irrigation systems are the most productive, producing around 5–8 t/ha, providing a good annual income. Chestnut trees can produce a good yield, both in quantity and quality, for more than a century, although it is good practice to prune trees to clear away dead branches and promote the growth of new ones.

## 4 Propagation

Chestnut is usually propagated by seeds in order to obtain seedlings to establish natural stands and orchards and for use as rootstocks to be grafted with selected cultivars. Seedlings differ in important traits on account of genetic recombination resulting from inevitable cross-pollination from originally highly heterozygous parents. Genetic variation in natural populations is essential to better adapt to climate change and to face new threats such as pests and diseases.

Vegetative (asexual) propagation offers considerable advantages in modern chestnut cultivation. There are four methods for vegetative propagation of chestnut:

1. grafting
2. layering
3. cutting
4. micropropagation

Grafting is the main method for *Castanea* spp. propagation for nut production, while layering and cutting are used to propagate Euro-Japanese hybrids or selected genotypes (both as rootstocks and as production trees). Micropropagation has great potential and is now beginning to be used commercially as well as for breeding programs to propagate improved genotypes.

### 4.1 Grafting

Grafting is the main propagation method for orchards. For European chestnut cultivars, grafting and budding are still the preferred propagation methods because large-scale self-rooting techniques have yet to be perfected (Mellano et al., 2012; Bounous, 2014). This is due to the juvenile phase affecting the other methods, which require rejuvenation in order to improve rooting. Moreover, when chestnut material is rejuvenated in order to root, a longer period is needed to overcome juvenility and reach the mature fruit-producing stage. This is clearly uneconomical in a commercial plantation, so grafting uses sticks or buds to form the aerial part or the crown where mature tissues are ready to bear fruit, which reduces the unproductive period. However, these parts of the plant are difficult to root in adult plants.

Grafting success requires the correct technique, skill, time and favourable weather conditions (mild with high moisture and little wind), suitable propagation materials, sterilization methods and storage conditions for scions. It also requires sufficiently mature rootstock as well as understanding of scion polarity and rootstock/scion genetic compatibility. Selection of the right rootstock is important for the success of grafting.

#### 4.1.1 Rootstocks

Chestnut rootstocks are produced from seed or clonally by vegetative propagation. Using genotypes with small-sized nuts (<10 g) for generative rootstocks may decrease the cost of seeds. However, seedlings obtained from small nuts generally grow poorly in comparison to those obtained from large nuts (Okada, 1969). Based on these results, if the cost of the rootstock is not important, bigger nuts (about 20 g) should be used to obtain seedlings of suitable quality. There have been many studies of rootstock selection criteria for chestnut (Solignat and Chapa, 1975; Ferrini et al., 1993; Pereira et al., 1993; Ertan, 1999; Soylu et al., 1999; Soylu and Serdar, 2000). Among the many rootstock characteristics, germination and sprouting ratios, growth by diameter and height, rooting capacity for clonal rootstocks and resistance to *Phytophthora* spp. and to drought are especially important. Most chestnut cultivars are propagated by grafting onto seedling rootstocks. However, some specific rootstocks tolerant to root rot (*Phytophthora*) have been bred from Euro-Asiatic crosses. Specific rootstocks must be propagated through vegetative methods to achieve clonal cultivation of these varieties.

Clonal rootstocks of 'Marsol', 'Maraval' and 'Marigoule', known to be tolerant of *Phytophthora* spp., are propagated by layering, cutting or micropropagation in France, Spain, Portugal and Italy. In Portugal, ColUTAD rootstock is also propagated by layering. Besides its tolerance to root rot, ColUTAD also displays tolerance of the gall wasp (*Dryocosmus kuriphilus* Yasumatsu). In Spain, the chestnut rootstocks used are derived from a Spanish breeding program developed in the mid-20th century. The reference clones 111-1, 7521, 2671 and 1483 are the most widely used for rootstock as they are very tolerant to root rot and show very high compatibility with fruit varieties. Some nurseries have their own rootstock collection. A new breeding program is currently being carried out by the Lourizan Research Center using nut varieties as *C. sativa* parietals in order to increase graft compatibility for hybrid generation.

In Australia, random seedlings were once used as nursery rootstocks. However, due to concerns about graft incompatibility, 'half sibling' seedlings (rootstock grown from seeds of the cultivar being propagated) were introduced. In the early 1990s, a trial was conducted to identify a rootstock resistant to the fungal root rot *Phytophthora*. The cultivar 'Menzies', thought to be a European/Japanese hybrid, proved to be the least susceptible to *Phytophthora* and is now widely used as rootstock (Casey and Casey, 2009). In Turkey, growers formerly used *C. sativa* seedlings as rootstocks; however, they are now encouraged to obtain 'Marigoule' seedlings, which are thought to be better for root rot.

In southwestern France, 'Marigoule' was grown on its own roots and was planted in large quantities between 1980 and 2000. The use of 'Marsol' as a rootstock makes it possible to limit the problem of ink disease. 'Marigoule' and especially 'Bouche de Bétizac' have proved to be less sensitive to canker than most *C. sativa* varieties. 'Bouche de Betizac' also displays tolerance to the gall wasp (Hennion, 2009).

In Japan, seedlings of 'Shibaguri', which are wild genotypes of Japanese chestnut, were traditionally used as rootstocks. However, it became difficult to obtain nuts because many 'Shibaguri' genotypes were devastated by the chestnut gall wasp (*D. kuriphilus*). Recently, seedlings of cultivated varieties have been used as rootstocks. In Italy, the most common rootstocks are seedlings coming from *C. sativa* or hybrids. Recently, the cultivar 'Marrone' has been grafted with success on 'Marsol' and 'Marlhac' clonal rootstocks (Beccaro et al., 2017).

#### 4.1.1.1 Propagation of seedling rootstocks

Seeds are used to propagate seedling rootstocks. In order to obtain a high germination ratio, seeds must be healthy. To avoid the development of mould, they should be collected immediately after they drop to the ground. Good germination occurs after stratification (for about 90 days) in humid conditions to remove growth-inhibiting hormones. During germination, chestnut seeds absorb high quantities of water and oxygen, after which carbohydrates stored in the cotyledons are transferred to the growing embryo. The primordial root (radicle and hypocotyl) and the young shoot (epicotyl) both emerge from the pointed tip of the nut. Pre-germinated nuts are sown in raised beds or pots. The soil substrate should be soft and well drained. The germinated chestnuts are planted with the flat part down to facilitate root penetration in the soil. In raised beds, the seedlings are 100–150 cm tall with a diameter of 8–12 mm at the end of the first growing season (August–September), and are ready to be budded in the autumn or grafted or budded in the following spring.

#### 4.1.1.2 Propagation of clonal rootstocks

The main objectives of a chestnut nursery are to produce as well-formed and genetically uniform trees as possible, avoid ink disease and blight infections and achieve intensive cultivation with high-density planting (Beccaro et al., 2018). Chestnut orchards employ the same technologies as modern fruit orchards, using plants with a vigorous juvenile rooting system (resistant clonal rootstock) as a mature crown (selected fruit variety) that will start producing nuts almost immediately. Vegetative production methods are essential for modern chestnut rootstock production, requiring propagation of self-rooted plants. Chestnut self-rooting was first studied in Spain from the middle of the 20th century and is an ideal way to clone rootstocks and produce *Phytophthora*-resistant hybrids (Bounous and Torello Marinoni, 2004). Thanks to collaboration with research centres, some Spanish, French, Italian and Portuguese nurseries also use vegetative propagation to produce rootstocks useful for intensive chestnut farming. Self-rooted plants and rootstocks can be obtained by using layering, soft and hardwood cuttings, or by micropropagation (Beccaro et al., 2018).

### 4.1.2 Grafting times and methods

There are two main periods for grafting: spring (early to mid-spring) and summer (end of August to the middle of September) depending on the grafting method and the local climate. Summer grafts benefit from the higher temperatures to heal the graft union in cooler areas. In spring grafting, scion woods are collected in winter during the full dormancy period and stored in cold rooms (+ 2 ± 1°C) until grafting takes place. In summer grafting, scions are collected on the same day as grafting and used immediately. In summer grafts, the same growing season's shoots are used. However, if they are insufficiently lignified, the rate of graft success can decrease.

Spring grafts are generally made over quiescent rootstock, when it is close to restarting vegetative growth and the buds from the scions are ready to burst when temperatures rise. However, if spring grafts are done early, spring frost may damage the graft union. On the other hand, if grafting is done late, the quality of the nursery tree is diminished. Lateral grafts (budding methods) can be delayed to mid-May, by keeping the scions refrigerated and performing the lateral graft on the rootstock without cutting back in order to allow normal movement of sap during healing. When the callus is formed, the rootstock is pruned heavily to promote the sprouting of buds from the scion. Summer grafts use quiescent buds that will become dormant if temperatures go down at the end of the summer; in consequence, they will sprout the following year.

The time chosen for grafting affects both graft success and nursery tree quality. Summer buddings generally have less success than spring buddings, except when patch and pipe budding methods are employed (Özkarakaş and Önal, 1997; Serdar and Soylu, 2005). In Japan, budding is rarely used for producing grafted plants, and grafting is usually carried out from March to April. In Turkey, summer buddings are made generally in the first half of September, though they can be done until the end of October in İzmir since this region has very suitable ecological conditions for production of nursery plants. Spring buddings generally have lower nursery tree quality than summer buddings, but high grafting success. Summer buddings start to sprout at the beginning of the next vegetation period and use the full vegetation period, so they are better-quality nursery trees overall.

There are many techniques for grafting chestnut, but some of them are used more often than others, depending on local traditions, the skill of the grafter and climatic conditions. For spring grafts, the most frequent grafting methods are cleft and splice and, for budding T, chip and pipe. In addition to these methods, cleft and whip-tongue grafting techniques are commonly used in Portugal, H budding in Spain and triangle grafting in Italy.

Whip grafts are widely preferred, with a high graft success and good resulting nursery tree quality, especially in the field (Serdar and Soylu, 2005). Cleft grafting is currently used at nurseries and in the field when the scion and the rootstock have the same diameter (and is sometimes performed by machine). Triangle (inlay) grafting is made either in the nursery or in the orchard, on rootstocks 2–3 cm in diameter; it involves creating a triangular wedge in the rootstock in which the scion is inserted (Mellano et al., 2015; Beccaro et al., 2017).

Budding is the main technique used for summer grafting. T-budding is the most favoured method, followed by the chip, patch and pipe methods. T-budding is a main summer graft in Spain and in the Aegean region of Turkey, primarily due to the ease with which it is possible to separate the bark and place the bud, a method which is much easier and faster than chip budding. The inverted T-budding method has a higher graft success than the T method, since it drains better during rainy periods (Serdar and Soylu, 2005). However, nursery workers prefer T-budding as an easier technique. Patch and pipe buds take more time to adjust to the rootstock, but these methods promote better healing in the graft area. On the other hand, the chip budding method has a longer period for budding than the T-budding method, because it is not necessary to separate the bark in order to insert the bud.

### 4.1.3 Seed grafts

Grafting can also be attempted on germinated chestnut seeds, the radicle of germinated seeds and the hypocotyls and epicotyls. These grafting methods offer several advantages over conventional field grafting. They obviate the need for stock trees, and grafting can be done indoors; furthermore, timing is not critical and scions that have a small diameter can be used. The method can also decrease costs because grafted plants can be produced within a year. Among these methods, better graft success has been obtained from the inverted radicle grafting method (Duman and Serdar, 2006). In this method, germinated seeds are used as rootstocks. The radicle tip is cut off at 4–6 cm and split. The bottom of the scion is then cut into a wedge-shape and inserted into the radicle as in a normal cleft (Duman and Serdar, 2006; Ozturk et al., 2009; Ozturk and Serdar, 2011).

A new chestnut orchard was established in 2013 with grafted plants produced by inverted radicle and whip grafting methods at Samsun, Turkey. So far, there have been no differences in survival ratio, tree growth and time to fruit bearing between these grafted plants and others. Moreover, this method may improve resistance to chestnut blight disease if the scion is obtained from resistant cultivars, because the graft union stays under the soil, protecting it from chestnut blight.

### 4.1.4 Green and in vitro grafts

Green or semi-soft scions are grafted onto rootstocks grown in pots during the vegetation period, using semi-herbaceous stems. *In vitro* grafts are made on *in vitro* micropropagated explants using scions also grown *in vitro*. Cleft grafting is common in this type of graft. When performed *in vitro*, a scalpel instead of a knife is used to cut the upper part of the shoot and the scion bud is laid on top; healing occurs in aseptic conditions inside the *in vitro* vessel (Errea et al., 2001). In green grafting, the union is maintained with a pin and grafts are set in a controlled atmosphere to avoid desiccation during healing, due to the herbaceous condition of both scion and rootstock and the presence of leaves in the graft (Fig. 1). In a nursery in Spain where both chip budding on hard tissue and green grafting on herbaceous tissue are undertaken, green grafting increases the success rate compared to chip budding due to faster healing of the soft green tissues (Cuenca et al., 2018).

**Figure 1** Example of green grafting in chestnut.

### *4.1.5 Care of grafts*

After grafting or budding, the shoots developed from the graft union must be removed. The growth of the scion shoot will be interrupted if more than one stem grows; only the best stem should be preserved. Growing stems are often staked to ensure regular development.

Although grafting is the most common method for propagating chestnut, it is expensive and gives rise to some problems owing to rootstock/scion incompatibility, blight and viral infection. The use of wax to protect grafting wounds prevents infection by blight and other diseases, as well as preventing drying of the scion wood. Attributes of a good wax include elasticity, impermeability, disinfectant action and the capacity to prevent or reduce infections, especially in the early years after grafting. *Cryphonectria parasitica* easily inoculate through the graft union, causing cortical tissue death. Grafting and budding are at risk of infection in the late spring, when the diffusion of pathogens is high (Bounous and Torello Marinoni, 2004). Canciani et al. (1993) have successfully tested some formulations including the fungicide carbendazim mixed with mineral oil, confirming the results of Jaynes and Anagnostakis (1971), Jaynes and Van Alfen (1977) and Elkins et al. (1978). In Turkey, copper compounds in the form of copper hydroxide or copper oxychloride were used for *C. parasitica* infections until 2015, but the transmission of blight to grafts could not be prevented. In recent years, some chemicals have been tested for *C. parasitica* infections and successful results for both sterilization and scion survival were obtained using 2.5% sodium hypochlorite (Serdar, personal communication).

### 4.1.6 Graft compatibility

Incompatibility has sometimes been observed within the same European and Chinese species (Santamour, 1988; Serdar and Soylu, 2005). However, incompatibility often occurs between European and Asian species. Weak unions have been described between European cultivars and the interspecific rootstock 'Marigoule', with fatal failure at the beginning of the third year of grafting (Pereira-Lorenzo and Ramos-Cabrer, 2004). In a study conducted on European chestnuts, the survival ratio was determined as 58.1% in the first year after spring grafts, and 29.5, 21.8 and 14.1% in the following years, respectively, in the combination of SA5-1 scion on 554-14 rootstock (Serdar et al., 2010). In another study carried out on *C. sativa*, the survival ratio was determined as 16.7–55.6 % for complex hybrids and 65.7% for 'Marigoule' in the second year after grafting (Serdar et al., 2014). In recent decades, Euro-Japanese hybrids have been preferred as a cultivar on account of their tolerance to chestnut blight, in contrast to *C. sativa* genotypes. 'Marigoule' has generally acceptable compatibility with *C. sativa*; however, root rot diseases prevent the use of *C. sativa* as a rootstock. 'Marigoule' is, therefore, generally grafted on its own seedlings in Turkey. However, 'Marigoule' seedlings' tolerance of root rot is not the same as that of clonal rootstocks. It is, therefore, very important to propagate rootstocks by vegetative methods. In Japan, Chinese chestnuts were first imported from China by some private chestnut growers in the 1930s. However, when seedlings of Japanese chestnut cultivars were used as rootstocks, some genotypes did not survive. From then on, attention was paid to graft compatibility between scion and rootstock. Torikata and Higuchi (1962) studied inter- and intra-specific graft compatibility between Japanese and Chinese chestnuts. Four years of observation of tree growth showed that Chinese scions achieved poor growth on the Japanese rootstock, while other combinations grew satisfactorily. Shimura et al. (1969) studied inter- and intra-specific graft compatibility by grafting Japanese cultivars on Japanese, Chinese and European rootstocks, respectively. Most scion/rootstock combinations made a successful union soon after grafting. However, many nursery trees of Japanese/Chinese combination died during the first growing season, while those of Japanese/Japanese and Japanese/European combinations grew well. Consequently, it was recommended that the scion and rootstock should belong to the same species.

Graft incompatibility has different symptoms. These include low graft success, abnormalities on graft union such as swelling or cracking, thicker bark tissue in the scion, cambial connectivity on the graft union, dying of graft shoots or grafted plants in the years following grafting, a tendency to develop excessive suckers from rootstocks, weak growth, early leaf yellowing and the falling of scion shoot leaves as well as direct flower formation from scion buds.

### 4.1.7 Grafting in orchards and natural coppices

It is still very common to plant rootstocks for grafting after 3–4 years. In this way, rootstocks spread roots and form the trunk, lifting the crown beyond the reach of grazing animals and making it easier to operate machinery. Bark and cleft grafting methods are the preferred ways to achieve this and are applied in the spring. When the rootstock is too thick to be split for cleft grafting, the bark grafting method is used. The bark grafting method should be performed at the beginning of vegetative growth when the bark separates easily. The placing of the scion under the bark can be a weak link to the rootstock in windy areas, especially when bark grafting is performed on large trees. In areas affected by high winds or snow, the scion is supported by a stick attached to the scion by wrapping. Damage caused by chestnut blight (*C. parasitica*) is a potential problem with this method. Bark grafting can also be used to graft the branches of the crown, but in windy areas pipe grafts or chip budding are preferred. Bark grafting can also be used for top working when part of the crown is renovated to maintain the quality of the nuts. That makes it possible for 300-year-old orchards to be kept productive and updated with new cultivars required for changing market conditions. In the past, easy peeling was an important quality characteristic for growers and consumers, but the large size and bright colour of the nuts are the most desirable characteristics.

## 4.2 Layering

Layering means the cutting down of a young tree to ground level so the stump will produce new shoots. The basal part of the new shoots is covered with soil to facilitate the formation of roots. Mother trees are cut every year and, when they sprout, the best shoots are selected, eliminating the others to avoid competition. In the spring, selected shoots are defoliated in the basal part of the shoot, scratched lightly and auxins applied over the scratch (usually a balsam with IBA 4–8%). Shoots are surrounded by a wire girdle under the area to which auxins have been applied in order to promote the accumulation of photosynthetic compounds that will aid root formation. Rooting response will be higher when juvenile shoots are selected and soil is not compacted or damp. The rooted plants are pulled out from mid-November when leaves fall. At this point, it is not uncommon that the root system is weak and the plants are curved where they join the mother plant. Both problems can be solved by reinforcing the rooted shoot with one more year's growth in the nursery, cutting the malformed shoot and choosing an upright one (Fig. 2).

Layering has several advantages. The technique does not need a high level of technology, highly skilled labour or complex facilities so it is cheaper than any other vegetative propagation method and plants can be produced within a year. However, it has many disadvantages: larger areas are required and there is a high demand for labour concentrated in a short period. Plants can only be commercialized during the dormant period, when their roots are bare; and any decision about changing the clones under production is difficult and expensive. Furthermore, the root system produced by the plant is poorer

**Figure 2** Stages of layering in chestnut.

than in other methods and reinforcement is frequently required (Cuenca and Majada, 2012).

### 4.3 Cutting

It was traditionally thought that chestnut was a difficult species to root by cuttings. However, the development of environmental humidification systems has changed this, allowing rooting of chestnut stems that are not lignified (semi-herbaceous). In the last few decades, the Forest Research Centre of Lourizán (CIF Lourizán) in Spain has developed a semi-herbaceous cutting methodology, which allows the rooting of cuttings using substrates that ensure adequate draining after their immersion in a concentrated auxin solution (Fernández-López et al., 1992).

Propagation through cuttings is currently used with some success in some industrial nurseries. Some of the advantages of this system include a better-quality root system and more cuttings obtained per mother plant related to stooling. It is also a more controlled process and, as plants are normally potted after rooting, there is more time to commercialize them. Nevertheless, this technique has some disadvantages, which mean it is not commonly used for mass propagation. These include the need for facilities to control humidity, the fact that plants may display the effects of plagiotropism and a poor root system compared to seed plants. It is slow and expensive to make changes in clones in production and there is a significant demand for labour concentrated in a very short time (Cuenca and Majada, 2012).

Biochemical, anatomical and environmental factors influence adventitious rooting. In general, young plants have higher quantities of root promoters and lack auxin-inhibiting factors. Juvenility, etiolation, exogenous promoters and mycorrhizal fungi promote rooting. This means that the success of rooting cuttings from young plants is higher than from adult plants (Bounous and Torello Marinoni, 2004). However, mother plants can be kept in a juvenile stage to maintain their ability to root. The juvenility grade of a meristem is inversely proportional to the distance (along the branches and trunk) between the root collar and the meristem (Bonga and Durzan, 1982). Therefore, cuttings collected from severely pruned plants will be ontogenically young and maintain their ability to root, even though they come from physiologically old mother plants.

Cuttings are collected from mother plants and usually grown in pots in the nursery, where it is easier to keep them healthy. The mother plants need to be true to genotype and free from chestnut mosaic virus (ChMV) (Mellano et al., 2018). The success of propagating chestnut by softwood cutting depends not only on genotype and stem juvenility, but also on the time of cutting and environmental factors such as the presence of growth regulators, soil, relative humidity, and temperature. *C. crenata* × *C. sativa* hybrids show a higher ability to self-root. Cuttings of some *C. sativa* genotypes, when taken from young mother plants (i.e., those generated by micropropagation), also show a high rooting ability (Cuenca et al., 2005). In Italy, the Chestnut Research and Development Center has developed an effective routine method for the rooting of clonal rootstocks ('Marsol' and 'Maraval'), with results that range from 40 to 80% rooting success (Mellano and Donno, 2015). In Spain, the company TRAGSA roots around 50,000 cuttings per year of Euro-Asiatic (European × Japanese) hybrids of French and Spanish varieties as well as one variety developed by the company's own genetic program for use as clonal rootstocks (Ocaña et al., 2001; Rodriguez et al., 2005).

The semi-herbaceous stems are collected in June from mother plants grown in big pots and are then cut into portions of 10–15 cm with 1–3 buds, with always one at the bottom and one at the top. The apical leaf, or at least part of it, is left attached to the cutting. Buds and leaves are points of auxin synthesis so new roots will always appear in the buried bottom bud. The cuttings are basally dipped in a concentrated auxin solution (1000–4000 ppm) for a few minutes. The cuttings are then laid in rooting trays filled with high-draining substrate in climate-controlled rooms or rooting tunnels in the greenhouse (Fig. 3). A fog or similar system to ensure high humidity during the first weeks is imperative, and should then be progressively reduced. Basal heating with heating wires or hot water pipelines is helpful during the night to maintain a minimum temperature of 15°C so the rooting period can be shortened.

Factors that reduce the rooting ability of cuttings include vanyllic, hydroxybenzoic, salicylic, syringic, aliphatic and ellagic acids, and some water-soluble compounds. Inhibiting factors in adult plants interfere with indoleacetic acid (IAA). Leaching by soaking in water promotes rooting. The anatomical structure of cuttings also influences rooting (Beakbane, 1961; Vazquez et al., 1978; Edwards and Thomas, 1980; Vieitez, 1981; Vieitez et al., 1987; Vieitez and Ballester, 1988, 1989). At the end of the growing season, when five continuous rings are developed, root emergence can be difficult. Etiolation of shoots increases the endogenous auxin content and phenolic cofactors, preventing the formation of inhibitors and modifying the histo-anatomical structure, thereby reducing the sclerenchymatic rings and promoting xylematic ray formation where former roots are initiated (Rinallo et al., 1987; Vieitez and Ballester, 1989). Treatments with fungicides avoid the risk of rot during the rooting process. Rooting usually occurs after 60–70 days, preferably under climate-controlled conditions. Good rooting substrates are porous and well drained, such as perlite and peat moss or peat moss and pine bark. Agronomic care for young rooted plants should include inter-row weeding, irrigation, phosphorus and potassium fertilization and the addition of low amount of nitrogen. The young rooted plants are transplanted to a nursery to increase root growth (Bounous, 2014; Donno

**Figure 3** Stages of cutting in chestnut.

et al., 2014). The rooting system of a rooted cutting is more developed and more balanced than that of a stooled plant although, in potted plants, it is limited by the pot size. Rooted cuttings transplanted to a 2 L pot can be grafted the following year.

### 4.4 Micropropagation

Micropropagation offers alternatives to traditional vegetative propagation systems by using *in vitro* tissue culture methods. Classic micropropagation involves the development of micro-cuttings in a semisolid aseptic culture media, confined in closed vessels. Plants developed in this way are heterotrophic, photosynthetically inactive and with dysfunctional stomata because they cannot perform gas exchange inside the vessel. If a semisolid culture is used, culture media should contain a source of carbohydrates as well as mineral salts, vitamins, growth regulators and agar. This method means that plants demonstrate higher development rates in the field compared with other propagation methods, due to the profuse and well-balanced rooting system developed by *in vitro* plants. It is an ideal technique to achieve rapid propagation of clonal rootstocks or self-rooted trees starting from very few initial materials, which can then produce chestnuts during the whole year. The production of plants free from viruses or diseases is another strength of micropropagation, as it is an aseptic process using controlled conditions. Problems related to the loss of rooting ability of adult genotypes can also be easily overcome as micropropagation can use young materials such as epicormics, ethiolated or basal sprouts and spheroblasts (lignified bulbs formed by shoots that did not emerge) that allow adult genotypes to be cloned (Sánchez et al., 1997). Furthermore, it is quick and easy to introduce changes in the genotypes produced by a nursery. On the other hand, the facilities needed, the high demand for skilled labour and the bottleneck of acclimatization make this method more expensive than traditional techniques.

Micropropagation protocols have been widely developed by different research teams (Viéitez and Viéitez, 1982; San-José and Viéitez, 1984; Vieitez et al., 1989; Sánchez and Viéitez, 1991; Ballester et al., 1992; Miranda and Fernández, 1992; Cuenca et al., 2005, 2009; Oakes et al., 2013; Vidal et al., 2015; Cuenca et al., 2017). Bearing in mind that *in vitro* establishment in chestnut is possible from both juvenile and mature materials, since clonal propagation is genotype-dependent, the optimum selection of culture media, carbon sources, rooting stages and acclimatization periods remains an area of research (Vieitez and Vietiez, 1980; Sánchez et al., 1997).

*In vitro* propagation of chestnut consists of a number of different steps, with different needs in terms of growth regulators, namely establishment, multiplication, elongation, rooting and acclimatization. A higher cytokinin/auxin ratio promotes multiplication, while the inverse ratio promotes rooting.

Micropropagation starts by establishing material of a certain genotype in *in vitro* culture. Juvenile winter material is collected, disinfected and forced to sprout in a controlled environment. The new sprouting is surface sterilized in a solution of hypochlorite and stabilized in culture. After contaminated or non-reactive explants are eliminated, multiplication rates become more regular and commercial propagation begins. In the multiplication phase, new shoots are generated from the initial shoot. During the elongation phase, the generation of new shoots stops and the existing shoots reach a suitable size for rooting. During the rooting phase, the elongated shoots are basally dipped in a sterile concentrated auxin solution (1 g/L) and laid to root. During the acclimatization phase, rooted explants switch from a heterotrophic to an autotrophic habit (Fig. 4).

**Figure 4** Stages of micropropagation in chestnut.

The most important steps of the whole process are rooting and acclimatization to *ex vitro* conditions. Although chestnut species and hybrids have been rooted and acclimatized with success (Vieitez et al., 1989; Seabra and Pais, 1998; Gonçalves et al., 1998; Oakes et al., 2013), some rooting and acclimatization rates have not been consistent. Roots produced *in vitro* frequently show low functionality and efficiency in uptake of water and nutrients. Problems with root induction and the physiological state of the shoot during and after the rooting process have made this stage the most critical step in the successful culture of mature chestnut material.

In classic micropropagation techniques, all these phases take place in semisolid media (Murashige and Skoog, 1962; Greshoff and Doy, 1972). However, as noted, roots produced *in vitro* frequently show low functionality and efficiency for uptake of water and nutrients, and plants inside the closed vessels fail to achieve gas exchange and photosynthesis. This results in low acclimatization and survival rates (Gonçalves et al., 1998; Tetsumura and Yamashita, 2004). Different approaches are under development to overcome this bottleneck. The Portuguese research team at INIAV have developed new genotypes with improved resistance to root rot (Costa et al., 2011). In doing so, they developed a more efficient protocol for mass propagation and efficient *in vitro* shoot rooting that could be used within the nursery industry for large-scale production of different species of *Castanea*, namely, *C. sativa*, *C. crenata*, *C. mollissima* and hybrid genotypes. They found that the medium used for inducing ectomycorrhizae, namely Modified Melin-Norkrans medium (MMN) developed by Marx (1977), with an equimolar solution of citric acid and calcium citrate (pH 5.4) before sterilization, showed itself to be an efficient inducer of root formation and development across different genotypes, functioning like a rooting booster. The morphology of the roots improved with the use of MMN are found having a higher number of roots per plant and numerous root hairs, traits that generally do not develop when the rooting is artificially induced *in vitro* with indolbutyric acid (IBA) . On the other hand, the use of zeatin in the culture medium, prior to rooting, favours shoot elongation, increasing the rooting percentage when compared to media without zeatin.

The use of zeatin to provide a good shoot system suggests that the elongation step, prior to root induction, is a key factor for producing functional rooting systems after IBA dipping. The key advantages that make this protocol attractive are (1) the production of whole plants with both vigorous shoots and roots; (2) reduction in propagation time when compared with traditional methods of chestnut propagation; and (3) provision of an easy and cost-efficient method for the production of improved *Castanea* genotypes for industry.

A Spanish research team from TRAGSA and IIAG has recently overcome the rooting and acclimatization problem by culturing in temporary immersion systems (TISs) using liquid media. These systems use a special vessel that contains the liquid media and pumps air into the vessel through a 0.2-μm filter, so it is sterile. The incoming air has a double function: it makes the liquid move upwards to the plants to feed them and it renews the atmosphere inside the vessel, allowing gas exchange to occur and therefore ensuring the stomata of the explant remain functional (Fig. 4) (Vidal et al., 2013, 2015).

The elongated shoots generated in this way are set to root *ex vitro* so rooting and acclimatization occur at the same time. Environmental control facilities, growing chambers or temperature-controlled greenhouses allow the maintenance of high humidity and a temperature around 25°C. Controlling temperature and humidity is essential for success. More recently, Spanish researchers have developed a photoautotrophic micropropagation (PAM) system for chestnut culture (Cuenca et al., 2017; Aldrey et al., 2018), in which a temporary or continuous immersion system is created under photosynthetic active radiation (PAR) lights and the incoming air is supplemented with $CO_2$. The PAR lights emit in the same spectral range as solar radiation, from 400 to 700 nm, which allows photosynthesis. $CO_2$ supplied between 2000 and 3000 ppm also allows young plants in culture to be photosynthetically active. The success of the process is particularly noticeable during the rooting process (Fig. 4). Explants set to root in rockwool block in this culture system show higher rooting rates. When the moment for acclimatization arrives, there is no stress for the explants as they are already autotrophic, and the roots that develop are normal and functional. The TRAGSA nursery in Spain is producing 30 000 plants a year in this way. This shows that chestnut micropropagation can now be conducted on a large industrial scale and not just for research purposes. Acclimatized *in vitro* plants are normally transplanted to bigger pots during the spring and grown until they are large enough for grafting. The success of the system suggests that the investment associated with the propagation of elite chestnut plant material, through *in vitro* tissue culture, is worthwhile, since the plants show higher development rates in the field compared with other propagation methods. This is due to factors such as more robust and functional rooting systems that are free of infection.

## 5 Orchard establishment and management

Chestnut orchards are typically established in degraded forest areas or agricultural land. Chestnut needs to not only avoid wet and cold positions, especially excessively wet forest soils, such as the upper parts of river streams, marshes, and flooded areas, but it also needs to avoid excessively dry and warm conditions, even though it tolerates them better than the wet conditions (Solar et al., 2009). Chestnut trees are susceptible to root rot and nut rot and both diseases are more severe in humid weather. Nut rot, in particular, can cause crop loss when there is high rainfall or foggy conditions during the growing season.

If it is a forest area or an area where different fruit species are cultivated, old trees must first be cut from the bottom and then the roots must be removed. After this operation, land should be deeply ploughed and the soil pretreated. Some growers just drill the planting holes with machines, without ploughing the soil deeply. However, in this case, depending on soil characteristics (heavy and impermeable), water can fill in the planting holes and cause root rot problems.

Good soil preparation is required to establish an orchard. Farmers have two options. Deep tillage is suitable for stony soils. Alternatively, holes can be dug for plants after stakeout to ensure good root system development. If tillage is undertaken, the entire land should be fertilized. In the second approach, each hole should be fertilized directly. Farmers can plant rootstocks or plants, which are already grafted. In both cases, they can use wild plants from *C. sativa* (which are more vigorous) or hybrid rootstocks.

Planting density varies according to variety, rootstock, climate and soil fertility. When wild *C. sativa* are used, the planting density should be 10 × 10 m, corresponding to a density of 100 plants/ha, while with Euro-Japanese hybrids, the plants are placed closer together (6–8 m in the row by 8–10 m between the rows), resulting in 170–180 plants/ha. In recent years, this has been reduced and densities of 6 × 6m or lower are more common in newer orchards in Australia, China, etc. (Casey and Casey, 2009; Qin and Feng, 2009). Since chestnut is self-sterilising, it is necessary to plant other varieties with long staminate catkins and overlapping the feminine flowering, in a percentage of 5–10% to obtain sufficient yield (Ohata and Sato, 1961; Soylu, 2004, Gomes-Laranjo, personal information).

The planting time can be varied according to the ecology and type of the nursery plant. Some compost should be mixed with soil before planting. If mulching (durable ground cover by polyethylene material) is done immediately after planting, weed control can be done more easily. However, the mulch should be removed after about 5 years because it can inhibit tillage and fertilization.

As noted earlier, freezing injury often occurs in young trees and has become one of the biggest problems for production in Japan. A combination of raised ridge planting and covering soil with polyethylene nonwoven fabric from mid-September has been tried successfully to prevent freezing by reducing soil moisture (Kamio and Mizuta, 2013).

### 5.1 Training and pruning system

Chestnut trees naturally grow up to 30 m in height depending on the species. American and European chestnuts have higher rates of tree growth. Although the Japanese chestnut has the smallest tree canopy among the four cultivated species in the world, trees grown without training or pruning can grow up to >10 m in height. The most common training system for European varieties is the vase shape; for Euro-Japanese hybrids and for those European varieties with an upright growth habit, the pyramid form is preferable (Fig. 5). Modified leader type training when trees reached about 5 m in height was initially undertaken in Japan, since chestnut trees usually show strong apical dominance and become tall when this technique is employed (Tsukamoto, 1985). However, this training method tended to produce taller trees than expected because since many (4–5) primary scaffold limbs (PSLs) were needed to cover the tree canopy and this was difficult for farmers to master. This increased the labour needed and reduced the nut size. In order to resolve this problem, new pruning methods were developed in the 1970s to make trees shorter (about 3.0–3.5 m in height) using modified open centre training (Hiyama, 1978; Araki and Nakaoka, 1982; Tsukamoto and Tanahashi, 1983). The common characteristics of these methods were to (1) train the form of a tree's central leader until around 5 years old and then shift to an open centre form by topping, (2) locate 2–3 PSLs and 2 secondary

**Figure 5** Trees (vr. *Longal* grafted on Ferrossacri rootstock, Minho Region, Portugal) growing in a pyramidal shape given by the central axis pruning.

scaffold limbs/PSLs, and (3) locate the bearing mother branch at a density of 6–8 shoots/square meter.

Making trees shorter and reducing tree canopies through training and pruning enable the production of large nuts and decrease the damage caused by disease or insects through improving the quality of mother shoots by reducing the wood/leaf ratio of each tree and increasing the amount of sunlight that reaches the canopy. Furthermore, reducing height and tree canopies extends the number of productive years per tree by controlling the occurrence of chestnut gall wasp, since tree vigour is maintained, and both yield and nut quality are kept high.

Research suggests that the most productive chestnut orchards have 80% of the ground area covered by the canopy with 25–30% sunlight penetration through the canopy in midsummer (Sakuma, 1988). There are good correlations between the length and thickness of bearing mother shoots and the number of female flowers, sets of burrs and total and average nut weight per bearing mother shoot. To obtain a high yield and larger nuts, it is preferable to locate the bearing mother shoot at a length of 30–50 cm with a basal thickness of 6–8 mm. It is recommended that 6–8 bearing mother shoots should be located per square meter to grow large nuts. The lateral branch should be renewed every third year since its productivity decreases with age. It is possible to keep the tree height short and tree canopy small by cutting back old branches. In the case of renovating a senescent orchard with low yields, or to change a variety, old branches should be pruned near the trunk, leaving only one with the cut ends protected. The next year, when renovated branches are available, this last older branch should also be cut (Fig. 6). The renovation of old chestnut orchards is carried out by severely pruning the trees right above their grafting point in order to re-form the trees from the juvenile buds, which will regrow. The trees begin bearing after 3 years (Hennion, 2009).

Further training methods to lower tree height have recently been developed to reduce labour, which is becoming an increasing problem as a result of an ageing farming population (Kamio et al., 2003; Mizuta, 2009). These methods limit tree height to 2.5 m after pruning and pruning times by 20–50% compared to earlier methods. This is because they reduce the need for stepladders to reach higher branches while maintaining both the yield per unit area and nut size. Moreover, they were able to decrease the amount of chemical spraying by 25% compared with conventional methods. As with conventional techniques, it is preferable to locate the bearing mother shoots at around 1 m length at a density of 3-5 per square meter.

**Figure 6** Wild trees of *C. sativa*, under renovating process in order to allow the grafting of trees (Sortegel Company, Bragança-Portugal).

## 5.2 Care activities

### 5.2.1 Fertilization

Depending on the results of physical and chemical analysis, manure and mineral fertilizers need to be incorporated in the soil. In *C. sativa* plantations, the organic matter should be equal to or higher than 2% of overall soil content and ploughed to a depth of 20–30 cm. Mineral fertilizers containing phosphorus (10–20 ppm) and potassium (100–150 ppm) should be added as mineral superphosphate and potassium sulphate (Bounous, 2014).

In Japan, for 5-year-old trees, fertilizer should be applied in amounts of 60–70, 40–50 and 40–50 kg/ha/year for N, $P_2O_5$ and $K_2O$, and 140–160, 130–140 and 130–140 kg/ha/year for trees over 10 years old (Araki, 2004). It is recommended that fertilizers are applied three times per year: in spring for pre-planting (50% of the total N required), in summer (20%) and in the autumn (30%) as side dressing. Sod culture (where fertilizer is combined with allowing grass or weeds to cover the soil) is an affective technique since it supplies organic substances to the soil, improves the physical condition of the soil and prevents soil runoff. In order to prevent competition for nutrients and soil moisture between chestnut trees and weeds, it is necessary to mow around trees three times between the flowering season and harvesting.

Chestnut is a siliceous species. Silicon (Si) fertilization is a good strategy to help protect trees against both biotic and abiotic stresses (Carneiro-Carvalho et al., 2017). The application of Si acts as a biological inducer of plant defense responses against *Phytophthora cinnamomi* and promotes a mechanical barrier, restricting lesion size and reducing the production of spores for secondary infection by enhancing the amounts of antifungal phytoalexins as well as by activating antioxidative enzymes after infection. The presence of Si bodies called phytoliths in the leaf mesophyll has been identified as an important factor in increasing trees' tolerance of heat and drought stress.

### 5.2.2 Irrigation

The chestnut tree is traditionally cultivated in mountainous areas with different types of soils and climatic conditions that usually include dry, warm summers and cold and wet winters. Irrigation has not traditionally been an issue for chestnut cultivation since there has usually been enough precipitation in these conditions. Recently, however, many problems associated with small nut size have been caused by insufficient rain, especially during the growing period.

To deal with changing climatic conditions (longer, drier and hotter summers), as well as improve yields and nut quality, irrigation in chestnut orchards has become a standard practice. Warmer and drier summers halt tree growth and fruit development and increase the incidence of fruit abortion. Under such conditions, trees also become less vigorous and more susceptible to diseases and pests. Precision irrigation systems have been developed based on monitoring levels of stem water. On adult *C. sativa* trees, the threshold level has been defined at -1.2 MPa. Whenever water status falls below that level, irrigation is initiated until the reference value is reached (Mota et al., 2014). Since nut growth begins close to 45 days after pollination, which corresponds in Europe to the end of August, it is essential to provide an adequate level of water at this point, as well as during flowering. Mota et al. (2014) determined a water requirement of about 1,600 square meters of water/ha between flowering and the beginning of fruit fall. The introduction of irrigation systems has led to an increase in production of about 35%.

### 5.2.3 Weed management

Most growers weed before harvesting the chestnuts. In some countries, weed management is done by using herbicides or by soil tillage. However, continuous use of herbicides may have harmful effects on the environment. On the other hand, tillage can spread root rot disease to clean areas. Orchard floor management with grass is a preferred method.

### 5.2.4 Harvesting

Chestnut burs begin to dehisce between mid-August and mid-November in the Northern Hemisphere, depending on the production area and the cultivar shedding the nuts. Most chestnut varieties produce more than one nut in each burr and can produce as many as five. In France, chestnuts that have more than 10% of nuts with multiembryos are called 'chataigne', while nuts that have a single embryo are called 'marron'. Marron is a large, sweet, easy-peeling nut variety and is prized in Europe for use in sweet desserts (Wilkinson, 2005).

Chestnuts that drop naturally should be picked up daily during the harvest season to minimize fungal infection and growth and the loss of quality due to excessive drying and/or sunburn. Harvesting nuts is traditionally done by collecting from the ground or using a pole to dislodge the nuts. However, harvesting by pole can damage the buds and increase blight infections. Chemicals can also be used for preharvesting treatment. Preharvest or postharvest exposure of burs to 500–1000 ppm ethephon (ethylene-releasing liquid) accelerates dehiscence (Soylu et al., 1987; Kader, 2003).

In European countries such as Portugal, there has been a growth in mechanical harvesting to offset labour shortages. Mechanical harvesting can reduce the visual appearance of nuts as well as shelf-life. Mechanical harvesting can be done using self-propelled harvesters, trailed aspiration harvesters or mechanical mounted harvesters. The most-used harvesters are trailed harvesters as they are cheaper and more appropriate for smaller areas, with a harvesting capacity of about 150–400 kg nuts/h. In France, it is common to cover the soil with nets in conjunction with a static husk remover.

#### 5.2.4.1 Water soaking and curing pre-treatments before storage

Thermo-hydrotherapy is now the most common procedure for treating nuts to kill larvae and eggs and other parasites inside nuts. The method involves soaking the nuts in water at room temperature for about 4–10 days (Bounous, 2002). This method is still used on a large scale to preserve nuts for several months. When the tubs are emptied, the cleaned product is piled up for a few days and then dried in well-aerated rooms with porous floors on which the nuts are spread in layers of 30–40 cm, where they are manually shovelled for several days. Today, this operation is increasingly being mechanized.

More recently, the use of hot water at a temperature of 48°C has been introduced, shortening the time needed to treat nuts. The nuts are immersed in a cylindrical drum equipped with a rotator. They are 'cured' for 45 minutes to kill larvae and eggs without denaturing the nut protein. Finally, the nuts are briefly immersed in cold water to cool and separate the floating chestnuts and dried in a ventilated cylindrical drum.

### 5.2.5 Storage

Chestnuts are traditionally stored in open areas as bur stacks and covered with ferns or other plant materials (Serdar et al., 2014). In this case, growers store their nuts and then sell them based on market demand. Chestnuts are generally consumed after roasting. For chestnut candy production, companies generally store peeled chestnuts in freezers. Growers generally sell chestnuts directly to merchants, customers or cooperatives (Serdar et al., 2018; Bozoğlu et al., 2019).

Chestnuts can be stored at 2°C and 90–95% HR under normal atmosphere (NA) or controlled atmosphere (CA) by enriching $CO_2$ content by 5–10%, or by ultra-freezing. Drying is an ancient technique still used in countries such as Italy. Dried chestnuts can be stored for more than 12 months without deteriorating. Generally, this method is used for small-sized nuts or those destined for fresh consumption. Besides traditional techniques, drying is also carried out in modern electric resistance-heated or convection ovens, which are able to reduce drying times without affecting the quality of the product.

## 6 Disease and pest management

### 6.1 Diseases

Chestnut blight is a plant disease caused by the ascomycete fungus *Cryphonectria parasitica*. It is the major disease of chestnut due to the damage it does to *C. dentata* and (to a lesser extent) to *C. sativa*. The blight penetrates into the bark and destroys the bark and cambium, causing the death of the branches or the tree (Anagnostakis, 1987). Conidia and ascospores of *C. parasitica* are spread by wind and rain or transmitted by beetles and birds. Cultural, chemical and biological methods have been used to manage chestnut blight. Chemical control techniques have had limited effect (Aksoy and Serdar, 2004). Chestnut blight has killed nearly all American chestnut species, but European chestnut species have started to recover due to the natural occurrence of hypovirulent races. This suggests the best method to control the pathogen is biological control, which depends on the transfer of hypovirulence into virulent populations of *C. parasitica* (Heiniger and Rigling, 1994). In North America, a breeding program is being implemented to reintroduce chestnut by crossing *C. dentata* with *C. mollissima*.

In Europe, hypovirulent strains have been successfully used to control chestnut blight, resulting in the recovery of chestnut trees. Hypovirulence is the result of a viral disease caused by Cryphonectria HypoVirus (CHV-1), which reduces the virulence and sporulation capacity of the infected fungal strains. This can occur naturally, but, in many regions, efforts must be made to introduce the hypovirus artificially through inoculation of cankers with hypovirus-infected *C. parasitica* strains (Robin and Heiniger, 2001). It has been found that Vc groups are important for biological control. However, biological control is complicated when the number of Vc groups increases.

The removal of infected bark followed by disinfestation with hypochlorite and coverage by copper paste is also an efficient control method. Several chemical products such as carbendazim may also be used. Another important method often used to manage plant disease is the development of tolerant or resistant selections or varieties. It has been reported that Chinese (*C. mollissima*) and Japanese (*C. crenata*) chestnut trees are more resistant than other species. 'Marigoule' and especially 'Bouche de Bétizac' are less sensitive to canker than most *C. sativa* varieties (Hennion, 2009).

Another important disease of chestnut trees is root rot (*Phytophthora cambivora* and *P. cinnamomi*). It kills the root and/or rots the collar of trees and seedlings in nurseries, plantations and forests (Vannini and Vettraino, 2001). Integrated management of root rot disease is difficult and must take into consideration several variables and factors associated with its spread, incidence and severity. Tree and soil treatments and the use of resistant rootstocks are important for controlling root rot. This disease is responsible for the disappearance of millions of trees every year in Europe. This soil oomycete attacks the root system of fragile trees during the summer when the combination of drought and heat limits tree vigour and increases the aggressiveness of the oomycete. There is currently no efficient measure to control the disease in infected tress. Currently, silicates (Carneiro-Carvalho et al., 2017), acilamines, phosphites, salts or esters of phosphorous acid are used with some success owing to their symbiotic relationship with the tree's biological resistance mechanisms.

Many breeding programs in Europe have started to produce hybrid plants of *C. sativa* with *C. crenata* or *C. mollissima* resistant to ink disease (Santos et al., 2015). Some of those selections are today used as rootstock for new plantations or as direct producers. Leaf spot (*Marssonina ochroleuca*), twig canker (*Cryptodiaporthe castanea*) and ChMV are also diseases affecting chestnut. However, they do minor damage to chestnut cultivation compared to diseases such as root rot.

### 6.2 Pests

Chestnut gall wasp (*D. kuriphilus*) is the most dangerous pest threatening chestnut production. In recent years, the European chestnut sector has also had to deal with the Asian chestnut gall wasp *D. kuriphilus* (DK), an exotic and invasive pest. The first reports of the pest in Europe were from Italy in 2002, in the Cuneo Region. Since then, DK has dispersed to cover almost all of the European chestnut-growing regions. The larvae of this species induce the formation of greenish-red galls, which develop at the time of budburst in the early spring on new shoots, suppressing shoot elongation and causing twig dieback (Fig. 7). Severe reduction of fruiting has resulted in estimated yield losses of up to 85% over 3–4 years. The annual rate of spread of the gall wasp is about 25 km; however, one means of transmission (and the most dangerous) is via the distribution of infected nursery plants (Rieske, 2007; Graziosi and Santi, 2008).

As a preventative measure, scion woods should be soaked in water of 52°C for 10 min (Warmund, 2014) and nursery plant production should be done in areas where gall wasp is not present, or tulles should be used, which cover the whole plant. Research suggests that insecticides have a limited impact on gall wasps, but that they have a negative effect on beneficial insects (especially bees) essential to pollination, as well as being costly (Bosio et al., 2009; Quacchia et al., 2014). The most effective control method for chestnut gall wasp is biological control by the parasitoid *Torymus sinensis* Kamijo (Hymenoptera: Torymidae) in the infected area. This method was first used in Japan in the 1980s and has been successfully applied in Asia, America and Europe (Quacchia et al., 2008). *T. sinensis* has, for example, become established in many parts of Italy and has

**Figure 7** Damage of gall wasp in chestnut.

been successful in suppressing the chestnut gall wasp (Quacchia et al., 2014). The parasitism rate has been recorded at over 80%, with a drastic decline in the presence of galls on the plants (Quachia et al., 2013). The other control method for chestnut gall wasp is to use resistant cultivars (Anagnostakis, 2014). However, very few genotypes of *C. sativa* and *C. crenata* have been found to be resistant. In a study conducted in Italy, 'Bouche de Betizac' was found to be resistant to the chestnut gall wasp (Sartor et al., 2009; Botta et al., 2015). Finally, studies conducted in Turkey found the A14 chestnut genotype to be resistant to this pest (Çil, 2018).

Carpophagous insects such as chestnut tortrix moths (*Cydia splendana*, *C. fagiglandana*, *Pammene fasciana*) and the chestnut weevil (*Curculio elephas*) are some important insect pests that also affect chestnut production. Since the larvae feed on the starch content of the seed, they cannot be detected until they come out of the chestnut fruit and the typical hole they form becomes visible. The damage can sometimes be very extensive, affecting up to 60-70%

of the crop. For these kinds of pests, control techniques include storing nuts at -2°C for 4 weeks, soaking nuts in water at 50°C for 30 min and chemical spraying (deltamethrine and diflubenzuron) in late August to early September. A combination of agronomical and chemical practices and mating disruption treatment methods are commonly used to control the pests. Promising results have been achieved with the use of entomopathogenic organisms such as bacteria, fungi and nematodes. However, each technique has its disadvantages in terms of cost, treatment capacity and degree of control.

Over the past 10 years, tree losses have been observed due to an infestation of ambrosia beetles *Anisandrus (Xyleborus) dispar* (Coleoptera) (Greece). In Turkey, ambrosia beetles were reported to cause significant damage in hazelnut orchards (Ak et al., 2005a,b,c). However, in the absence of hazelnut trees, they can damage drupe, pome, chestnut and also some forest trees (Mani et al., 1986, 1990; Tsankov and Ganchev, 1988; Kaya, 2004; Ak et al., 2006; Pereira-Lorenzo et al., 2006; Toksöz et al., 2017). *X. dispar* damages trees by opening up galleries within trees. They mate inside and then leave their eggs in the galleries. The plant continuously leaks from these holes, causing withering and weaker growth (Fig. 8). Since ambrosia beetles remain in these galleries, chemical control is not feasible. In recent years, some growers have started to use traps. Ambrosia beetles tend to attack weaker or damaged plants, so growers hang some branches to the sides of the orchards before the vegetation period begins. Ambrosia beetles attack these branches instead of main plants. Then these branches are collected and burned. These traps can be placed until the end of the vegetation period. However, the best solution is to use resistant cultivars. To this end, breeding studies need to be conducted in order to obtain cultivars resistant to ambrosia beetles.

Peach moth (*Dichocrocis punctiferalis*) and goat moth (*Cossus cossus*) are currently important insect pests of chestnut in Japan and Turkey, respectively. Spraying with insecticides such as Cypeimethrin may be undertaken when adults emerge, using pheromone traps to identify when emergence occurs. White striped longicorn beetles and swift moths are also serious insect pests of the trunk and branch. Since white longicorn beetles usually lay eggs at the basal part of trunk from late July to the middle of August, an effective method is to target oviposition scars using hammers and to inject insecticide directly into oviposition holes during that period. Chemical control of moths is very difficult. Mowing regularly effectively reduces damage by swift moths by disrupting their habitat. An alternative method to chemical control is to strengthen the immune system of trees. Some chestnut producers in Turkey use a mixture of soil regulator to provide better resistance to pests and diseases, as well as to obtain better growth in chestnut seedlings and trees. Potato leafhopper (*Empoasca fabae*), Japanese beetle (*Popillia japonica*), Rose chafer (*Macrodactylus subspinosus*) and spider mites are less damaging than the pests mentioned earlier.

**Figure 8** Damage of ambrosia beetle in chestnut.

## 7 Future trends

Climate change is affecting agricultural system such as chestnut orchards and forest stands. Solutions are needed to adapt agricultural practices to rising temperatures and drought as well as the increasing occurrence of extreme weather events. Research is needed to better understand how genotype, soil biota and animals adapt to environmental constraints and why certain genetic resources are tolerant or resilient to biotic and abiotic stresses such as summer drought and warm winters. Rootstock breeding to improve dwarfing and resistance to root rot is a top priority. Graft incompatibility is a serious problem in introducing new cultivars and combining them with rootstocks. Early selection methods for graft compatibility need to be investigated. There also needs to be more research on pruning systems, bud management and high-density planting.

## 8 Where to look for further information

The best sources of information on chestnut research and cultivation are reported below:

- International Society for Horticultural Science (ISHS) (http://www.ishs.org/) in the Section Nuts and Mediterranean Climate Fruits and related Chestnut Working Group and in the peer reviewed series of the Proceedings of the International Congresses on Chestnut.
- Michigan State University (MSU) "Chestnut and related Extension Services" (https://www.canr.msu.edu/chestnuts/establishing_orchards/)
- European Chestnut Network (http://www.eurocastanea.org/)
- International Nut and Dried Fruit Council: INC and related publications (https://www.nutfruit.org/)

## 9 References


Ak, K., Uysal, M. and Tuncer, C. (2005a), 'Bark beetle (*Coleoptera: Scolytidae*) species which are harmful in hazelnut orchards, their short biology and densities in Giresun, Ordu and Samsun provinces of Turkey', *J. Fac. Agric.*, 20(2), 37–44.

Ak, K., Uysal, M. and Tuncer, C. (2005b), 'The injury level of bark beetles (*Coleoptera:Scolytidae*) in hazelnut orchards in Giresun, Ordu and Samsun Provinces of Turkey', *J. Fac. Agric.*, 22(1), 9–14.

Ak, K., Uysal, M., Tuncer, C. and Akyol, H. (2005c), 'The bark beetle species (*Col.: Scolytidae*) harmful on hazelnut in middle and east Black Sea Region of Turkey and their control strategies', *J. Fac. Agric.*, 19(37), 37–9.

Ak, K., Uysal, M. and Tuncer, C. (2006), 'Harmful bark beetles (*Coleoptera: Scolytidae*) species and struggle in kiwi in the Black Sea Region', Gaziosmanpaşa University, Agriculture Faculty, II. *National Kiwi and Grapes Fruit Symposium*, pp. 365–70.

Aksoy, H. M. and Serdar, U. (2004), 'A research on chemical control against chestnut blight (*Cryphonectria parasitica* (Murill) Barr)', *Plant Pathol. J.*, 3, 44–7.

Aldrey, A., Blanco, B., Bogo, B., Cuenca, B., Sánchez, C., Luquero, L., Ocaña, L., Mandujano, M. and Vidal, N. (2018), 'Photomixotropic and photoautotrophic micropropagation of *Phytophthora* resistant chestnut genotypes using liquid media', *Acta Hortic*, 1220, 177–84.

Anagnostakis, S. L. (1987), 'Chestnut blight - the classical problem of an introduced pathogen', *Mycologia*, 79, 23–37.

Anagnostakis, S. L. (2014), 'Asian chestnut gall wasp on Connecticut chestnut trees', http://www.ct.gov/caes/lib/caes/documents/publications/fact_sheets/plant_pathology_and_ecology/asian_chestnut_gall_wasp_on_connecticut_chestnut_trees.pdf.

Araki, H. (2004), *Chestnut Practical Handbook*, Noubunkyo, Tokyo (In Japanese).

Araki, H. and Nakaoka, T. (1982), 'Effects of severity of renovation pruning on tree growth and nut yield and size in chestnut tree (*Castanea crenata* Sieb. et Zucc.)', *J. Jpn. Soc. Hortic. Sci.*, 51, 278–85.

Ballester, A., Sanchez, M. C. and Viéitez, A. M. (1992), 'New strategies for *in vitro* propagation of chestnut', *Proceedings of the World Chestnut Industry Conference*, Morgantown, USA, pp. 32–40.

Beakbane, A. B. (1961), 'Structure of the plant stem in relation to adventitious rooting', *Nature*, 192, 954–5.

Beccaro, G. L., Mellano, M. G. and Bounous, G. (2017), 'Advances in high density chestnut plantations and clonal rootstocks: re-thinking a *Castanea sativa* cultivation model', *6th International Chestnut Symposium - Book of Abstracts*, p. 7.

Beccaro, G. L., Mellano, M. G., Rossi, J., Chiavazza, P. M. and Donno, D. (2018), 'Evoluzione della castanicoltura intensiva con portinnesti ibridi clonali', *Rivista di Frutticoltura e di Ortofloricoltura*, 3, 60–2.

Bonga, J. M. and Durzan, D. (Eds). (1982), *Tissue Culture in Forestry*, Springer Science & Business Media.

Bosio, G., Gerbaudo, C. and Piazza, E. (2009), *Dryocosmus kuriphilus* Yasumatsu: an outline of seven years after the first report in Piedmont (Italy)', in *A Global Serious Pest of Chestnut Trees, Dryocosmus Kuriphilus: Yesterday, Today and Tomorrow*, Proceedings of the Japan-Italy Joint International Symposium, Tsukuba, Japan, 24–25 November, pp. 3–13.

Botta, R., Sartor, C., Morinani, T., Mellono, M. G., Beccoro, G. L., Alma, A., Dini, F. and Quacchia, A. (2015), 'Impact of the Asian wasp *Dryocosmus kuriphilus* (Yasumatsu) oncultivated chestnut: yield loss and cultivar susceptibility', *Scientia Horticulturae*, 197, 454–60.

Bounous, G. (2002), *Il Castagno*, Coltura, ambiente e utilizzazioni in Italia e nel mondo.

Bounous, G. (2014), *Il Castagno*, Edagricole.

Bounous, G. and Torello Marinoni, D. (2004), 'Chestnut: botany, horticulture, and utilization', in *Horticultural Reviews ISHS* (vol. 31), Wiley and Sons, New York, pp. 291–347.

Bozoğlu, M., Başer, U., Eroğlu Alhas, N. and Topuz Kılıç, B. (2019), 'Developments in the chestnut market of Turkey', *KSU J. Agric. Nat.*, 22(1), 19–25.

Canciani, L., Dellavalle, E., Zambonelli, A. and Zecchini D'Aulerio, A. (1993), 'Prove di protezione chimica su innesti di castagno', *Proc.* Intl. Cong., *Chesnut*, Spoleto (PG), Italy, pp. 235–8.

Carneiro-Carvalho, A., Pereira, C., Marques, T. and Martins, L. (2017), 'Potential of silicon fertilization in the resistance of chestnut plants to ink disease (*Phytophthora cinnamomi*)', *International Journal of Environment, Agriculture and Biotechnology (IJEAB)*, 2(5), 2740–53.

Casey, J. and Casey, B. (2009), 'Australia', in Avanzato, D. (Ed.), *Following Chestnut Footprints (Castanea spp.) Cultivation and Culture, Folklore and History, Tradition and Uses*, ISHS, Leuven, pp. 14–19.

Çil, Y. (2018), 'Determination of sensitivities of some chestnut cultivars and genotypes to chestnut gal wasp', Master thesis (in Turkish), Ondokuz Mayıs University.

Costa, R., Santos, C., Tavares, F., Machado, H., Gomes-Laranjo, J., Kubisiak, T. and Nelson, C. D. (2011), 'Mapping and transcriptomic approches implemented for understanding disease resistance to *Phytophthora cinammomi* in *Castanea spp.*', *BMC Proc.*, 5.

Cuenca, B. and Majada, J. (2012), '*Castanea sativa* Mill', in García, J. P., Navarro Cerrillo, R., Nicolás Peragón, J. L., Prada Sáez, M. A. and Serrada Hierro, R. (Eds), Producción *y manejo de semillas y plantas forestales. Tomos I y II. Organismo autónomo Parques Nacionales*, Ministerio de Agricultura, Alimentación y Medio Ambiente.

Cuenca, B., Rodriguez, L., Cámara, M. J. and Ocaña, L. (2005), 'Micropropagación de ejemplares adultos de *Castanea sativa* Mill. seleccionados por resistencia a

*Phytophthora cinnamomi*', in *Actas de la VI Reunión de la Sociedad Española de Cultivo in vitro de Tejidos Vegetales*, pp. 11-13.

Cuenca, B., Ocaña, L., Salinero, C., Pintos, C., Mansilla, J. P. and Rial, C. (2009), 'Selection of *Castanea sativa* Mill. for resistance to *Phytophthora cinnamomi*: micropropagation and testing of selected clones', in *I European Congress on Chestnut-Castanea 866*, pp. 111-19.

Cuenca, B., Sánchez, C., Aldrey, A., Bogo B., Blanco, B., Correa, B. and Vidal, B. (2017), 'Micropropagation of axillary shoots of hybrid chestnut (*Castanea sativa* × *C.* crenata) in liquid medium in a continuous immersion system', *Plant Cell Tiss. Organ Cult.*, 131, 307-20.

Cuenca, B., Lario, F. J., Luquero, L., Ocaña, L. and Mandujano, M. (2018), 'Early grafting of chestnut by green grafting', *Acta Hortic.*, 1220, 141-8.

Diamandis, S. (2009), 'Greece', in Avanzato, D. (Ed.), *Following Chestnut Footprints (Castanea spp.) Cultivation and Culture, Folklore and History, Tradition and Uses*, ISHS, Leuven.

Donno, D., Beccaro, G. L., Maghenzani, M., Mellano, M. G., Cerutti, A. K. and Bounous, G. (2014), 'Nuove tecniche per la propagazione agamica mediante talea di ibridi di *Castanea* spp', *Corylus & Co.*, 1, 33.

Duman, E. and Serdar, U. (2006), 'A research on shortening the nursery period in grafted chestnut', *Horticultural Science (Prague)*, 33(1), 16-22.

Edwards, R. A. and Thomas, M. B. (1980), 'Observations on physical barrier to root formation in cuttings', *Plant Propagat.*, 26, 6-8.

Elkins, J. R., Griffin, G. J. and Stipes R. J. (1978), 'Blight development and methyl-2-benzimidazole carbammate levels in bark tissues of American chestnut trees following sol injection of benomyl', in *Proc. Am. Chestnut Symp.*, W. V. University Books, Morgantown, WV, pp. 73-9.

Errea, P., Garay, L. and Marin, J. A. (2001), 'Early detection of graft incompatibility in apricot (*Prunus armeniaca*) using *in vitro* techniques', *Physiol. Plant.*, 112, 135-41.

Ertan, E. (1999), 'Investigations on the determination of rootstock characteristics of aegean region chaestnut (*Castanea sativa* mill.) types obtained selection', PhD thesis (in Turkish), Adnan Menderes University.

FAOSTAT. (2019), 'World chestnut production numbers', http://www.fao.org/faostat/en/#data (accessed on 01 February 2019).

Fernández-López, J. and Alia, R. (2003), *EUFORGEN Technical Guidelines for Genetic Conservation and Use for Chestnut (Castanea sativa)*, International Plant Genetics Research Institute, Rome.

Fernández-López, J., Miranda-Fontaíña, M. E. and Pereira-Lorenzo, S. (1992), 'Fog and substrate conditions for chestnut propagation by leafy cuttings', En: *Proceeding Mass Production Technology for Genetically Improved Fast-growing Forest Tree Species I*, AFOCEL/IUFRO, Bordeux, pp. 379-83.

Ferrini, F., Mattii, G. B., Nicese, F. P. and Pisani, P. L. (1993), 'A breeding program for chestnut rootstocks: preliminary results', *Proc. International Congress on Chestnut*, 20-23 October, Spoleto, Italy, pp. 365-7.

Gomes-Laranjo, J., Dinis, L. T., Marques, T., Mota, M., Carvalho, A., Pinto, T., Anjos, R., Martins, L., Marques, G., Gaspar, M. J., Gonzalez Pereira, M., Raimundo, F. and Ferreira-Cardoso, J. (2018), 'Increasing chestnut resilience to climate change with innovative management practices', *Acta Hortic.*, 1220, 163-76.

Gonçalves, J. C., Diogo, G. and Amâncio, S. (1998), '*In vitro* propagation of chestnut (*Castanea sativa* x *C. crenata*): effects of rooting treatments on plant survival, peroxidase activity and anatomical changes during adventitious root formation', *Sci. Hortic. (Amsterdam)*, 72, 265–75.
Graziosi, I. and Santi, F. (2008), 'Chestnut gall wasp (*Dryocosmus kuriphilus*): spreading in Italy and new records in Bologna province', *Bull. Insectol.*, 61, 343–8.
Gresshoff, P. M. and Doy, C. H. (1972), 'Development and differentiation of haploide *Lycopersicum esculentum*', *Planta*, 107, 161–70.
Heiniger, U. and Rigling, D. (1994), 'Biological control of chestnut blight in Europe', *Annu. Rev. Phytopathol.*, 32, 581-99.
Hennion, B. (2009), 'France', in Avanzato, D. (Ed.), *Following Chestnut Footprints (Castanea spp.) Cultivation and Culture, Folklore and history, Tradition and Uses*, ISHS, Leuven.
Hiyama, H. (1978), 'Training and pruning', in Isaki, M. (Ed.), *Theory and Practice of Chestnut Cultivation*, Hakuyusya, Tokyo, 263-96 (In Japanese).
Horimoto, M. and Araki, H. (1999), 'Effect of restricted water uptake on freezing injury of Japanese chestnut (*Castanea crenata* Sieb. et Zucc.) trees (in Japanese with English summary)', *J. Agr. Meteorol.*, 55, 329-36.
Inoue, H., Kusaba, S., Sakamoto, D., Mizuta, Y. and Kamio S. (2014), 'Influence of nitrogen fertilization during winter on freezing injury of chestnut trees', *Hortic. Res. (Japan)*, 13(2), 373 (In Japanese).
Jaynes, R. A. and Anagnostakis, S. R. (1971), 'Inhibition of *Endothia parasitica* in fieldgrown American chestnut trees' *Plant Dis. Rptr.*, 55, 199–200.
Jaynes, R. A. and Van Alfen, R. K. (1977), 'Control of the chestnut blight fungus whit injected methil-2-benzimidazolecarbamate', *Plant Dis. Rptr.*, 61, 1032–6.
Kader, A. A. (2003), 'A perspective on postharvest horticulture (1978-2003)', *HortScience*, 38(5), 1004–8.
Kamio, S. and Mizuta, Y. (2013), 'Study on inhibitory effect of the freezing injury of young chestnut tree by high ridge and sheet mulching', *Hortic. Res. (Japan)*, 13(1), 98 (In Japanese).
Kamio, S., Taguchi, M. and Yanase S. (2003), 'Studies on super low height cultivation of chestnuts (1) Effect of pruning on the nut yields and size', *Bull. Gifu Pref. Res. Inst. Agr. Sci. Hill. Mount. Are.*, 2, 27–32 (In Japanese).
Kaya, M. (2004), 'Studies on population fluctuations of adults of *Xyleborus dispar* (F.) (*Coleoptera: Scolytidae*) on different fruits in Bursa Province', *J. Agric. Sci.*, 14, 113–17.
Mani, E., Remund, U. and Schwaller, F. (1986), 'Alkolfalle Zur Flugkontrolle und Befallsreduktion Beim Unleichen Holzbohrer', *Sparatdruck Aus Der. Zeitschrift für Obst-und Weinbau*, 122(7), 203–7.
Mani, E., Remund, U. and Schwaller, F. (1990), 'Der Ungleiche Holzbohrer, *Xyleboryus dispar* F. (*Coleoptera: Scolytidae*) im Obst-und Weinbay', *Landwirtschaft Schweiz Band*, 3(3), 105–12.
Marx, D. H., Hatch, A. B. and Mendicino, J. F. (1977), 'High soil fertility decreases sucrose content and susceptibility of loblolly pine roots to ectomycorrhizal infection by *Pisolithus tinctorius*', *Can. J. Bot.*, 55(12), 1569–74.
Mellano, M. G. and Donno, D. (2015), 'Chestnut Regional Center: new techniques for the propagation', *Castanea*, 3, 8–9.
Mellano, M. G., Beccaro, G. L., Donno, D., Torello Marinoni, D., Boccacci, P., Canterino, S., Cerutti, A. K. and Bounous, G. (2012), '*Castanea* spp. biodiversity conservation:

collection and characterization of the genetic diversity of an endangered species', *Genet. Resour. Crop Evol.*, 59, 1727–41.
Mellano, M. G., Donno, D. and Beccaro, G. L. (2015), 'Innovations in nursery: the micrografting of *C. sativa* and hybrids', *Castanea*, 4, 10–11.
Mellano, M. G., Torello Marinoni, D., Boccacci, P., Donno, D., Cerutti, A. K. and Beccaro, G. L. (2018), '*Ex situ* conservation and characterization of the genetic diversity of *Castanea* spp.', *Acta Hortic.*, 1220, 1–6.
Miranda, E. and Fernandez, J. (1992), 'Micropropagation as nursery technique in chestnut compared with stooling', in Wallace, R. D. and Spinella, L. G. (Eds), *Conference: World Chestnut Industry*, Morgantown, WA. Chestnut Marketing Association, USA, pp. 54–60.
Mizuta, Y. (2009), 'Super shrub training cultivation of chestnut by Hyogo method. Saishin nougyou gijutsu (The latest agricultural technique)', *Fruit Tree*, 2, 315–26. Nobunkyo, Tokyo (In Japanese).
Mizuta, Y. and Oribe, F. (2015), 'Relationship between locational conditions and the occurrence of freezing injuries in Japanese chestnuts in Hyogo prefecture', *Hyogo. Pre. Tech. Cent. Agr. Forest. Fish. (Agriculture)*, 63, 20–24 (In Japanese with English summary).
Mota, M., Pinto, T., Marques, T., Borges, A., Raimundo, F., Veiga, V., Caço, J., Martins, A. and Gomes-Laranjo, J. (2014), 'Watering chestnuts orchards in Northeast Portugal - preliminary study', *Acta Hortic.*, 1043, 129–33.
Murashige, T. and Skoog, F. (1962), 'A revised medium for rapid growth and bio assays with tobacco tissue cultures', *Physiologia Plantarum*, 15(3), 473–97.
Oakes, A. D., Powell, W. A. and Maynard, C. A. (2013), 'Doubling acclimatization survival of micropropagated American chestnuts with darkness and shortened rooting induction time', *J. Environ. Hortic.*, 31, 77–83.
Ocaña, L., Santos, M. I., Gómez, J. A. and Cuenca, B. (2001), 'Reproducción en vivero de castaños híbridos resistentes a la tinta mediante estaquillado semiherbáceo', in *Congresos Forestales*.
Ohata, T. and Sato, T. (1961), 'Studies on pollinations of chestnut trees', *Bull. Nat. Inst. Agric. Sci., Ser. E.*, 9, 185–93.
Okada, Y. (1969), 'Investigetion on the grafting rootstock of chestnut', *Bull. Okayama Forest. Stn.*, 9, 128–31 (In Japanese).
Özkarakaş, İ. and Önal, M. K. (1997), 'Kestane (*Castenea sativa* Mill.) Çoğaltımında En Uygun Göz Aşı Yöntemi ve Zamanının Belirlenmesi Üzerine Araştırmalar', *Anadolu. J. AARI*, 7(2), 74–9.
Ozturk, A. and Serdar, U. (2011), 'Effects of different nursery conditions on the plant development and some leaf characteristics in Chestnuts (*Castanea sativa* Mill.)', *Austral. J. C. Sci.*, 5(10), 1218–23.
Ozturk, A., Serdar, Ü. and Balcı, G. (2009), 'The influence of different nursery conditions on graft success and plant survival using the inverted radicle grafting method on the chestnut', *Acta Hortic.*, 815, 193–8.
Pereira, J. G., Pinto de Abreu, C. and Valdiviesso, T. (1993), 'Chestnut selection in Portugal evaluation of some characteristics', in *Proceedings of the International Congress on Chestnut*, Spoleto, Italy, pp. 357–60.
Pereira-Lorenzo, S. and Ramos-Cabrer, A. M. (2004), 'Chestnut, and old crop with future', in Dris, R. and Mohan Jain, S. (Ed.), *Production Practices and Quality Assessment of Food Crops, Volume 1, Preharvest Practice*, Kluwer Academic Publishers, pp. 151–61.

Pereira-Lorenzo, S., Ramos-Cabrer, A. M., Díaz-Hernández, M. B., Ciordia-Ara, M. and Ríos-Mesa, D. (2006), 'Chemical composition of chestnut cultivars from Spain', *Sci. Hortic.*, 107(3), 306–14.

Qin, L. and Feng, Y. (2009), 'China', in Avanzato, D. (Ed.), *Following Chestnut Footprints (Castanea spp.) Cultivation and Culture, Folklore and History, Tradition and Uses*, ISHS, Leuven.

Quacchia, A., Moriya, S., Bosio, G., Scapin, I. and Alma, A. (2008), 'Rearing, release and settlement prospect in Italy of *Torymus sinensis*, the biological control agent of the chestnut gall wasp *Dryocosmus kuriphilus*', *BioControl*, 53, 829–39.

Quacchia, A., Ferracini, C., Nicholls, J. A., Piazza, E., Saladini, M. A., Tota, F., Melika, G. and Alma, A. (2013), 'Chalcid parasitoid community associated with the invading pest *Dryocosmus kuriphilus* in North-western Italy', *Insect Conserv. Diver.*, 6, 114–23.

Quacchia, A., Moriya, S. and Bosio, G. (2014), 'Effectiveness of *Torymus sinensis* in the biological control of *Dryocosmus kuriphilus* in Italy', in *II European Congress on Chestnut*, 1043, pp. 199–204.

Rieske, L. K. (2007), 'Success of an exotic gallmaker, *Dryocosmus kuriphilus*, on chestnut in the USA: a historical account', *EPPO Bull.*, 37, 172–4.

Rinallo, C., Gellini, R. and Fabbri, A. (1987), 'Studies on rhizogenesis in *Castanea sativa* Mill. cuttings', *Adv. Hort. Sci.*, 1, 27–33.

Robin, C. and Heiniger, U. (2001), 'Chestnut blight in Europe: diversity of *Cryphonectria parasitica*, hypovirulence and biocontrol', *For. Snow Landsc. Res.*, 76, 361–7.

Rodriguez, L., Cuenca, B. and Pato, B. (2005), 'Producción *in vitro* y mediante estaquillado semiherbáceo de híbridos de castaño en un vivero comercial: Coste y eficiencia de ambos sistemas', in *Congresos Forestales*.

Sakai, A. (1967), 'Freezing injury of fruit trees occur near the ground', *Agr. Hortic.* 42(11), 49–53 (in Japanese).

Sakamoto, D., Inoue, H. and Kusaba S. (2015), 'Effect of soil moisture conditions during the period from late autumn to early spring on the freezing tolerance of the Japanese chestnut (*Castanea crenata* Sieb. et Zucc)', *Bul1. NARO Inst. Fruit Tree Sci.*, 20, 21–8.

Sakuma, F. (1988), 'Current situation and issues of low tree height training cultivation of chestnut', *Res. J. Food Agr.*, 11(4), 49–53 (In Japanese).

Sánchez, M. C. and Viéitez, A. M. (1991), '*In vitro* morphogenetic competence of basal sprouts and crown branches of mature chestnut', *Tree Physiol.*, 8, 59–70.

Sánchez, M. C., San-José, M. C., Ferro, E., Ballester, A. and Vieitez A. M. (1997), Improving micropropagation conditions for adult-phase shoots of chestnut', *J. Hortic. Sci.*, 72, 433–43.

San-José, M. C. and Viéitez, E. (1984), 'Morfogénesis en secciones de plántulas de castaño cultivadas *in vitro*', *Anales Edafología y Agrobiología*, 43, 587–97.

Santamour, F. S. (1988), 'Graft incompatibility related to cambial peroxidaz isozymes in Chinese chestnut', *J. Environ. Hortic. (USA)*, 6(2), 33–9.

Santos, C., Machado, H., Correia, I., Gomes, F., Gomes-Laranjo, J. and Costa R. (2015), 'Phenotyping *Castanea* hybrids for *Phytophthora cinnamomi* resistance', *Plant Pathol.*, 64, 901–10.

Sartor, C., Torello Marinoni, D., Quacchia, A. and Botta, R. (2009), 'Genes involved in chestnut response to infestation by *Dryocosmus kuripilus* (Yasumatsu, *Hymenoptera:Cynipidae*)', in *Proceedings of the 53rd Italian Society of Agricultural Genetics Annual Congress*, Torino, Italy.

Seabra, R. C. and Pais, M. S. (1998), 'Genetic transformation of European chestnut', *Plant Cell Rep.*, 17, 177–82.

Serdar, U. and Soylu, A. (2005), 'The effect of grafting time and methods on chestnut nursery tree production', *Acta Hortic.*, 693, 187–94.

Serdar, U., Beyhan, N., Demirsoy, L. and Demirsoy, H. (2010), 'Some phenological properties of chestnut genotypes in the Black Sea Region, Turkey', *Acta Hortic.*, 866, 135–41.

Serdar, U., Akyuz, B. and Fulbright, W. D. (2014), 'Graft success of hybrids on European chestnut rootstock and development of chestnut blight disease', in *2nd Symposium of Turkey Forest Entomology and Pathology*, 7–9 April, Antalya, Turkey. Symposium Proceeding, pp. 127–31.

Serdar, Ü., Akyüz, B., Ceyhan, V., Hazneci, K., Mert, C., Er, E., Ertan, E., Çoskuncu Savaş, K. S. and Uylaşer, V. (2018), 'Horticultural characteristics of chestnut growing in Turkey', *Erwerbs-Obstbau*, 60(3), 239–45.

Shimura, I., Kanato, K. and Yasuno, M. (1969), 'Study on Japanese chestnut rootstock. The graft compatibility between Japanese chestnut and other species rootstock', *Agr. Hortic.*, 44, 1285–6 (In Japanese).

Solar, A., Grecs, Z., Seljak, G., Osterc, G., Stampar, F. and Jurc D. (2009), 'Slovenia', in Avanzato, D. (Ed.), *Following Chestnut Footprints (Castanea spp.) Cultivation and Culture, Folklore and History, Tradition and Uses*, ISHS, Leuven.

Solignat, G. and Chapa, J. (1975), *Les porte-greffes du chataignier C. sativa*, INRA, France, No. 393.

Soylu, A. (2004), *Kestane Yetiştiriciliği ve Özellikleri (Chestnut Cultivation)* (in Turkish), Hasad Publishing, İstanbul.

Soylu, A. and Serdar, U. (2000), 'Rootstock selection on chestnut (*Castanea sativa* Mill.) in the middle of Black Sea Region in Turkey', *Acta Hortic.*, 538, 483–7.

Soylu, A., Eriş, A. and Sermenli, T. (1987), *Kestanelerde hasatın kolaylaştırılması amacıyla ethephon'dan (2-(2-Chloroethylphosphonicacid) yararlanma imkanları üzerinde bir araştırma*, Uludağ Üniversitesi Basımevi.

Soylu, A., Eris, A., Ozgur, M. and Dalkilic, Z. (1999), 'Researches on the rootstock potentiality of chestnut types (*Castanea sativa* Mill.) grown in Marmara Region', in *Proc. 2nd Int. Symp. on Chestnut.*, Acta Hortic., 494, 213–21.

Tetsumura, T. and Yamashita, K. (2004), 'Micropropagation of Japanese chestnut (*Castanea crenata* Sieb and Zucc.) seedlings', *HortScience*, 39(7), 1684–7.

Toksöz, S., Akyüz, B., Saruhan, İ., Akça, İ., Serdar, U. (2017), 'Determination of the damage ratios of Ambrosia beetles (*Coleoptera: Curculionidae: Scolytinae*) on hybrid chestnuts', *J. Zool. Stud.*, 4(5), 19–22.

Torikata, H. and Higuchi, H. (1962), 'Studies on the grafting congeniality in chestnut trees. I. Influence of rootstocks on the growth and mineral composition (in Japanese with English summary)', *J. Jpn. Soc. Hort. Sci.*, 31, 115–22.

Tsankov, G. and Ganchev, P. (1988), 'Attack by Dryocoetes (Anisandrus) dispar on *Castanea mollissima* in Bulgaria', *Gorsko Stopanstvo*, 44(5), 16–17.

Tsukamoto, N. (1985), *Tree Form and Training. Compendium of Fruit Trees. Chestnut, Walnut, Cherry and Apricot*, Noubunkyo, Tokyo, pp. 163–4 (In Japanese).

Tsukamoto, N. and Tanahashi, H. (1983), 'Study on the establishment of low tree height training cultivation of chestnut', *Res. Bul. Gifu. Chu-sankanchi Agr. Exp. Stn.*, 3, 18–34 (In Japanese).

Vannini, A. and Vettraino, A. M. (2001), 'Ink disease of chestnut: impact on European chestnut', *For. Snow Landsc. Res.*, 76, 345–50.

Vazquez, A., Gesto, M. D. V. and Vieitez, E. (1978), 'A growth inhibitor from *Castanea sativa* Mill, cuttings', *Biol. Plant.*, 20, 146–8.

Vidal, N., Correa, B., Rial, E., Regueira, M., Sánchez, C. and Cuenca, B. (2013), 'Comparison of temporary and continuous immersion systems for micropropagation of axillary shoots of chestnut and willow', in *VIII International Symposium on In Vitro Culture and Horticultural Breeding 1083*, pp. 227–33.

Vidal, N., Blanco, B. and Cuenca, B. (2015), 'A temporary immersion system for micropropagation of axillary shoots of hybrid chestnut', *Plant Cell Tiss. Organ Cult.*, 123(2), 229–43.

Vieitez, E. (1981), 'Current knowledge of the physiology of the vegetative propagation of chestnut', in *International Union of Forest Research Organizations XVII IUFRO World Congress Proceeding*, 17(2), pp. 61–71.

Vieitez, A. M. and Vieitez, E. (1980), 'Plantlet formation from embryonic tissue of chestnut grown *in vitro*', *Physiol. Plant.*, 50, 127–31.

Viéitez, F. J. and Ballester, A. (1988), 'Effect of etiolation and shading on the formation of rooting inhibitors in chestnut trees', *Phyton*, 48(1–2), 13–19.

Vieitez, F. J. and Ballester, A. (1989), 'Effect of etiolation and shading on the formation of rooting inhibitors in chestnut trees', *Hort. Abst.*, 59, 4604.

Vieitez, M. L. and Vieitez, A. M. (1982), 'Observaciones sobre el injerto juvenil del castano', *Anales de Edafologia y Agrobiologia*, 41(9–10), 1999–2002.

Vieitez, J., Kingston, D. G. I., Ballester, A. and Vieitez, E. (1987), 'Identification of two compound correlated with lack of rooting capacity of chestnut cuttings', *Tree Physiol.*, 3(3), 247–55.

Vieitez, A. M., Sanchez, C. and San-José, C. (1989), 'Prevention of shoot - tip necrosis in shoot cultures of chestnut and oak', *Sci. Hortic. (Amsterdam)*, 41, 151–9.

Villani, F., Sansotta, A., Cherubini, M., Donatella, C. and Valerio, S. (2001), 'Genetic structure of natural populations of *Castanea sativa* in Turkey: evidence of a hybrid zone', *J. Evol. Biol.*, 12, 233–44.

Warmund, M. R. (2014), 'Disinfestation of *Dryocosmus kuriphilus* Yasumatsu in *Castanea* scion wood', *Acta Hortic.*, 1019, 243–7.

Wilkinson, J. (2005), *Nut Grower's Guide: the Complete Handbook for Producers and Hobbyists*, Landlinks Press, Australia.

# Chapter 14

# Advances in breeding of pistachio

*Salih Kafkas, University of Çukurova, Turkey*



## 1 Introduction

The genus *Pistacia* L. is a member of the Anacardiaceae family that includes mango, cashew, poison oak, poison ivy, pepper tree and sumac. The genus consists of 11 or more species (Kafkas, 2006a) and is estimated to be approximately 80 million years old (Parfitt and Badenes, 1997). *P. vera* L. is the only commercially important species due to its edible nuts. Other species grow in natural areas and several have been used as rootstock seed sources for *P. vera*, with most also being used as forest trees. Iran, the United States, Turkey and Syria are currently the main pistachio producers in the world, contributing over 90% of the worldwide production (FAOSTAT, 2018).

*Pistacia* species are dioecious and wind pollinated; however, monoecious individuals within *P. atlantica* have also been reported (Kafkas et al., 2000). *P. vera* is believed to be the most ancestral species in the genus and other species are probably derived from it (Zohary, 1952). There are two centres of diversity of cultivated pistachios. The first comprises the Mediterranean region of Europe, Northern Africa and Middle East countries. The second comprises the Eastern part of the Zagros Mountains from Crimea to the Caspian Sea. Pistachio cultivation extended westward from its centre of origin to Italy, Spain and other

http://dx.doi.org/10.19103/AS.2018.0042.17

Mediterranean regions of Southern Europe, North Africa and the Middle East, in addition to China and more recently to the United States and Australia (Maggs, 1973; Hormaza et al., 1994a, 1998).

The primary difficulties encountered during breeding and genetic studies of *Pistacia* include its alternate bearing, dioecious character and an extensive juvenile period. As a result of the latter two factors, large investments of time and land are required for the characterization and evaluation of progenies in a breeding program. Furthermore, genetic control of the most economically significant pistachio traits is not clearly understood, including disease and pest resistance, yield, nut quality characteristics and alternate bearing. However, current technologies such as molecular markers are able to overcome such difficulties in breeding programs.

Deoxyribonucleic acid (DNA) markers have played a major role in breeding programs for several decades. Numerous molecular markers such as randomly amplified polymorphic DNA (RAPD; Kafkas and Perl-Treves, 2001), simple sequence repeat (SSR; Zaloglu et al., 2015), sequence-related amplified polymorphism (SRAP; Ahmad et al., 2005), amplified fragment length polymorphism (AFLP; Kafkas, 2006b), inter-SSR (ISSR; Kafkas et al., 2006a), selectively amplified microsatellite polymorphic loci (SAMPL; Karimi and Kafkas, 2011) and single nucleotide polymorphisms (SNP; Kafkas et al., 2015) have been used to assess genetic diversity, fingerprinting, phylogenetic relationships, germplasm characterization, sex determination and genetic linkage mapping, as well as enabling quantitative trait locus (QTL) analysis, in cultivated and wild *Pistacia* species.

Most pistachio cultivars used in production are earlier selections by growers or breeders. Pistachio cultivars from controlled crosses are limited in number, and selected genotypes in current breeding programs are under evaluation in different countries. Using molecular tools in conventional breeding programs could enable the advancement of pistachio production. Initial molecular studies involving pistachios were mainly focused on germplasm characterization using different molecular marker techniques. The identification of molecular markers linked to sex determination that allowed the early selection of females in a cultivar breeding program has also been studied (Hormaza et al., 1994b; Kafkas et al., 2015; Khodaeiaminjan et al., 2017). Recently, the markers linked to sex have been used in a cultivar breeding program in Turkey (Kafkas et al., 2017a).

The pistachio is a diploid plant which has a haploid chromosome number of $n = 15$ (Ila et al., 2003). It is also a highly heterozygous species due to its dioecious character (Motalebipour et al., 2016). The first genetic linkage map was constructed by Turkeli and Kafkas (2013) who used an inter-specific F1 population derived from a cross between *P. vera* cv. *Siirt* and *P. atlantica* (monoecious genotype Pa-18). Recently, Khodaeiaminjan et al. (2018)

constructed the first SSR-based linkage map of pistachios using an intra-specific F1 population by developing numerous novel SSR markers in silico. In addition, Motalebipour et al. (2018) constructed a genetic linkage map and performed the first QTL analysis of pistachios using an inter-specific F1 population.

Next-generation sequencing (NGS) has provided a new perspective for research, owing to its high throughput and speed of data generation, which has enabled low-cost whole-genome and transcriptome sequencing in a short time. Motalebipour et al. (2016) performed a genome survey study of pistachios and estimated the genome size to be 600 Mb. NGS technology was also used to identify the whole genome (Kafkas et al., 2017b) and transcriptome (Jazi et al., 2017) sequences of pistachios. In this chapter, germplasm resources, taxonomy and phylogenetic analysis, marker development, uses of molecular markers in germplasm characterization, achievements in cultivar and rootstock breeding, as well as advances in micropropagation of pistachios will be discussed.

## 2 Breeding objectives

The progress of traditional pistachio breeding is very slow and is longer than other tree-crop species due to a very long juvenile period. The pistachio is a dioecious species, and breeding objectives differ for female and male cultivars in a breeding program. Furthermore, the seeds of pistachio cultivars and other *Pistacia* species have been used as rootstock sources for cultivated pistachios; therefore, the breeding objectives for female, male and rootstock cultivars are discussed separately in this section.

### *2.1 Female cultivar breeding*

#### *2.1.1 Yield*

As with other plant species, a high and regular yield is a prerequisite for a cultivar in a pistachio breeding program. Obtaining a regular yield in each growing season is a major problem as all pistachio cultivars have a tendency for alternate bearing. Nut production fluctuates between an 'On' year of high yields and an 'Off' year of low or no yields. Pistachio growers perform regular orchard management in both 'On' and 'Off' years but they obtain lower or no yield and income in an 'Off' year. It also causes higher prices for customers in an 'Off' year. The mechanism of alternate bearing in pistachios is different to that of other fruit tree species. As with other deciduous tree species, pistachio inflorescence buds are formed one year before. However, they abscise during the summer season of an 'On' year rather than a lack of flower bud formation (Crane and Nielson, 1971). There are many studies on alternate bearing in pistachios and two main hypotheses have been proposed for inflorescence bud abscission:

(1) the carbohydrate-nitrogen sink hypothesis states that carbohydrate-nitrogen competition between nuts and buds causes inflorescence bud abscission (Crane et al., 1976; Crane and Al-Shalan, 1977; Crane and Iwakiri, 1981; Weinbaum et al., 1994; Nzima et al., 1997) and (2) endogenous plant growth regulators (PGR) produced in the nuts or in the leaves lead to pistachio bud abscission (Monselise and Goldschmidt, 1982; Pontikis, 1990). Although many studies have attempted to explain the mechanism of irregular bearing in pistachios, inflorescence bud abscission requires elucidation.

### 2.1.2 Nut quality

As with other nut crops, nut quality traits such as nut size, weight and shape, splitting rate, kernel colour and flavour, and the chemical composition of the kernel are very important in the scion breeding of pistachios. Of these traits, nut size and splitting rate are the primary characteristics. Progenies with large nuts and high splitting rates are preferred in breeding program selection and by consumers. With regard to the splitting characteristics of the nuts, wide opening of the suture is preferred by customers, as this enables easy consumption; however, if the suture is very narrow, it might be very hard to open. Splitting is mostly a genetically controlled character, however growing conditions and pollinators also have an effect. A greenish kernel colour and good flavour are also very important traits, especially for industrial use, for example such as in the sweet, chocolate and ice cream industries. The main Turkish and Italian cultivars have a green kernel with a good flavour. It is obvious that the chemical composition of the kernel, such as high protein and tocopherol content, could be selection criteria in future scion pistachio breeding programs.

### 2.1.3 Blank nut percentage

The pistachio nut and kernel development process is unusual. The kernels start to grow after several weeks of flowering when ovary growth is complete. Nut set typically follows successful pollination in pistachio, whereas blank nuts (unfilled nuts) result when there is fruit set and ovary growth, but the embryo fails to grow, leaving the nut shell empty or blank. Blanking can occur during two different phases of pistachio nut development: (1) nut setting and (2) nut filling. Blank nuts occur during nut setting when pollination occurs but fertilization fails (parthenocarpy). Blanks during nut filling can occur during kernel development when a certain percentage of the fertilized embryos fail to enlarge to fill the shell. It can also be affected by crop load and production practices (Ferguson et al., 2005; Ferguson and Polito, 2016). Hence, in breeding programs a crucial criterion includes the selection of progenies that have lower rates of blank nuts under the same growing conditions.

### 2.1.4 Flowering and harvest time

Late flowering, in particular, is an essential selection criterion in regions where late spring frost is common, as it can affect nut production and springtime fungal infections. Early harvesting is also a desirable characteristic in a pistachio cultivar to reduce the cost of pest and disease management in the orchard. An early harvest is a very important trait that enables pistachio growing and production in regions where the summer period is short. The number of days from flowering to harvest is also an important factor in breeding.

### 2.1.5 Early bearing (precocity)

Early bearing indicates the period from the grafting of a pistachio plant to the first flowering of a scion cultivar, with a short period being desirable, and provides earlier harvesting after orchard establishment. Genetic and environmental factors also influence this period. Hence, progenies that are early bearing are a desirable factor in a breeding program.

### 2.1.6 Chill requirement

Climate change and global warming have become one of the main problems in temperate fruit and nut growing, as specific hours of chilling accumulation are required to produce normal bud break and inflorescences, viable pollen, adequate nut set and vegetative growth. When the cumulative hours fall below the desired level, the bloom and foliation are irregular and delayed, leaves are deformed and the yield is reduced. Therefore, the breeding of pistachio cultivars that have a low chill requirement might be one of the main goals in future pistachio scion breeding programs.

### 2.1.7 Resistance to pests

There are a few diseases that damage pistachios and their impact is highly variable and dependent upon environmental conditions as well as the cultivar. Several diseases are very common in pistachios such as leaf spot (*Pseudocercospora pistacina* Allesch), panicle and shoot blight (*Botryosphaerias dothidea*), alternaria late blight (*Alternaria alternate*), botrytis blossom and shoot blight (*Botrytis cinerea*), with pysillid (*Agonoscena* spp.) and navel orangeworm (*Amyelois transitella*) being other important pistachio pests (Michailidies et al., 2016). It is essential to screen different pistachio germplasms for these pests prior to starting the breeding program; hence, upcoming pistachio scion breeding programs might focus on these pests.

### 2.2 Male cultivar breeding

Most of the effort in pistachio scion breeding has focussed on the fruiting-related problems of female cultivars. Half of the progeny in a typical cross are male plants and this allows the selection of the best males for major female cultivars. Besides the above criteria for females, the main traits of interest in male cultivars include (1) strong growth, (2) no alternate bearing, (3) producing a large amount of male inflorescence, viable and fertile pollen, (4) a long flowering period that overlaps with the blooming of a certain female cultivar under differing ecological conditions. Several experiments have demonstrated that nut quality is negatively affected by using the pollen of wild *Pistacia* species (Kaska and Ak, 1996a,b; Hormaza and Herrero, 1998; Ak, 2001). Therefore, it is desirable to select a male cultivar for the cultivated pistachio from *P. vera*.

### 2.3 Rootstock breeding

Several wild species such as *P. atlantica*, *P. terebinthus*, *P. integerrima*, *P. khinjuk* and *P. eurycarpa*, as well as cultivated pistachios, are used as rootstocks for *P. vera*, either by establishing new orchards using their seeds or by top-grafting them into cultivated pistachios. The latter situation is very common, especially in Turkey and Italy (Kaska et al., 1996; Barone and Marra, 2004). Seeds or micropropagated plants of a hybrid rootstock, UCB1 (*P. atlantica* × *P. integerrima*), are commonly used in California (Ferguson et al., 2016).

#### 2.3.1 Vigour

Unlike some other fruit species, highly vigorous rootstock seed sources are a desirable characteristic in pistachio, for example, *P. integerrima* is a highly vigorous species in the genus.

#### 2.3.2 Uniformity

The pistachio is a highly heterozygous species due to its dioecious character. Therefore, seeds of an open-pollinated *Pistacia* tree are usually not uniform and result in great variability in the nursery, and affects performance of the scion cultivar on various traits such as yield, nut quality and soil-borne disease resistance.

#### 2.3.3 Clonal propagation

The ability of a rootstock to be clonally propagated is of great demand in rootstock breeding, as its suitability for mass propagation by tissue culture is desired for commercial production.

### 2.3.4 Resistance to soil-borne diseases

There are several soil-borne pathogens that affect pistachio rootstocks. The most important include *Verticillium* wilt (caused by *Verticillium dahliae* Kleb.), *Armillaria* root rot (caused by *Armillaria mellea* (Vahl.) *Quel.*), *Phytophthora* crown and root rot (caused by *Phytophthora* spp.) and seedling blight (caused by *Rhizoctonia solani* Kühn.) (Holtz and Teviotdale, 2016).

### 2.3.5 Tolerance of different soil conditions

Salinity tolerance of a rootstock is very important especially in the Iranian and Californian pistachio industry where salinity is a serious problem. Sanden et al. (2004) conducted a salinity trial in California using four different rootstocks and concluded that pistachio can tolerate up to 8 dS/m without significant yield reduction. Rootstocks are also required to tolerate high pH levels and calcareous soils, especially in Turkey.

### 2.3.6 Tolerance of cold

Most *Pistacia* species are tolerant to winter temperatures that reach −25°C (Ozbek, 1978). *P. integerrima* was found to be the most sensitive rootstock, while *P. terebinthus* and *P. atlantica* were the most tolerant during the winters of 1989 to 1990 when temperatures ranged from −11°C to −15°C in California (Ferguson et al., 2016).

### 2.3.7 Effects of rootstock on scion cultivar

Rootstocks may affect yield and nut quality characteristics as well as nutrient uptake efficiency. For example, it has been demonstrated that trees on the UCB1 rootstock produced approximately 45% more marketable yield than trees on *P. atlantica* (Ferguson et al., 2016).

## 3 Genetic resources

### 3.1 Female cultivars

As there are only a certain number of pistachio-growing countries in the world, the number of cultivars in production is limited. A very long juvenile stage and dioecious character are also limiting factors to the development of new pistachio cultivars. Most of the pistachio cultivars in production or in germplasm collections are selections made by growers.

Iran has a large pistachio germplasm and many of the Iranian cultivars have relatively larger nuts with higher splitting rates than other cultivars in the world.

The main germplasm collection is located in the Rafsanjan Pistachio Research Institute and was established at the beginning of the 1980s. The earlier selections were performed by curious pistachio growers, which other growers took for commercial production. The cultivars were divided into two categories based on their shape: round and long. The original wild *P. vera* varieties usually have a long nut shape and the Ohadi cultivar was the first round-shaped cultivar in Iranian pistachio production, which was developed by a farmer from Rafsanjan, named Mr Ohadi, about 75 years ago. Sefid Peste and Momtaz are the oldest cultivars and are not currently used for production. There are about 70 pistachio cultivars growing in different parts of Iran, and the current main pistachio cultivars in production are Fandoghi, Ohadi, Kalehghouchi, Ahmad Aghaei and Akbari. Fandoghi is the most widely available cultivar and is grown in most pistachio-growing areas of Iran. It is a round-type cultivar, has small nuts and a low yield. Kalehghouchi is one of the best cultivars in production with a high splitting percentage, good yield and high commercial value of its nuts. However, due to the growth habit of the trees, there are no new plantations with this cultivar in Iran. It was also tested under Californian conditions and was found to be unsuitable for mechanical harvesting due to the production of excessive vegetative growth on mature trees. It was also tested in the Gaziantep province in Turkey, and the climatic conditions were found to be unsuitable for this cultivar due to it requiring longer periods to mature and completely fill its nuts. Akbari is one of the newest cultivars in Iranian pistachio production. It has long large nuts with a high splitting percentage and high yield. Its nuts have a wide suture opening which increases its commercial value. Ahmad Aghaei is also a new cultivar in Iran and is in high demand by growers due to its short flowering precocity. It has long-type nuts. Among the four cultivars, Akbari and Ahmad Aghaei are preferred in new plantations (Esmail-pour, 2001; Iran Pistachio Association, 2017).

Several cultivars are used for pistachio production in Turkey, and currently, the main cultivars are Uzun, Kirmizi and Siirt. Uzun and Kirmizi cultivars have similar nut characteristics and both have a strong alternate bearing tendency. Their nuts are smaller with lower splitting percentages than Iranian and Californian pistachio cultivars. However, the growers in Turkey still prefer to establish their orchards with these cultivars as their green kernels have a great flavour; hence these nuts are preferred in the sweet industries such as baklava, chocolate, ice cream and so on. Turkey exported 575 tons of baklava mainly to the United States, Saudi Arabia, United Arab Emirates, Switzerland and Germany, in 2016 (TUIK, 2017). Therefore, baklava is a highly developed sector in Turkey and mainly uses nuts of Uzun and Kirmizi pistachio cultivars. The Siirt cultivar is another main cultivar and is mostly cultivated in the Siirt province of Turkey. Its nuts are larger and have a higher splitting percentage than Uzun and Kirmizi cutivars. The kernel colour is yellowish and is not preferred for industrial purposes. Halebi, Ketengomlek, Sultani, Beyaz Ben and Degirmi are also Turkish cultivars and are not currently preferred for new plantations by growers. There are two main germplasms of pistachio cultivars in Turkey: one is in the Pistachio Research Institute in Gaziantep province and the second is in the Ceylanpinar State Farm in Sanliurfa province (Kaska, 1995; Ak and Acar, 2001; Ak et al., 2016).

Kerman, Joley and Lassen were the seedling cultivars developed by chance at the beginning of pistachio cultivation in California. Kerman was collected in 1929 by W.E. Whitehouse in Iran, selected as a seedling in 1936 and released in 1957 (Whitehouse, 1957). 'Lassen' and 'Damghan' were developed from the same seed lot as 'Kerman'. The Joley cultivar was another open-pollinated seedling introduced from the Damghan county of Iran. However, 'Lassen', 'Damghan' and 'Joley' were not used for production due to low yield or other factors. Kerman is now the main female cultivar in California and is responsible for more than 90% of pistachio production. It matures late, requires a large number of heat units, has a good yield and produces large nuts. It has a yellowish kernel colour with minimal flavour. The late maturing of this cultivar has been a serious problem due, in particular, to navel orangeworm infestations that may result from aflatoxin contamination in California. Therefore, a cross-breeding program was started at the end of the 1980s in California, and is discussed in the breeding achievements section. During the twentieth century, the US Department of Agriculture introduced pistachio cultivars from different countries such as Italy, Syria, Tunisia and Israel, and they were collected in a germplasm repository at Davis in California during the 1980s (Parfitt et al., 2012).

Syria is one of the oldest pistachio-growing countries, and Ashoury (Red Aleppo) is the main cultivar making up 85% of Syrian production. 'Red Oleimy' (5%) and 'White Batoury' (5%) result in approximately 10% of pistachio production. Several other cultivars are also used for pistachio cultivation in Syria including Ajamy, Red Jalab, Bundouky, Marawhy, Lazwardy, White Oleimy, Nab Al-Djamal, White Jalab, Antaby and Ein El-Tainah (Hadj-Hassan, 2001).

Pistachios were introduced to Italy from Syria in AD 30, at approximately the same time as they were introduced in Spain. Currently, Italian pistachio production is focussed in Sicily, and Bianca (syn. Napoletana) is the main cultivar. Other cultivars such as Femminella, Natalora, Agostana, Silvaan, Insolia, Cerasola and Cappuccia represent less than 5% of production. Italian cultivars have low yields, small nuts with low splitting percentages, but they have a highly appreciated greenish kernel with a good flavour which is preferred in industries such as confectionery and ice cream (Barone and Marra, 2004).

Mateur, Sfax and Guettar are the main cultivars in Tunisia, while Batoury, Mateur, Ashoury, Bayadi and Oleimy are cultivars in Morocco and lead to a limited amount of production (Ghorbel et al., 2001; Loudyi, 2001). Aeginea is the major cultivar in Greece and Pontikis is a new cultivar in cultivation. The main pistachio germplasm collection in Greece is that of the Vardates Research Station of NAGREF, Lamias. The main cultivar in Australia is Sirora, while in Cyprus it is 'Larnaca'.

One of the largest germplasms was established at IRTA in Spain in 1976 (Vargas et al., 1996) and includes 45 female and 27 male cultivars collected from Turkey, Iran, Syria, Israel, Italy, the United States, Tunisia, Iran, Cyprus, Australia and Greece, which were characterized by Batlle et al. (2001, 2006). Some cultivars are vigorous, such as Ajamy, Bondoki, Marawhy, Mateur, Oleimy, Red Aleppo and Silvana, while some are less vigorous such as Bronte, Kerman, Sirora, Sfax, Joley, Lassen and White Oleimy. The blooming time is an important issue and Batoury, Aeginea, White Oleimy, Mateur and Marawhy are early blooming cultivars, whereas Joley, Oleimy, Kerman and Bianca are late blooming cultivars. Precocity is an important trait in pistachio breeding, as well as for growers, in order to reach a productive stage at an earlier time. The cultivars Batoury, Kerman and Sfax have the shortest unproductive periods, while Bianca, Bondoki, Bronte, Cappuccia, Red Jalab and Silvana have the longest. 'Kerman' produces larger nuts along with 'Ajamy', 'Batoury' and 'Oleimy', whereas 'Avidon', 'Cappuccia' and 'El Guetar' has smaller nuts. A new cultivar breeding program was started at the end of the 1980s in Spain and the results are discussed in the breeding achievements section.

### 3.2 Male cultivars

The importance of male cultivars in some pistachio-growing countries was not well understood when cultivation first began. Later, the necessity of males in the pistachio orchard was understood; however, growers budded males randomly without taking into consideration if their blooming period overlapped with the females or their potential effect on pollen quantity and quality. Less attention has been paid to males than females, with most of the male selections historically performed by growers; however, the recent cultivar breeding programs consider male selection in combination with females.

There are three main male cultivars in Turkey and they are selections from commercial pistachio orchards released by Atli et al. (2006). The authors screened pistachio orchards to find the best males in the Gaziantep, Sanliurfa, Kahramanmaras and Adiyaman provinces of Turkey during 1980–7. Fifty male genotypes were then selected and grown in one location along with major female cultivars. The Uygur male cultivar was selected for Uzun and Halebi female cultivars, the Atli male cultivar was chosen for Kirmizi and Siirt cultivars, and the Kaska male cultivar was suggested for a late flowering cultivar such as Ohadi. Acar and Ak (1998) also characterized 24 males during five consecutive years (1992–6) in the Ceylanpinar State Farm in Sanliurfa, Turkey, which contains approximately 3500 ha of pistachio orchards. The authors suggested Male-1,

Male-3, Male-4, Male-15 and Male-22 for the Kirmizi cultivar, Male-7, Male-9 and Male-23 for the Siirt cultivar, and Male-17 and Male-21 for the Ohadi cultivar.

In Iran, many female cultivars are grown and each of them has its own blooming period, which differentiates from early to late. Sheibani (1995) reported a survey study in Iran to find males with a very long blooming period to overlap most of the female cultivars, and one cultivar was reported. In California, there is only one main male cultivar, Peters, which was discovered in the early 1900s by A.B. Peters in Fresno, California. Two selections, 02-16 and 02-18, bloom earlier and later than 'Peters', respectively, and have been used to match either early or late stages of the 'Kerman' blooming period in California (Parfitt et al., 2016).

The most comprehensive study on the characterization of male cultivars was performed by Batlle et al. (2001, 2006) in Spain. The authors collected 27 male cultivars from different countries such as Greece, Tunisia, Italy, the United States, Syria and Israel, and compared their performance by growing them in one location. 'Peters' (USA), 'B' (Greece) and 'M-P3' (Italy) were the most vigorous, while 'Enk' (Israel), 'M-47' (Syria), '02-18' (USA) and '40A' (Tunisia) exhibited weaker growth. Peters, A (Greece) and B cultivars are early blooming, while M-P9 (Italy), M-9 (Italy) and M-5 (Italy) are late blooming cultivars. Peters had the longest blooming period among the tested male cultivars. The juvenile period was shorter in Peters, M-502 (Italy), M-25A (Tunisia), B, C (Greece) and Enk than the other male cultivars.

## 4 Cytogenetics

The results of cytogenetic studies may be useful in taxonomy and for the phylogenetic analysis of *Pistacia* species, in addition to elucidating their genetic structure at the chromosomal level. Firstly, Zohary (1952) reported that chromosome numbers of *P. vera*, *P. atlantica* and *P. lentiscus* species varied and are 30, 28 and 24, respectively. By studying the meiosis in pollen mother cells, Ozbek and Ayfer (1957) found chromosome numbers of *P. vera*, *P. khinjuk* and *P. terebinthus* species to be $2n = 30$, and *P. atlantica* to be $2n = 28$.

Harandi and Ghaffari (2001) performed meiosis studies using pollen mother cells and mitotic studies using meristematic cells of root tips and the results showed that the chromosome number of *P. vera* is $2n = 30$. Laggard chromosomes were observed in some cells, both at meiosis and mitosis, and the authors suggested that they are probably candidates for sex chromosomes. Ghaffari and Harandi (2002) and Ghaffari et al. (2005) performed studies on chromosome counts and karyotype analysis of Iranian *Pistacia* species and subspecies using mitotic cells. The chromosome number of *P. vera* was found to be $2n = 30$ with four metacentric, eight submetacentric and three acrocentric pairs. The karyotype of *P. khinjuk* ($2n = 24$) exhibited seven submetacentric and five metacentric pairs. Chromosome counting in *P. atlantica*, including subspecies cabulica, kurdica and mutica, resulted in $2n = 28$ with three metacentric, nine submetacentric and two acrocentric pairs. One pair of metacentric chromosomes was heterochromatic in three species and the authors suggested the possible existence of an XY chromosomal mechanism of sex determination in *Pistacia*. The chromosomes of *P. khinjuk* were symmetric with a median or submedian centromere position, whereas the others were asymmetric varying with centromere positions. According to the authors, basic chromosome numbers are $x = 12$, 14, 15, and $x = 12$ with $x = 15$ derived from $x = 14$ in *Pistacia*.

Ila et al. (2003) counted the chromosomes of four major *Pistacia* species in Turkey (*P. vera*, *P. eurycarpa*, *P. atlantica*, *P. terebinthus*) using their root tips and found $2n = 30$ for all four species. The sizes of a couple of homolog chromosomes were found to be larger than the others and were hypothesized to be sex chromosomes. Furthermore, the authors were in disagreement as a result of unequal number of chromosomes of different closely related *Pistacia* species due to successful hybridizations and subsequent nut set found among them in nature. The authors noted that the differences in the chromosome numbers of *P. atlantica* and *P. eurycarpa* between the studies may come from the very small sizes of *Pistacia* chromosomes that led to error when counting.

Ayaz and Namli (2009) used *P. vera* roots obtained in vitro from mature seeds for karyotype analysis. The chromosome number was found to be $2n = 30$ at the c-metaphase of mitosis cells. The centromere types of all chromosomes were designated as median, submedian, subtelocentric and telocentric, and the total lengths of chromosome pairs were found to be between 35.4 and 5.97 μm. Al-Saghir et al. (2014) used fluorescent microscopy for karyotyping analysis in *Pistacia* species. Leaves from the herbarium specimens of *P. atlantica*, *P. khinjuk*, *P. lentiscus* and *P. terebinthus* were used. The chromosome number was found to be $2n = 30$ for all the analysed species.

Sola-Campoy et al. (2015) reported the first molecular cytogenetic study in cultivated pistachios. The karyotype was characterized by fluorescent in situ hybridization (FISH) with probes for 5S and 45S rDNAs, and two pistachio-specific satellite DNAs, PIVE-40 (43 bp repeat) and PIVE-180 (176–178 bp repeat), together with 4′,6-diamidino-2-phenylindole (DAPI) staining. Most chromosomes of the *P. vera* genome were found to be highly heterochromatic. Although classifying chromosomes according to morphology was difficult, using a combination of FISH probes, 10 out of 15 pairs could be distinguished by their unique banding patterns. The largest pair was designated HC1 (heteropycnotic chromosome I), which was strongly heteropycnotic and showed massive enrichment in PIVE-40 repeats. HC1 chromosomes with a differing PIVE-40 hybridization signal were divided into two types, type-I having homozygote individuals and type-II having both heterozygote and homozygote individuals. The authors hypothesized the involvement of HC1 chromosomes in sex determination and confirmed a ZW sex-determining system in pistachios suggested by Kafkas et al. (2015).

## 5 Use of molecular markers

### *5.1 Development of simple sequence repeat markers*

SSR is a very useful molecular marker system due to being abundant in genomes and its codominant inheritance. SSR also have a high level of transferability between closely related species. They have been used for assaying diversity in natural populations, marker discovery, germplasm characterization, parental identification, genetic linkage mapping and evolutionary studies. Until a few years ago there existed a very limited number of SSR markers for *Pistacia* species, however, NGS technology has allowed the development of many SSRs since 2016 in *Pistacia*, which are discussed in this section.

The first pistachio SSR markers were developed by Ahmad et al. (2003) in *P. vera*, using an enrichment method from the Kerman pistachio cultivar. Dinucleotide (CT and CA)- and trinucleotide (CTT)-enriched genomic DNA libraries were constructed. A total of 25 primer pairs were designed, 14 (56%) SSR loci were amplified and produced polymerase chain reaction (PCR) products in tested commercially important American, Iranian, Turkish and Syrian pistachio cultivars (Ahmad et al., 2003). Kolahi-Zonoozi et al. (2014) designed 42 SSR primer pairs and tested them in 45 Iranian pistachio cultivars for amplification and polymorphism. Sixteen were successfully amplified and 12 were polymorphic in the characterization of 45 pistachio cultivars. A total of 32 alleles were obtained with an average of 2.7.

Zaloglu et al. (2015) developed SSR markers from *P. vera* cv. Siirt using genomic libraries enriched with repeats of 'CA', 'GA', and 'AAC' and 'AAG'. Fifty-seven clones were sequenced from each library, 84 primer pairs were designed and 59 bands generated (70.2%), of which 47 were polymorphic, during the characterization of seven diverse pistachio cultivars. The number of alleles among 47 polymorphic loci ranged from two to nine with an average of 3.6. The SSR loci from a GA-enriched library was found to be more polymorphic and informative than the others. Fifty-nine SSR primer pairs were tested for amplification in eight *Pistacia* species, and 54 were transferable to at least one of the *Pistacia* species. Topcu et al. (2016) sequenced more clones from genomic libraries enriched with (GA)n and (AAG)n motifs constructed by Zaloglu et al. (2015). A total of 192 clones were sequenced from each library, 135 primer pairs were designed, 110 were amplified and 64 were polymorphic in 12 diverse pistachio cultivars. From the polymorphic loci, 264 alleles were produced with an average of 4.13 alleles/locus. The observed (Ho) and expected heterozygosity (He) values were 0.52 and 0.56, respectively, while the polymorphism information content was 0.51. One hundred out of 110 SSR primer pairs were transferable to at least one of the ten tested *Pistacia* species.

Motelebipour et al. (2016) conducted a genome survey study of pistachios by obtaining 40× depth sequences of the Siirt cultivar by next-generation Illumina sequencing to develop SSR markers; a total of 59 280 SSR motifs were detected with a frequency of 8.67 kb. A total of 2036 alleles were produced by 206 SSR loci during the testing of 24 pistachio cultivars and 20 wild genotypes belonged to five wild *Pistacia* species. In *P. vera*, 897 alleles were obtained from 206 loci with an average of 4.5 alleles/locus. The novel SSR loci developed from cultivated pistachios were highly transferable to wild *Pistacia* species ranging from 73.3% in *P. lentiscus* to 97.1% in *P. atlantica*. A total of 136 loci were successfully amplified and 41 were polymorphic in all tested *Pistacia* species. Jazi et al. (2017) searched SSRs in transcriptome sequences of pistachio, and 11 337 SSRs were defined as di- to hexa nucleotide motifs in 11 130 contigs. A total of 7605 primer pairs were designed for further studies involving pistachios.

Khodaeiaminjan et al. (2018) designed in silico polymorphic 750 SSR primer pairs by comparing 10× sequences of three *P. vera* (Siirt, Ohadi and Bağyolu) cultivars and one monoecious *P. atlantica* genotype (Pa-18). SSR-PCR analysis of all the primer pairs in 18 *P. vera* cultivars and 6 *P. atlantica* genotypes produced 625 polymorphic loci from 618 primer pairs. A total of 3947 alleles were obtained from 625 loci with an average of 6.2 alleles/locus. A total of 613 (98.1%) SSR loci in 18 *P. vera* cultivars and 544 SSR loci (87.0%) in six *P. atlantica* genotypes were polymorphic. Finally, an in silico approach provided a high number of polymorphic pistachio SSR markers.

There are also several SSR development studies on wild *Pistacia* species. Albaladejo et al. (2008) developed eight polymorphic nuclear microsatellite markers from *P. lentiscus* using an enrichment method. Forty-two individuals from two populations sampled in southern Spain were used for the characterization of eight loci. A total of 59 alleles were obtained, ranging from 3 to 13 per locus. Arabnezhad et al. (2011) constructed two enriched DNA libraries with dinucleotide (AG) and trinucleotide (ATG) microsatellite motifs in *P. khinjuk*. A total of 27 primer pairs were developed and they were successful in cross-species amplification within the *Pistacia* genus. Chen et al. (2011) developed SSRs from the genomic DNA of *P. weinmannifolia*. A total of 147 sequences contained SSRs from 205 sequenced clones, and 94 allowed primer designs with sufficient flanking regions. Fourteen polymorphic microsatellite loci screened on 12 individuals from two wild *P. weinmannifolia* populations had an average of 4.1 alleles/locus. *P. chinensis* and *P. mexicana* were used to test the transferability of the 14 primer pairs. Nine (64.2%) loci in *P. chinensis* and four (28.6%) loci in *P. mexicana* were successfully transferable.

## 5.2 Phylogenetic analysis of *Pistacia* species

Before reviewing phylogenetic analysis at the molecular level, it may be beneficial to present some information on morphological description in order to provide a better understanding of the classification of *Pistacia* species. The first complete taxonomic study was performed by Zohary (1952), who divided the *Pistacia* genus into four sections including 11 species: Lentiscella, Eu-Lentiscus, Butmela, and Eu-Terebinthus, *P. mexicana* and *P. texana*. Swingle in the section Lentiscella occupy a strikingly isolated position both taxonomically and geographically. These species are easily distinguishable from all other species by having smaller and more numerous leaflets. In the section Eu-Lentiscus, *P. lentiscus* is one of the Mediterranean species in the genus and a number of varieties such as var. *chia* Duham. have been described by various authors mainly based on size, shape and number of leaflets in this species. It is easily distinguishable from the original type as it has more leaflets and considerably broader winged leaves. *P. saportae* was described in this section, which is a hybrid between *P. lentiscus* and *P. palaestina*. In the Butmela section, only one species, *P. atlantica*, with two varieties, was described. Zohary (1952) did not find justification to retain *P. mutica* F&M and *P. cabulica* stocks, neither as species nor as variety, and they correspond well to *P. atlantica* var. *latifolia* DC. The author noted that var. *kurdica* Zoh. is easily differentiated from the original form and var. *latifolia* by its large, broad, ovate leaflets and by its nuts, which are considerably broader than long. The Eu-Terebinthus section was divided into two groups: Khinjuk (*P. vera*, *P. terebinthus* and *P. khinjuk*) and Chinensis (*P. palaestina* and *P. chinensis*). The Khinjuk group is discernible due to its imparipinnate leaves with a terminal leaflet as large or larger than the lateral ones, whereas the Chinensis group is differentiated by a reduction of the terminal leaflet or by having paripinnate leaves with an acuminate apex. The author stated that *P. palaestina* is a very close relative of *P. terebinthus* but it is diverged by the absence or presence of the terminal leaflet and leaflet apex shape. In this section, *P. integerrima* and *P. falcata* Beccari were retained as a variety of *P. chinensis*, although their distribution areas are entirely different (Zohary, 1952).

Yaltirik (1967) described six *Pistacia* species in Turkey: *P. lentiscus*, *P. atlantica*, *P. eurycarpa*, *P. khinjuk*, *P. vera* and *P. terebinthus*. The author classified *P. palaestina* as a subspecies of *P. terebinthus*. He added a new species in *Pistacia* and named it *P. eurycarpa*. Zohary (1952) treated this plant as a variety of *P. atlantica* (var. *kurdica*), but it differs from the latter in at least two characteristics and it seemed justifiable to treat it as a different species. It differs from *P. atlantica* by having leaves that are light green on both sides (instead of being dark green above and pale below) and in its depressed nuts. Furthermore, the leaflets are usually thicker and never as numerous or narrow as they often are in *P. atlantica*, and the wing of the rachis is narrower or even absent. An attempt was made by Grundwag and Werker (1976) to characterize *Pistacia* species growing in Israel and to clarify the origin of *P. saportae* according to wood anatomy. They proved, based on several anatomical characteristics, that *P. saportae* is a hybrid between *P. palaestina* and *P. lentiscus* or vice versa.

The first molecular study of the *Pistacia* genus was performed by Louskas and Pontikis (1979) who investigated the pollen isozyme patterns of nine different enzymes in four genotypes of *P. vera* and also in *P. terebinthus* and *P. lentiscus*, in an attempt to study the inter- and intra-specific relationships. The authors found a closer phylogenetic relationship between *P. vera* and *P. terebinthus* than between *P. vera* and *P. lentiscus*. PCR-RFLP analysis using chloroplast DNA was used to study the phylogenetic relationships of ten *Pistacia* species by Parfitt and Badenes (1997) who divided the genus into two sections, Lentiscus and Terebinthus rather than the four sections described by Zohary (1952). The Terebinthus group included *P. vera*, *P. khinjuk*, *P. atlantica*, *P. terebinthus*, *P. integerrima* and *P. chinensis*, while the Lentiscus group consisted of *P. mexicana*, *P. texana*, *P. lentiscus* and *P. weinmannifolia*. All of the species in Lentiscus are evergreen with a paripinnate leaflet arrangement, whereas Terebinthus species lose their leaves in the autumn and have imparipinnate leaves. The seed size in Lentiscus species is considerably reduced compared with that of the Terebinthus species. Two pairs of species could not be separated: *P. mexicana*:*P. texana* and *P. khinjuk*:*P. vera*. *P. integerrima* was clearly diverged from *P. chinensis* and this result differed from the classification performed by Zohary (1952) who described *P. integerrima* as a more recently diverged subspecies of *P. chinensis*.

Kafkas and Perl-Treves (2001) studied the taxonomic relationship of the wild *Pistacia* germplasm in Turkey using morphological data and RAPD analysis. The study included *P. vera*, *P. atlantica*, *P. terebinthus* and *P. eurycarpa* (locally identified as *P. khinjuk*) species. Cluster analysis based on morphological data revealed that the closest species to *P. vera* is *P. eurycarpa*, followed by *P. atlantica* and *P. terebinthus*. In the resulting molecular phylogeny, *P. terebinthus* appeared to be the most diverged species, and the closest pair of species was found to be *P. atlantica* and *P. eurycarpa*. This supported the classification of trees that had been identified by local growers as *P. khinjuk* to be *P. eurycarpa*. This hypothesis was proved by including a reference *P. khinjuk* sample in the analysis, which clustered further away from *P. atlantica* and from the trees identified as *P. eurycarpa*, and closer to *P. vera*.

Phylogenetic relationships among nine *Pistacia* species were examined using RAPD analysis by Kafkas and Perl-Treves (2002). Phylogenetic analysis yielded a cladogram featuring two main groups: *P. vera*, *P. khinjuk*, *P. eurycarpa*, *P. atlantica* and *P. integerrima* were included in one group, while *P. terebinthus*, *P. palaestina*, *P. mexicana* and *P. lentiscus* formed the second group. The first group included species with single-trunked big trees, whereas the species included in the second group mostly grow as shrubs or small trees. The cladogram showed that the closest pairs of species were *P. terebinthus – P. palaestina*, *P. eurycarpa – P. atlantica*, *P. vera – P. khinjuk*, and *P. mexicana – P. lentiscus*. This study suggested that among the wild species, *P. khinjuk* is the closest relative of cultivated pistachios and is a different species rather than the same species of *P. vera* suggested by Parfitt and Badenes (1997). Kafkas and Perl-Treves (2002) found that *P. vera* is more closely related to *P. integerrima* than to *P. terebinthus*, which is not in agreement with the results of Parfitt and Badenes (1997). The authors suggested that *P. palaestina* is in fact a variety of *P. terebinthus* in view of the small genetic distance between them. The study also showed that *P. eurycarpa* is a distinct species from *P. atlantica*, rather than a variety within the same species suggested by Zohary (1952). Karimi et al. (2009) obtained similar results by AFLP analysis of several *Pistacia* species concluding *P. eurycarpa* to be a different species.

Werner et al. (2001) identified *P. saportae* morphologically and RAPD analysis determined it to be a hybrid between *P. lentiscus* and *P. terebinthus*; in addition, RAPD analysis produced several species-specific RAPD markers. Katsiotis et al. (2003) characterized five *Pistacia* species (*P. vera*, *P. palaestina*, *P. terebinthus*, *P. chinensis* and *P. lentiscus*) and one variety of *P. lentiscus* (var. *chia*) in Greece by AFLP and RAPD analysis. According to their results, two main clusters were produced: one containing the evergreen species *P. lentiscus* and the resin-producing *P. lentiscus* var. *chia* (cultivated only on the island of Chios), and the other cluster containing the deciduous species *P. vera*, *P. palaestina* and *P. terebinthus*. Two major subgroups were formed within the deciduous group: one containing *P. vera* and another containing *P. palaestina* and *P. terebinthus*. *P. chinensis* was clustered either with the evergreen species by RAPD analysis or with the deciduous species by AFLP data. Katsiotis et al. (2003) also suggested considering *P. palaestina* as a subspecies of *P. terebinthus* due to the close genetic similarity value between them. Golan-Goldhirsh et al. (2004) studied the taxonomic relationships of six Mediterranean *Pistacia* species and accessions within species using RAPD and AFLP markers. The authors were in agreement with Parfitt and Badenes (1997) on dividing the genus into two sections, and suggested recognizing *P. palaestina* as a subspecies of *P. terebinthus* due to the high similarity between them.

Kafkas (2006b) performed a phylogenetic study on ten *Pistacia* species using AFLP markers. *P. atlantica*, *P. khinjuk*, *P. vera*, *P. mutica*, *P. integerrima* and UCB1 (*P. atlantica* × *P. integerrima*) were placed in one group, and *P. lentiscus*, *P. palaestina*, *P. terebinthus* and *P. mexicana* were in the second group. *P. atlantica – P. mutica* and *P. terebinthus – P. palaestina* pairs were the closest species, therefore *P. mutica* was classified as *P. atlantica*, and *P. palaestina* as *P. terebinthus*. Karimi and Kafkas (2011) also classified *P. mutica* as a subspecies of *P. atlantica* by using SAMPL markers.

Salehi Shanjani et al. (2009) and Pazouki et al. (2010) used AFLP and SSR markers and characterized 216 and 304 *Pistacia* accessions, respectively, belonging to *P. khinjuk*, *P. vera* and *P. atlantica* subsp. *kurdica*. The authors indicated a lower level of polymorphism and variation within *P. atlantica* subsp. *kurdica* than *P. vera* and *P. khinjuk*. Iranjo et al. (2016) studied the genetic diversity of *Pistacia* species in Iran by RAPD analysis, and *P. vera* had the highest genetic variability, while *P. khinjuk* had the lowest variability among Iranian populations. Motalebipour et al. (2016) used 1505 alleles amplified by 136 SSR primer pairs for the phylogenetic analysis of six *Pistacia* species. The closest species to *P. vera* was *P. atlantica*, and *P. integerrima*, *P. chinensis*, *P. terebinthus* followed it, while *P. lentiscus* was the most diverse species to the cultivated pistachio. The structure analysis also confirmed the cluster analysis.

## 5.3 Characterization of pistachio cultivars

The natural distribution of wild *P. vera* is centered in Tajikistan, Kirgizia and north Afghanistan, and extends westward to the northern part of the Khorasan district in Iran, and the Kopet mountain range of southern Turkmenistan (Zohary, 1996). The Badgtis region of southeastern Turkmenistan and northeastern Afghanistan is an area where significant undomesticated *P. vera* forests remain (Maggs, 1973; Popov, 1994; Kaska et al., 2006) and appears to be the centre of diversity for this species (Whitehouse, 1957; Khanazarov et al., 2009).

There are a few studies describing the *P. vera* germplasm at the molecular level. The isozyme polymorphism of pistachio cultivars in Sicily was studied and the offspring obtained by controlled pollination of *P. vera* cv. Bianca with the pollen of *P. atlantica*, *P. vera*, *P. terebinthus* and Santangilisi, a Sicilian pollinator that was previously published to be a hybrid between *P. vera* and *P. terebinthus*, was analysed. The results suggested that Santangilisi is a hybrid between *P. vera* varieties, and the *P. vera* cv. Insolia is a hybrid of *P. vera* × *P. terebinthus* (Dollo, 1996). Isozyme analyses (esterase, peroxidase and acid phosphatase) of leaf samples were used to identify eight male pistachio selections and ten female pistachio cultivars in Italy. Isozyme analysis showed a higher degree of polymorphism in the males compared with the females. Hence, using only three enzymes, it was possible to identify all of the male selections, but only 50% of the females (Barone et al., 1996).

Hormaza et al. (1994a) used RAPD markers to examine patterns of relatedness among 15 pistachio cultivars. The results revealed two major clusters of the *P. vera* germplasm: a Mediterranean cluster, which includes cultivars originating from the Mediterranean region of Europe, North Africa and the Middle East; and an Iranian-Caspian cluster, comprising the germplasm originating from locations east of the Zagros Mountains. The molecular data in combination with historical and geographical records support the hypothesis that pistachio cultivation originated within, or near, the present natural range of the species and was spread by cultivation to the Mediterranean region of the Middle East (Hormaza et al., 1994a). Hormaza et al. (1998) added 16 more accessions to check the consistency of this hypothesis, and the conclusions resembled those obtained with 15 cultivars.

Morphological description and RAPD fingerprinting analysis were conducted on 24 cultivars of *P. vera* (8 males and 16 females) collected from Italy, Greece, Morocco, Spain and Turkey (Caruso et al., 1998). A high degree of polymorphism was detected both at the phenotypic and molecular levels. Among the female accessions, cluster analyses of both morphological and bio-molecular characteristics did not separate the Mediterranean from the Iranian-Caspian genotypes, in contrast to the study by Hormaza et al. (1994a). Seedlings of *P. vera* developed from the seeds of two separate populations in Turkmenistan, Kepele and Agachli, were characterized by Barazani et al. (2003) using the RAPD technique. Cluster analysis divided the accessions into two main groups according to their origins. Kafkas et al. (2006b) characterized 17 wild pistachio accessions, collected from natural pistachio populations of Afghanistan, by AFLP markers and found high genetic diversity in the population.

One of the most detailed germplasm characterization studies of *P. vera* was performed by Kafkas et al. (2006a) who analysed 69 pistachio cultivars originating in Iran, Turkey, the United States, Syria, Greece, Italy, Israel, Cyprus and Tunisia by AFLP, ISSR and RAPD analysis. Cluster analysis of the combined data mainly formed two groups according to the geographical origin of the pistachio genotypes: one group contained the cultivars originating from Iran and the second group included cultivars originating from Turkey, Syria, Greece, Italy, Cyprus and Tunisia. The Siirt cultivar (originating in the southeast of Turkey) and its variants were located between the two main groups. The authors hypothesized that the Siirt cultivar is in a transition subcluster between the Iranian and Mediterranean cultivars and pistachio cultivation diffused from its centre of origin, the Iranian-Caspian region, via southeastern Turkey to Syria, the Mediterranean region of Europe and North Africa. Motalebipour et al. (2016) used 1505 alleles from 136 SSR primer pairs for the genetic diversity analysis of 24 pistachio cultivars, and similar results were obtained with the study performed by Kafkas et al. (2006a). Iranian cultivars were well separated from the other pistachio cultivars and the Siirt cultivar was placed between the Iranian and Mediterranean cultivars. The structure analysis confirmed the results of cluster analysis and male cultivars in Turkey such as Kaska, Uygur, Atli and Bagyolu also had shared alleles from wild *Pistacia* species. Khodaeiaminjan et al. (2018) analysed the genetic relatedness of 18 pistachio cultivars from different origins by 2631 SSR alleles from 625 SSR loci producing a very robust dendrogram that is in agreement with the above hypothesis on the diffusion of pistachio cultivars suggested by Kafkas et al. (2006a). Salehi-Shanjani et al. (2009) characterized 43 Iranian and 23 Mediterranean pistachio cultivars using AFLP markers, and the cultivars were well separated based on their origins and the results were similar to Kafkas et al. (2006a).

Mirzaei et al. (2006) characterized 22 pistachio cultivars grown in Iran using RAPD markers. Momtaz and Gholamrezaei were the closest pair of cultivars. The authors speculated that Iranian pistachio cultivars originated from the Sarakhs cultivar, and Khanjary Damghan, Ghazvini Zoodras, Italiaei Zoodras and Badami Nishkalaghi are a bridge between Sarakhs and commercial cultivars. Talebi et al. (2016) supported this hypothesis by conducting chloroplast DNA analysis in non-coding regions of pistachio. Afzadi et al. (2007), Tagizad et al. (2010), Baghizadeh et al. (2010) and Aliakbarkhani et al. (2015) studied the genetic diversity of Iranian pistachio cultivars using various marker systems such as RAPD, ISSR, AFLP and SSR. The authors detected a high level of polymorphism and genetic diversity in the Iranian *P. vera* germplasm. Noroozi et al. (2009) studied the genetic diversity of Iranian pistachio cultivars using ISSR markers, with a low level of genetic diversity detected in Iranian pistachio cultivars and several pairs of cultivars couldn't be distinguished.

Basha et al. (2007) determined the genetic relationships of 25 *P. vera* cultivars in Syria using AFLP markers. The cultivars separated into three groups: Ashoury, Red Jalab, Entaby 2, Red Oleimy, Batoury, Nab Al-Djamal, White Ashoury, Ajamy and Sen Al Feel were in group I, while Baidy, White Turkey, Marawhy, Boundouky, Entaby 1, Bayd At-Tair, Wardany and Ain Al-Arab cultivars formed group II. Group III contained Lazwardy, Botmy, Abu-Rieha, Al-Grahy, White Oleimy Zaroory and Ain Al-Tinaha cultivars. Kebour et al. (2012) characterized ten Syrian pistachio cultivars grown in Algeria using ISSR analysis and indicated a low level of polymorphism. Fares et al. (2009) and Guenni et al. (2016) characterized the Tunisian pistachio germplasm using ISSR and SRAP markers, respectively. Meknessy and Red Aleppo were the closest cultivars. A high level of genetic diversity was detected in the Tunisian pistachio germplasm.

## 5.4 Sex-linked marker development

Marker-assisted selection may greatly facilitate pistachio rootstock breeding, as well as cultivar improvement, because of the long juvenile period of *Pistacia* species. In addition, due to the dioecious character of *Pistacia* species, early diagnosis of the seedling sex type would assist breeding, nursery management and germplasm collections of the genus (Hormaza et al., 1994b; Kafkas et al., 2001).

There are two main studies on sex-linked marker development in *P. vera*. The first attempt was performed by Hormaza et al. (1994b) who used bulk segregant analysis in combination with the RAPD technique. A single female-specific sex marker ($OPO08_{945}$) was found by screening 700 decamer primers, and the authors suggested that the low frequency of sex-linked bands may indicate that the DNA segment involved in sex determination is small and probably represents a single gene or very few genes. Kamiab et al. (2014) tested the female-specific sex marker ($OPO08_{945}$) in female and male genotypes in Iran and found a 15% error rate in discriminating males. Yakubov et al. (2005) converted the $OPO08_{945}$ sex marker to SCAR (sequence characterized amplified region) and used it in combination with touchdown PCR for specific amplification of the sex-linked marker. The primer OPO08 amplified a 905 bp fragment in all female trees, but also in several male trees. SCAR primer pairs amplified a 905 bp band in females and a 909 bp band in males. A second internal set of SCAR primer pairs were designed after the sequencing of both bands. All females amplified a specific 297 bp product, while males had no amplification product in combination with the touchdown PCR technique. However, the SCAR marker has been ineffective for gender determination in other *Pistacia* species. The authors suggested the use of the new set of SCAR primer pairs with touchdown PCR to determine females in a breeding program or in a germplasm collection of *P. vera*.

Restriction site-associated DNA (RAD) markers were used to identify sex-linked markers and to elucidate the sex determination system in pistachios by Kafkas et al. (2015). Eight male and eight female F1 progenies from a *P. vera* L. Siirt × Bagyolu cross, along with the parents, were subjected to RAD sequencing in two lanes of a Hiseq 2000 sequencing platform. There were 33 757 polymorphic SNP loci between the parents. Thirty-eight of these, from 28 RAD reads, were detected as putative sex-associated loci in pistachio. Primer designs were successful in 13 loci and their validation was performed using SNaPshot analysis in 42 mature F1 progenies and in 184 cultivars and genotypes in a germplasm collection. Eight loci could distinguish the sex of the pistachio with 100% accuracy. To ascertain the cost-effective application of markers in a breeding program, high-resolution melting analysis was performed; four markers were found to perfectly separate sexes in pistachio. Due to the female heterogamety in all candidate SNP loci, the authors reported for the first time that pistachio has a ZZ/ZW sex determination system. Khodaeiaminjan et al. (2017) tested 25 remaining sex-associated SNP loci from the study by Kafkas et al. (2015). The authors extended the RAD sequences to allow primer design for validation by SNaPshot analysis. Of the 25 markers, the authors found nine more SNP markers linked to sex in pistachios. All the markers were heterozygote in females and homozygote in males, supporting a ZW/ZZ sex determination system in pistachios. However, 17 SNP markers were ineffective for sex identification in other *Pistacia* species.

There are several studies on sex determination in wild *Pistacia* species. Kafkas et al. (2001) searched for RAPD markers associated with sex in *P. atlantica*,

*P. terebinthus* and *P. eurycarpa*, the major wild species in Turkey that are used as rootstocks for *P. vera*. The authors found, by screening 472 RAPD primers, one sex marker ($BC156_{1300}$) in *P. eurycarpa*, one ($OPAK09_{850}$) in *P. atlantica* and none in *P. terebinthus*. Such low frequency is consistent with the results of Hormaza et al. (1994b). Sun et al (2014) aimed to find sex markers in *P. chinensis* Bunge using RAPD and ISSR techniques. Among 21 RAPD primers and 20 ISSR primers, two sex-specific primers (S1 and S281) were identified which were able to amplify female-specific fragments of 473 and 1242 bp, respectively. However, only one fragment (S281) was successfully converted into a SCAR marker. A 636 bp fragment was amplified by the SCAR primer pairs in all females while it was absent in all male samples at an annealing temperature of 64°C. Xiong et al. (2013) searched differentially expressed proteins between female and male plants using the leaf and stem of *P. chinensis*. Phosphoglycerate kinase (PGK) was in high abundance in stem phloem in the females, while it was almost undetectable in the males. Ascorbate peroxidase (APX) and temperature-induced lipocalin (TIL) were highly abundant in male stems, while they were found in lower amounts in females. The authors assumed that APX, PGK and TIL might be promising candidates to serve as protein molecular markers for sex determination in *P. chinensis*.

## 5.5 Genetic linkage maps and QTL analysis

*Pistacia* species are suitable for genetic linkage mapping studies in which inter- and intra-specific hybrids can be easily produced (Kafkas, 2003). F1 progenies derived from crosses between two heterozygous parents are typically used for genetic linkage mapping studies in outbreeding tree species like pistachio. The double pseudo-testcross strategy is applied in linkage map construction (Grattapaglia and Sederoff, 1994).

There are three genetic linkage mapping studies of pistachio. The first one was performed by Turkeli and Kafkas (2013) who used 16 ISSR primers, 27 SRAP and 27 AFLP primer combinations to construct a genetic linkage map of pistachios. An inter-specific population of a cross between the Siirt (*P. vera*) cultivar and monoecious Pa-18 (*P. atlantica*) genotype was used. The maternal Siirt map included 165 markers along with 17 linkage groups (LGs) spanning 1237.3 cM (centiMorgan) with a mean distance of 7.50 cM between markers, while the paternal monoecious *P. atlantica* (Pa-18) map comprised 156 markers in 19 LGs covering 1337.5 cM with an average of 8.57 cM between the markers. The first map in pistachio was slightly fragmented and it was impossible to construct the consensus map due to the use of only dominant marker systems that produced an insufficient number of shared markers.

Khodaeiaminjan et al. (2018) constructed the first SSR-based genetic linkage map of pistachios using F1 progenies derived from an intra-specific cross between Siirt and Bagyolu cultivars. The consensus and parental genetic linkage maps spanned into 15 LGs. A total of 384, 317 and 341 markers were mapped in consensus, Siirt and Bagyolu genetic linkage maps. A high percentage of the mapped markers were common markers (71.4%), while the remainder were parental markers (28.6%). The total length of the consensus map was 1511.3 cM with an average of 25.6 SSR markers per LG, and the average distance between the markers was 3.9 cM. The maternal Siirt map had an average of 21.1 markers per LG with a total length of 1427.0 cM and had an average distance of 4.5 cM between the markers. The Bagyolu paternal linkage map comprised an average of 22.7 markers per LG, with the total length of 1453.4 cM and had an average distance of 4.2 cM between SSR markers. Finally, the authors constructed a moderately dense SSR-based genetic linkage map of pistachios, which can be deemed as a framework map for further genetic studies such as genomics, QTL mapping, genome-wide association studies and marker-assisted selection of pistachios.

Motalebipour et al. (2018) performed the first QTL study involving pistachios by constructing a genetic linkage map using an inter-specific F1 population derived from *P. vera* 'Siirt' and monoecious *P. atlantica* (Pa-18 genotype). A total of 388 SSR markers were mapped. The length of the consensus map was 1492 cM along with 15 LGs with an average marker distance of 3.7 cM. A total of 306 SSR markers were mapped in the Siirt female parent with an average number of 20.4 markers per LG. The Siirt total map length was 1410.4 cM with an average of 4.6 cM between the markers. The Pa-18 male parent map included 285 SSR markers with an average of 19.0 markers per LG. The Pa-18 map length was 1362.5 cM with an average distance of 4.8 cM. The authors performed the first QTL analysis of pistachios for five morphological traits including leaf length, leaf width, number of leaflets, young shoot colour and leaf colour, and 17 stable QTLs during two consecutive years were identified in both parental maps.

The recently released SSR-based genetic linkage maps and reported QTLs will be useful genetic resources for future genetic linkage maps and QTL analysis, and for identification of important genes in *Pistacia* as well as genome sequencing projects.

## 6 Genome and transcriptome sequencing

Over the last decade, NGS technology has enabled the release of whole genome sequences, and the rapid release of transcriptome sequences, of different plant species. Pistachio is an interesting species to study using whole genome sequencing due to its dioecious character. Motalebipour et al. (2016) was the first to elucidate the genome structure of pistachio by performing a genome survey study. The authors obtained 26.77 G Illumina data and estimated the genome size to be approximately 600 Mb by K-mer analysis. The GC content was about 37.1% in pistachios. The K-mer analysis also showed that the pistachio genome is highly heterozygous. The assembly of 26.77 G Illumina data produced a highly fragmented genome that was used to screen SSR motifs, and a total of 59280 SSR motifs were detected. The dinucleotide motifs were the most abundant type of repeats (63.2%) in pistachio, followed by tri- (18.0%), tetra- (12.8%), penta- (3.8%) and hexanucleotide motifs (2.2%). The most abundant repeats were AT/AT (23.0%) and TA/TA (21.6%), followed by AG/CT (7.6%) and GA/TC (6.2%), AAT/ATT (4.1%), CA/TG (2.6%) and TAA/ATT (2.6%). The distribution of SSRs in the pistachio genome was calculated as one SSR per 8.67 kb. Kafkas et al. (2015) detected an average of one SNP for each 449 bp in pistachios using RAD sequencing.

Kafkas et al. (2017b) gave a presentation on genome sequencing of the Siirt pistachio cultivar at the VII International Almond and Pistachio Symposium. The authors used an Illumina sequencing platform and obtained a fragmented genome assembly with a scaffold size of N50 = 208.7. The addition of PacBio long read sequencing in combination with the Illumina assembly resulted in an N50 of 1.51 Mb with 1787 scaffolds. In the third step, the authors used Hi-C scaffolding technology that generated a chromosome scale assembly with N50 = 30.6 Mb and 1193 scaffolds of which 16 scaffolds contained the majority of the genome. More information will be available when the pistachio genome is published. The genome assemblies may provide breeders with foundational information, new tools and methodologies to accelerate cultivar and rootstock breeding programs.

Although pistachio is a salt-tolerant species, a medium to high level of soil salt content adversely affects the pistachio yield and nut quality. Therefore, Jazi et al. (2017) performed a study to find salinity-related genes of pistachios by transcriptome sequencing. Sarakhs and Ghazvini, the most salt-sensitive and salt-tolerant cultivars, respectively, were grown and subjected to salt stress. The leaf, stem and root samples from sensitive and salt-tolerant cultivars after 0, 6, 24 and 48 h salt treatment (24 samples in total) were subjected to transcriptome sequencing by using an Illumina Hiseq 2000 platform, and 13097 transcripts were identified as candidate stress-responsive genes. The expressions of some of these genes were experimentally validated through quantitative real-time PCR (qRT-PCR), and the contrasting expression pattern of NCED3 and SOS1 genes were observed between salt-sensitive and salt-tolerant cultivars. The authors also searched SSR repeats in the pistachio transcriptome, and found that di- and trinucleotide repeats were the most abundant accounting for 40–44% of total SSRs, followed by tetra- (9.5%), penta- (3.1%) and hexanucleotide repeats (2.2%). The pistachio transcriptome was found to be rich in GA/TC (12.1%), AG/CT (11.0%), AT/AT (8.3%), TA/TA (8.0%), GAA/TTC (5.0%) and AGA/TCT (4.0%) repeats.

## 7 Micropropagation

Pistachios are one of the most difficult species to micropropagate among fruit and nut tree species (Joley, 1979). Pistachio rootstocks have generally been propagated by seeds (sexual propagation), with propagation by cuttings proving to be rather difficult compared with those of other tree species (Almehdi et al., 2002). Like other heterozygous plants, clonal propagation techniques, in particular the micropropagation technique, are important and offer significant advantages to obtain uniform pistachio rootstocks.

There are many reports on the micropropagation of pistachio cultivars and wild *Pistacia* species. Organogenesis has been the main method for micropropagation of pistachios by using shoot tips (Barghchi, 1986a,b; Barghchi and Alderson, 1983a,b; Martinelli and Loretti, 1988; Parfitt and Almehdi, 1994). Embryo culture (Benmahioul et al., 2009), callus culture (Alderson and Barghchi, 1982) and meristem culture (Cetiner et al., 1996) were the other methods applied to *Pistacia* species.

There are several stages in micropropagation, and the first step is the surface sterilization of tissues. The major factor in the success of cultures is to have a good surface sterilization method, as contamination by bacteria and fungi may cause failures. The plants grown under greenhouse conditions may have less contamination than plants grown outside, and mature plants may exhibit more contamination than juvenile plants. Furthermore, explants from current year expanding shoots may have less contamination than explants obtained from old shoots. Therefore, an effective surface sterilization method is necessary, and plant genetic material and its physiological stage, age of plant and explant, and growing conditions should be taken into consideration when choosing a sterilization method (Onay et al., 2004a; Benmahioul et al., 2009, 2016; Benmahioul, 2017). Tilkat et al. (2013) suggested the use of small meristematic tips from actively growing shoots of mature *P. vera* to eliminate the risk of contamination. Recently, Benmahioul et al. (2016) suggested a sterilization protocol for pistachios. The authors obtained the best results when all expanded leaves were removed from the shoots, washed under running water and soaked in 70% ethanol for 1 min and then disinfected by immersion for 10 min in a solution of sodium hypochlorite (NaOCl, 2.6% available chlorine).

Shoot tip necrosis (STN) is one of the most common problems during the large-scale in vitro propagation of pistachios. Some reasons that cause STN include nutrient deficiency such as Ca (Calcium) and B (Bor), concentration of cytokinins, medium type and its nutrient concentration, aeration, gelling agent, pH of the medium, and subculture period (Bairu et al., 2009). Barghchi and Alderson (1996) found that Ca and B deficiency are the most common reasons for STN. Srivastava and Joshi (2013) revealed that increasing the calcium chloride ($CaCl_2$) concentration to 18 mM in Murashige and Skoog (MS) medium decreased in vitro STN with no effect on the growth of the explants. In order to prevent STN, Nezami et al. (2015) treated the explants with varying concentrations of $CaCl_2$ (1×, 1.5× and 2×), and boric acid (1×, 2× and 3×), and the lowest necrosis rate (17%) was obtained in the treatment containing 3× $CaCl_2$ in combination with 2× boric acid.

The type, mixing rate and PGR concentration affect growth and differentiation of the tissues. Several authors have reported that cytokinin alone or in combination with an auxin is necessary for shoot proliferation (Benmahioul et al., 2009). Benmahioul (2017) reported that the highest shoot proliferation frequency was obtained with nodal explants on MS basal salts containing Gamborg (B5) vitamins and supplemented with 4 mg $L^{-1}$ 6-benzyladenine (BA). The addition of 2 mg $L^{-1}$ meta-topolin (mT) generated an optimal number of shoots with suitable morphological features, while kinetin (KIN) was found to be unsuitable for pistachio shoot proliferation. In another study, mT was also suggested for the effective in vitro proliferation of axillary shoots (Benmahioul et al., 2012a, 2016) and the same authors reported that the best results were obtained with 2 mg $L^{-1}$ mT (2.5 shoots/explant, 2.3 cm shoot height).

The rooting of pistachio under in vitro conditions is also difficult. Tilkat et al. (2008) compared the effects of washing basal cut ends in sterile distilled water and dipping the basal cut ends in different indole-3-butyric acid (IBA) solutions at different explant sizes (1.0, 2.0, 3.0 and 4.0 cm long) for root induction. The highest rooting frequency (92%) of microshoots was recorded with 4.0 cm long explants and twice-washed microshoots, and rooting was apparent within two weeks. Benmahioul et al. (2012b) studied ex vitro rooting and the best results were obtained from shoot explants treated with 2% active charcoal. Micrografting of *P. vera* was previously reported by Onay et al. (2003), however more detailed studies are necessary in order to use it on a commercial level.

Somatic embryos and/or the callus is necessary for the first step of the genetic engineering of pistachios. The somatic embryogenesis of pistachios was first achieved using immature kernels as explants (Onay et al., 1995). Similar reports on the somatic embryogenesis of several genotypes has been reported by several authors using immature and mature zygotic embryos, cotyledons, flowers and juvenile leaf tissues as explants (Taskin et al., 1996; Onay and Namli, 1998; Onay and Firat, 1999; Onay and Jeffree, 2000; Onay et al., 2004b, 2007).

Recently, Delijam et al. (2016) developed a complete micropropagation protocol for the Ghazvini pistachio cultivar, which is an important pistachio rootstock source in Iran. The authors generated a new medium called GNH (Garoosi, Nezami and Haddad), and compared it with standard media including MS medium, Juglans Medium (DKW) and McCown Woody Plant Medium. The most suitable concentration for Ca and vitamin sources for shoot multiplication were 3.0 mM Ca gluconate and DKW vitamins. The most suitable PGRs were a combination of 0.5 or 1.0 mg $L^{-1}$ BAP (6-benzylaminopurine) and 0.1 mg $L^{-1}$ IBA. The highest rooting parameters were obtained when 3.0 mM Ca gluconate or Fe-EDDHA (Etilendiamin di (ohidroksifonil asetik) asit) (with 0.2 mM Fe) were incorporated into the GNH medium containing 2.0 mg $L^{-1}$ α-naphthalenacetic acid (NAA).

Although there are many studies on pistachio micropropagation, only a hybrid UCB1 (*P. atlantica* × *P. integerrima*) rootstock is micropropagated on a commercial level by different companies in California, and recently in Spain and Turkey. These companies have their own UCB1 plants for clonal propagation, while the parents of these clones are the same.

# 8 Breeding achievements

## 8.1 Cultivars

Breeding new cultivars is of great interest in pistachio-growing countries. The breeding period of pistachios is longer than for other fruit and nut tree species, and the slow progress limits achieving the results in a timely manner. There have been several cultivar breeding programs in pistachio-growing countries worldwide, and the major cross-breeding programs are in the United States, Turkey and Spain. The breeding program in Iran is mainly focussed on selecting from natural populations rather than crosses between female and male cultivars.

Several cultivars have been obtained from known open-pollinated cultivars. Pontikis is a descendent of the Aeginea cultivar in Greece (Pontikis, 1986), whereas Sirora is a chance seedling of Red Aleppo in Australia (Maggs, 1982).

The first major scion program was initiated in California in 1989 by Dr Parfitt (Chao et al., 1998). The program was started by developing an initial set of 1940 progeny from 78 different crosses. In 1990, an additional 5470 seedlings were produced from 176 controlled crosses. The progenies were planted at three locations, Davis-Winters, Kearney Agricultural Center and near Bakersfield. The nuts were evaluated for number of nuts per tree, % splits, % blanks, wet and dry nut weight, and kernel weight. Replicated advanced selection trials were established in 1997 to test nine selected genotypes for yield and performance on two different rootstocks. Three cultivars were initially released in 2005 from this program namely, Golden Hills, Lost Hills and Randy (Parfitt et al., 2006a,b,c, 2007, 2008).

Kerman is a primary cultivar in California pistachio production and growing one cultivar dominantly limits the efficient use of harvesting and processing equipment. The early nut maturity of a pistachio cultivar permits pistachio growers to extend their harvest period. The earlier maturing cultivars are also less susceptible to insect damage from navel orangeworm, a major pest of pistachios implicated in the occurrence of aflatoxin contamination. Therefore, an early harvest is one of the most important traits for California pistachio production. Golden Hills is a female cultivar that flowers 5–7 days before Kerman, and its harvest time is 15 days earlier than Kerman. Golden Hills has a higher yield and split nut percentage, and lower blank nut percentage than Kerman, whereas the nut size and weight are similar to Kerman. Lost Hills is also a new female cultivar, and the flowering and harvesting time are similar to Golden Hills. It has a higher yield, nut size, nut weight and nut splitting percentage than Kerman. Recently, a new female cultivar, namely Gumdrop, was released by Kallsen and Parfitt (2016a,b). It matures earlier than all commercial cultivars with equivalent yield and nut quality to Kerman. Gumdrop can be harvested about 10–12 days before Golden Hills and 24 days before Kerman. Gumdrop blooms about 5 days before Golden Hills and 10–11 days before Kerman. Randy is a male cultivar that flowers 1–2 weeks earlier than Peters. The most important characteristic of this cultivar is a long blooming period which is twice that of Peters or most of the male cultivars. However, it flowers 1–2 weeks earlier than Kerman, and it can only be used as a pollinator for early flowering female cultivars. Recently, two new male cultivars were released. Famoso (Kallsen et al., 2016) is well-suited to serve as a pollenizer of the Kerman female pistachio variety and has a better overlap in the flowering period of Kerman than Peters. Famoso is also more precocious than Peters and produces flowers three to four years after budding compared with Peters, which typically takes 4–5 years to produce flowers after budding. Famoso also has a flowering period that is more coincident with the flowering period of Kerman, while Peters often flowers later than Kerman and has been reported to have poor blooms in some years. Tejon is another new male cultivar in the Californian breeding program released by Kallsen and Parfitt (2018). This variety exhibits 6–10 day earlier flowering than the male pistachio variety Randy. It also produces large quantities of relatively durable and viable pollen. The flowering time of Tejon also overlaps with the bloom period of the female pistachio variety Gumdrop.

In Turkey, the cross-breeding program was started in 1996 (Uzun et al., 2011), and the crosses were performed between five female (Siirt, Ohadi, Haciserifi, Kalehghouchi and Sel-14) and three male cultivars (Kaska, Atli and Uygur). Open-pollinated seeds of Kalehghouchi and Akbari cultivars were also used in the cultivar breeding program. Approximately 3500 progenies were obtained and selections have been evaluated. In Turkey, commercial pistachio orchards were screened during the 1987–9 growing season and 16 female genotypes were selected and grown in the Gaziantep location along with major cultivars, and two cultivars were released (Tahtaci et al., 2011). Tekin is a new cultivar that is a close relative of the Siirt cultivar (Motalebipour et al., 2016), however it has a better yield and nut characteristics such as higher nut weight and splitting percentage with a light endocarp colour. It also exhibits less alternate bearing tendency than Siirt. Barak Yildizi is another newly released cultivar and it has a low chill requirement and low total heat accumulation, thereby maturing very early. Another program was initiated by Kafkas et al. (2004) to breed monoecious pistachio cultivars. Kafkas et al. (2000) found several monoecious *P. atlantica* trees in the Yunt Mountains of the Manisa province of Turkey. One genotype (Pa-18) was fully monoecious and other genotypes have male branches or male inflorescence in several female trees. Kafkas et al. (2004) used monoecious *P. atlantica* trees as male parents, while Siirt and Ohadi cultivars as female parents. Several other crosses were also performed between monoecious, female and male *P. atlantica* trees to understand sex inheritance in *Pistacia*. Most of the progenies are still in the juvenile stage, and no progeny has bisexual flowers among mature ones. About 20% of the progenies belong to *P. vera* × monoecious *P. atlantica* crosses, and about 60% of the progenies belonging to crosses between monoecious *P. atlantica* trees have died. There were also progenies obtained from selfing of the fully monoecious *P. atlantica* tree (Pa-18), and all the selfed progenies died after only a few years of planting. The remaining progenies in the program need several more years to pass the juvenile stage and first flowering. The last cultivar breeding program in Turkey was conducted in 2016 by Kafkas et al. (2017a), and is the first pistachio breeding program to use marker-assisted breeding technology. In total, about 17000 progenies were obtained from crosses between four female and 11 male cultivars along with several backcross and several pseudo-F2 populations. Female-specific sex markers developed by Kafkas et al. (2015) were used to eliminate males. Finally, approximately 9000 progenies were transplanted to the breeding orchard. All the segregating populations in this program may serve to develop new cultivars, as well as help develop novel markers linked to economically important characteristics in pistachios in the future.

A pistachio scion breeding program has been underway in Spain since 1989 (Vargas et al., 1996). Ten female and 12 male cultivars were used to obtain 31 controlled crosses at IRTA Mas de Bove, and about 2000 progenies were obtained. The main female parents were Aeginea (Greece), Batoury and Ouleimy (Syria), Kerman (USA), Larnaca (Cyprus) and Mateur (Tunisia), while B and C (Greece), M-36 and M-38 (Syria), M-502 (Italy) and Nazar (Israel) were the main male cultivars in the breeding program. Several traits have been studied in these progenies such as progeny vigour, blooming and leafing time, flowering precocity and nut traits (Vargas et al., 1996, 2001, 2002; Vargas and Romero, 1998a,b, 2005). The leafing date was correlated with the full blooming date ($r = 0.67$). The selected ten male and eight female genotypes are still under replicated trial. Two males have been released, namely Chaparillo and Guerrero. They are late flowering male cultivars overlapping with late flowering female cultivars such as Kerman (Guerrero et al., 2014).

There is no report of a pistachio cross-breeding program in Iran. There are about 70 registered pistachio cultivars in Iran and they have been selected from the natural *P. vera* population in the northeast of Iran (Safarnejad, 2012). The most important registered cultivars are Akbari, Kalehghouchi, Ahmad Aghaie, Ohadi, Zarand, Momtaz, Damghan, Shahpasand, Fandoghi and Ghazvini. Of them, Akbari has better nut and kernel weights, splitting percentage and nut size than the other cultivars. A detailed description of several major female pistachio cultivars in Iran was performed by Javanshah et al. (2014). Kamiab et al. (2006) characterized ten male pistachio genotypes for major cultivars in Iran including Kalehghouchi, Ohadi, Ahmad Aghaei and Akbari. The flowering period of male genotypes P1 and P2 overlaps with Kalehghouchi, P3, P4 and P5 with Ahmag Aghaei, P6 and P7 with Ohadi, and P8, P9 and P10 overlaps with Akbari. Kashanizadeh et al. (2006) selected a few male seedling trees in pistachio orchards in Gazvin and Buin-Zahra. The authors identified early, mid-season and late blooming male genotypes for major female cultivars. The flowering period changed between 7 and 14 days, and the authors suggested their use in new pistachio plantations for their high pollen quality and long overlapping periods with major female cultivars in Iran.

## 8.2 Rootstocks

Most *Pistacia* species can be used as a rootstock for cultivated pistachio, *P. vera*, as they are mostly graft-compatible. The worldwide pistachio production is achieved in two ways: one is by establishing commercial orchards as in Turkey, California, Iran and Syria, and the second way is by top-budding of *Pistacia* plants in nature as found in Turkey and Italy.

To establish commercial orchards, the seeds of cultivated pistachios have been used as rootstock seed sources in Iran, Turkey and Syria. In Iran, the seeds of the Badami cultivar have been used due to its better salinity tolerance. In Turkey, the seeds of several cultivars were used at the beginning of cultivation, and the Siirt cultivar has been used for several decades due to its more vigorous growth than other cultivars, such as Uzun and Kirmizi. Some growers prefer *P. eurycarpa* (buttum) seeds for rootstock. Its seedlings grow slowly in the first year, however they become a similar size to the seedlings of the Siirt cultivar in the second year.

Currently, there is only one clonal rootstock in the world that is used in pistachio production. UCB1 is the short name for the University of California Berkeley #1. It is a hybrid of *P. atlantica* × *P. integerrima*, invented by Dr Lee Ashworth of the UCB1 Department of Pathology. UCB1 is a superior rootstock for its resistance to *Phytophthora* spp., Verticillium wilt diseases, salt and cold tolerance, and extreme vigour (Epstein et al., 2004; Sanden et al., 2004, 2010; Spann et al., 2007; Ferguson et al., 2016). However, *Rhodococcus* sp.

bacterial infections have become a serious problem in clonally propagated UCB1 rootstocks in California since 2010 (Stamler et al., 2015). Each company has its own UCB1 clone for tissue culture propagation; however, all these new rootstocks are still the progeny of the same *P. atlantica* and *P. integerrima* parents. For example, the Paramount company patented different UCB1 clones as 'Paramount Pistachio rootstock #1' (MacIlvaine et al., 2015a) and 'Paramount Pistachio rootstock #2' (MacIlvaine et al., 2015b), the Duarte company patented it as 'UCB1-D110' (Sharma et al., 2017) and 'UCB1-D154' (Ash et al., 2017), and the Wonderful Orchards company patented it as 'Wonderful Pistachio Rootstock #2' (MacIlvaine et al., 2016a) and 'Wonderful Pistachio Rootstock #4' (MacIlvaine et al., 2016b).

Recently, a new hybrid rootstock cultivar (P.N.B.1) was released by Blackwell (2016). Both its parents are open-pollinated progeny between *P. integerrima* and *P. atlantica*. The P.N.B.1 rootstock cultivar is suitable for commercial propagation by vegetative cuttings. The author indicated that its micropropagation is easier than known rootstocks, highly vigorous and influences the scion yield by approximately 1.5–2.0%.

## 9 Future trends and conclusion

Large germplasms with all *Pistacia* species and most of the cultivars should be established in different pistachio-growing countries. In addition, it is also necessary to obtain samples from Irano-Turanian, East African and Sino-Japanese regions. Furthermore, it is very important to have a wild *P. vera* gene pool from Turkmenistan, Kazakhstan, Kyrgyzstan, Afghanistan and Uzbekistan. The rich pistachio germplasm may help to breed superior cultivars and rootstocks with better traits such as resistance to pests, high and stable yield, and nut characteristics. Marker-assisted selection is definitely very important in *Pistacia*, where the life cycle is very long, that is up to 10–15 years. Therefore, many controlled crosses with high numbers of progeny should be produced to study the genetics of commercially important characteristics, and to develop markers for breeding programs. It is very difficult to find continuous funding for the maintenance of collections and evaluation of breeding plots due to the relatively large size of pistachio plants and very long breeding period. The development of molecular markers linked to economically important traits may help reduce the number of progeny as a result of breeding. Recent technologies such as genome editing may provide rapid advances in pistachio breeding. Therefore, much more research is necessary to produce scientific knowledge regarding the pistachio genome and finally, it is obviously necessary to have interdisciplinary breeding programs bringing together conventional breeding and molecular techniques to make qualitative advances in pistachio production.

## 10 Where to look for further information

Key meetings for pistachio breeding are as follows:

1 the International Symposium on Almonds and Pistachios organized by the International Society for Horticultural Science, and the articles are published in *Acta Horticulturae* (https://www.actahort.org)
2 GREMPA (Group de Recherches et d'Études Méditerranéen pour l'Amandier) Meeting on Almonds and Pistachios organized by CIHEAM (Centre International de Hautes Etudes Agronomiques Méditerranéennes), and the articles are published in OPTIONS Méditerranéennes (https://www.ciheam.org/en/publications/options_mediterraneenes) every four years.

The NUCIS newsletter is the information bulletin of the FAO-CIHEAM network on nuts including pistachio (http://networks.iamz.ciheam.org/nuts/nucis_newsletter.html).

There are also published reviews on pistachios by Kafkas (2006a), Hormaza and Wünsch (2007) and Parfitt et al. (2012).

General information regarding pistachio production can be obtained from Pistachio Production Manuals (Ferguson et al., 2005, 2016) published by the University of California, and from the Fruit and Nut information Center at the University of California (http://fruitsandnuts.ucdavis.edu).

## 11 References

Acar, I. and Ak, B. E. (1998), 'An investigation on pollen germination rates of selected some male trees at Ceylanpinar state farm', *Options Médit.*, 33, 63–6.
Afzadi, M. A., Tabatabaei, B. E. S., Mohammadi, S. A. and Tajabadipur, A. (2007), 'Comparison of genetic diversity in species and cultivars of pistachio (*Pistacia* sp. L.) based on amplified fragment length polymorphism (AFLP) markers', *Iran. J. Biotech.*, 5(3), 147–52.
Ahmad, R., Ferguson, L. and Southwick, S. M. (2003), 'Identification of pistachio (*Pistacia vera* L.) nuts with microsatellite markers', *J. Am. Soc. Hortic. Sci.*, 128, 898–903.
Ahmad, R., Ferguson, L. and Southwick, S. M. (2005), 'Analyses of pistachio rootstocks by SSR and SRAP molecular markers', *J. Hortic. Sci. Biotechnol.*, 80(3), 382–6.
Ak, B. E. (2001), 'Effects of different *Pistacia* species pollen on fruit dimension and weight in the Kirmizi variety', *Options Médit.*, 56, 311–14.
Ak, B. E. and Acar, I. (2001), 'Pistachio production and cultivated varieties grown in Turkey', *in* S. Padulosi and A. Hadj-Hassan (Eds), *Towards a Comprehensive Documentation and Use of Pistacia Genetic Diversity in Central and West Asia, North Africa and Europe*, IPGRI, Roma, Italy, pp. 27–34.
Ak, B. E., Karadag, S. and Sakar, E. (2016), 'Pistachio production and industry in Turkey: Current status and future perspective', *Options Médit.*, 119, 323–9.
Albaladejo, R. G., Sebastiani, F., Aparicio, A., Buonamici, A., Gonzalez-Martinez, S. C. and Vendramins, G. G. (2008), 'Development and characterization of eight

polymorphic microsatellite loci from *Pistacia lentiscus* L. (Anacardiaceae)', *Mol. Ecol. Resour.*, 8, 904-6.

Alderson, P. G. and Barghchi, M. (1982), 'The potential for *Pistacia* propagation through shoot, leaf and callus culture', *in* A. Fujiwara (Ed.), *Plant Tissue Culture*, 11-16 July 1982, Maruzen, Tokyo, pp. 739-40.

Aliakbarkhani, S. T., Akbari, M., Hassankhah, A., Talaie, A. and Moghadam, M. F. (2015), 'Phenotypic and genotypic variation in Iranian pistachios', *J. Genet. Eng. Biotechnol.*, 13(2), 235-41.

Almehdi, A. A., Parfitt, D. E. and Chan, H. (2002), 'Propagation of pistachio rootstock by rooted stem cuttings', *Sci. Hortic.*, 96, 359-63.

Al-Saghir, M., Shadi Abu Baker, S. and Pusok, R. (2014), 'Effective method to resolve the chromosome numbers in *Pistacia* species (Anacardiaceae)', *Am. J. Plant Sci.*, 2913-16.

Arabnezhad, H., Bahar, M. and Pour, A. T. (2011), 'Evaluation of genetic relationship among Iranian pistachios using microsatellite markers developed from *Pistacia khinjuk* Stocks', *Sci. Hortic.*, 249-54.

Ash, K., Castillon, J. and Sharma, D. (2017), 'Pistachio rootstock named clone UCB1-D110', U.S. Patent No. 2017/0094852.

Atli, H. S., Arpaci, S. and Uygur, N. (2006), 'Selection of pistachio pollinators in Turkey', *Acta Hortic.*, 26, 417-22.

Ayaz, E. and Namli, S. (2009), 'The karyotype analysis of *Pistacia vera* L. from Turkey', *Nat. Prod. Res.*, 23(9), 866-70.

Baghizadeh, A., Noroozi, S. and Javaran, M. J. (2010), 'Study on genetic diversity of some Iranian pistachio (*Pistacia vera* L.) cultivars using random amplified polymorphic DNA (RAPD), inter sequence repeat (ISSR) and simple sequence repeat (SSR) markers: A comparative study', *Afr. J. Biotechnol.*, 9(45), 7632-40.

Bairu, M. W., Jain N., Stirk W. A., Doležal K. and Van Staden, J. (2009), 'Solving the problem of shoot-tip necrosis in *Harpagophytum procumbens* by changing the cytokinin types, calcium and boron concentrations in the medium', *S. Afr. J. Bot.*, 75, 122-7.

Barazani, O., Atayev, A., Yakubov, B., Kostiukovsky, V., Popov, K. and Golan-Goldhirsh, A. (2003), 'Genetic variability in Turkmen populations of *Pistacia vera* L.', *Genet. Resour. Crop Evol.*, 50, 383-9.

Barghchi, M. (1986a), '*In vitro* propagation of *Pistacia* rootstocks', *Proc. Int. Plant Pro. Soc.*, 35, 336-7.

Barghchi, M. (1986b), '*In vitro* culture of mature commercial varieties of *Pistacia vera* L.', *Proc. Int. Plant Prop. Soc.*, 35, 331-3.

Barghchi, M. and Alderson, P. G. (1983a), '*In vitro* propagation of *Pistacia vera* L. from seedling tissues', *J. Hortic. Sci.*, 58, 435-45.

Barghchi, M. and Alderson, P. G. (1983b), '*In vitro* propagation of *Pistacia* species', *Acta Hortic.*, 131, 49-60.

Barghchi, M. and Alderson, P. G. (1996), 'The control of shoot tip necrosis in *Pistacia vera* L. *in vitro*', *Plant Growth Regul.*, 20, 31-5.

Barone, E. and Marra, F. P. (2004), 'The pistachio industry in Italy: Current situation and prospects', *Nucis*, 12, 16-19.

Barone, E., Di Marco, L., Marra, F. P. and Sidari, M. (1996), 'Isozymes and canonical discriminant analysis to identify pistachio (*Pistacia vera* L.) germplasm', *HortSci.*, 31, 134-8.

Basha, A. I., Padulosi, S., Chabane, K., Hadj-Hassan, A., Dulloo, E., Pagnotta, M. A. and Porceddu, E. (2007), 'Genetic diversity of Syrian pistachio (*Pistacia vera* L.) varieties evaluated by AFLP markers', *Genet. Resour. Crop Evol.*, 54, 1807–16.
Batlle, I., Romero, M. A., Rovira, M. and Vargas, F. J. (2001), 'Pistacia conservation, characterization and use at IRTA: Current situation and prospects in Spain', *in* S. Padulosi and A. Hadj-Hassan (Eds), *Towards a Comprehensive Documentation and Use of Pistacia genetic diversity in Central and West Asia, North Africa and Europe*, IPGRI, Rome, Italy, pp. 77–87.
Batlle, I., Romero, M. A., Rovira, M. and Vargas, F. J. (2006), '*Pistacia* species genebank, conservation, characterization and use at IRTA: Current situation and prospects in Spain', *Nucis*, 13, 47–53.
Benmahioul, B. (2017), 'Factors affecting *in vitro* micropropagation of pistachio (*Pistacia vera* L.)', *Agric. For. J.*, 1(1), 56–61.
Benmahioul, B., Kaid-Harche, M., Dorion, N. and Daguin, F. (2009), '*In vitro* embryo germination and proliferation of pistachio (*Pistacia vera* L.)', *Sci. Hortic.*, 122, 479–83.
Benmahioul, B., Dorion, N., Kaid-Harche, M. And Daguin, F. (2012a), 'Micropropagation and *ex vitro* rooting of pistachio (*Pistacia vera* L.)', *Plant Cell Tiss. Org. Cult.*, 108, 353–8.
Benmahioul, B., Kaid-Harche, M. and Daguin, F. (2012b), 'Influence of activated charcoal on *in vitro* germination and growth of plantlets of pistachio (*Pistacia vera* L.)', *Int. J. Sci. Nat.*, 3, 613–16.
Benmahioul, B., Kaid-Harche, M. and Daguin, F. (2016), '*In vitro* regeneration of *Pistacia vera* L. from nodal explants', *J. For. Sci.*, 62, 198–203.
Blackwell, B. (2016), '*Pistacia* plant named P.N.B.1', U.S. Patent No. PP27,319.
Caruso, T., Iannini, C., Barone E., Marra, F. P., Sottile, F., Greco, C. I., Sabina, M. R., Martelli, G., Monastra, F., Batlle, I., Vargas, F., Romero, M., Ak, B. E., Zakynthinos, G., Rouskas, D., Padulosi, S. and Laghezali M. (1998), 'Genetic and phenotypic diversity in pistachio (*P. vera* L.) germplasm collected in Mediterranean countries', *Acta Hortic.*, 470, 168–78.
Cetiner, S, Yucel, H, Aka-Kacar, Y. and Yalcin-Mendi, Y. (1996), 'Clonal propagation of *Pistacia* rootstocks by meristem and shoot culture', *Acta Hortic.*, 441, 333–6.
Chao, C. T., Parfitt, D. E., Ferguson, L., Kallsen, C. and Maranto, I. (1998), 'Breeding and genetics of pistachio: The Californian program', *Acta Hortic.*, 470, 152–61.
Chen, S., Wu, X., Ji, Y. and Yang, J. (2011), 'Isolation and characterization of microsatellite loci in *Pistacia weinmannifolia* (Anacardiaceae)', *Int. J. Mol. Sci.*, 12(11), 7818–23.
Crane, J. C. and Al-Shalan, I. (1977), 'Carbohydrate and nitrogen levels in pistachio branches as related to shoot extension and yield', *J. Am. Soc. Hortic. Sci.*, 102, 396–9.
Crane, I. C. and Iwakiri, B. T. (1981), 'Morphology and reproduction of pistachio', *Hortic. Rev.*, 3, 376–93.
Crane, I. C. and Nielsen, M. M. (1971), 'The unusual mechanism of alternate bearing in pistachio', *HortSci.*, 6, 489–90.
Crane, J. C., Catlin, P. B. and Al-Shalan, I. (1976), 'Carbohydrate levels in the pistachio as related to alternate bearing', *J. Am. Soc. Hortic. Sci.*, 101, 371–4.
Delijam, M., Garoosi, G., Nezami-Alanagh, E. and Hosseini, R. (2016), 'Improving *Pistacia vera* micropropagation: With emphasis on the efficiency of minerals, vitamins and PGRs', *J. Plant Mol. Breed.*, 4(1), 43–54.
Dollo, L. (1996), 'An isozyme study of Sicilian *Pistacia* species, varieties and offspring from artificial pollination', *in Proceedings of the IX GREMPA Meeting on Almonds and Pistachios*, 20–21 May 1993, Bronte-Sciacca, Italy, pp. 112–18.

Epstein, L., Beede, R., Kaur, S. and Ferguson L. (2004), 'Rootstock effects on pistachio trees grown in *Verticillium dahliae*-infested soil', *Phytopathology*, 94(4), 388–95.

Esmail-pour, A. (2001), 'Distribution, use and conservation of pistachio in Iran', *in* S. Padulosi and A. Hadj-Hassan (Eds), *Towards a Comprehensive Documentation and Use of Pistacia genetic diversity in Central and West Asia, North Africa and Europe*, IPGRI, Roma, Italy, pp. 16–26.

FAOSTAT (2018), 'Crops', viewed 20 February 2018, from http://www.fao.org/faostat/en/#data/QC

Fares, K., Guasmi, F., Touil, L., Triki, T. and Ferchichi, A. (2009), 'Genetic diversity of pistachio tree using inter-simple sequence repeat markers ISSR supported by morphological and chemical markers', *Biotechnology*, 8, 24–34.

Ferguson, L. and Polito, V. S. (2016), 'Alternate bearing, nut blanking and shell splitting', *in* L. Ferguson and D. R. Haviland (Eds), *Pistachio Production Manual*, University of California, Davis, CA, pp. 313–21.

Ferguson, L., Beede, R. H., Freeman, M. W., Haviland, D. R., Holtz, B. A. and Kallsen, C. E. (2005), *Pistachio Production Manual*, 4th edition, University of California, Davis, CA.

Ferguson, L., Beede, R. H., Reyes, H., Sanden, B. L., Grattan, S. R. and Epstein, L. (2016), 'Pistachio rootstocks', *in* L. Ferguson and D. R. Haviland (Eds), *Pistachio Production Manual*, University of California, Davis, CA, pp. 65–74.

Ghaffari, S. M. and Harandi, O. F. (2002), 'Chromosome counts and assessment of two heterochromatic chromosomes in some species of *Pistacia* L. from Iran', *Acta Hortic.*, 591, 389–93.

Ghaffari, S. M, Shabazaz, M. and Behboodi, B. S. (2005), 'Chromosome variation in *Pistacia* genus', *Options Médit.*, 63, 347–54.

Ghorbel, A., Ben Salem-Fnayou, A., Chatibi, A. and Twey, M. (2001), 'Genetic resources of *Pistacia* in Tunisia', *in* S. Padulosi and A. Hadj-Hassan (Eds), *Towards a Comprehensive Documentation and Use of Pistacia Genetic Diversity in Central and West Asia, North Africa and Europe*, IPGRI, Roma, Italy, pp. 62–71.

Golan-Goldhirsh, A., Barazani, O., Wang, Z. S., Khadka, D. K., Saunders, J. A., Kostiukovsky, V. and Rowland, L. J. (2004), 'Genetic relationships among Mediterranean *Pistacia* species evaluated by RAPD and AFLP', *Plant Syst. Evol.*, 246, 9–18.

Grattapaglia, D. and Sederoff, R. (1994), 'Genetic linkage maps of *Eucalyptus grandis* and *eucalyptus urophylla* using a pseudo-testcross: Mapping strategy and RAPD markers', *Genetics*, 137, 1121–37.

Grundwag, M. and Werker, W. (1976), 'Comparative wood anatomy as an aid to identification of *Pistacia* species', *Isr. J. Bot.*, 25, 152–67.

Guenni, K., Aouadi, M., Chatti, K. and Salhi-Hannachi, A. (2016), 'Analysis of genetic diversity of Tunisian pistachio (*Pistacia vera* L.) using sequence-related amplified polymorphism (SRAP) markers', *Genet. Mol. Res.*, 15 (4), 1–15.

Guerrero, J., Memmi, H., Lopez, P. D., Couceiro, J. F., Moriana, A., Martinez, E. and Gijon, M. C. (2014), 'Phenological behavior of two new male cultivars of pistachio (*Pistacia vera* L.): Chaparrillo and Guerrero', *Acta Hortic.*, 1028, 297–302.

Hadj-Hassan, A. (2001), 'Cultivated Syrian pistachio varieties', *in* S. Padulosi and A. Hadj-Hassan (Eds), *Towards a Comprehensive Documentation and Use of Pistacia Genetic Diversity in Central and West Asia, North Africa and Europe*, IPGRI, Roma, Italy, pp. 1–12.

Harandi, O. F. and Ghaffari, M. (2001), 'Chromosome studies on pistachio (*Pistacia vera* L.) from Iran', *Options Médit.*, 56, 35–9.

Holtz, B. A. and Teviotdale, B. L. (2016), 'Soilborne diseases', *in* L. Ferguson and D. R. Haviland (Eds), *Pistachio Production Manual*, University of California, Davis, CA, pp. 257–64.

Hormaza, J. I. and Herrero, M. (1998), 'Pollen effects on fruit and seed characteristics in pistachio', *Ann. Appl. Biol.*, 132, 357–64.

Hormaza, J. I. and Wünsch. A. (2007), 'Pistachio, in genome mapping and molecular breeding in plants', *in* C. Kole (Eds), *Fruits and Nuts*, Springer, Berlin, Germany, 4, pp. 243–9.

Hormaza, J. I., Dollo, L. and Polito, V. S. (1994a), 'Determination of relatedness and geographic movements of *Pistacia vera* (Pistachio; Anacardiaceae) germplasm by RAPD analysis', *Econ. Bot.*, 48(4), 349–58.

Hormaza, J. I., Dollo, L. and Polito, V. S. (1994b), 'Identification of a RAPD marker linked to sex determination in *Pistacia vera* using bulked segregant analysis', *Theor. Appl. Genet.*, 89(1), 9–13.

Hormaza, J. I., Pinney, K. and Polito, V. S. (1998), 'Genetic diversity of pistachio (*Pistacia vera* L., Anacardiaceae) germplasm based on randomly amplified polymorphic DNA (RAPD) markers', *Econ. Bot.*, 52, 78–87.

Ila, H. B., Kafkas, S. and Topaktas, M. (2003), 'Chromosome numbers of four *Pistacia* (Anacardiaceae) species', *J. Hortic. Sci. Biotechnol.*, 78, 35–8.

Iranjo, P., Nabatiahmadi, D., Sorkheh, K., Memeari, H. R. and Ercisli, S. (2016), 'Genetic diversity and phylogenetic relationships between and within wild *Pistacia* species populations and implications for its conservation', *J. For. Res.*, 27(3), 685–97.

Iran Pistachio Association (2017), 'Iranian pistachio varieties', viewed 27 December 2017, from https://www.iranpistachio.org/

Javanshah, A., Tajabadipour, A., Shuraki A. D., Sherafati, A., Mortazavi, A., Kashanizadeh, S. and Esmaeilpour, A. (2014), 'Identification and registration of Iranian commercial pistachio by morphological characteristics', *Acta Hortic.*, 1028, 289–95.

Jazi, M., Seyedi, S., Ebrahimie, E., Ebrahimi, M., Moro, G. and Botanga, C. (2017), 'A genome-wide transcriptome map of pistachio (*Pistacia vera* L.) provides novel insights into salinity-related genes and marker discovery', *BMC Genom.*, 18 (1), 627.

Joley, L. E. (1979), 'Pistachios', *in* R. A. Jaynes (Ed.), *Nut Tree Culture in North America'*, Northern Nut Growers Association, Hamden, CT, pp. 163–74.

Kafkas, S. (2003), 'Developing of monoecious pistachio (*P. vera* L.) populations and the sex determination mechanism in *Pistacia* by cross-breeding', *Acta Hortic.*, 591, 179–85.

Kafkas, S. (2006a), 'Phylogeny, evolution and biodiversity in the genus *Pistacia* (Anacardiaceae)', *in* A. K., Shama and A. Sharma (Eds), *Plant Genome, Biodiversity and Evolution, Volume 1. Part C: Phanerogams (Angiosperm-Dicotyledons)*, Science, Enfield, NH, pp. 525–57.

Kafkas, S. (2006b), 'Phylogenetic analysis of the genus *Pistacia* by AFLP markers', *Plant Syst. Evol.*, 262(1–2), 113–24.

Kafkas, S. and Perl-Treves, R. (2001), 'Morphological and molecular phylogeny of *Pistacia* species in Turkey', *Theor. Appl. Genet.*, 102, 908–15.

Kafkas, S. and Perl-Treves, R. (2002), 'Interspecific relationships in *Pistacia* based on RAPD fingerprinting', *HortSci.*, 37, 168–71.

Kafkas, S., Perl-Treves, R. and Kaska, N. (2000), 'Unusual *Pistacia atlantica* Desf. (Anacardiaceae) monoecious sex type in the Yunt Mountains of the Manisa province of Turkey', *Isr. J. Plant Sci.*, 48, 277–80.

Kafkas, S., Cetiner, S. and Perl-Treves, R. (2001), 'Development of sex-associated RAPD markers in wild *Pistacia* species', *J. Hortic. Sci. Biotech.*, 76, 242–6.

Kafkas, S., Acar, I., Gozel, H. and Eti, S. (2004), 'Breeding monoecious pistachio cultivars', *Nucis*, 12, 21–3.

Kafkas, S., Ozkan, H., Ak, B. E., Acar, I., Atli, H. S. and Koyuncu, S. (2006a), 'Detecting DNA polymorphism and genetic diversity in a wide pistachio germplasm: Comparison of AFLP, ISSR and RAPD markers', *J. Am. Soc. Hortic. Sci.*, 131(4), 522–9.

Kafkas, S., Kafkas, N., Wassimi, A. N. and Padulosi, S. (2006b), 'Molecular characterisation of Afghan pistachio accessions by amplified fragment length polymorphisms (AFLPs)', *J. Hortic. Sci. Biotechnol.*, 81, 864–8.

Kafkas, S., Khodaeiaminjan, M., Guney, M. and Kafkas, E. (2015), 'Identification of sex-linked SNP markers using RAD sequencing suggests ZW/ZZ sex determination in *Pistacia vera* L.', *BMC Genom.*, 16(1), 98.

Kafkas, S., Gozel, H., Bozkurt, H., Karci, H., Topcu, H., Paizila, H., Kafkas, E., Khodaeiaminjan, M., Guney, M., Paizila, A., Coban, N., Kefayati, S., Islam, R., Zhaanbaev, M., Kafkas, N. E., Acar, I. and Uzun, M. (2017a), 'Marker-assisted cultivar breeding in pistachio', *in 7th International Symposium on Almonds and Pistachios*, 5–9 November 2017, Adelaide, Australia.

Kafkas, S., Michelmore, R., Tang, H., Ming, R., Navajas-Pérez, R., Wai, J., Palmer, W., Ozkan, H., Zhang, X., Val Muñoz, C. D., Min, X., Liao, Z., Jacygrad, E., Rwahni, M., Golino, D., Preece, J., Zhang, X., Robles, F., Herran, R., Campoy P., Garcia, A., Kafkas, E., Khodaeiaminjan, M., Guney, M., Topcu, H., Motalebipour, E., Paizila, A., Gozel, H., Coban, N., Killian, A., Song, J., Carling, J., Kefayati, S., Karci, H., Islam, R. and Zhaanbaev, M. (2017b), 'Pistachio genome sequencing and assembly', *in 7th International Symposium on Almonds and Pistachios*, 5–9 November 2017, Adelaide, Australia.

Kallsen, E. C. and Parfitt, D. E. (2016a), 'Gumdrop, a new early harvest pistachio cultivar' *HortSci.*, 52, 310–12.

Kallsen, C. E. and Parfitt, D. E. (2016b), 'Female pistachio variety named Gumdrop', U.S. Patent No. 2016/0309632.

Kallsen, C. E. and Parfitt, D. E. (2018), 'Female pistachio tree named Tejon', U.S. Patent No. PP28,931.

Kallsen, C. E., Parfitt, D. E. and Maranto, J. (2016), 'Male pistachio variety named Famoso', U.S. Patent No. 2016/0309631.

Kamiab, F., Vesvaei, A. and Panahi, B. (2006), 'Male performance in pistachio (*Pistacia vera* L.)', *Acta Hortic.*, 726, 133–8.

Kamiab, F., Ebadi, A., Panahi, B. and Tajabadi, A. (2014), 'RAPD analysis for sex determination in *Pistacia vera* L.', *J. Nuts*, 5(1), 51–5.

Karimi, H. R. and Kafkas, S. (2011), 'Genetic relationships among *Pistacia* species studied by SAMPL markers', *Plant Syst. Evol.*, 297, 207–12.

Karimi, H. R., Kafkas, S., Zamani, Z., Ebadi, A. and Fatahi, M. R. (2009), 'Genetic relationships among *Pistacia* species using AFLP markers', *Plant Syst. Evol.* 279, 21–8.

Kashanizadeh, S., Gholipour, Y. and Mohammadbeigi, A. (2006), 'Identification and collection of pistachio male cultivars in Qazvin region', *Acta Hortic.*, 726, 459–60.

Kaska, N. (1995), 'Pistachio nut growing in Turkey', *Acta Hortic.*, 419, 161–4.

Kaska, N. and Ak, B. E. (1996a), 'Effect of pollens of different *Pistacia* species on nut set of pistachios', *Proceedings of the IX GREMPA Meeting on Almonds and Pistachios*, 20–21 May 1993, Bronte-Sciacca, Italy, pp. 85–9.

Kaska, N. and Ak, B. E. (1996b), 'Effect of pollens of different *Pistacia* species on some physiological features of pistachios', *Proceedings of the IX GREMPA Meeting on Almonds and Pistachios*, 20–21 May 1993, Bronte-Sciacca, Italy, pp. 43–7.

Kaska, N., Caglar, S. and Kafkas, S. (1996), 'Genetic diversity and germplasm conservation of *Pistacia* species in Turkey', *in* S. Padulosi, T. Caruso and E. Barone (Eds), *Taxonomy, Distribution, Conservation and Uses of Pistacia Genetic Resources. Report of a Workshop*, 29–30 June 1995, Palermo, Italy, International Plant Genetic Resources Institute, Rome, Italy, pp. 46–50.

Kaska, N., Kafkas, S., Padulosi, S., Wassimi, N. and Ak, B. E. (2006), 'Characterization nut species of Afghanistan: II-Pistachio', *Acta Hortic.*, 726, 157–64.

Katsiotis, A., Hagidimitriou, M., Drossou, A., Pontikis, C. and Loukas, M. (2003), 'Genetic relationships among species and cultivars of *Pistacia* using RAPDs and AFLPs', *Euphytica*, 132, 279–86.

Kebour, D., Boutekrabt, A., Mefti, M. and AbdulHussain, A. S. (2012), 'Use of the ISSR markers for the study of genetic polymorphism of the pistachio fruit (*Pistacia vera* L.) in Algeria', *UPB Sci. Bull Series B*, 74, 191–200.

Khanazarov, A. A., Chernova, G. M., Rakmonov, A. M., Nikolyi, L. V., Ablaeva, E., Zaurov, D. E., Molnar, T. J., Eisenman, S. W. and Funk, C. R. (2009), 'Genetic resources of *Pistacia vera* L. in Central Asia', *Genet. Resour. Crop Evol.*, 56(3), 429–43.

Khodaeiaminjan, M., Kafkas N. E., Guney, M. and Kafkas, S. (2017), 'Development and genetic linkage mapping of novel sex-linked SNP markers in pistachio', *Mol. Breed.*, 37 (98).

Khodaeiaminjan, M., Kafkas, S., Motalebipour, E. and Coban, N. (2018), '*In silico* polymorphic novel SSR marker development and the first SSR-based genetic linkage map in pistachio', *Tree Genet. Genom.*, 14, 45.

Kolahi-Zonoozi, S. H., Mardi, M., Zeinalabedini, M., Pirseyedi, S. M., Mahmoodi, P., Tabatabaei, I, Mosavi-Derazmahalleh, S. M., Farsi, M., Ebrahimi, M. A., Khayam-Nekoui, S. M. and Ahmadi, K. (2014), 'Development of 12 new SSR markers for genetic diversity and structure analysis in pistachio (*Pistacia vera* L.)', *J. Hortic. Sci. Biotechnol.*, 89, 707–11.

Loudyi, W. (2001),'*Pistacia* genetic resources and pistachio nut production in Morocco', *in* S. Padulosi and A. Hadj-Hassan (Eds), *Towards a Comprehensive Documentation and Use of Pistacia Genetic Diversity in Central and West Asia, North Africa and Europe*, IPGRI, Roma, Italy, pp. 56–61.

Louskas, M. and Pontikis, C. A. (1979), 'Pollen isozyme polymorphism in types of *Pistacia vera* and related species as aid in taxonomy', *J. Hort. Sci.*, 54, 95–102.

MacIlvaine, J., Mercure, E. and Etchamendy, J. (2015a), 'Pistachio rootstock tree named Paramount pistachio rootstock #1', U.S. Patent No. 2015/0156935.

MacIlvaine, J., Mercure, E. and Etchamendy, J. (2015b), 'Pistachio rootstock tree named Paramount pistachio rootstock #2', U.S. Patent No. 2015/0156936.

MacIlvaine, J., Mercure, E. and Etchamendy, J. (2016a), 'Pistachio rootstock tree named Wonderful pistachio rootstock #2', U.S. Patent No. PP26,915.

MacIlvaine, J., Mercure, E. and Etchamendy, J. (2016b), 'Pistachio rootstock tree named Wonderful pistachio rootstock #4', U.S. Patent No. PP27,319.

Maggs, D. H. (1973), 'Genetic resources in pistachio', *Plant Genet. Resour. Newsl.*, 29, 7–15.

Maggs, D. H. (1982), *An Introduction to Pistachio Growing in Australia*, CSIRO Publishing, Adelaide, Australia.

Martinelli, A. and Loretti. F. (1988), 'Use of in vitro techniques for selection and cloning of different *Pistacia* species', *Acta Hortic.*, 227, 436-7.

Michailides, T. J., Morgan, D. P. and Doster, M. A. (2016), 'Foliar, fruit and branch diseases', *in* L. Ferguson and D. R. Haviland (Eds), *Pistachio Production Manual*, University of California, Davis, CA, pp. 265-92.

Mirzaei, S., Bahar M. and Sharifnabi, B. (2006), 'A phylogenetic study of Iranian wild pistachio species and some cultivars using RAPD markers', *Acta Hortic.*, 726, 39-43.

Monselise, S. P. and Goldschmidt, E. E. (1982), 'Alternate bearing in fruit trees', *in* J. Janic (Ed.), *Horticultural Reviews*, Volume 4, The AVI Publishing Company, Westport, CT, pp. 128-73.

Motalebipour, E., Kafkas, S., Khodaeiaminjan, M., Coban, N. and Gozel, H. (2016), 'Genome survey of pistachio (*Pistacia vera* L.) by next generation sequencing: Development of novel SSR markers and genetic diversity in *Pistacia* species', *BMC Genom.*, 17, 998.

Motalebipour, E. Z., Gozel, H., Khodaeiaminjan, M. and Kafkas, S. (2018), 'SSR-based linkage map construction in pistachio using an interspecific F1 population and QTL analysis for leaf and shoot traits', *Mol. Breed.* (*accepted manuscript*).

Nezami, S. R., Yadollahi, A., Hokmabadi, H. and Eftekhari, M. (2015), 'Control of shoot tip necrosis and plant death during *in vitro* multiplication of pistachio rootstock UCB1 (*Pistacia integrima* × *P. atlantica*)', *J. Nuts*, 6(1), 27-35.

Noroozi, S., Baghizadeh, A. and Javaran, M. J. (2009), 'The genetic diversity of Iranian pistachio (*Pistacia vera* L.) cultivars revealed by ISSR markers', *Biol. Divers. Conserv.*, 2(2), 50-6.

Nzima, M., Martin, G. and Nishijima, C. (1997), 'Seasonal changes in total non-structural carbohydrates within branches and roots of naturally "off" and "on" "Kerman" pistachio cultivar', *J. Am. Soc. Hortic. Sci.*, 122, 856-62.

Onay, A. and Firat, M. Z. (1999), 'Analysis of binominal data obtained from plant tissue culture studies using generalised linear models', *Turk. J. Biol.*, 23, 261-7.

Onay, A. and Jeffree, C. E. (2000), 'Somatic embryogenesis in pistachio', *in* S. M., Jain, P. K. Gupta and R. J. Newton (Eds), *Somatic Embryogenesis in Woody Plants*, Kluwer Academic Publishers, Dordrecht, the Netherlands, pp. 361-90.

Onay, A. and Namli, S. (1998), 'Antepfistiginin yapraklarindaki somatik embriyogenesis', (Somatic embryogenesis in pistachio leaf', *in XIV Ulusal Biyoloji Kongresi*, 7-10 September 1998, Samsun-Turkey.

Onay, A., Jefree, C. E. and Yeoman, M. M. (1995), 'Somatic embryogenesis in cultured immature kernels of pistachio, *Pistacia vera* L.', *Plant Cell Rep.*, 15, 192-5.

Onay, A., Pirinc, V., Isikalan, C., Adiyaman, F., Tilkat, E. and Baflaran, D. (2003), '*In vivo* and *in vitro* micrografting of pistachio, *Pistacia vera* L. cv. Siirt', *Turk. J. Biol.*, 27, 95-100.

Onay, A., Pirinc, V., Yildirim, H. and Basaran, D. (2004a), '*In vitro* micropropagation of pistachio (*Pistacia vera* var. Siirt)', *Plant Cell Tissue Organ Cult.*, 77, 215-19.

Onay, A., Pirinc, V., Yildirim Tilkat, E., Akturk, Z. and Yildirim, H. (2004b), 'Somatic embryogenesis of pistachio from female flowers', *J. Hortic. Sci. Biotechnol.*, 79(6), 960-4.

Onay, A., Tilkat, E., Yildirim, H. and Suzerer, V. (2007), 'Indirect somatic embryogenesis from mature embryo cultures of pistachio, *Pistacia vera* L.', *Propag. Ornam. Plants*, 7(2), 68-74.

Ozbek, S. (1978), 'Ozel Meyvecilik' (Fruit Growing), University of Cukurova. Agr. Dept. No. 128 Adana.

Ozbek, S. and Ayfer M. (1957), '*Pistacia* turleri uzerinde sitolojik arastirmalar' (Studies on cytology of *Pistacia* species), *Ankara Universitesi Ziraat Fakultesi Yilligi*, 3, 203-22.

Parfitt, D. E. and Almehdi, A. (1994) 'Use of high $CO_2$ atmospheric and medium modifications for the successful micropropagation of pistachio', *Sci. Hortic.*, 56, 321-9.

Parfitt, D. E. and Badenes, M. L. (1997), 'Phylogeny of the genus *Pistacia* as determined from analysis of the chloroplast genome', *Proceedings of the National Academy of Sciences*, 94, 7987-92.

Parfitt, D. E., Maranto, J. and Kallsen, C. E. (2006a), 'Golden Hills female pistachio', U.S. Patent No. 2006/0212980.

Parfitt, D. E., Maranto, J. and Kallsen, C. E. (2006b), 'Lost Hills female pistachio', U.S. Patent No. 2006/0212981.

Parfitt, D. E., Maranto, J. and Kallsen, C. E. (2006c), 'Randy male pistachio', U.S. Patent No. 2006/0212979.

Parfitt, D. E., C. Kallsen, J. Maranto and B. Holtz. (2007). 'Golden Hills Pistachio', *HortSci.*, 42(3), 694-6.

Parfitt, D. E., Kallsen C. E., Holtz, B. and Maranto, J. (2008), 'Lost Hills a new pistachio cultivar for California', *HortSci.*, 43(1), 247.

Parfitt, D. E., S. Kafkas, I. Batlle, F. J. Vargas and C. E. Kallsen (2012), 'Pistachio', *in* M. L. Badenes and D. H. Byrne (Eds), *Fruit Breeding*, Springer, New York, NY, pp. 803-26.

Parfitt, D. E., Kallsen, C. E. and Maranto, J. (2016), 'Pistachio cultivars', *in* L. Ferguson and D. R. Haviland (Eds), *Pistachio Production Manual*, University of California, Richmond, CA, pp. 59-64.

Pazouki, L., Salehi, Shanjani, P., Hagidimitriou, M., Pirseyedi, S. M., Naghavi, M. R., Avanzato, D., Quarta, R., Kafkas, S., Ghareyazie, B., Ghaffari, M. R., Khayam, Nekoui, S. M. and Mardi, M. (2010), 'Genetic diversity and relationship among *Pistacia* species and cultivars', *Conserv. Genet.*, 11, 311-18.

Pontikis, C. A. (1986), '"Pontikis" pistachio', *HortSci.*, 21(4), 1074.

Pontikis, C. A. (1990), 'Effect of 2-naphthalene acetic acid on alternate bearing of pistachio', *Fruits*, 45, 281-5.

Popov, K. P. (1994), 'Trees, shrub, and semi shrubs in the mountains of Turkmenistan', *in* V. Fet and K. I. Atamuradov, (Eds), *Biogeography and Ecology of Turkmenistan*, Springer, Kluwer, Netherlands, pp. 173-86.

Safarnejad, A. (2012), 'Characterization of different *Pistacia vera* varieties in Khorasan province of Iran', *Int. J. Adv. Biol. Res.*, 2(4), 636-40.

Salehi Shanjani, P., Mardi, M., Pazouki, L., Hagidimitriou, M., Avanzato, D., Pirseyedi, S. M., Ghaffari, M. R. and Khayam, Nekoui, S. M. (2009), 'Analysis of the molecular variation between and within cultivated and wild *Pistacia* species using AFLPs', *Tree Genet. Genom.*, 5, 447-58.

Sanden, B. L., Ferguson, L., Reyes, H. C. and Grattan, S. C. (2004), 'Effect of salinity on evapotranspiration and yield of San Joaquin Valley pistachios', *Acta Hortic.*, 664, 583-9.

Sanden, B. L., Ferguson, L., Corwin, D. and Kallsen, C. E. (2010), 'Using saline groundwater for large-scale development and irrigation of pistachio interplanted with cotton', *Options Médit.*, 94, 31-6.

Sharma, D., Castillon, J. and Ash, K. (2017), 'Pistachio rootstock named clone UCB1-D110', U.S. Patent No. 2017/0094851.

Sheibani, A. (1995), 'Pistachio production in Iran', *Acta Hortic.*, 419, 14-15.

Sola-Campoy, P. J., Robles, F., Schwarzacher, T., Rejon, C. R., Herran, R. and Navajas-Perez, R. (2015), 'The molecular cytogenetic characterization of pistachio (*Pistacia vera* L.) suggests the arrest of recombination in the largest heteropycnotic pair HC1', *PLoS One*, 10, e0143861.

Spann, T. M., Beede, R. H. and DeJong, T. M. (2007), 'Preformation in vegetative buds of pistachio (*Pistacia vera* L.): Relation to shoot morphology, crown architecture and rootstock vigor', *Tree Physiol.*, 27, 1189-96.

Srivastava, A. and Joshi, A. G. (2013), 'Control of shoot tip necrosis in shoot cultures of *Portulaca grandiflora* hook', *Not. Sci. Biol.*, 5(1), 45-9.

Stamler, R. A., Kilcrease, J., Kallsen, C., Fichtner, E. J., Cooke, P., Heerema, R. J. and Randall J. J. (2015), 'First report of *Rhodococcus* isolates causing pistachio bushy top syndrome on "UCB-1" rootstock in California and Arizona', *Plant Dis.*, 99, 1468-76.

Sun, Q., Yang, X. and Li, R. (2014), 'SCAR marker for sex identification of *Pistacia chinensis* Bunge (Anacardiaceae)', *Genet. Mol. Res.*, 13(1), 1395-401.

Tagizad, A., Ahmadi, J., Haddad, R. and Zarrabi, M. (2010), 'A comparative analysis of ISSR and RAPD markers for studying genetic diversity in Iranian pistachio cultivars', *Iran. J. Genet. Plant Breed.* 1(1), 6-16.

Tahtaci, S. A., Arpaci, S., Gozel, H., Bilim, H. C., Atli, H. S. and Tekin, H. (2011), 'Breeding selection on pistachio', *Acta Hortic.*, 912, 427-33.

Talebi, M., Akbari, M., Zamani, M. and Sayed-Tabatabaei, E. B. (2016), 'Molecular polymorphism in *Pistacia vera* L. using non-coding regions of chloroplast DNA', *J. Genet. Eng. Biotechnol.*, 14, 31-7.

Taskin, T., Cobanoglu, D., Colak, G., Namli, S., Basaran, D. and Gaspar, T. (1996), 'Plant regeneration through somatic embryogenesis in *Pistacia vera* L.', *Saussurea* 27, 59-65.

Tilkat, E., Onay, A., Yildirim, H. and Ozen, H. C. (2008), 'Micropropagation of mature male pistachio *Pistacia vera* L.', *J. Hortic. Sci. Biotech.*, 83(3), 328-33.

Tilkat, E., Suzerer, V., Akdemir H., Tilkat, E. A., Ozden C. Y. and Onay A. (2013), 'A rapid and effective protocol for surface sterilization and *in vitro* culture initiation of adult male pistachio (*Pistacia vera* L. cv."Atli")', *Acad. J. Sci. Res.*, 1(8), 134-41.

Topcu, H., Coban, N. and Kafkas, S. (2016), 'Novel microsatellite markers in *Pistacia vera* L. and their transferability across the genus *Pistacia*', *Sci. Hortic.*, 198, 91-7.

TUIK (2017), Viewed on 25 December 2017, from http://www.tuik.gov.tr/PreTablo.do?alt_id=1046.

Turkeli, Y. and Kafkas, S. (2013), 'First genetic linkage map in pistachio constructed using an interspecific cross between *Pistacia vera* L. and monoecious *Pistacia atlantica* Desf', *Sci. Hortic.*, 151, 30-7.

Uzun, M., Acar, I., Atli, H. S., Arpaci, S., Sarpkaya, K. and Gozel, H. (2011), 'Breeding of new pistachio cultivars by hybridization in Turkey', *Acta Hortic.*, 912, 321-6.

Vargas, F. J. and Romero, M. A. (1998a), 'Vigour and juvenile stage in pistachio progenies', *Options Médit.*, 33, 105-11.

Vargas, F. J. and Romero, M. A. (1998b), 'Vigour in pistachio progenies', *Acta Hortic.*, 470, 162-7.

Vargas, F. J. and Romero, M. A. (2005), 'Fruit characters in pistachio', *Options Médit.*, 63, 49-56.

Vargas, P. J., Romero, M., Rovira, M. and Batlle, I. (1996), 'Pistachio cultivar improvement at IRTA-Mas Bove', *in* T. Caruso, E. Barone and F. Sottile (Eds), *Proceedings of the IX GREMPA Meeting-Pistachio*, 20-21 May 1993, Bronte - Sciacca, Italy, pp. 15-19.

Vargas, F. J., Romero, M. A. and Vargas, I. (2001), 'Blooming and leafing time in pistachio progenies', *Options Médit.*, 56, 41–6.

Vargas, F. J., Romero, M. A. and Vargas, I. (2002), 'Flowering precocity in pistachio progenies', *Acta Hortic.*, 591, 297–303.

Weinbaum, S. A., Picchioni, G. A., Muraoka, T. T., Ferguson L. and Brown, P. H. (1994), 'Fertilizer nitrogen and boron uptake, storage and allocation vary during the alternate bearing cycle in pistachio tree', *J. Jpn. Soc. Hortic. Sci.*, 119, 24–31.

Werner, O., Sánchez-Gómez, P., Guerra, J. and Martínez, J. F. (2001), 'Identification of *Pistacia × Saportae* Burnat (Anacardiaceae) by RAPD analysis and morphological characters', *Sci. Hortic.*, 91, 179–86.

Whitehouse, W. E. (1957). 'The pistachio nut – a new crop for the Western United States' *Econ. Bot.* 11, 281-321.

Xiong, E., Wu, X., Shi, J., Wang, X. and Wang, W. (2013), 'Proteomic identification of differentially expressed proteins between male and female plants in *Pistacia chinensis*', *PLoS One*, 8(5), e64276.

Yakubov, B., Barazani, O. and Golan-Goldhirsh, A. (2005), 'Combination of SCAR primers and touchdown-PCR for sex identification in *Pistacia vera* L.', *Sci. Hortic.*, 103, 473–78.

Yaltirik, F. (1967) *Anacardiaceae*, *in* P. H. Davis (Ed.), *Flora of Turkey*, Edinburgh University Press, Edinburgh, UK, pp. 544–8.

Zaloglu, S., Kafkas, S., Dogan, Y. and Guney, M. (2015), 'Development and characterization of SSR markers from pistachio (*Pistacia vera* L.) and their transferability to eight *Pistacia* species', *Sci. Hortic.*, 189, 94–103.

Zohary, M. (1952), 'A monographical study of the genus *Pistacia*', *Palestine Journal of Botany*, Jerusalem Series 5, 187-228.

Zohary, D. (1996), 'The genus *Pistacia* L.', *in* S. Padulosi, T. Caruso and E. Barone (Eds), *Taxonomy, Distribution, Conservation and Uses of Pistacia Genetic Resources. Report of a Workshop*, 29–30 June 1995, Palermo, Italy, International Plant Genetic Resources Institute, Rome, Italy, pp. 1–11.

# Chapter 15

## Advances in cultivation of pistachio

*Bekir Erol Ak, University of Harran, Turkey*



## 1 Introduction

The pistachio tree is native to western Asia and Anatolia, from Syria to the Caucasus and Afghanistan. Archaeological evidence in Turkey indicates that the nuts were being used for food as early as 7000 BC (Anon., 1997). Anatolia is the major centre of diversity for the genus *Pistacia*. Its species are naturally distributed throughout the country, with the exception of very cold and humid areas. *Pistacia* is present in the form of bushes or trees of different plant sizes that differ due to genetic factors, soil type and the climate conditions that characterize the region. In the Mediterranean and Aegean regions, *Pistacia atlantica* trees reach gigantic dimensions. In the transitional zone between the Mediterranean and central Anatolia, where rainfall is less, a gradual reduction in the height of these trees is observed. *Pistacia terebinthus*, which grows in the form of relatively big bushes in the Mediterranean region, has a reduced plant size in Southeast Anatolia. Because of its wide adaptability to a variety of soils and climates, *P. terebinthus* thrives well from the coastal zone at sea level

http://dx.doi.org/10.19103/AS.2018.0042.23

to up 2000 m into the mainland, this taxon being the most widespread among all *Pistacia* species (Kaska and Bilgen, 1992; Kaska et al., 1996).

The pistachio was introduced to Italy from Syria in the early first century AD at approximately the same time as it was introduced to Spain. Subsequently, its cultivation spread to other Mediterranean countries. Its introduction to Sicily probably occurred some time later, but it started to be grown on a large scale only during the Arab rule (AD 827–1040) (Barone and Caruso, 1996). The tree was first introduced into the United States in 1854 by Charles Mason, who distributed seeds for experimental plantings in California, Texas and some southern states. In 1875, a few small pistachio trees, imported from France, were planted in Sonoma, California. In the early 1900s, the US Department of Agriculture assembled a collection of *Pistacia* species and pistachio nut varieties at the Plant Introduction Station in Chico, California. Commercial production of pistachio nuts began in the late 1970s and rapidly expanded to become a major operation in the San Joaquin Valley. Other major pistachio-producing areas are Iran and Turkey and to a lesser extent, Syria, India, Greece, Pakistan and elsewhere (Anon., 1997).

Areas of pistachio plantation have increased from 407.9 thousand ha in 1994 to 526.6 thousand ha in 2004. The main producing country, Iran, has a pistachio plantation area that has increased (by 53%) to 280 thousand ha in 2004 (Razavi, 2006). Turkey has a big plantation area, but its yield is very low when compared to other production areas (Arpaci et al., 2005). Pistachio is grown most intensively in Iran, Syria, Turkey and the United States. The other pistachio-producing countries are in the Near East, North Africa and the Southern Europe. According to a 4-year average (2002–2005), the production value in Iran was over 242,000 tons, which is nearly half of the world production (47.46%). The second major producer, the United States, produces 119,000 tons (23.40%), and Turkey has a yield of 54000 tons, 10.52 % of the total world production.

## 2 Pistachio genetics

The pistachio plant belongs to the genus *Pistacia*, which is a member of the family Anacardiaceae, which also includes cashew, mango, poison ivy, poison oak, pepper tree and sumac. The genus consists of 11 or more species (Zohary, 1952; Whitehouse, 1957). There are only a few chromosome count studies relating to *Pistacia*, some of which are rather old. Zohary (1952) reported in his study of the *Pistacia* species that the chromosome numbers of the *P. vera*, *P. atlantica* and *P. lentiscus* species are different at 30, 28 and 24, respectively. Ozbek and Ayfer (1957) then found the chromosome numbers of *P. vera*, *P. khinjuk* and *P. terebinthus* species to be $2n = 30$, and that of *P. atlantica* to be $2n = 28$ by studying meiosis in pollen mother cells.

Harandi and Ghaffari (2001) performed a chromosomal study in *P. vera*; meiosis in pollen mother cells and mitotic studies in meristematic cells of root tips showed that the chromosome number of *P. vera* is $2n = 30$. Laggard chromosomes were observed in some cells both at meiosis and mitosis, and the authors suggested that they are probably candidates for sex chromosomes. Chromosome counting in *P. atlantica*, including the subspecies *cabulica*, *kurdica* and *mutica*, gave $2n = 28$ with three metacentric, nine submetacentric and two acrocentric pairs. One pair of metacentric chromosomes was heterochromatic in three species and the authors suggest the possible existence of an XY chromosomal mechanism for sex determination in *Pistacia*. The chromosomes of *P. khinjuk* were found to be symmetric, with a median or submedian centromere, whereas the others were asymmetric, varying with the centromere positions. According to the authors, basic chromosome numbers are $x = 12, 14, 15$ and $x = 12$ with $x = 15$ derived from $x = 14$ in *Pistacia* (Kafkas, 2006).

According to Zohary (1952), the distribution range of the *Pistacia* species comprises five different regions. These regions are (1) Mexican region (which may be revised to the Northwestern America region in the future), (2) Mediterranean region, (3) Irano-Turanian region, (4) Sino-Japan region and (5) Northeastern Africa region. Zohary (1952) also indicated several isolated areas such as the Saudi Arabian desert (*P. palaestina*), Egypt (*P. khinjuk*) and southern Crimea (*P. atlantica*).

As members of the second section of the species, *P. mexicana* and *P. texana* have a more or less continuous area of distribution, but are entirely isolated from all other species in the genus. The distribution area of *P. weinmannifolia* is China, Taiwan and Malaysia. *P. lentiscus* extends to the whole of the Mediterranean region and the Canary Islands and, with the exception of Egypt, from Portugal to Southeastern Turkey, Syria, Jordan, Israel and Morocco to Libya. *P. sported* is a hybrid between *P. lentiscus* and P. *terebinthus* and extends from the south of France, Spain, Italy, Israel and Jordon to Morocco and Algeria. *P. aethiopica* in this section is an endemic typical Northeast African species and its area comprises Somalia, Ethiopia, Tanzania, Eritrea, Kenya and Uganda (Kafkas, 2006; El-Oqlah, 1996; Kokwaro and Gillett, 1980; Yaltirik, 1967; Zohary, 1952, 1972). *P. falcata* (syn. *P. chinensis* var. falcata Zoh.) shares some of the distribution area of *P. aethiopica* and includes Somalia and Eritrea; however, this species is deciduous and may be considered to be within the first section, named the Lentiscella section (Zohary, 1952).

## 3 Climatic requirements

*P. vera* is grown at an altitude of 2–1220 m. Pistachio can be grown at latitudes of 36°–37° and, being a deciduous tree, is drought-resistant and tolerant to hot and dry summer weather. Areas that are suitable for olive cultivation are generally satisfactory for pistachios, which need cold conditions to satisfy chilling requirements and hot and dry summers to produce fruits. Pistachio trees are frost hardy, normally withstanding temperatures from −14 to −16°C. During some winters in Turkey, it has been observed that pistachio trees have even been able to withstand temperatures as low as −30°C without injury. In Iran, it has been observed that pistachios can tolerate temperatures as low as −20°C (Kaska, 1990). *P. vera* usually blooms in the first week of April, preventing the flowers from becoming damaged from late spring frosts, although plantations established in areas where frost pockets appear may suffer frost damage. Sometimes, the flowers of the pistachio burn from dry and hot winds during the spring.

As with other temperate fruit trees, pistachio requires a period of chilling during the winter period. Inadequate chilling creates problems including very low yield, unsuitable and low percentage of bud break, late flowering, the pre-death of stigma, very late vegetative and generative development, fewer leaves, reduced pollen production, irregular sprouting of the buds and late ripening (Kuden et al., 1995). Crane and Takeda (1979) found that the 'Kerman' cultivar needs about 1,000 h of chilling. Kuden et al. (1995) reported that the male trees require less chilling than females. The 'Uzun' cultivar needs 600 h of chilling and the 'Ohady' cultivar requires more than 1,000 h, while the 'Siirt' cultivar needs 700 h. Summer temperatures are another very important climatic factor. If the summer temperatures are low, kernel development will not be sufficient, splitting will not occur properly and the hard shell might not be easily dehulled.

The pistachio is a very hardy and drought-resistant tree because of its deep tap root system. It can be grown even on rocks and stony soils without irrigation. If trees are irrigated over many years, the tap root system may not develop to be very deep. If irrigation, fertilization and good culturing practices are provided, then the yield is better and the fruits have a better quality compared to non-irrigated trees. If the trees are grown without irrigation, young trees should still be irrigated to help them to establish before they are able to grow independently. The pistachio tree is also resistant to more alkaline soils, being the most salt-resistant tree after the date palm tree. The roots are very active in taking minerals and water and, therefore, survive well in harsh conditions. If grown using irrigation, soil drainage should be good; otherwise, mortality will be high due to *Verticillium* wilt caused by a soil-borne fungus (Tareen et al., 1989).

## 4 Flower and fruit characteristics

Pistachio trees (*P. vera* L.) are dioecious. This means that the male and female flowers occur on separate trees and that pollination and fertilization are necessary to obtain seeded fruits (Whitehouse and Stone, 1941). Pollination occurs by the transport of pollen from male to female trees by wind. Pistachio flowers have no petals, which attract insects, and therefore pistachio orchards must contain male trees with a ratio of male-to-female trees of 1/8 or 1/11 (Kaska, 1990, 1995; Ak and Kaska, 1990; Kaska et al., 1989; Ozbek, 1978; Ak, 1992) (Fig. 1). It is necessary to have enough male trees to ensure adequate pollination as insufficient pollen is a primary cause of crop yield failure. On the other hand, the pollen of the wild species can be used in artificial pollination of *P. vera*. Protandry is common in pistachios (Ak et al., 1998).

Flowers are small and apetalous. In the male flowers, there are long and pointed sepals, and each male flower contains 5 anthers. The female flowers have short styles and greenish ovaries. Stigma is divided into three long and tongue-shaped parts with uneven surfaces and an area of 3 $mm^2$. The length of one flower is 2 mm. Fertilization occurs 20–24 h after pollination. Chalazogamy is noted in *P. vera* L. The inflorescence buds expand at the end of March and wind pollination of the 100–300 flowers per panicle occurs during the first 2 weeks of April (Woodroof, 1982).

The size of the pericarp increases very rapidly from fertilization to the beginning of embryo development. The zygote persists in a resting stage for 50–55 days and, during this period, the flowers and fruits drop, so that there is no relationship between the drops and embryo abortion. If embryo abortion occurs, the fruit can continue to develop without seed. However, embryo abortion and abnormal embryo formation are very rare and there is no seedless fruit problem for this reason (Ayfer, 1964).

Male trees in an orchard should have certain characteristics. A good male tree should have some properties (Ak et al., 1998; Ak, 2001) such as growth must be strong and upright, the flowering period must be synchronized (i.e., flowering periods should be at the same time as in females) (Fig. 2), the flowering period should be long enough to overlap with the flowering period of the females, the number of clusters must be high, flower clusters must be

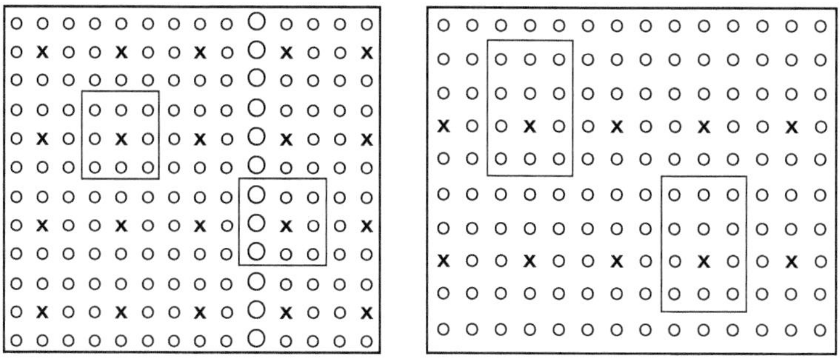

**Figure 1** Distribution of male and female trees in the pistachio orchard. 8 or 11 Female:1 Male Plantation System (x: Male Tree, o: Female Tree)

**Figure 2** Male tree between female trees, with the male tree flowering before the females.

large in size, the amount of pollen production in each cluster must be high, the germination rate of pollen must be high, the pollen's viability in vivo must be long and the plant must not show alternate bearing.

The trees of *P. vera* L. are rather large in size and they can grow up to 10 m in height, generally. They have a wider spread than other *Pistacia* spp. trees. Compound leaves are formed of 3–5 (rarely 1–4 or 7) roundish leaflets.

The fruit is a drupe with a fleshy mesocarp and the shell (endocarp) is stone and ivory in colour. The exocarp and mesocarp become dark red and dark green-purple, respectively. When the nuts are matured and harvested, they can be easily dehulled. Normally, the shells split longitudinally along their sutures when mature. Under unfavourable conditions during nut growth, the shells may not split, although normally the fruits have a split rate of 70% or higher (Ak and Acar, 2001). The colour of the kernel varies from yellowish through to shades of green, which extends throughout the kernel. In general, the deeper the shade of green is, the higher is the perceived quality. Pistachio nuts are rich in oil, with an average content of about 55%. The trees begin bearing fruit after 5–8 years, but full bearing is not normally attained until the 15th or 20th year. Pistachios tend towards biennial bearing, producing heavy crop 1 year followed by little or none the next (Anon., 1997).

## 5 Cultivars

Pistachio trees are grown in approximately 15 countries according to the FAO records; however, this fruit species is grown in a number of other countries for production and research, including Spain, Afghanistan, Chile, Australia, and South Africa. In these countries, different cultivars from different producing countries have been collected for trial purposes. Spain has 29 female and 24 male cultivars or types (Vargas et al., 1995) and new cultivar development studies are ongoing with the breeding of male and female cultivars (Vargas et al., 2001; Vargas and Romero, 2005). The largest collection exists in Iran. The pistachio cultivars of Iran, which have been described are numerous, with characteristics determined for more than 60 named varieties. However, the main cultivars produced are 'Ohady' and 'Kaleh Ghochi' (Sheibani, 1995). However, 'Akbari', 'Ahmad Aghaii' and 'Badami Zarand' have also become common in commercial orchards.

There are major cultivars in each of the main producing countries. In Turkey, 'Kirmizi', 'Uzun' and 'Siirt' are the major cultivars. Turkish cultivars are preferred in many European and US markets due to their good taste and uniformly green kernels. In the United States, the main cultivar is 'Kerman'. In Syria, different cultivars are common, such as 'Ashoury', 'Oleimy', and 'White Batoury', but the main cultivar is 'Ashoury'. Selection studies are underway in Turkey, Syria and Iran and breeding studies have been performed on rootstock development in the United States and Turkey.

### *5.1 Turkish cultivars*

Some pistachio-producing countries prefer Turkish cultivars because of the kernel colour and taste. Some Turkish cultivars also have non-splitting traits. This is an important feature in maintaining a yellow or yellowish kernel and for protection from aflatoxin contamination. While many varieties of pistachio nut have been developed throughout the world, few cultivars have adapted well to the conditions in Turkey. These pistachio cultivars include 'Uzun', 'Kirmizi', 'Siirt', 'Halebi' and the Iranian cultivar 'Ohady'. Of these cultivars, 'Uzun', 'Kirmizi' and 'Halebi' have longer nuts, while 'Siirt' and 'Ohady' have round-shaped nuts (Fig. 3). Cultivars with longer nuts are preferred on account of their kernel colours, while cultivars with round nuts may be preferred by some growers due to their larger size and higher split rates (Anon., 1993). Commonly propagated pistachio cultivars in Turkey are 'Kirmizi' and 'Uzun'; however, these cultivars are

**Figure 3** Fruit colours of two Turkish pistachio cultivars. 'Kirmizi' (left) and 'Siirt' (right).

less popular in international markets due to their lower split rates and kernel yields. Other pistachio cultivars such as 'Keten Gomlegi', 'Beyaz Ben', 'Degirmi', 'Cakmak', 'Sultani', 'Vahidi', 'Mumtaz', 'Sefidi' and 'Haci Serifi' are also grown in Turkey.

Turkish pistachios are thinner and smaller than Iranian pistachios. Only the 'Siirt' cultivar is similar to Iranian pistachios, with shapes somewhat in between Gaziantep and Iranian pistachios. 'Siirt' pistachios, representing about 15% of the total production, are bigger and fetch higher prices in the market than traditional Turkish pistachios. In Turkey, size is directly related to the quality; nuts of 90 or less than 100 g are considered the highest quality, 90–100 g nuts are second, 100–120 gram nuts are of third quality and nuts above 120 g are considered to be of the lowest quality. New Turkish cultivars are developed and registered by the Gaziantep Pistachio Research Institute.

## 5.2 European and North African cultivars

'Napoletana' is the predominant variety in Sicily. Less often planted are 'Agostana', 'Girasola', 'Notaloro', 'Cappuccia' and 'Femminello'. 'Trabonella' and 'Bronte' are Sicilian cultivars with similar characteristics. The nut colour is greenish and the nut shape is longer and thinner than the 'Kerman' or Iranian cultivars. Nut size is considerably smaller than the 'Kerman' and nut quality under local conditions is poor, with a high level of non-splits in some seasons and significant disease and pest problems (Parfitt et al., 2005).

'Sfax', 'Mateur' and 'El Guettar' are grown in Tunisia. 'Mateur' may be the best of these and is currently being tested in Spain. 'Sfax' produces large tight nut clusters, but nut size, yield and per cent splits are inferior to 'Kerman' nuts. 'Aegina' ('Aegenes') and the more recently released 'Pontikis' are grown in Greece. 'Aegina' appears to be very susceptible to *Botryosphaeria dothidea*, perhaps due to its very early flowering in the spring. 'Pontikis' was developed and tested by the C.A. Pontikis Agricultural College of Athens, Greece (Pontikis, 1986). It has a moderately large fruit, the kernel weight is at 55% of the fruit weight and has an oblong-ovate kernel. It splits much better than 'Aegenes', with 90–98% splits. Blank nut percentage is about 5–10%, which is similar to 'Aegenes'. However, the adaptation of 'Pontikis' outside of the Athens area has not been determined. 'Lamarka' is the main cultivar in Cyprus. 'Sirora' was developed in Australia (Maggs, 1990) although, since the industry has remained small in Australia, this cultivar has not been widely planted in Australia or in Greece. 'Kastel' and 'Rashti', grown in Israel, are similar in some aspects to the 'Kerman' variety. Neither 'Kastel' nor 'Rashti' have been evaluated in comparison with 'Kerman' in Californian yield trials. 'Rashti' has large nuts, a high split percentage and a good flavour. The tree structure is similar to 'Kerman' as is its alternate bearing characteristic. It is a late maturing cultivar, probably maturing several weeks after 'Kerman'. This has been an issue during years when a late summer and an autumn have been very cool. Under these conditions, it may not ripen before the start of the winter rains. 'Kastel' seems to be very similar to 'Kerman' in most characteristics (Parfitt et al., 2005).

## 5.3 Iranian cultivars

The history of pistachio cultivation in Iran dates back to around 4,500 years. Since the first trees grew in ancient Persia, it is likely that Iranians were the first consumers of pistachio. It is generally believed that pistachio was introduced from Iran to other parts of the world and there is evidence that the name used in other parts of the world is derived from its Persian name. The history of the cultivation of Pistachio trees in the cities of Damghan, Semnan and Ghazvin goes back to the thirteenth century and in the Kerman Province to the eighteenth century. In other parts of Iran, cultivation of the tree goes back 150–200 years. In spite of this long history, cultivation of pistachio trees in Iran has only developed significantly within the recent decades (Hokmabadi et al., 2008).

There are many pistachio cultivars in Iran. However, there are 10 main commercial cultivars adopted and grown in different provinces. One of the major production provinces is Kerman, growing the cultivars 'Kalleh-ghouchi', 'Ohady', 'Ahmad Aghaii', 'Akbari', 'Momtaz' and 'Foundoghi 48'. Other provinces and their cultivars are Semnan ('Abbas Ali', 'Shah pasand'), Ghazvin ('Kalleh bozi') and Khorasan razavi ('Badami Sefid'). Among these cultivars, 'Ohady' is the oldest cultivar. New commercial cultivars in each province are becoming popular.

'Ohady' is an important cultivar, suitable for cultivation in most parts of the Rafsanjan area. It was selected first in Rafsanjan during 1941–1951. Its cultivation has increased during the past 40 years and now occupies probably more than 60–70% of the cultivated pistachio orchards in the Rafsanjan area. 'Kalleh-ghouchi' is one of the important commercial cultivars and is widespread in both Rafsanjan and Kerman provinces. Its high yield and large fruits are the main reasons for its popularity. It has a clearer habit growth when compared with 'Ouhadi', with stronger branches. It fruits on the third year after grafting (Hokmabadi et al., 2008).

## 5.4 Californian cultivars

California is unique among the world's commercial pistachio industries in that it is comprised solely of two cultivars; the female 'Kerman' and the male 'Peters'. Beginning in the late 1800s and early 1900s, a small number (less than 20) of pistachio cultivars were imported into the United States from Syria, Sicily, Tunisia, Israel and Iran. The current commercial Californian female pistachio cultivar,

'Kerman', named for its area of origin in Iran, was among them. It was collected by W. E. Whitehouse in 1929 and selected from seed lot P.I. 86327 in 1936 by Dr Lloyd Joley of the United States Department of Agriculture (USDA) at the Plant Introduction Station in Chico, CA. It was released for trial in 1957. 'Kerman' is pollinated primarily by 'Peters', a seedling of unknown parentage identified by A.B. Peters of Fresno, California, USA. These two cultivars comprise over 99% of the commercial plantings in California. Three recent breeding releases by D. A. Parfitt of University of California Davis, the male 'Randy', and the females 'Lost Hills' and 'Golden Hills', are now being planted in limited amounts ( Parfitt, 1997; Ferguson, 2008).

The release of 'Kerman' occurred just prior to a period of major growth of the industry in California. 'Kerman' is now the primary female cultivar commercially grown in all regions in California and produces high yields of large and attractive nuts, especially in the primary growing areas of the southern San Joaquin valley. It has a strong alternate bearing tendency, high percentage of blank nuts in some years, a relatively high level of non-split nuts and a light greenish-yellow kernel with almost minimal flavour when dried at commercial temperatures. Kernel colour and flavour have not been an issue with American consumers, who have not been exposed to European cultivars with different flavour characteristics (Parfitt et al., 2005).

'Lassen' was developed by Whitehouse from the same seed lot as 'Kerman' (P.I. 86372) and released in 1962 from the USDA Plant Introduction Station at Chico CA. It is very similar to 'Kerman' with respect to nut characteristics. 'Damghan' was also developed from this seed collection (Joley, 1979). It has very large nuts, but appears to be very low yielding under Californian conditions. Two new female cultivars were released from the pistachio-breeding program by Parfitt et al. in 2005. 'Golden Hills' produced 46% higher yield than 'Kerman' during the first 3 years that harvestable yields were collected from a replicated trial in the southern San Joaquin Valley. Nut size and weight were similar to 'Kerman', but the percentage of blank nuts was lower than 'Kerman' during the 3 years of tests. 'Golden Hills' has a balanced shell-hinge strength that provided for a split nut percentage of about 5% greater than 'Kerman' during the 3 years of data collection, without excessive loss of shells and loose kernels. 'Golden Hills' has more, but smaller scaffold branches than 'Kerman', producing a smaller more bushy tree after 3-4 years of training (Parfitt et al., 2005).

'Lost Hills' is a new female cultivar from the breeding program of Parfitt et al. that flowers 1-2 weeks before 'Kerman' and matures 2-4 weeks earlier than 'Kerman'. As with 'Golden Hills', earlier maturity results in less navel orange worm damage at harvest. This cultivar produced 26% higher grower paid yield than 'Kerman' during the first 3 years of harvestable yield collection. Nut size and weight were larger than 'Kerman'. Percentage splits were about

7% higher than for 'Kerman'. 'Lost Hills' produced more lost shells and loose kernels than 'Kerman' during hulling as well. Flowering was more uniform than 'Kerman' for both the cultivars during low chilling seasons. This translated into a more uniform maturity and less difficulty in determining the correct time at which to harvest for the maximum splits and minimum staining (Parfitt et al., 2005).

## 6 Rootstocks

Rootstocks are very important for pistachio nut production. Choice of rootstock depends on resistance to biotic and abiotic stresses such as insect pests and diseases, occurrence of low and high temperatures, drought, low soil moisture, salinity and its effects on the size of the crown, alternate bearing, yield, quality, shell splitting and blank nut production. Apart from *P. lentiscus*, almost all of the *Pistacia* species can be used as rootstock for *P. vera*. Although in Iran and Turkey *P. vera* seedlings have been used as common rootstock for centuries, seedlings of *P. terebinthus* in Turkey and *P. mutica* in Iran have also been used during the last few decades. In Greece and Italy (Sicily), the use of *P. terebinthus* is also common. In North Africa, *P. atlantica* is used. In California, although both *P. atlantica* and *P. terebinthus* were used in the early days of the pistachio industry, now *P. atlantica* is the preferred rootstock because of its greater success in budding and faster growth than *P. terebinthus*. However, since both species are very susceptible to *Verticillium* wilt, a new rootstock, *P. integerrima*, has now been recommended to growers because of its vigour and resistance to the disease (Maranto and Crane, 1982; Needs and Alexander, 1982; Crane and Maranto, 1988; Kaska, 1990). Some important characteristics of the main rootstocks are summarized later (Kaska and Bilgen, 1988; Kaska, 1990).

### 6.1 Pistacia vera

The trees of *P. vera* are large in size and around 10 m tall. They have wider crowns than other pistachios (Bilgen, 1968). *P. vera* seedlings are stronger and more homogeneous than other rootstocks. Budded trees on this rootstock grow slowly during the first year, but growth becomes faster in the following years. There are no incompatibility problems in terms of budding with the pistachio (Ozbek, 1978; Ozbek and Ayfer, 1959). In the budded trees, the juvenility period lasts for a long period of time; therefore, the trees bear fruit quite late. In arid zones, the trees reach the bearing stage 15–20 years after planting. They can withstand drought, high lime content and salinity in the soil, but are sensitive to high soil moisture, as a result of which drainage is necessary. *P. vera* seedlings are sensitive to *Melodogyne* nematodes and some soil-borne diseases (Kaska, 1990).

## 6.2 Pistacia terebinthus

The trees of this species are grown generally as bushes. The plant has a strong and deep root system (Ayfer, 1964) and, therefore, it can be grown in poor, rocky and stony soils. It is a kind of dwarfing rootstock for *P. vera* L., so the trees budded on this rootstock can be easily recognized (Ozbek, 1978). *P. terebinthus* is as hardy as *P. vera*, but hardier than *P. atlantica*; therefore, it is preferred to *P. atlantica* in cold areas. In Turkey, it can be grown successfully at elevations of up to 1,200 m. It is also resistant to *Melodogyne* nematodes, but very susceptible to *Verticillium* wilt. *P. terebinthus* is suitable for planting in calcareous soils along the Mediterranean coast and grows very well in places where the annual precipitation is about 400–600 mm. It can also grow in somewhat heavy soils. According to Joley (Woodroof, 1982), pistachio varieties budded on *P. terebinthus* and *P. atlantica* grow faster than those budded on *P. vera*.

## 6.3 Pistacia atlantica

The trees of *P. atlantica* are rather tall. Mature trees may reach 7–20 m. *P. atlantica* is an excellent rootstock for *P. vera*. Since they are open pollinated and compatible with other pistachios, the seedlings may not be homogeneous, although this problem can be overcome by discarding the non-homogeneous materials. The seedlings grow slowly and come to the budding stage within 2–3 years on dry sites. Taproots of *P. atlantica* are not as strong as those of *P. vera;* therefore, they are not as difficult to transplant as the latter. This rootstock can be recommended for irrigated plantations and is also successful in somewhat heavier soils. In such soils, it gives better results than *P. vera*. (Ozbek, 1978; Spiegel-Roy et al., 1972; Needs and Alexander, 1982; Crane and Maranto, 1988). *P. atlantica* is sensitive to temperatures below −10°C (Maranto and Crane, 1982), for which reason it is widespread along the Mediterranean coastline. *P. atlantica* is resistant to *Melodogyne* nematodes and some soil-borne diseases such as *Phytophtora parasitica*. However, the seedlings are sensitive to *Verticillium* wilt and, therefore, they are not recommended for planting in *Verticillium*-infested soils where cotton was previously grown.

## 6.4 Pistacia khinjuk

The trees of *P. khinjuk* look similar to those of *P. vera*; however, they are taller in height and more compact in shape (Ozbek, 1978). This rootstock is hardier and more resistant to drought than *P. terebinthus* and is therefore grown at much higher altitudes (2,500–3,000 m) and in arid areas. When trees of *P. khinjuk* are top budded, they bear fruit within 5–6 years. Detailed studies are needed regarding the value of this rootstock when it is grown from seed. Moreover, the behaviour of different pistachio varieties budded on this rootstock needs to be investigated. In recent decades, *P. khinjuk* has shown better performance according to rootstock x scion trails in Turkey.

### 6.5 Pistacia integerrima

*P. integerrima* is considered to be a variety of *P. chinensis* (Joley, 1979), but Monastra et al. (1988) have stated that it is in fact a variety of *P. khinjuk*. In recent years, it has been used in California as a rootstock for pistachio, as it grows vigorously and is resistant to *Verticillium* wilt (Needs and Alexander, 1982). It is recommended for both re-planting and new plantations in cotton areas where the soils are infested with wilt-causing fungus. According to Maranto and Crane (1982), *P. integerrima* is less tolerant to cold temperatures than either *P. atlantica* or *P. terebinthus* (Kaska, 1990)

After *P. integerrima* was discovered to be tolerant to *Verticillium* wilt, it became the most widely planted rootstock in California. It is vigorous, buds easily and results in a uniform tree stand. It has the least cold tolerance of the commonly used rootstocks and can be damaged by winter freezes and early frosts, particularly if they occur when the tree is young. Results of the state-wide rootstock trials conducted in 2002 indicated that its growth rates and early production are superior to *P. atlantica*, 19.1% lower than UCB I and equal to or slightly less than the two interspecific hybrids. Nutrient studies have shown *P. integerrima* to be less efficient than *P. atlantica* for boron, zinc and copper uptake (Ferguson et al., 2005).

## 7 Propagation techniques

### *7.1 Propagation by seed*

Pistachio trees are propagated by budding of *P. vera* varieties on the seedlings of *P. vera, P. atlantica, P. khinjuk, P. terebinthus, P. integerrima,* etc. All the seeds of the different *Pistacia* species require some chilling for germination. The seeds of the pistachio should be stratified for 40 or 50 days at +4°C after soaking in water overnight. If this period is extended, germination may start after 50 days or more. After germination, when the radicle has grown to 0.5–1.0 cm, the tip of the radicle can be removed by scissors to induce secondary roots and to stop tap root formation (Ak, 1988). The seeds are usually sown in polythene bags at least 40 cm long with a diameter of 20 cm and filled with mixed material (1:1:1 soil:sand:farm manure). Stratified seeds may be sown in seedbeds at distances of 10 cm in the row with 50 cm to 60 cm between rows. The seeds should be chosen from the current year's nuts because the germination capacity of old seeds is low (Ak, 1988; Kaska, 1990). The growth regulator gibberellic acid ($GA_3$) increases the seed germination rate and growth of the seedlings and, to obtain the best results using this product, the pistachio seeds should be soaked in 250–500 ppm $GA_3$ for 24 h (Ak et al., 1995).

Split seeds of *P. vera* germinate more quickly and homogeneously than unsplit seeds. Among the unsplit seeds, the number of blanks may be rather high and, in such cases, they may be separated by soaking the nuts in water and discarding the floating ones. The separation of split and unsplit nuts at the time of soaking is also beneficial to obtaining an even germination in seed. In continental areas, where the winters are cold, direct sowing of the seeds in nurseries in November or December is recommended. If sowing is delayed for some reason, then stratification of the seeds is the only solution to obtain a more homogeneous germination (Kaska, 1990). The seeds of *P. khinjuk, P. atlantica* and *P. terebinthus* should be dehulled (separation of the mesocarp) before they are sown or stratified, otherwise their germination percentages will be low due to infestation by mould (Kuru et al., 1988). In the seedbeds, precautions must be taken against pests such as crows, rats and moles.

The seedlings continue to grow during late spring, summer and early autumn. They should be irrigated every 15–20 days in places where irrigation facilities are available. In non-irrigated nurseries, their growth rate is very low. All the seedlings of *Pistacia* species are taprooted and, since they are transplanted with a bare root, they cannot withstand a limited water supply. Therefore, lateral root formation of the seedlings should be stimulated by root pruning. This can be done in the nursery in June or July by using root cutters at a depth of 40–50 cm in the soil. This is not a common procedure in Turkey. Instead, seedlings uprooted in late autumn are either transplanted in budding plots in nurseries or directly in orchards and they are prepared for budding either in June–July or September. The transplanted seedlings must be irrigated 3–4 times during the hot summer months (Kaska, 1990).

In the United States and more recently in Turkey, container-grown pistachio nut seedlings have become popular. In this method, the seeds are sown first in a jiffy bag, tin pot or black plastic bag and, 2–4 weeks after emergence, the seedlings are shifted to bigger (15 cm in diameter) and taller (50–60 cm) containers. These containers can be made of pressed paper (papier-mache) or fashioned from plastic or tin pots. If the seeds are sown in December or January and kept in a heated greenhouse until spring, they can attain budding size by May or June. They are either chip or shield budded and, at the end of the season, they can be transplanted into the orchard as 1-year-old nursery stock. If the containers are kept in the open, the growth rate is slow and, in this case, they may attain budding size in the late autumn or the following spring. If the pots are kept on a cement floor, taproot growth is prevented and lateral root formation is stimulated. Any taproot emerging from the bottom of the pot should be cut and protruding lateral roots should also be pruned.

Transplanting of the bare-rooted seedlings should be done in late autumn if the soil conditions are suitable for planting. In arid areas, spring planting should be avoided because the seedlings cannot establish a water balance between the organs above and below ground. Percentage survival is high when container-grown seedlings are planted and, although these plants can be transplanted any time in the season, the best planting times are November and December. Whenever they are planted, they must be irrigated at the time of planting and 3–4 times during the summer months.

## 7.2 Budding of pistachio

The budding operation can be carried out when the bark of the rootstock easily separates from the wood at a time when sufficiently developed buds of the desired cultivar are obtainable. Depending on the region, budding can be carried out from June until September, depending on the growing conditions for different fruit trees. The best time to bud fruit stocks is dependent on certain growth characteristics of the stock and, with the pistachio, generally 'T' budding occurs in June. Budding can be initiated at the end of May until the end of June, with the start time depending on the readiness of the scion or bud, and the completion time depending on the ease of separation of the bark from the stocks. When the plants lose their water and the bark sticks tightly and does not separate from the wood, this means that the time for 'T' budding is over.

'T' budding is easy to learn and apply. The most important point for 'T' budding success is that the knife blade should not touch the trunk, otherwise resins come between the bud and stock and the budding fails. The blade of budding knives must be very sharp and the budder should be experienced (Ak, 2011). During pistachio budding, extra care should be taken to prevent wounding the core or wood part of the stock while the 'T' is incised. The main reason for this is the mastic gum. This gum oozes behind the bud and wood and the result is a failed operation. The 'T' budding steps are pictured step by step in Figs. 4 and 5 (Ak, 2015).

Successful budding occurs with satisfactory union of the cells in a graft and the formation of callus and common tissues. It can be established only 10–15 days after budding. If the bud has sprouts, the colour of the bark is bright and the petiole drops off when touched, then the budding has been successfully completed. If the bark is shrunken and the petiola is dried and attached, the budding has failed. In this case, if the season is suitable and the stock grows in a healthy manner, a second budding attempt can be made at the same spot on the other side or on another part of the trunk. Chip budding is an alternative approach, which can be successful when scions are in the rest period, kept in

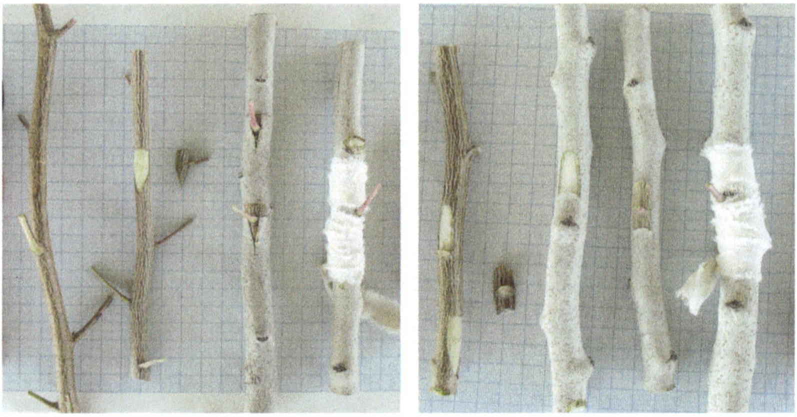

**Figure 4** 'T' Budding steps on the left and 'Chip' budding on the right.

**Figure 5** Pistachio seedlings produced and budded, sprouted plants in a greenhouse in Turkey.

cold store at 2–4°C and is undertaken in April just at the beginning of spring growth (Fig. 4) (Kaska et al., 1990).

Generally, farmers use bare-rooted plantlets to establish pistachio orchards. After 4–5 years, the plants are budded in the field. More recently, many farmers have started to used potted plants (unbudded) that have hairy or more roots developed in the pot. Such plants, when transplanted in field conditions, bud fast because of the root systems.

## 8 Establishing and planting of pistachio orchards

Planting of pistachio trees differs according to whether irrigation is available or not. In new orchards in arid areas of Turkey and Syria, trees are planted 8 × 10 m, i.e., 100–160 nursery plants per ha are required. In irrigated lands, the planting distances of 6 × 6 m are used. Under arid conditions, pistachio trees start bearing fruit at the age of 15–20 years In order to obtain some income

**Figure 6** Pistachio plantation in Iran.

during this long interval, farmers inter-cultivate grapevines with pistachios (Kaska, 1990).

A high-density plantation system is practised in Iran (Fig. 6). The plants are planted very near to each other and the roots and the trunk do not develop as they would in a normal tree. A channel is opened and the seeds are sown 40–50 cm apart. The seedlings that emerge are budded. In such a high-density system, with the maximum number of plants in a limited area, the roots compete with each other so the plants are dwarf plants and irrigation is essential. The dwarf growth provides small plants, very high yield per area and maintenance and harvesting are easy.

Modern plantations should be operated under irrigated conditions. The interval until production of fruit can be shortened in this manner compared to non-irrigated conditions, although it should be kept in mind that the pistachio does not like excessive water. Distances between the rows and plants are usually 6–8 m and the rows should be arranged according to optimize mechanization. Pistachio is resistant to dry conditions due to its quite strong tap root system. When the plantation is done under dry, non-irrigated conditions, the interval between the rows should be large. This distance can be changed according to the rootstocks and may be 6–12 m.

Planting of seedlings is generally done in autumn or spring time before bud bursting time. If the winter conditions are harsh, then spring planting will be better, otherwise the seedlings can be planted in autumn. The holes are dug to around 40–50 cm diameter and 60–80 cm depth. Around 500 g of sunburned farm manure mixed with soil should be placed at the bottom of the hole and then the seedlings are planted and irrigated.

## 9 Orchard management

Pistachio orchards have been established in dry and unirrigated lands in Turkey and Syria. In order to preserve the water from the winter and spring rains and snows, the soils are ploughed in the autumn. On the other hand, in order to prevent evaporation, the soils are cultivated by chisel ploughs during the entire spring and summer months. This type of soil management destroys the weeds, which are in competition with the pistachio trees and which are also the host plants for many diseases and insects (Kaska, 1995).

### 9.1 Soil cultivation

The pistachio is a very hardy and drought-resistant tree due to its deep tap root system. It can be grown even in rocky and stony soils without any irrigation. If irrigation is carried out for a number of years, then the tap root system may not be well developed, which means irrigation must be continued. With irrigation, fertilization and good culturing practices, a higher yield of good-quality nuts is produced compared to that of non-irrigated trees. Trees should be irrigated while young in order to aid establishment even if they are not to be irrigated when mature. If grown in irrigated conditions, the soil drainage should be good, otherwise mortality will be high due to Verticillium wilt caused by soil-borne fungi (Tareen et al., 1989).

Soil cultivation is a very important factor in pistachio production, especially under dry conditions. It helps preserve water in the soil and prevents its evaporation. Under dry conditions, soil cultivation should be practised three times a year. First in the autumn, to help the rain water penetrate deep into the soil, second in the spring time, when the cultivation should not be very deep, and third in the summer, before harvesting, to cope with dried weeds and the nuts that have dropped onto the soil and been lost among the weeds. This is important to prevent a fire hazard because pistachio-growing areas have very dry conditions, high temperatures and very low relative humidity (Fig. 7).

### 9.2 Irrigation

Pistachio trees are commonly irrigated in Iran and the United States; it is well known that irrigation is a very important factor in obtaining high yields and good-quality fruit. The use of irrigation increases the leaf size, and the number and length of the shoots. Irrigation also increases yields, nut size, splitting percentage and decreases blank nut rate and degree of alternate bearing (Fig. 8).

Pistachio cultivation is generally practised in water-stress areas. This situation, along with some other factors, determines the irrigation method employed for pistachio production (Figs. 9–11). In Iran, the furrow irrigation method is very common, although this is now being converted to the drip irrigation method. Pistachio does not like excessive water as this causes some soil-borne disease problems. Drip irrigation is the best method to irrigate pistachio orchard plantations provided enough irrigation water is available. This

**Figure 7** Pistachio orchard with good weed control in Turkey.

**Figure 8** Non-irrigated grown pistachio, current shoots are small and, at harvest time, the leaves and next year's flower buds will drop.

**Figure 9** Furrow irrigation in Iran and drip irrigation in Greece.

**Figure 10** Drip irrigation system (the pipeline is near the trees or in the middle of the rows).

**Figure 11** Pistachio plantation and irrigation system in California (USA).

underground irrigation method has some major advantages. In this system, the pipelines are laid down under the soil; irrigation by underground drip systems is currently used in olive plantations in Spain. This system has recently been adopted for pistachio and almond orchards in Turkey (Ak, 2004). However, this system is still not widely accepted or used by farmers.

Water use in pistachio production may involve several field operations, but especially includes irrigation and the application of pesticides and fertilizers. There is no water used in the fields during harvesting operations, but water is used during the hulling/drying process. Water use in these off-farm harvest operations will be covered in a later section. Water can be a direct source of microbial contamination and it can spread a localized contamination source; its potential as a direct source of contamination is related to the source and its quality (Anon., 2009, Ak, 2015).

Pistachio growers routinely treat water to ensure its suitability for use in micro-irrigation systems. While this treatment may not address microbial contamination, water seldom arrives at the farm in a condition to be used directly and treating water to ensure its suitability is an accepted practice. Checking the water quality at the end of the growing season allows the maximum time in which to correct the problem. Treating wells to ensure that the water meets potable standards with respect to microbial contaminants is advisable (Anon., 2009).

Pistachio orchards are commonly irrigated through mini-sprinkler systems and furrow irrigation in the United States and Iran, respectively. The best irrigation methods include drip irrigation and subsurface drip irrigation systems. The amount of water, irrigation interval or the depth of irrigated soil should be determined before choosing the irrigation method and amount of water to be used. Generally, these conditions depend on the soil infiltration rate. Irrigation intervals and amounts are very important under heavy soil conditions. Pistachio growers prefer drip irrigation systems, either on the surface or under the soil (Arpaci et al., 1995). The water should not touch the trunk; otherwise, diseases such as *Phytophythora* spp. may occur.

## 9.3 Fertilization

If fertilization programmes are applied regularly in pistachio orchards, vegetative growth is better. Adequate plant nutrition is essential for commercial pistachio production with the degree of fertilization largely dependent on the level of irrigation and the pH of the soil. The pH of soils in the main production areas of Iran is between 7.2 and 8.5, with an average of 7.9. Pistachios are tolerant to salinity, and studies have shown that they can tolerate an electrical conductivity (EC) of 8.0 dS/m without any considerable decrease in yield (Sheibani, 1995).

Soils of pistachio orchards often have inadequate levels of nitrogen, phosphorus, potassium and organic matter. In a survey conducted in 30 pistachio orchards in Southeast Anatolia, Tekin et al. (1985) found that, in many orchards, the trees were markedly deficient in phosphorus and zinc and slightly deficient in nitrogen, iron and manganese. The level of potassium was found to be adequate in many orchards although there were some districts where trees showed slight deficiencies. In such regions, the soil pH varied between 7.5 and 9.3 and the organic matter content was very low.

Crane and Maranto (1988) have claimed that the pistachio is not a significant consumer of nitrogen when it is abundantly available in the soil. Chemical forms of nitrogenous fertilizers should be chosen depending on the soil pH. For instance, in some GAP guidelines (Southeast Anatolia Region), only ammonium sulphate is recommended as a source of nitrogen because of the alkalinity of the soils. Nitrogenous fertilizers should be applied at the end of February or in the beginning of March at a rate of 1.5–4 kg per tree (Kaska, 1995). In the United States, boron (B) plays a unique role in pistachio production. Responses to foliar B application include increased pollen viability and germination rate, increased fruit set and yield, decreased blanking percentage and increased leaf B concentration (Brown, 1995a).

Generally, the nutrient status of pistachios should be calculated according to leaf analyses. Leaf samples were collected in the second half of July by Tekin et al. (1990) and the levels of nutrients in pistachio trees grown under

**Table 1** Critical nutrition levels of pistachio leaves grown under dry conditions

| Macronutrient | % | Micronutrient | ppm |
|---|---|---|---|
| Nitrogen (N) | 1.80–2.40 | Iron (Fe) | 43–170 |
| Phosphorus (P) | 0.06–0.14 | Zinc (Zn) | 10–25 |
| Potassium (K) | 0.80–1.20 | Manganese (Mn) | 25–50 |
| Calcium (Ca) | 2.30–3.00 | Copper (Cu) | 6–90 |
| Magnesium (Mg) | 0.50–0.90 | Boron (B) | 100–180 |

**Table 2** Nutrition levels of 'Kerman' pistachio leaves under irrigated conditions

| Macronutrient | % | Micronutrient | ppm |
|---|---|---|---|
| Nitrogen (N) | 2.50–2.90 | Zinc (Zn) | 7–15 |
| Phosphorus (P) | 0.14–0.17 | Manganese (Mn) | 30–80 |
| Potassium (K) | 1.00–2.00 | Copper (Cu) | 4–10 |
| Calcium (Ca) | 1.30–4.00 | Boron (B) | 90–250 |
| Magnesium (Mg) | 0.60–1.20 | | |

dry conditions determined by the authors are reported in Table 1 (Tekin, 2002).

On the other hand, the critical levels of nutrients of pistachio leaves grown under irrigated conditions in 'Kerman' cultivars are reported in Table 2 (Brown, 1995b).

Soil analyses of pistachio orchards have indicated that there are enough micro and macro nutrients available in the soil but that these cannot be taken up by the leaves due to antagonism. Analyses of the soil and leaves are required to determine any deficiency of plant nutrition and to help maintain balanced nutrition of pistachio plants. The best time to take soil samples is after the harvest of pistachio, which occurs in the middle of March. The sampling of leaves is carried out in August, when spots appear on the leaves. Some nutrients cannot be taken up by tree roots so foliar application may be practised. If lime content is very high, zinc, an important element for pistachio, will not be taken up by the tree. It should then be sprayed, although present in sufficient quantities in the soil. Any deficiencies will be determined by leaf analyses.

The following key points should be followed by the growers and processors involved in pistachio production (Ak, 2015):

- The soil analyses for nutrients and some physical traits should be conducted on the soil before any decision is made to establish a pistachio orchard. The fertilizer applications should be organized on the basis of these analyses results.

- Farm yard manure is used by the farmers to increase the organic matter content of the soil; any rotten manure should be used. In dry areas, however, this is not very useful because of the high and dry temperature as it can only correct the physical traits of the soil.
- For the adult trees, leaf analyses should be added along with soil analyses. Excessive application of nutrients is uneconomical, can lead to lush growth that is more susceptible to disease and pests and can cause ground and surface water pollution.
- Application of fertilizers, manures, biosolids and soil amendments are subject to a number of different regulatory schemes. Fertilizer and manure nutrient contents should be documented. Biosolids must be tested for pathogens and treated to reduce or eliminate them. Pistachio processors may have additional requirements or exclusions for biosolids, manures and fertilizers. Growers need to follow good agricultural practices for handling manure and biosolids to minimize microbial hazards.
- Pathogens may be introduced into the orchard through water, runoff and air pollution or through animal and human manure and faeces. Apply only actively treated certified manures whenever possible. Obtain records and analyses from the supplier regarding treatment parameters, nutrient and pathogen tests, treatment dates and other analytical tests

Growers must have documentation for the materials applied to their orchards. Application of biosolids requires documentation and a certificate of analysis. The nutrient content of any applied manure should be documented. While there are currently no restrictions on soil amendments, it is advisable to keep the same records as required for other applications. These include the supplier of the material, the nature of the material, relevant test data, dates of active/passive treatments, date of application to the orchard, date of other activities done to incorporate the material and expected/actual dates of harvest (pre-harvest interval) (Anon., 2009).

## 9.4 Training and pruning

Normally pistachio trees are trained with an open-centre (vase) system or modified central leader system. If it is not compulsory, the tip of the main branch should not be cut (Fig. 12).

Pistachio trees should be trained through two different systems:

- The modified central leader or delayed open-centre system: The modified central leader system is used mainly on the higher headed nut trees. With this system, 4–6 primary scaffolds are developed from the main

**Figure 12** Pistachio trees in Kyrgyzstan before pruning and after pruning.

trunk before the open centre is allowed to develop. This system should be practised in humid pistachio-growing areas because these areas have high light density and sunshine can damage the branches exposed to sunshine directly. The centre of the tree is attached with a central leader branch. This will protect branches from sunburn.

- The open-centre system: For this aim, three to five major scaffold limbs with no central leader are chosen. This system can be used in areas where humidity is high with limited sunshine.

Pruning of pistachio trees grown under dry and poor soil conditions should be done carefully; otherwise, the tree might be seriously damaged. Pruning plays an effective role in obtaining high yields. Many farmers, however, either do not prune pistachio trees at all or undertake incorrect pruning. Research has been performed on three levels of pruning; the shoot lengths were 14 cm on the heavily pruned, 12 cm on the lightly pruned and 10 cm on the location (farmer) pruned trees. This research has indicated that the lightly pruned trees have a lower canopy volume (25%) than the location pruned trees (farmer practice), although the lightly pruned trees have a 12% higher yield than the location pruned trees (Arpaci et al., 2001). Proper pruning and training of pistachios may also reduce the tendency for biennial bearing and reduce incidence and spread of certain diseases.

The pruning operation begins in autumn and continues until March or until 1 or 2 weeks of flowering time. The tree age, trunk thickness and crown width should be taken into consideration and pruning should be made in alternating rotations over 1–3 years. In pistachio trees, summer pruning or training is not applicable because of the wax (resinous) of trees. Extensive pruning should not be done during the growing season except to remove diseased or damaged wood. The dried or dead branches should be cut out soon after the harvesting period in pistachio trees because the colour may be confused after leaves are shed in pistachio trees.

The following points should be taken into consideration when pruning:

- The pistachio tree is resinous. In deep cuts, it oozes resin and, due to the slow development of the tree, closes the wound with difficulty. Therefore, deep cuts should be avoided as much as possible.
- After the crown has developed from three or four, or sometimes two, main branches, care should be taken to keep the vegetative and generative development in balance.
- Terminal buds develop strongly. When these buds are cut, the branch dries towards the rear to the point of junction with another branch. Therefore, care should be taken not to prune terminal buds unnecessarily. During cutting of branches, no stub should be left.
- Pruning of the pistachio tree after trimming for shape should be in the form of the ridding of dry branches on young trees. Thinning out of the crowded and interfering branches and not too deep pruning of the older trees should occur every 2 or 3 years in order to facilitate the growth of new shoots. Light exposure must reach into the crown, otherwise some very crowded branches inside and below the crown will dry and will host insect eggs.

It should not be forgotten that the pistachio tree must be lightly pruned. In very old trees left unpruned for a long period of time, some of the branches should be pruned to provide normal life to the tree (Fig. 12). At the beginning, one-third of the wood should be removed. Pruning tools should be clean and hygienic and the pruner should be an expert in the pistachio pruning system as the pistachio tree is different than other fruit trees.

## 9.5 Harvesting and post-harvest management

The pistachio is harvested from late August to late October depending on cultivar maturation. During maturation, the under colour of the exocarp changes from a light green to a pale straw or whitish, opaque appearance. The hulls can be removed from the nuts by being squeezed between the fingers. Another sign of maturity is splitting of the shell, a decrease in fruit removal force. Kernel dry weight and crude fat content are increased (Ak, 1998a).

In Turkey and Syria, pistachio nuts are harvested manually as clusters, which can easily be separated from the branches by simply bending them back. They drop onto the canvas spread out under the tree. All the clusters are put into sacks and brought to drying yards made of concrete or hardened soil and covered with canvas and left to dry in the sun. The dried nuts in the hull are separated from their clusters and left again in the sun for further drying.

In Turkey, all the cultivars, except 'Siirt' and 'Ohady', are stored with their hulls. The nuts are kept in the sun until they are completely dry, then they are put into sacks and stored (Kaska, 1995). In the United States and Iran, the system is different. Harvesting is done in a manner similar to Turkey in Iran, but machines are used to harvest in the United States. Post-harvest practices of dehulling, drying, separation of blank nuts and separation of split nuts are mechanized.

## 10 Pests and diseases

Pistachios are susceptible to a number of insects and diseases that affect the yield and quality of the crop. Some of the most widespread and/or serious pest problems are listed later (Barone, et al., 1997).

### *10.1 Fungi*

All fungi affect pistachio production, with some being very important for one producing country, but not being a problem for another country. The first record of Alternaria blight of pistachio is from Egypt. Symptoms on the leaves of both male and female trees are circular, dark brown to black necrotic lesions, and later in the season, they are usually with black sporulation, especially in the centre of the lesions. Propagules of Alternaria alternata are present on the developing fruit and leaves from early April until harvest. Septoria leaf spots are caused by *Septoria pistaciae* Desm., which is controlled with preventative fungicide sprays. Dithiocarbamates (zinep, mancozeb) are recommended. Copper fungicides are also effective, but they should only be applied after the fruit has reached 1 cm size, since they might be phytotoxic to the very young fruit. Applications should be begin after the first leaves have unfolded and, if necessary, be repeated monthly up to the beginning of June (Michailides and Young, 1989; Michailides, 1998).

Important fungal diseases are: *Alternaria alternata* (Alternaria late blight), *Alternaria tenuissima* (Alternaria blight), *Armillaria mellea* (Armillaria root rot), *Aspergillus candidus, Aspergillus clavatus, Aspergillus flavus, Aspergillus parasiticus (Seed rot (aflatoxins)), Aspergillus fumigatus, Aspergillus ochraceus* (Ochratoxins), *Aspergillus niger, Aspergillus* spp. (Aspergillus blight), *Asteromella pistaciarum, Aureobasidium pullulans* (Stigmatomycosis), *Botryosphaeria dothidea* (Botryosphaeria panicle and shoot blight), *Botryosphaeria obtusa* (Steam canker), *Botryodiplodia pistaciae, Botryotinia fuckeliana* (syn. *Botrytis cinerea)* (Botrytis blossom, shoot blight) ,*Cenangium vagabundum, Camarosporium pistaciae* (Camarosporium shoot) ,*Chaetomium* spp., *Cladosporium herbarum, Cylindrosporium garbowskii, Camarosporium pistaciae, Cytospora terebinthi* (Gum canker), *Cochliobolus spielfer, Epicoccum purpurascens, Eutypa lata, Fomes rimosus, Fusarium equiseti, F. roseum, F.*

*solan, F. Oxysporum, Fusarium* spp. (Root and stem rot), *Monilia pistaciae, Melampsora pistaciae, Mycosphaerella pistacina, Nematospora coryli* (Stigmatomycosis), *Paecilomyces variotii* (Die-back of young shoots), *Pestalotia breviseta, Phytophthora* spp. (Gommosis, crown and root rot), *Phyllosticta terebinthi, P. lentisci, Phymatotrichum omnivorum* (Texas root rot), *Penicillium camemberti, P. decumbens, Penicillium* spp., *Pestalotiopsis* spp., *Phyllactinia suffulta, Phomopsis* sp. (Phomopsis shoot blight), *Pleospora montemartinii, P. pistaciae, Pileolaria terebinthi* (Rust), *Pleurotus ostreatus, Rhizoctonia bataticola, Rhizoctonia solani* (AG-4) (Nursery seedling blight), *Rhizopus* sp., *Rosellina necatrix, Schyzophyllum commune (*Schyzophyllum wood decay), *Sclerotinia selerotiorum (*Scelerotinia shoot blight*) Septogloeum pistaciae, Septoria pistaciae, S. pistacina* (syn. *Mycospharella pisticina), S. pistaciarum* (syn. *Mycospherella pistacearum)*(Septoria blight), *Septoria* spp. (Septoria leaf and fruit spot), *Sphaerella pistaciae, Stemphyllium botryosum, Tetracoccosporium* sp., *Trichoderma harzianum, Trichothecium roseum, Tzavella roumbos* (Panicle blight), *Uromyces terebinthi* (syn. *Pileolaria terebinthi,*(Rust), *Verticillium albo-atrum, V. dahliae* (Verticillium wilt).

## *10.2 Bacteria*

Pistachio dieback is caused by *Xanthomonas translucens* (Xtp) (Facelli et al., 2002). Symptoms of this disease include decline, dieback, internal staining, trunk and limb lesions and, in some instances, tree death. Bacteria are normally found in the sapwood of trunks, limbs and twigs, and less frequently in current season growth and leaves. The main mode of transmission of the bacteria from tree to tree appears to be through pruning tools (Taylor et al., 2005; Facelli et al., 2006).

## *10.3 Insects*

Numerous insects have been reported to feed on the foliage, shucks, nuts, twigs, wood and roots of pistachio (Barone, et al., 1997). Some pistachio nut insects (Sheibani, 1996) are summarized later.

Pistachio psylla (*Agonoseena targionii* Licht) is the most destructive pest for pistachio trees grown in Turkey and Iran. The damage caused by this insect includes direct injury as a result of sucking the plant sap and stunting of the tree, shedding of leaves, fruits and buds, increasing the rate of nut blanks, lowering the quality of nut, hindering transpiration and photosynthesis due to the closure of stomata by honey-dew secreted from the insect larvae and/or sooty mould growing on leaves if conditions become favourable.

Pistachio Leafhopper (*Idiocerus stali* Fieb.): This pest overwinters as the adult insect. In early spring after the formation of fruit clusters, the adult insect penetrates its ovipositor into young tissues and deposits eggs. The eggs are often laid at the junction of fruit clusters and rarely on petioles. After 7 days, the larvae emerge and start feeding on the plant sap. The larva is often found on fruit clusters. Feeding is associated with abundant honey-dew secretion. Damage includes stunted trees, shedding, drying and shrinking of fruits. Reduced transpiration through leaves and the growth of sooty mould are other injuries.

Pistachio wood boring moth (*Kermania pistaciella* Amsel): The pest is a small dark coloured moth 11–13 mm in length. It overwinters as full-grown larva inside pistachio woody branches. As soon as the average environment temperature reaches 10°C, it comes out and, after mating, the females deposit their eggs singly on the clusters axis and twigs of the same year. Then the larvae enter the tree tissues and bore either into the clusters axis or the centres of young shoots and discharge dark excrements. They cause falling or drying of the clusters fruit in the mid-season. This pest has one generation per year.

Foliage-feeding insects: *Agonoscena targionii* (Pistachio psylla), *Anapulvinaria pistaciae* (Pistachio cushion scale ), *Ceroplastes rusci* (Homoptera), *Eulecanium rugulosum* (Scale), *Idiocerinus stali* (Pistachio leafhopper), *Melanaspis inopinatus* (Pistachio trunk scale), *Pistaciaspis pistaciae* (Pistachio scale), *Saissetia oleae* (Homoptera), *Slavum wertheimae* (Gall-forming aphid), *Suturaspis pistaciae* (Pistachio white scale), *Thaumetopoea solitaria* (Pistachio bud moth), *Tenuipalpus* (Pistachio mites), *Tetranychus* (Mites).

Flower-feeding insects: *Anthascia* sp. (Coleoptera), *Eriophyte pistacia*, (Mites)

*Frankliniella occidentalis* (Western flower thrips), *Polydrosus davatchi (Coleopptera), Telphusa pistaciae* (Pistachio bud moth), *Thrips iranicus, Thrips pistacia* (Thrips).

Fruit-feeding insects: *Acrostemum hegeriir* (Stink bug), *Amyelois transitella* (Naval orangeworm), *Apomyelois ceratoniae* (Carob moth), *Arimania komaroffi* Ragonot, (Lepidoptera) *Brachynema* (Common stink bug), *Brevipalpus lewisi* (Citrus flat mite), *Calocoris norvegicus* (Gmelin bug), *Chlorochroa uhleri, C. ligata* (Stink bug), *Dinarmus pistacia* (Hymenoptera), *Ephestia elutella* (Lepidoptera), *Eurytoma plotnikovi* (Pistachio fruit wasp), *E. pistachiae* (Pistachio wasp), *Gonocerus acuteangulatis, Graphosoma semipunctatum* (Heteroptera), *Leptocoris trivittatus* (Boxelder bug), *Leptoglossus dypealis,* L. *occidentalis* (Leaf-footed bug), *Liorhyssus hyalinus* (Epicarp lesion), *Lygaeus hesperus* (Hemiptera), *Lygaeus panderus* (Pistachio red bug), *Megastigmus pistaciae* (Pistachio seed chalcid, Pistachio golden fruit wasp), *Neurocolpus califomicus, Psallus vaceinola, P ancorifer* (Epicarp lesion), *Nezara viridula* (Green stink bug), *Plodia interpunctella* (Lepidoptera) *Phytocoris* spp. (Miridae) (Epicarp lesion), *Recurvaria pistaciicola* (Pistachio fruit moth), *Solenostedium bilunatum,* **Thyanta pallidovirens* (Red-shouldered stink bug, Pistachio nut worm).

### 10.4 Nematodes

*Heterodera mediterranea, Heterodera marioni* (Cyst nematode), *Meloidogyne* spp. (Root knot nematode), *Pratylenchus hamatus, P. neglectus* ((Root lesion nematode), *Xiphinema* spp. (Dagger nematode).

## 11 Improving yields

Yield is affected by some physiological problems such as alternate bearing and blank fruit production. Alternate bearing is one of the important features of pistachio trees. Alternate bearer cultivars produce heavy crops in 'on' years and little or no crop in the 'off' years. In contrast to other fruit species, such as olive and apple, (Ak and Kaska, 1992) this situation can be change from one cultivar to another one. For example, the 'Siirt' variety is a less alternating one (Ak, 1998b). This means that the tendency of alternate bearing of cultivars is different and can be decreased by irrigation and fertilization.

Although blank nut formation is a varietal character, it is significantly affected by climate, efficiency of pollination, fertilization, irrigation and rootstocks. Blank nuts found in the Turkish cultivars are due to inefficient pollination and lack of irrigation. Beside this, vegetative or stimulative parthenocarpy and embryo abortion have other impacts on blank nut formation. The tendency to parthenocarpy and percentage of abortive embryos can be changed from one cultivar to another (Caglar et al., 1997).

As mentioned earlier, pistachio orchards have been established under dry and very bad soil conditions; however, high yields can be obtained in irrigated and fertile soil conditions. According to Ak and Agackesen (2003), pistachio trees should be established under irrigated and fertile or good soil conditions to obtain high yield and quality. The yield of 'Kirmizi' pistachio cultivar either as fresh weight or dried weight in good conditions, meaning irrigated and fertile soil, is higher than that obtained under other conditions. Fertile soil without irrigation is better than poor soil with irrigation. Fresh fruit weight was determined in arid soil irrigated conditions to be 21.80 kg, and in unirrigated conditions, 15.50 kg. The yield obtained under fertile soil irrigated conditions was 31.00 kg, and under unirrigated conditions it was 25.87 kg; 1 kg fresh fruit was determined to weigh 559.63 g after being sun dried. According to this calculation, the highest dry weight was determined at 17.043 kg per tree grown under fertile soil at irrigated conditions (Ak and Parlakci, 2006). Some researchers have shown the following effects on the yield of pistachios trees: irrigation 70%, fertilization 50%, variety 45% and light pruning 17%.

## 12 Future trends

Pistachio nut culturing is centuries old in Iran, Turkey and Syria. In these countries, the areas given over to pistachio nut cultivation are the largest and the number of trees the highest in the world. In spite of these facts, yield is very low in Turkey and Syria; although the taste is good, the nuts are small and their splitting percentage is low. However, one should bear in mind that Turkey is the only country in the world in which pistachio nuts are grown in such marginal lands with a dry climate and poor, rocky and calcareous soils.

On the other hand, Turkey has very important advantages in terms of aflatoxin. Because of the low humidity and high temperature during growing and the harvest season, the fruits are free from aflatoxin, which can cause contamination during processing. Processing systems are also modernizing, meaning that Turkish pistachios have no aflatoxin and possess a high-quality flavour. Turkish cultivars are preferred in many European and US markets due to their good taste and uniformly green kernels.

The most important characteristics of the pistachio nuts desired by markets are large size, high percentage of shell splitting, low percentage of blank nuts, high oil and protein contents, regular bearing and high percentage of green kernels. In pistachios, a green kernel is one of the most desired characteristics and green kernel nuts are always at a premium. Although a green kernel is a varietal characteristic, it is also related to altitude and time of harvest. Generally, the nuts of early harvested trees and those of plants grown on high plateaus produce greener kernels than those of late harvested and low land grown plants. On the other hand, the green colour is affected by the pollen source.

## 13 Where to look for further information

Ministry of Agriculture. Gaziantep Pistachio Research Institute, Turkey. https://arastirma.tarimorman.gov.tr/afistik

Pistachio Research Institute, Rafsanjan-Iran. https://www.gfar.net/organizations/pistachio-research-institute-rafsanjan

California Pistachio Research Bopard- USA. https://calpistachioresearch.org/

Acta Horticulturae books Volumes of ISHS Symposium on Pistachios and Almonds

Pistachio Health Institute, Iran. https://pistachiohealthinstitute.org/

University of California, Division of Agriculture and Natural Resources, Fruit & Nut Research & Information Center. http://fruitsandnuts.ucdavis.edu

Prevention of aflatoxin in pistachios. http://www.fao.org/3/w9474t/w9474t06.htm

## 14 References

Ak, B. E. (1988), An investigation on some *Pistacia* spp. seed germination. MSc thesis. Publication Code Number; 297. Department of Horticulture, Cukurova University Institute of Natural and Applied Science, Adana, Turkey, 110p. (In Turkish with an English summary).

Ak, B. E. (1992), Effects of different *Pistacia* spp pollens on the fruit set and quality of pistachios. PhD thesis. Cukurova University Institute of Natural and Applied Science, Department of Horticulture, Adana, Turkey, 210p. (In Turkish with an English summary).

Ak, B. E. (1998a), Harvest systems: pistachio harvest in Turkey and Iran. In: *Advanced Course Production and Economics of Nut Crops*. Organized by CIHEAM-IAMZ, FAO and Univ. Cukurova Fac. of Agr. Course Notes: 5p.

Ak, B. E. (1998b), The yield and fruit quality of *Pistacia vera* cv. 'Siirt' grown at the Ceylanpınar State Farm. *Acta Horticulturae*, 470, 510–15.

Ak, B. E. (2001), Selection criteria of the best pistachio male trees. *Nucis*, 10, 24–5.

Ak, B. E. (2004), Underground drip system: the new irrigation method for pistachio and almond orchard in Turkey. *Nucis*, 12, 24–5.

Ak, B. E. (2011), Pistachio and walnut development project. TCP/KYR/3203. FAO, Sub Regional Office for Central Asia (SEC). Consultancy report. Mission carried out from 17–31 March 2011, Kyrgyzstan, 26p.

Ak, B. E. (2015), *Pistachio GAP Manual. Good Agricultural Practices for Pistachio Cultivation*. Food and Agriculture Organization of the United Nations, Rome, 90p.

Ak, B. E. and Acar, I. (2001), Pistachio production and cultivated varieties grown in Turkey. In: *International Workshop on Pistachio: Towards a Comprehensive Documentation of Distribution and Use of Its Genetic Diversity in the CWANA Region*. Report of the IPGRI Workshop, 14–17 December 1998, Irbid, Jordan, pp. 27–34.

Ak, B. E. and Agackesen, N. (2003), An investigation on determination of yield and quality on pistachio grown at Birecik province. Project Number: 58 final report. University of Harran, 59p. (In Turkish with English summary).

Ak, B. E. and Kaska, N. (1990), Morphological differences between the organs of male and female pistachio (*Pistacia vera* L.) trees. In: *XXIII. International Horticultural Congress*. Abstracts of Contributed papers 2. Poster, 3012.

Ak, B. E. and Kaska, N. (1992), Alternate bearing problem in pistachios and its reasons, and the sitiuation in different varieties (in Turkish). In: *First Turkish National Horticultural Congress*, 13–16 October 1992, vol. I (Fruit), pp. 67–72.

Ak, B. E. and Parlakcı, H. (2006), The low yield reasons and solution of pistachio nut in turkey. *Nucis* 13, 37–40.

Ak, B. E., Ozguven, A. I. and Nikpeyma, Y. (1995), The effect of $GA_3$ applications on pistachio nut seed germination and seedling growth. *Acta Horticulturae*, 419, 115–20.

Ak, B. E., Acar, I. and Kaska, N. (1998), An investigation on the male determination for some female varieties throughout five years (1992–1996) grown at Ceylanpinar State Farm in Sanliurfa conditions. In: *Proceedings of the X. GREMPA Seminar*, 14–17 October 1996, Meknes (Morocco). Cahiers Options Mediterraneennes (vol. 33), pp. 99–104.

Anon. (1993), Pistachio variety catalog. Occupational Publications Series No: 20. Turkish Ministry of Agriculture and Rural Affairs, Department of Production Agriculture and Development, Ankara, Turkey.

Anon. (1997), Pistachio. Available at: http://www.crfg.org/pubs/ff/pistachio.html.

Anon. (2009), Good agricultural practices manual. Guidelines for California pistachio growers. California Pistachio Research Board, 27p.

Arpaci, S., Akkök, F. and Tekin, H. (1995), Determination of yield and quality changes of under dry and irrigated conditions. In: *Türkiye II. Ulusal Bahçe Bitkileri Kongresi*, Adana, Cilt I, pp. 429–33. (In Turkish with English summary).

Arpaci, S., Atli, H. S. and Tekin, H. (2001), Improvement of pruning techniques for bearing pistachio nut trees. In: *GAP II. Tarim Kongresi, Harran Universitesi*, 24-26 Ekim 2001, Sanliurfa, pp. 31-38. (In Turkish with English summary).

Arpaci, S., Acar, I. Atli H. S. and Karadag, S. (2005), Comparison of pistachio production in Turkey and World. In: *GAP IV. Tarim Kongresi, Harran Üniversitesi Ziraat Fakültesi*, 21-23 Eylül 2005, Sanliurfa, pp. 238-43. (In Turkish with English summary).

Ayfer, M. (1964), Pistachio nut culture and its problems with special reference to Turkey. *University of Ankara, Faculty of Agriculture Yearbook*, pp. 189-217.

Barone E. and Caruso T. (1996), Genetic diversity within *Pistacia vera* in Italy. In: Padulosi, S., Caruso, T. and Barone, E. (Eds), *Taxonomy, Distribution, Conservation and Uses of Pistacia Genetic Resources*. Project on underutilized Mediterranean species. Report of a Workshop 29-30 June 1995. IPGRI, Palermo, Italy, pp. 20-8.

Barone, E., Padulosi, S. and Van Mele, P. (1997), *Descriptors for Pistachio (Pistacia vera L.)*. International Plant Genetic Resources, Rome, Italy, 53p.

Bilgen, A. M. (1968), *Memleketimizde bulunan antepfistiği anacları ve asilama teknigi*. Tarim Bakanlıgı, Ziraat Isl. Genel Mud. Yayinlari, Ankara, 35s.

Brown, P. (1995a), Nutrition of pistachio. *Acta Horticulturae*, 419, 77.

Brown, P. (1995b), Diagnosing and correcting nutrient deficiencies. Pistachio Production, Sects. 95-100.

Caglar, S., Nikpeyma, Y. and Ak, B. E. (1997), The causes of blank nut formation in pistachio. University of Sütcüimam. *Journal of Science and Engineering*, 1(1), 37-41.

Crane, J. C. and Maranto, J. (1988), Pistachio production. Univ. of California, Publication 2279, 15p.

Crane, J. C. and Takeda, F. (1979), The unique response of the pistachio tree to inadequate winter chilling. *HortScience*, 14(2), 135-8.

El-Oqlah, A. A (1996), Biosystematic research on the genus *Pistacia* in Jordan. In: Padulosi, S., Caruso, T. and Barone, E. (Eds), *Taxonomy, Distribution, Conservation and Uses of Pistacia Genetic Resources*. Project on underutilized Mediterranean species. Report of a Workshop 29-30 June 1995. IPGRI, Palermo, Italy, pp. 12-19.

Facelli, E., Taylor, C., Scott, E., Emmett, R., Fegan, M. and Sedgly, M. (2002), Bacterial dieback of pistachio in Australia. *Australian Plant Pathology*, 31, 95-6.

Facelli, E., Scott, E., Taylor, C., Nancarrow, N., Emmett, R., Rodriguez, E. and Sedgly, M. (2006), Pistachio bacterial dieback in Australia - characteristics and pathways of the infection. *Acta Horticulturae*, 726, 613-14.

Ferguson, L. (2008), Pistachios in California. Following pistachio footprints (*Pistacia vera*, L.) cultivation and culture, folklore and history, traditions and uses. *Scripta Horticulturae*, 7, 112-16.

Ferguson, L., Sanden, B., Grattan, S., Epstein, S. L. and Krueger, B. (2005), Pistachio rootstocks. The orchard. In: Ferguson, L. (Ed.), *Pistachio Production Manual* (4th edn.), pp. 67-73.

Harandi, O. F. and Ghaffari, M. (2001), Chromosome studies on pistachio (*Pistacia vera* L.) from Iran. *Cahiers Options Mediterraneennes*, 56, 35-9.

Hokmabadi, H., Tajapadipour, A., Shaker Ardakani, A. and Javanshah, A. (2008), Pistachios in Iran. Following pistachio footprints (*Pistacia vera*, L.) cultivation and culture, folklore and history, traditions and uses. *Scripta Horticulturae*, 7, 39-49.

Joley, L. E. (1979), Pistachios. In: Jaynes, R. A. (Ed.), *Nut Tree Culture in North America*. Northern Nut Growers Association, Handen, CT, pp. 163-74.

Kafkas S. (2006), Phylogeny, evolution and biodiversity in the genus *Pistacia* (Anacardiaceae). In: Sharma, A. K. and Sharma, A. (Eds), *Plant Genome: Biodiversity and Evolution. Volume 1C: Phanerogams (Angiosperm - Dicotyledons)*. Science Publishers Inc, Eufield, NH, pp. 525-57. Chapter 20.

Kaska, N. (1990), Pistachio research and development in the Near East, North Africa and Southern Europe. *Nut Production Industry in Europe, Near East and North Africa*. Reur Technical Series 13, pp. 133-60.

Kaska, N. (1995), Pistachio nut production growing in Turkey. *Acta Horticulturae*, 419, 161-4.

Kaska, N. and Bilgen, A. M. (1988), Top working of wild pistachios in Turkey. In: Grassely, C. (Ed.), *Programme de recherche Agrimed*. Rapport EUR 11557, pp. 317-25.

Kaska, N., Eti, S. and Ak, B. E. (1989), An imagination of artificial pollination by plane in Pistachio nut. In: *Second National Symposium on Agricultural Aviation*, 11/13 January 1989, pp. 127-33. (In Turkish with an English summary).

Kaska, N., Ak, B. E. and Nikpeyma, Y. (1990), Comparison of chip and shield budding in top-worked *Pistacia terebinthus* L. in Çukurova Region. In: *XXIII. International Horticultural Congress*. Abstracts of Contributed papers 2. Poster, 4035.

Kaska, N., Caglar, S. and Kafkas, S. (1996), Genetic diversity and germplasm conservation of *Pistacia* species in Turkey. In: Padulosi, S., Caruso, T. and Barone, E. (Eds), *Taxonomy, Distribution, Conservation and Uses of Pistacia Genetic Resources*. Project on underutilized Mediterranean species. Report of a Workshop, 29-30 June 1995. IPGRI, Palermo, Italy, pp. 46-50.

Kokwaro, J. O. and Gillett, J. B. (1980), Notes on the *Anacardiaceae* of Eastern Africa. *Kew Bulletin* 34, 745-60.

Kuden, A. B., Kaska, N., Tanriver, E., Ak, B. E. and Tekin, H. (1995). Determining the chilling requirements and growing degree hours of some pistachio nut cultivars and regions. *Acta Horticulturae*, 419, 85-90.

Kuru, C., Karaca, R., Tekin, H., Akkok F. and Uygur, N. (1988), *Antepfistigi Yetistiriciligi*. Tarim Orman ve Köyisleri Bakanligi, Ankara, 54p.

Maggs, D. H. (1990), The Australian pistachio Sirora. *Fruit Varieties Journal*, 44, 178-9.

Maranto, J. and Crane, J. C. (1982), Pistachio production. Division of Agr. Sciences. Univ. of California leaflet 2279, 18pp.

Michailides, T. J. (1998), Fliar and fruit fungal diseases of pistachio. In: *Advanced Course Production and Economics of Nut Crops*. Organized by CIHEAM-IAMZ, FAO and Univ. Cukurova Fac. of Agr. Course Notes: 29p.

Michailides, T. J. and Young, D. J. (1989), Septoria leaf spot of pistachio caused by Septoria pistaciarum in Arizona. *Annual Pistachio Res. Report Crop Year 1988-89*, California Pistachio Industry, 112.

Monastra, F., Avanzato, D. and Lodoli, E. (1988), 11 pistachio del Monto confronto tra la Pistacchicoltura delle aree tradizionali e quella eraergente degli Stati Uniti. In: Grassely, C. (de), *7 Colloque du Grempa Groupe de Recherche et d'Etude Mediterraneen pour le plstachier et l'amandier*. Rapport EUR 11557, pp. 271-88.

Needs, R. A. and Alexander, M.E. (1982). Pistachio. A technique for chip budding. *Australian Horticulture*, 80(10), 87-9.

Ozbek, S. (1978), Decidious fruits. Çukurova Univ. Publicatiion, Number 128, Reference Book no: 11, 486p. (In Turkish).

Ozbek, S. and Ayfer, M. (1957), *Pistacia* türleri üzerinde sitolojik arastirmalar. Ankara Universitesi Ziraat Fakultesi Yilligi, 3, pp. 203-22.

Ozbek, S. and Ayfer, M. (1959), M. Türkiye'de Antepfistigi (*Pistacia vera*, L.) Anaçlari ve Asi Tekniği A. U. Ziraat Fak. Yilligi 4: 189-214.

Parfitt, D. E. (1997), Pistachio. In: *The Brooks and Olmo Register of Fruit and Nut Varieties*. ASHS Press, Alexandria, VA, pp. 581-2.

Parfitt, D. E., Kallsen, C. ve Maranto, J. (2005), Pistachio cultivars. In: Ferguson, L. (Ed.), *Pistachio Production Manual* (4th edn.). Fruit & Nut Research and Information Center, University of California, USA, pp. 62-6.

Pontikis, C. A. (1986), 'Pontikis' pistachio. *HortScience*, 21, 1074.

Razavi, S. (2006), Pistachio production, Iran vs. the world. *Acta Horticulturae*, 726, 689-94.

Sheibani, A. (1995), Pistachio production in Iran. *Acta Horticulturae*, 419, 165-8.

Sheibani, A. (1996), Pistachio production in Iran. Pistachio Research Institute, Ministry of Agriculture of Iran, 28p.

Spiegel-Roy, P., Assaf, R. and Garm, I. (1972), Essais d'acclimatation et de culture du pistachier en Israel. *Fruits*, 27(9), 619-25.

Tareen, M. R., Kakar, M. A. and Marri, M. I. (1989), Report on the training course on the nut propagation and production held in Adana/Turkey from April to July 1989 Sponsored by FAO, 105p.

Taylor, C., Facelli, E., Scott, E., Emmett, R. and Sedgly, M. (2005). Spread and control of pistachio dieback in Australia. *XIII GREMPA Meeting on Almonds and Pistachios*. Options Mediterraneennes Serie A, No. 63, pp. 183-5.

Tekin, H. (2002), Deficiencies of nutrients and fertilization in pistachio. (in Turkish), Pistachio Research Institute Publication No:14, 33p.

Tekin, H., Genc, C., Kuru, C. and Akkok, F. (1985), Investigations in the determinations of nutrient contents of pistachio nuts. *Bahce*, 14(1-2), 47-57. (In Turkish).

Tekin, H., G. Caglar., Kuru C., and Akkok F. (1990), Determination of nutrients for pistachio trees and the best leaf samples collection periods. In: *First National Symposium on Pistachio Nut*, 11-12 September 1990, pp. 120-38. (In Turkish with an English summary).

Vargas, F. J. and Romero, M. A. (2005), Fruit characters in pistachio progenies. In: Ak, B. E. (Ed.), *Proceedings of The XIII GREMPA Seminar*, 1-5 June 2003, Mirandela, Portugal. Cahiers Options Mediterraneennes (vol. 63), pp. 49-56.

Vargas, F. J., Romero, M. A., Plana, J., Rovira, M. and Battle, I. (1995), Characterization and behaviour of pistachio cultivars in Catalonia (Spain). *Acta Horticulturae*, 419, 181-8.

Vargas, F. J., Romero, M. A. and Vargas, I. (2001), Blooming and leafing time in pistachio progenies. In: Ak, B. E. (Ed.), *Proceedings of The XI GREMPA Seminar*, 1-4 September 1999, Sanliurfa, Turkey. Cahiers Options Mediterraneennes (vol. 56), pp. 41-6.

Whitehouse, W. E. (1957), The pistachio nut-a new crop for the western United States. *Economic Botany*, 11(4), 281-321.

Whitehouse, W. E. and Stone, C. L. (1941), Some aspects of dichogamy and pollination in Pistache. *Proceedings of the American Society for Horticultural Science*, 39, 95-100.

Woodroof, J. C. (1982), *Tree Nuts. Production, Processing, Products* (2nd edn.). AVI Publishing Company, Inc., Westport, CT, 731pp.

Yaltirik, E. (1967), *Anacardiaceae*. In: Davis, P. H. (Ed.), *Flora of Turkey, vol. 2*. Edinburgh University Press, Edinburgh, UK, pp. 544-8.

Zohary, M. A. (1972), *Pistacia* L. *Flora Pa1estine*, 2, 297-300.

Zohary, M. A. (1952), A monographical study of the genus *Pistacia*. *Palestine Journal of Botany, Jerusalem series*, 5, 187-228.

# Chapter 16

# Developing hazelnuts as a sustainable and industrial crop

*Valerio Cristofori, Stefano Speranza and Cristian Silvestri, University of Tuscia, Italy*



## 1 Introduction

The European hazelnut (*Corylus avellana* L.) is a major species of interest for nutritional use and is the primary economically valuable commercial tree nut crop within the *Betulaceae* family. *C. avellana* is a temperate, diploid ($2n = 2x = 22$), monoecious, wind-pollinated, dichogamous species, exhibiting sporophytic incompatibility and a relatively small genome size (1C around 385 Mb) (Mehlenbacher, 2014). Its geographical distribution ranges from Asia Minor and the Caucasus region to Europe and North Africa (Boccacci et al., 2013). Approximately 660000 ha of the world's surface is used to cultivate hazelnuts (FAOSTAT, 2018), with an average in-shell annual production of approximately 835000 tons recorded between 2012 and 2016 (FAOSTAT, 2018). Production is mainly concentrated in two countries: Turkey (563000 tons) and Italy (112000 tons); however, other significant producers include the United States (34000 tons), Georgia (32000 tons), Azerbaijan (31000 tons) and Spain (15000 tons). This nut crop has recently been cultivated in the Southern Hemisphere in countries such as Chile, South Africa and Australia. In general, world production shows an increasing trend from the beginning of

http://dx.doi.org/10.19103/AS.2018.0042.21

the last century, in response to the demand of the confectionery industry which processes about 90% of the harvested nuts, while the remaining 10% is sold as in-shell nuts.

One of the major concerns for hazelnut growers is the suckering aptitude of the species, which varies from cultivar to cultivar but needs to be kept under control annually, and has two negative consequences: increased costs and spread of diseases. The selection and use of non-suckering rootstocks (*C. colurna*) proved to be a solution.

In order to reduce the initial low cropping of new orchards, increasing plant density has proved to be a positive option (Cristofori et al., 2017). The multi-system bush is the most common training system, whereas in recent plantings the open vase and hedgerow (V-shaped hedge) have also been adopted. Mineral fertilization has been the object of considerable research activity and in recent years, foliar fertilization has become common in the specialized orchards in Italy, Turkey, Oregon and elsewhere. The spread of the hazelnut industry from traditional areas, the related necessity to increase productivity and climatic changes have increased the irrigation need and related research. A rainfall of 800 mm/year, well distributed throughout all seasons, is needed for good hazelnut growth and production. In recent years, considerable work has been undertaken to improve mechanical harvesting using vacuum machines and sweepers and pick-up machines.

Important progress has also recently been achieved for the use of different hazelnut biomass (wood and shell) as biofuel and feedstock.

## 2 Propagation

The advances of European hazelnut propagation have only partially been achieved, as the species is usually propagated using own-rooted suckers collected from young plants directly in the orchards (Bignami et al., 1999; Cristofori et al., 2010). The use of this kind of plant material can negatively affect new plantations due to the absence of genetic fidelity and health guarantees.

Hazelnut is quite commonly propagated by layering and some progress has recently been made in grafting.

Contrariwise, the development of efficient and reliable techniques for hazelnut propagation using cuttings or micropropagation from mature tissues are still lacking, as the species is considered recalcitrant to induced rhizogenesis.

However, the increasing interest in hazelnut cultivation, which has led to the need for a large number of plants, and the presence of some diseases recently encountered in some hazelnut districts (i.e. dieback of hazelnuts in Italy and Eastern filbert blight in Oregon, USA), emphasize the risk connected with suckers and layering propagation, which is unable to satisfy the massive demand for high-quality plant material.

## 2.1 Propagation by cuttings

Propagation by cuttings, a technique widely used for many fruit species, has not produced protocols able to ensure the large-scale propagation on hazelnut (Kantarci and Ayer, 1994; Solar et al., 1994; Contessa et al., 2014; Tombesi et al., 2015).

Some researchers have shown that although many factors such as collection time, type of cutting, bottom heating and rooting chemicals affect rooting, the level of rooting obtained on hardwood cuttings from the Turkish cultivar 'Tombul' was not suitable for commercial multiplication (Soylu and Erturk, 1997).

Furthermore, a strong enhancing effect of polyamines on the microshoot rooting of Turkish hazelnut cultivars has been observed. Putrescine, spermidine and spermine improved the rooting of indole-3-butyric acid (IBA)-treated microshoots in a synergistic fashion, probably favouring a better induction of roots, whereas polyamines had a limited positive impact on rooting when applied without IBA (Rey et al., 1994).

More recently, Cristofori et al. (2010) achieved encouraging results with higher rooting success using semi-hardwood cuttings of some Italian hazelnut varieties ('Tonda di Giffoni', 'Tonda Gentile Romana' and 'Nocchione'), confirming that rooting ability was strongly influenced by shoot age and collection time. Similarly, Contessa et al. (2011a,b) obtained a high degree of rooting on the cultivar 'Tonda Gentile delle Langhe' confirming that the rooting response is markedly affected by the donor genotype, hormonal dosage and shoot diameter.

Many attempts have been made to increase the rooting efficiency of cuttings by using polyamine (putrescine) in combination with different auxin concentrations, which were found to increase rooting efficiency (Cristofori et al., 2010). In addition, a trial demonstrated the use of 1-methylcyclopropene (1-MCP) or silver nitrate ($AgNO_3$) in combination with IBA at 1000 mg/L improved the rooting percentage by reducing bud abscission and leaf retention (Contessa et al., 2011b).

Factors such as light intensity and cutting lignification can influence the success of cuttings during the acclimatization phase. Herbaceous cuttings (not lignified) were easier to root than the lignified explants, and a reduced light intensity (~600/μmol/$m^2$/$s^2$) promoted the photosynthetic activity of the remaining leaves (Tombesi et al., 2015).

However, the recalcitrance of this species to cutting propagation has recently been confirmed by Ozdemir and Dumanoglu (2018). In their trials,

**Table 1** The main recent advances obtained using specific Italian hazelnut cultivars

| Cultivar | Treatments | Response | Reference |
|---|---|---|---|
| Tonda Gentile Romana | Putrescine 1600 mg/L + IBA 1000 mg/L | After 2 months 60% of the cuttings showed callus formation with 76% rooting | Cristofori et al. (2010) |
| Tonda Gentile delle Langhe | IBA 1000 mg/L ± 1-MCP or $AgNO_3$ | The use of ethylene inhibitors induced a higher bud retention without a negative effect on rooting | Contessa et al. (2011b) |
| Tonda Gentile delle Langhe | Vermiculite/perlite or top soil/perlite | Best results were obtained using vermiculite/perlite (1:1) | Contessa et al. (2014) |

they obtained the highest root rates (43%) for 'Tombul' with softwood cuttings treated using the IBA quick-dip method and with different concentrations of polyvinyl alcohol (PVA); rooting was not achieved in 'Tombul' hardwood cuttings using different concentrations of IBA with or without PVA.

In general, it is clear that annual weather conditions and the health and vigour of donor plants are crucial for the success of cuttings, as well as setting up an efficient and reliable protocol suitable for a wide range of commercial varieties. Table 1 details the main results recently obtained for European hazelnut cutting propagation.

## 2.2 Propagation by grafting

Grafting is not largely used in hazelnut production, as it is mainly adopted for graft cultivars on clonal rootstocks in major fruit species to obtain high-density orchards and promote mechanized orchard management (Rovira et al., 2014).

In Serbia, grafting European hazelnut cultivars on Turkish hazelnut (*Corylus colurna* L.) seedlings is routine and the English grafting method is largely used due to the high incidence of graft success (Cerovic et al., 2009).

Currently, some trials are in progress to test the half-sib seedlings of the Turkish hazelnut, clonal rootstocks of *C. avellana* L. and interspecific crosses between *C. colurna* and *C. avellana* (named Dundee and Newberg) as no-sucker formation rootstocks (Lagerstedh, 1993; Tous et al., 2009). As for other nut trees, specialized procedures are required for successful callus formation at the grafted point, for instance, through a hot-callus graft maintaining the graft union at +27°C.

Encouraging results were recently obtained using etiolated hypocotyls of the American native hazelnut (*C. americana* Marshall) collected in southern Ontario (USA) and used as rootstocks. The 2–3 mm diameter and four bud scions of the European hazelnut (variety Carmela) were used for grafting on these etiolated hypocotyls and the rate of graft success was more than 80% when the scion was coated with paraffin (Rahemi et al., 2016).

These techniques need further investigation but it would be a successful way to establish new orchards grown in a single-trunk training system and for reducing the use of herbicide sprays in sucker control.

### 2.3 Micropropagation

Micropropagation represents a relatively new approach for hazelnut multiplication, as well as an efficient tool to promote *in vitro* tissue culture and genetic improvement (Bacchetta et al., 2008). Although it is a technique routinely used by some nurseries, micropropagation starting from mature tissues and organs still presents recalcitrance to *in vitro* culture, especially for the main hazelnut cultivars (Nas and Read, 2004; Damiano et al., 2005; Hand et al., 2014).

To date, successful protocols for hazelnut micropropagation have been published for both embryonic and juvenile material (Anderson, 1984). Numerous micropropagation protocols for different hazelnut varieties have also been proposed (Garrison et al., 2013; Bacchetta et al., 2008; Damiano et al., 2005; Nas and Read, 2004; Yu and Reed, 1993), but a suitable protocol for a wide range of cultivars has not yet been identified. Many problems still affect hazelnut micropropagation including the lack of multiplication, short shoot production, chlorosis and a milky white exudate (Hand et al., 2014). Furthermore, bacteria and fungi, combined with extensive damage as a result of sterilizing agents, are normally the main problems associated with the *in vitro* establishment of aseptic culture. Recently, Hand and Reed (2014), and Akin et al. (2017) employed a computer-assisted technique and surface design statistical software with the aim of modelling multiple factors and their influence on the overall outcome based on plant response. They concluded that the overall quality of the varieties used in the experiment was influenced by ammonium and nitrate nitrogen, micro- and macronutrients. The authors concluded that the results were genotype dependent and all cultivars require minor nutrient concentrations compared with the media commonly used (Bacchetta et al., 2008; Hand et al., 2014; Hand and Reed, 2014). In addition, growth regulators are crucial for hazelnut micropropagation and are the main components in the culture medium that determine proliferation success. The length of the shoots and the multiplication rate are influenced by cytokinin concentration. The most employed cytokinin is BAP (6-benzylaminopurine) at various concentrations and with different results for each cultivar and/or type of explant.

The chemical composition of the medium plays an important role in the success of micropropagation, as a suboptimal culture medium results in physiological disorders of the tissues (Nas and Read, 2004). In order to formulate an optimized culture medium in recalcitrant species, a novel approach based on the mineral composition of the seeds has been successfully applied (Bacchetta et al., 2008; Nas and Read, 2004), as the seeds store minerals and food reserves essential for starting vegetative growth if factors preventing germination are eliminated.

*In vitro* technologies, such as micropropagation, can be used to produce existing and new varieties, which can be multiplied exponentially in a short time. These techniques can also be used as a tool for implementing plant-breeding biotechnological approaches in woody species, particularly if mature tissues from valuable cultivars are used (Silvestri et al., 2016, 2018a,b; Karasawa et al., 2016).

## 3 Orchard site selection, cultivar choice and plantation

The richness of genetic variability ensures that European hazelnut have a remarkable adaptability to different climates. However, the optimum environmental conditions for cultivars are climates with mild and humid winters and cool summers. These conditions can be seen in the main hazelnut-producing areas of the world, located mainly between the 40th and 45th latitude: in Turkey and Georgia near the Black Sea, in Azerbaijan on the Caspian Sea, in Italy and Spain due to the presence of the Mediterranean basin and in Willamette Valley (Oregon), near the Pacific Ocean coast.

### *3.1 Orchard site selection*

Climate is the main factor limiting the expansion of hazelnut orchards and strongly influences its profitability.

Introducing cultivars into different environments from their original areas can reduce their productivity or can lead to ecophysiological disorders related to the lack of resistance to low temperatures or inadequate chilling (male flowers: 100–1200 h, female inflorescence: 500–1200 h, vegetative buds: 600–1200 h) that influences the correct development of the vegetative and floral parts of the plant (Mehlenbacher, 1991a).

During winter, unelongated hazelnut catkins (male inflorescence) can resist temperatures of −15/−20°C. During the elongation phase, just before pollen release, catkins become sensitive to temperatures of around −7°C. The vegetative and fertile buds, bearing female inflorescence, can tolerate temperatures of −20°C but, when the stigmas are in full bloom, these can be damaged by temperatures below −5°C (Hummer et al., 1986; Krpina et al., 1994).

In the bud break phase, cold damage occurs with temperatures between −3.5°C and −4°C at the first-leaf stage and at −2.5°C at the three-leaf stage; during orchard site selection, the land on the valley floor should be avoided as spring frosts may occur more frequently.

In general, the most suitable temperature for plant growth ranges from +23°C to +27°C during the day and +16°C to +18°C during the night. Temperatures above +35°C accentuate stomatal closure with a consequent growth break and photo-oxidation induction.

The amount of annual precipitation required is 800–1000 mm, well distributed throughout the year and the water requirement of the plant is between 80 and 100 mm/month from April to August; irrigation is appropriate below these levels (Tombesi and Rosati, 1997).

The hazelnut leaf surface, which is on average 90–100 $cm^2$ (Cristofori et al., 2007), has high transpiration when the relative humidity of the air is less than 70%. Under these conditions, vegetative development and the water reserve of the ground are affected, even if water is supplied via irrigation.

The species is adaptable to different soil conditions, avoiding clay types due to the risk of root asphyxia, and very sandy types as they tend to be poor and too permeable. The ideal soil must be homogeneous and fertile to at least 0.5 m in depth, where most of the hazelnut root system develops.

In general, deep, well-drained soils with an adequate amount of organic matter promote vegetative growth in the first years of planting and later, the precociousness of fruiting and higher productivity. The optimum soil pH for hazelnut is between 6 and 7.8; values higher than 7.8 may lead to the phenomena of ferric chlorosis and poor mobilization of microelements, while values below 6 may lead to magnesium deficiencies, reduced availability of phosphorus and reduced decomposition of organic matter. With regard to the limestone content, it is estimated that it should not exceed 8–10%.

## 3.2 Cultivar choice

The main demand for hazelnuts is a roundshape with well-blanched kernels for use in chocolates and other confectionery. High-quality kernels of the Italian cultivars 'Tonda Gentile delle Langhe', 'Tonda di Giffoni' and 'Tonda Gentile Romana', Turkish cultivars 'Tombul' 'Palaz', 'Çakıldak' and 'Fosa', and the Spanish cultivar 'Negret' set quality standards for the industry.

Despite hazelnuts having a long history of utilization (Boccacci et al., 2006), with over 400 cultivars worldwide (Mehlenbacher, 1991b), global production is still based on the use of a few dozen cultivars selected from local and wild populations, mainly in Europe and Turkey (Thompson et al., 1996). Hazelnut breeding is still poor compared with other domesticated crops and the most important public breeding programme is conducted in Oregon State University (OSU). Over the last 40 years, OSU has accumulated dozens of advanced selections from cultivars, pollinizers and ornamentals, which are listed in Table 2. Many of these releases have total or quantitative resistance to Eastern filbert blight. This fungal disease is caused by *Anisogramma anomala* (Peck) E. Miller,

**Table 2** New cultivars, pollinizers and ornamentals released by the hazelnut breeding programme at Oregon State University (Corvallis, Oregon, USA)

| Name | Type of selection | Year of cross | Year of release | Reference |
|---|---|---|---|---|
| Willamette | Cultivar | 1973 | 1990 | Mehlenbacher (2009) |
| VR 04-31 | Pollinizer | 1976 | 1990 | Mehlenbacher (2009) |
| VR 11-27 | Pollinizer | 1976 | 1990 | Mehlenbacher (2009) |
| VR 20-11 | Pollinizer | 1976 | 1990 | Mehlenbacher (2009) |
| VR 23-18 | Pollinizer | 1976 | 1990 | Mehlenbacher (2009) |
| Lewis | Cultivar | 1981 | 1997 | Mehlenbacher (2009) |
| Clark | Cultivar | 1982 | 1999 | Mehlenbacher (2009) |
| Rosita | Ornamental | 1984 | 1999 | Mehlenbacher (2009) |
| Gamma | Pollinizer | 1991 | 2002 | Mehlenbacher (2009) |
| Delta | Pollinizer | 1989 | 2002 | Mehlenbacher (2009) |
| Epsilon | Pollinizer | 1992 | 2002 | Mehlenbacher (2009) |
| Zeta | Pollinizer | 1992 | 2002 | Mehlenbacher (2009) |
| Santiam | Cultivar | 1989 | 2005 | Mehlenbacher (2009) |
| Sacajawea | Cultivar | 1990 | 2006 | Mehlenbacher (2009) |
| Yamhill | Cultivar | 1990 | 2008 | Mehlenbacher (2009) |
| Red Dragon | Ornamental | 1997 | 2008 | Mehlenbacher (2009) |
| Jefferson | Cultivar | 1993 | 2009 | Mehlenbacher et al. (2011a) |
| Eta | Pollinizer | 1999 | 2009 | Mehlenbacher et al. (2012) |
| Theta | Pollinizer | 1999 | 2009 | Mehlenbacher et al. (2012) |
| Tonda Pacifica | Cultivar | 1981 | 2010 | Mehlenbacher et al. (2011b) |
| Dorris | Cultivar | 1997 | 2011 | Mehlenbacher et al. (2013) |
| Wepster | Cultivar | 1997 | 2013 | Mehlenbacher et al. (2014) |
| McDonald | Cultivar | 1997 | 2014 | Mehlenbacher et al. (2016) |

which is spread in Oregon and Washington states (USA), and represents the main concern for local growers due to its high pressure on hazelnut orchards (Mehlenbacher et al., 2004). The new cultivars recently released in the breeding programme are mainly planted in the United States and Chile under legal protection agreements between OSU and the nurseries. Contrarily, in other countries the local cultivars, mainly with round-shaped nuts, are still extensively used to establish new plantations.

Considering the high industrial demand for hazelnuts, certain nut physical characteristics are preferred for market. The preferred nut shape is spherical which facilitates shelling operations. The industry requires a high percentage of kernel cultivars to have a kernel/nut ratio above 45% and most kernel sizes need to be 11–15 mm. The double kernel is considered a defect. Good pellicle removal after kernel roasting is also highly desired. These characteristics contribute to determining the industrial nut quality parameters and subsequently the price of hazelnut production.

On the basis of these qualitative parameters, the cultivars 'Barcelona' (sin. 'Castanyera'; 'Fertile de Coutard'), 'Ennis', 'San Giovanni' and 'Corabel' are often sold in-shell for fresh consumption, while the round-shaped cultivars 'Tonda Gentile delle Langhe', 'Tonda Gentile Romana', 'Tonda di Giffoni', 'Tombul' and 'Negret' are destined for market.

Currently, the industry uses 'Tonda Gentile delle Langhe', 'Tonda Gentile Romana', 'Tonda di Giffoni', 'Mortarella' and 'San Giovanni', which are mainly produced in Italy; 'Negret' and 'Pauetet' produced in Spain; 'Tombul', 'Palaz' and 'Fosa' produced in Turkey; 'Barcelona' and the OSU releases produced in Oregon and Chile (Tombesi, 2005).

The cultivars mainly spread in traditional hazelnut countries are discussed briefly.

### 3.2.1 Turkish cultivars

There are 18 standard registered cultivars in Turkey mainly obtained from natural hybridization (*C. avellana* × *C. maxima*), with three recently registered ('Okay 28', 'Giresun Melezi', 'Allahverdi') as new releases (Anil et al., 2016). Turkish varieties are grouped as 'Giresun' and 'Levant quality'. The first group produces high-quality nuts and includes 'Tombul' and the main varieties used in the oldest hazelnut area of the Turkish Black Sea (Giresun, Ordu and Trazbon provinces). 'Levant quality' group varieties are mainly cultivated in younger hazelnut provinces (Turan and İslam, 2018).

Turkish hazelnut varieties are also grouped into three classes depending on their nut shape, for example round-shaped group ('Allahverdi', 'Cavcava', 'Çakıldak', 'Foşa', 'Giresun Melezi', 'Kalınkara', 'Kan', 'Kara', 'Kargalak', 'Mincan', 'Okay 28', 'Palaz', 'Tombul', 'Uzunmusa'); pointed-shaped group ('Sivri', 'İncekara') and the long-shaped group ('Yuvarlak badem', 'Yassıbadem').

*'Tombul'*. Highly cultivated in Giresun, Trabzon and Ordu. The nuts are of medium to small size (2.0–2.2 g), with a kernel/nut ratio of approximately 50%, spherical shape, excellent pellicle removal and excellent taste and aroma. Generally, a single husk has three to four nuts and its length is twofold longer than the nut. For this reason, nuts are often collected by hand directly from the tree. $S_4S_{12}$ are alleles of incompatibility.

*'Çakıldak'*. Widely grown in middle to high altitude in Ordu. The cultivar is tolerant to spring frosts due to its late bud burst. The nuts are medium sized (2.1–2.5 g) with a kernel/nut ratio of approximately 55%. It has very high yield and large nut size. The alleles of incompatibility are unknown.

*'Sivri'*. This cultivar is highly productive but it is sensitive to drought, and as a result there is an unacceptably high pre-harvest fruit drop in dry seasons. The nuts are medium to small size (2.0 g). The nut/kernel ratio is 50% and the alleles of incompatibility are $S_4S_{\underline{12}}$.

*'Foşa'*. A large and attractive hazelnut cultivar mainly grown in the Trabzon province. Its shell is reddish-brown and the nuts are small sized (1.7 g). The nut/kernel ratio is 50–53% and the alleles of incompatibility are unknown.

*'Palaz'*. This cultivar is commonly grown in the Ordu and Samsun provinces, it can be easily distinguished by its tick-leaf tissue and long, split husk. It is sensitive to many diseases and pests. The nuts are of medium to small size (2.1 g) with a nut/kernel ratio of 51.4%. The alleles of incompatibility are $S_{\underline{2}}S_{4.}$

### 3.2.2 Italian cultivars

*'Tonda Gentile'* (sin: 'Tonda Gentile delle Langhe'). The nuts are medium sized (2.6 g), with a kernel/nut ratio of approximately 45%, spherical shape, good pellicle removal, excellent taste and aroma and good shelf life. This cultivar is of medium productivity and ripens early, has intermediate vigour and is characterized by poor adaptability to different environments from the area of origin. The alleles of incompatibility are $S_2S_{\underline{7}}$.

*'Tonda Gentile Romana'*. The nuts are medium sized (2.5 g), with a kernel/nut ratio of about 45%, spherical shape, low pellicle removal and good taste and aroma. The cultivar is characterized by a medium to high productivity, intermediate vigour, late bud break and medium to late ripening. The alleles of incompatibility are $S_{\underline{10}}S_{\underline{20}}$.

*'Tonda di Giffoni'*. The nuts are of medium to large size (2.7 g), with a medium-to-high kernel/nut ratio (46%), spherical shape, excellent pellicle removal and good taste and aroma. It is a cultivar characterized by high yield, intermediate to high vigour, early bud break, intermediate ripening and a good adaptability to different environments. The alleles of incompatibility are $S_{\underline{2}}S_{23}$.

*'Nocchione'* (sin. 'Montebello'). The nuts are of medium to large size (3 g), with a low kernel/nut ratio (38%), spherical shape, good pellicle removal and excellent taste and aroma. This cultivar displays high productivity, intermediate to high vigour, early bud break, early to intermediate ripening and good adaptability to different environments. The alleles of incompatibility are $S_{\underline{1}}S_2$.

*'San Giovanni'*. The nuts are medium sized (2.6 g), with a medium-to-high kernel/nut ratio (46%), semi-elongated shape, good pellicle removal and good taste and aroma. It is a cultivar characterized by high yield and vigour, early

bud break, early ripening and good adaptability to different environments. The alleles of incompatibility are $S_2S_{\underline{8}}$.

*'Mortarella'*. The nuts are of medium to low size (2.2 g), with a medium-to-high kernel/nut ratio (46%), semi-extended shape, excellent pellicle removal and excellent taste and aroma. This cultivar displays medium to high productivity, intermediate vigour, medium to late bud break and intermediate ripening. The alleles of incompatibility are $S_2S_{\underline{17}}$.

### 3.2.3 Spanish cultivars

*'Negret'*. This is the main variety cultivated in Tarragona and Girona and is of unknown origin. It has medium-to-small-sized nuts, with a kernel/nut ratio of 47–49%. It is highly appreciated by the industry for its good organoleptic characteristics and ease of pellicle removal after toasting. Its agronomic characteristics are not ideal; it shows weak vigour and sensitivity to iron chlorosis, produces a large number of shoots and has high chilling requirements to produce pollen. The alleles of incompatibility are $S_{\underline{10}}S_{22}$.

At the end of the 1980s, the Institute of Agrifood Research and Technology (IRTA) started a clonal selection of this cultivar, the free apple tree virus (ApMV) clone 'Negret-IRTA-N-9' was selected (Rovira et al., 2006) and was more productive than the standard 'Negret' variety infected by this virus, which affects hazelnut (Aramburu and Rovira, 1998).

*'Pauetet'*. The nuts are medium to small, oval in shape and are similar to those of 'Negret'. However, pellicle removal after roasting is lower than 'Negret'. It has a kernel/nut ratio of 46–49%. It is a very vigorous tree, more productive and resistant to compact and calcareous soils than 'Negret' and produces fewer suckers. This cultivar also performs well on rainfed land in mountain areas. The alleles of incompatibility are $S_{\underline{18}}S_{22}$. Since the 1990s, 'Pauetet' has been the most widely planted variety in France for the industrial hazelnut market (Germain and Sarraquigne, 2004).

*'Gironell'*. The nuts are small and round and peel poorly after roasting. It has a low kernel/nut ratio (40–42%). It is a vigorous cultivar, very productive, produces large numbers of suckers and old orchards show biennial bearing. The alleles of incompatibility are $S_{\underline{1}}S_2$. It is a good pollinizer as it produces large amounts of pollen.

*'Barcelona'* (sin: 'Castanyera'; 'Fertile de Coutard'). This cultivar is of unknown origin. It has round large nuts, with a kernel/nut ratio of 43%, intended for the table hazelnut market. The alleles of incompatibility are $S_{\underline{1}}S_2$. It is an old variety cultivated in some countries under different names: 'Barcelona' (USA), 'Fertile de Coutard' (France), 'Grada' (Portugal) and 'Grande' (Asturias, Spain) (Köksal, 2002).

### 3.3 Plantation

In suitable hazelnut districts, such as the Mediterranean area, Oregon (USA), and the new hazelnut-growing countries, such as Chile, the orchards are trained as a multi-stem bush or vase shape (Fig. 1) at a density of 400–500 plants/ha (Bignami et al., 2005a).

The advised spacing for new orchards is still 5 × 5 m (400 plants/ha) or 5 × 4 m (500 plants/ha) depending on the vigour of the cultivar, to enable mechanical working of the orchard in two directions. For some high vigour cultivars, such as Barcelona, 6 × 6 m tree spacing (about 278 plants/ha) is still quite commonly adopted. In new hazelnut areas, the orchard layout should incorporate at least 10% of pollinizers placed every third tree in every third row.

Dynamic spacing of the orchard has recently been suggested for fruit tree species with a long unproductive period after planting, such as olive and hazelnut (Riggert and McDonald, 1987; Sarraquigne, 1989; Tous et al., 1994a). This plantation technique consists of planting double the number of plants (i.e. 800 plants/ha, spaced 5 × 2.5 m) with the aim of enhancing the yield during the first 10–12 years of the plantation and then eliminating 50% of the plants to achieve the final orchard design (i.e. 400 plants/ha, spaced 5 × 5 m). Nevertheless, this system has not been applied on a large scale because of

**Figure 1** Hazelnut trees trained to a multi-stem bush, which is a growing system mainly adopted in Italy (a), Turkey (b) and Spain (c), in comparison with trees trained as a single trunk, which is the growing system adopted in Oregon (d).

the higher plant costs, the tendency of the farmer to delay the timing of tree spacing and the risks of competition among the plants (Bignami et al., 2005a). High-density systems have received little attention, and the new orchards are still mainly created by hand planting three to four rooted suckers per hole; however, this technique is labour intensive. In Turkey, growers have recently preferred to adopt the row system with spacings of 5–6 m between the rows and 1–2 m between contiguous plants to replace the traditional 'ocak system' that consists of planting at least four rooted suckers around a circle 1 m in diameter (Fig. 1b).

Considering that the hazelnut is a self-incompatible species, the orchard layout needs to include 10% pollinizers. Recently, Differential Global Navigation Satellite System – Real Time Kinematic (DGNSS-RTK) technology has been introduced to design and mechanically plant single-trunk high-density new orchards spaced at 4.5–5.0 × 3.0–2.5 m (Fig. 2), which can be managed with a high level of mechanization of future pruning operations. There are companies that will work with the grower to mark the exact spot for each tree according to the expected spacing and tree arrangement. The economic advantages derived from the adoption of DGNSS-RTK technology include a significant cost reduction for mechanized setting of 30 ha and over, indicating a good opportunity for the entire hazelnut agro-food chain (Cristofori et al., 2017).

Young plants can be planted anytime from autumn through spring to ensure tree roots do not dry out. Planting during the autumn is preferred to give the plant more efficient seasonal time to develop a working root system

**Figure 2** Differential Global Navigation Satellite System-RTK mechanized transplanting operation in the field.

before it leafs out in the spring. In a temperate climate, hazelnut root growth is active in late January whenever soil temperatures are higher than +4°C.

Manual planting holes need to be 50–60 cm wide and 30 cm deep. During planting, portions of the roots are lost, reducing water absorption and potentially slowing tree growth. For this reason, irrigation interventions during spring and summer are recommended. After planting, it is advisable to prune back the tops of the tree, leaving 80–90 cm of the trunk (particularly for the vase shape), and if trees are branched the laterals should be cut back to a few buds or removed. This top reduction should offset root damage that occurred during digging and handling.

## 4 Orchard management: irrigation, nutrition and pruning

In the most important hazelnut production areas, with the exception of Turkey where orchard management is mostly conducted manually, the agronomic management of specialized orchards is mainly mechanized.

Among orchard operations, soil management is crucial to maintain and improve fertility, to contain erosion in sloping areas, to facilitate the mechanical harvesting of nuts and to facilitate the access of agricultural machinery avoiding soil compaction.

Soil management techniques vary from milling to growing cover crops and using herbicides, the latter are also used in combination.

Soil management through controlled grassing using spontaneous flora, widely diffused in the main hazelnut orchards, is one example of the effective application of an integrated fruit-growing technique, which partially offsets the problem of dust rising during mechanical harvesting of the nuts and at the same time attenuates erosion and hydrogeological instability.

Managing soil with grassing involves the periodic shredding of the turf with shredders during two to three seasonal interventions based on the spring-summer rainfall, which influences the growth of the grass. Chemical control through spreading herbicides, mainly along the row, represents an economically sustainable method of grass control in the orchard, with positive effects also on the conservation of the water resource, otherwise transpired by the flora. The orientation towards sustainable and 'low-impact' soil management, as well as restrictions on the use of herbicides mainly required for organic and integrated production regulations, place some limits on the use of herbicides.

The European hazelnut is a species with an aptitude for high sucker emission and the quantity of suckers emitted varies among the cultivars (Cristofori et al., 2014a). Starting from the second leaf on the ground, sucker control is an essential practice. During the juvenile phase of the plant, the suckers are removed manually, while from the fourth year the control strategies range from manual or mechanical elimination in summer, to chemical control with the use of specific herbicides as the active principles 'Carfentrazone etile' and 'Pyraflufen etile' or proper nitrogen and salt solutions. To ensure the effectiveness of these chemical applications, the suckers must be in the herbaceous stage (10–15 cm in height).

In small orchards, sucker management is mostly achieved by hand using sharp tools such as pruning shears. Managing this operation by hand enables the farmer to leave some suckers to rejuvenate the bush. The method is environmentally friendly but is labour intensive and hence expensive (Serdar and Akyuz, 2018).

The use of non-sucker emission rootstocks are under evaluation in trials and in some commercial orchards in Serbia and Spain (Cerovic et al., 2009; Tous et al., 2009).

## 4.1 Irrigation

The irrigation of hazelnut is widely applied due to its positive effects on productivity both in areas and periods with limited rainfall, and in soils with low water capacity. The hazelnut is generally considered a species sensitive to water stress, with a low capacity for stomatal regulation. Several negative effects of water stress have been observed in this species, such as a decrease of photosynthetic activity, early break fruit growth and early leaf fall, a higher susceptibility to diseases (Girona et al., 1994; Tombesi, 1994; Bignami and Natali, 1997; Diaz et al., 2005) and an increase in blank nuts and decrease of per cent kernel (Bignami et al., 2009). Vegetative and productive activity is affected by water availability during the growing season, with consequences on production quality and quantity. The complex annual cycle, characterized by the overlapping of different vegetative and reproductive processes (end of leaf area development, fruit set, kernel filling, flower bud initiation and differentiation) from June to August, makes the availability of water an important factor to overcome competition. Usually, a hazelnut needs at least 800 mm of rain well distributed throughout the year.

Irrigation systems have been introduced in the main production areas with limited water availability, irregular rain distribution during the year and for young plants. Currently, drip irrigation is the most utilized technique in hazelnut orchards and is applicable in its variants according to the soil and orchard design.

It is necessary to calculate the water balance in order to calculate the volumes of water supply. Crop evapotranspiration is calculated using the formula ETc = ETo  Kc, where ETo represents the estimated daily water loss (for instance determined using a class A evaporimeter) and it is multiplied by the crop coefficient (Kc), which is specific for each development phase. The only current hazelnut Kc available in the literature were determined in a Spanish environment using 'Negret' cultivar plants (Tous et al., 1994b) reported as: April: 0.30, May: 0.40, June: 0.62, July: 0.70, August: 0.55 and September: 0.35.

Localized irrigation, as well as ensuring a high efficiency of water use, also allows the administration of nutrients via fertigation, after the installation of appropriate fertilizer mixers and injection systems. Drip irrigation with drip-wing laying on the row or on the canopy is highly spread in specialized orchards and water is distributed through anti-occlusion and self-compensating drips (with a variable flow rate of 4 and 8 L/h). Normally, two drips per plant are applied, one on each side, at a distance of about 1 m from the trunk.

Micro-jets, also called micro-sprayers, are very common in new high-density systems and provide large volumes of water in less time than drip systems. Micro-jets require the structuring of irrigation systems with a higher flow rate and require maintenance of the micro-sprayers, which can easily occlude. In areas with low relative humidity or, in particular, drought, they have the advantage of partially raising moisture in the foliage of the plant.

Sub-irrigation is a good application in flat and large plantations. This technique, based on the administration of irrigation water by capillary ascent, is more efficient due to the absence of evaporative phenomena. The installation of self-compensating drip-off wings, buried at a depth of 20–50 cm, is performed with special machines and affects only the centre line of each inter-row in adult systems, while in young orchards the laying of a dripping wing buried per single inter-row is carried out at distances of about 80–100 cm from the row.

Recent experiments conducted in Italy on 'Tonda Gentile Romana' have shown that irrigation aimed at returning 75% of the crop evapotranspiration allows a suitable balance between vegetative growth and production (Cristofori et al., 2014b).

## 4.2 Nutrition

The complex overlapping of vegetative and reproductive activity in a short period of the annual cycle makes hazelnut yield and quality highly dependent on correct nutritional support. Nitrogen is the most important component in the mineral nutrition of hazelnut (Bignami et al., 2005b); in addition, the foliar application of calcium plays an important role during the kernel filling stage (Cakca and Sanguankeo, 2014). Trials were recently carried out to develop optimized protocols of hazelnut nutrition including foliar applications, such as in Italy for the cultivar 'Tonda Gentile delle Langhe' (Roversi, 2011; Pansecchi et al., 2014), in Oregon on different cultivars (Olsen and Cakca, 2009) and in Chile on the 'Barcelona' cultivar (Ellena et al., 2014). The principle of foliar nutrition carried out in these experiments is based on the proven ecophysiological condition that the elements are absorbed, digested and used by the tree and are not dispersed in the environment. However, knowledge of the nutrition of the main hazelnut cultivars is still incomplete.

In general, the hazelnut has a high demand for nitrogen and potassium, while phosphorus requirements are slightly lower. The literature agrees that in soils of medium fertility and with mature plants in balance for vegetative and productive activity, the correct ratio for the three macroelements is 1:0.5:1 (N: P: K) (Olsen, 1997). The hazelnut also requires calcium and magnesium and their application is advised every 3–4 years during the spring according to soil and foliar analysis. The need for magnesium application is greater where potassium and calcium levels in the soil are high (Cacka and Sanguankeo, 2014).

Nitrogen demands are high in the juvenile phase because this element is required for vegetative development of the plant, but a balanced contribution of nitrogen is also fundamental in the production phase, considering that the hazelnut is a species that produces nuts exclusively on the vigorous 1-year-old branches and adequate nitrogen availability favours this development. Nitrogen is also in abundance in mature nuts (about 3% of kernel dry weight) and shows the highest content during kernel filling (5% of kernel dry weight), indicating that this growth phase has a high requirement for suitable nitrogen availability (Cristofori et al., 2015).

Potassium is necessary throughout the biological cycle both for vegetative and reproductive phases, and the availability of this macronutrient also favours the absorption of nitrogen, magnesium and manganese, the latter two elements are particularly important for the accumulation of fatty substances in the kernel.

Phosphorus is an essential element in many physiological functions of the plant including cellular respiration and sugar and protein metabolism. It is present in the kernel and reproductive organs.

For young plants, nitrogen fertilization can be performed in early spring using granular formulations (i.e. ammonium nitrate) starting from the second leaf in the field. The doses to be administered per plant (referred to as fertilizing units) are 50 g in the second year, 90–130 g from the third to the fifth year and 150–220 g from the sixth to the seventh year. In soils that have ample phosphorus and potassium and through adequate pre-implantation administration, the contributions of these elements can start from the third or fourth year.

In adult and specialized orchards, the quantities of macronutrients should be annually integrated to about 90 units of nitrogen per year (considering an average production per ha of 2–2.5 t for in-shell nuts) divided into 2–3 epochs (vegetative bud break stage, kernel filling and post-harvest). The nitrogen post-harvest intervention allows the development of the root system and the accumulation of reserves in the trunk and should be performed before the leaves have lost their photosynthetic functionality. For phosphorus and potassium, the yearly integration consists of giving 40 and 80 units/ha, respectively, in spring and in a single solution.

The administration of other elements is not necessary in soils in which the chemical analysis shows adequate levels and the plants do not exhibit symptoms of deficiency. Moreover, the periodic administration of organic matter is generally enough to cover microelement requirements. The administration of organic matter should be carried out every 2–3 years during the autumn in poor soils and every 4–5 years in those with a high silt and clay content. This intervention provides a certain amount of nitrogen, phosphorus and potassium, which must be taken into account at the time of mineral fertilization.

The foliar administration of macro- and microelements is an effective practice to compensate for any nutritional deficiencies that may occur during the economic cycle of the orchard and currently available formulations are particularly suited to hazelnut requirements.

Correct foliar application should however be accompanied by diagnostic observations to determine the actual nutritional status of the plant. Foliar applications of boron, iron, calcium and magnesium have shown positive effects in various environments and for different cultivars, for example in Italy for 'Tonda Gentile Romana', in Spain for the 'Negret' cultivar and in Oregon for 'Barcelona' (Olsen and Cacka, 2009).

Three to four proper foliar applications (macro- and micronutrients), sometimes combined with fungicide or pesticide applications, are recommended for adult hazelnut orchards.

Table 3 summarizes some recent findings of the micro- and macroelement content in adult leaves observed in different environmental conditions and are useful for correct diagnostic provision.

### 4.3 Pruning

Hazelnut orchards are mainly trained to a multi-stem bush system and in some traditional areas the new plantations are still created by hand planting three to four rooted suckers per hole. Alternatively, the plantations are made using one well-rooted young plant per hole and the trees are grown as a vase shape and

**Table 3** Minimum and maximum content of the main macro- and microelements present in adult hazelnut leaves considered 'standard' by the international literature, and average values observed in adult leaves of Italian cultivars 'Tonda Gentile Romana', 'Nocchione' and 'Tonda Gentile delle Langhe' sampled at the end of July in Latium and Piedmont

| Cv | Nitrogen % dw (dry weight) | Phosphorus % dw (dry weight) | Potassium % dw (dry weight) | Calcium % dw (dry weight) | Magnesium % dw (dry weight) | Boron ppm dw (dry weight) | Iron ppm dw (dry weight) |
|---|---|---|---|---|---|---|---|
| Standard (Olsen and Cacka, 2009) | 2.21-2.50 | 0.14-0.45 | 0.81-2.00 | 1.01-2.50 | 0.25-0.50 | 31-75 | 51-400 |
| Average Italian cultivars | 1.82-2.60 | 0.14-0.32 | 0.79-1.40 | 1.31-1.95 | 0.16-0.44 | 40-85 | 50-300 |

single trunk, as in the United States and Chile. Recently, mechanized planting systems have been adopted for hazelnut, mainly to establish large orchards on flat ground.

The pruning of adult trees is usually performed during vegetative rest and is generally limited to the removal of suckers, dead wood and old and diseased or broken branches. A reduction of shoot vigour, low light penetration within the tree canopy and yield decline are long-term consequences observed in hazelnut orchards (Cristofori et al., 2009). As a result, canopy density increases and over the years partial crossing of branches, reduction of shoot growth and lessening of fruiting potential occur especially in the lowest and inner parts of the canopies (Me et al., 1994; Roversi and Mozzone, 2005). Romisondo et al. (1983) showed that hazelnut productivity is positively related to the strength of mixed 1-year-old shoots, and indicated 15–20 cm as the optimum shoot length without cultivar interaction. Considering that hazelnut pruning carried out manually is labour intensive, a new trend to manage this operation mechanically is being developed in some areas, such as Italy and Spain, based on hedging and topping cutting combinations.

The pruning of young bush-grown plants is performed starting from the second to third leaf in the field and is limited to the selection of four to five main lignified suckers, thinning the apical portion and eliminating other suckers produced at the base. In single-trunk plants, a single and vigorous sucker is selected and thinned at heights of 60–80 cm for setting up the new vase shape and promoting the development of lateral shoots to generate new primary branches. During the first following season, four to five well-oriented main branches will be selected eliminating the remaining ones.

Annual pruning of adult trees is mainly focused on removing 15–20% of wood according to the age and vigour of the cultivar and soil fertility. In some areas, 'green pruning' techniques have been tried on adult plants, with interventions carried out in May, which seems to provide positive results in terms of plant productivity; however, the use of this technique needs further investigation before being applied on a large scale (Ughini et al., 2009).

Mechanical pruning was recently applied to hazelnut in both trials and commercial orchards and was performed using a rotating blade bar carried by a tractor (Fig. 3). The effectiveness of this operation, advisable for large orchards and in medium- to high-density plantations, is confirmed by the positive effects on production obtainable in the medium period.

Mechanical pruning is performed with side-cutting interventions along the row (hedging) and cutting the top of the plants (topping). The first intervention can occur when the branches of the two contiguous rows come to meet on the half of the row, on average from years 10 to 11 of the plantation depending on cultivar vigour. To avoid the pruning of all trees, which can drastically reduce productivity in the first year of application, hedging interventions can be

**Figure 3** Mechanical pruning of the hazelnut: hedging cut demonstration carried out in an adult orchard in Italy.

foreseen over some years (from 3 to 5). Figure 4 shows a hedging scheme that is adoptable every 3 years in the same row side.

Topping may be less frequent than hedging and is performed on the whole orchard every 5–8 years. The more promising technique involves cutting one side of two rows facing each other; this type of cut is carried out during the first year on an inter-row every three, leaving the two following inter-row spaces unaltered. The following year the first pruned inter-row is not cut, but the second one is, leaving the two following ones unchanged and resuming the sequence. The third year ends by pruning all the inter-row threads not yet pruned in previous years. The rotation every 3 years of pruned rows allows diluting the loss of nuts production due to pruning and maintaining a constant rhythm of renewal of the plantation. To complete mechanical pruning, it is advisable to proceed with the manual cutting of those branches that exhibit any mechanical damage or that are badly positioned. The clippings are transported out of the orchard and are eliminated or can be collected and treated for biomass energy production.

## 5 Integrated pest management

The hazelnut industry and several agricultural management laws in different countries encourage farmers to use integrated pest management (IPM) to

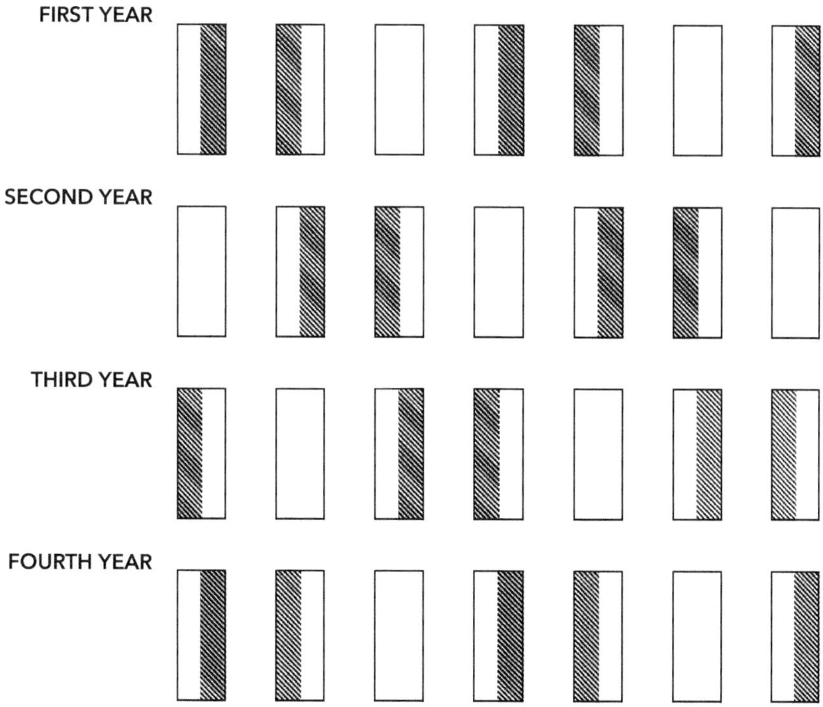


**Figure 4** The hedging scheme for mechanical pruning adopted in large commercial orchards. The same row side is cut every 3 years. Alternatively, the intervention can be extended to 4 or 5 years.

control phytophagous insects, certain diseases and weed pests that are found in hazelnut orchards to reduce the amount of pesticides used.

Adult orchards tend to have few weed problems because shading inhibits most weed establishment. Younger orchards have more light penetration and herbicides are often used mainly along the rows, but growers also manage weeds mechanically. Maintaining a vegetation strip between the tree rows and carefully timing mowing events provides habitat for beneficial insects that aid in conventional IPM.

The IPM techniques used for weed management not only provide weed control but also help maintain biological diversity in the orchard (De Francesco, 2006).

In Oregon (USA), advances in IPM practices have included the development of monitoring techniques and treatment thresholds for 'filbert aphid' (*Myzocallis coryli* (Goetze)) and 'filbert leafroller' (*Archips rosanus* (L.)), which have reduced prophylactic insecticide applications. Advances have also included the use of pheromone traps and a degree-day model for 'filbert worm' (*Cydia latiferreana*

(Walsingham)), which allow precise, targeted treatments for this pest (Olsen, 2002). The most valuable result of this IPM approach has been the successful identification and introduction of a filbert aphid parasitoid, *Trioxys pallidus* (Haliday). This parasitoid, a small Braconid wasp, has provided nearly complete biological control of the filbert aphid and has significantly reduced insecticide use for this pest. This approach, based on parasitoid distribution, should be proposed for the biological control of other hazelnut pests following a similar programme, as the elimination of aphicides improves the survival of predators and parasitoids that attack leaf rollers and filbert worms.

The effective monitoring of pests is the basis of pest management for IPM production and organic growing. Monitoring reveals the current pests and their levels so that decisions with regard to pest control can be made. In addition, timing of control is very important for many pests and this is determined by monitoring pest biology. Several methods can be applied to monitor hazelnut orchards, with the main monitoring methods including pheromone traps for lepidopteran pests, visual counts and beating sheets. In Turkey, current management programmes in hazelnut orchards are based on regular monitoring, field counts and phenology models (Tuncer et al., 2009).

In addition, beneficial organisms must be considered in hazelnut orchards as there are hundreds of natural enemies of the major hazelnut pests and diseases.

More generally, many adverse events contribute to the reduced quantity and quality of hazelnut production. The factors at higher risk for production quality are the infestations due to certain specialized arthropods. Various phytophagous arthropods are now associated with the hazelnut, as shown in Table 4.

The main phytophagous arthropods commonly recognized to be most harmful for hazelnuts are the *Phytoptus avellanae* (Nalepa) mite, hazelnut true bugs, hazelnut weevil and hazelnut scolitids. Each biological cycle, the damage caused to hazelnut and some guidelines of IPM will now be discussed.

## 5.1 Phytoptus avellanae

*Phytoptus avellanae* causes serious damage to hazelnut crops via bud deformation due to the development of hypertrophies that inhibit the normal development of the bud and inflorescence. The mite overwinters as an adult in hypertrophic buds and lays from the end of February to the end of March. It migrates between the scales of the hypertrophic bud, between April and June during leaf development and then infests new buds. During the autumn phase, there is an explosion in population density within the new infested buds, which will become hypertrophic. There are differing degrees of susceptibility to these infestations and the most sensitive cultivar is 'San Giovanni', while the

**Table 4** Phytophagous arthropods associated with the hazelnut[1]

| Pest | Order | Species | Damage | Country |
|---|---|---|---|---|
| Insects | *Isoptera* | *Kalotermes flavicollis* (Fabr.) | Roots and stumps | Europe, North Africa |
| | *Hemiptera* | *Acanalonia conica* (Say) | Leaves and sprouts | Italy, Romania, United States |
| | | *Oxycarenus lavaterae* (Fabricius) | Sprouts and leaves | Europe |
| | | *Halyomorpha halys* (Stäl) | Nuts | Europe, Turkey, Georgia, South Russia, China, United States |
| | | *Gonocerus acuteangulatus* (Goeze, 1778) | Nuts | Europe |
| | | *Nezara viridula* (Linnaeus, 1758) | Nuts | Europe |
| | | *Palomena prasina* (Linnaeus, 1761) | Nuts | Europe, North, South and East European Russia |
| | | *Piezodorus lituratus* (Fabricius, 1794) | Nuts | Europe, North, South and East European Russia |
| | | *Rhaphigaster nebulosa* (Poda, 1761) | Nuts | Italy |
| | | *Pantilius tunicatus* (Fabricius, 1781) | Meadows | Italy |
| | *Homoptera* | *Metcalfa pruinosa* (Say, 1830) | Leaves | Europe, North America, Korea |
| | | *Myzocallis coryli* (Goetze) | Sprouts and leaves | North America, Italy, Spain, Turkey |
| | | *Mytilococcus ulmi* (Monti, 1956) | Branches | Italy |
| | *Coleoptera* | *Anoplophora chinensis* (Foster) | Trunk | Italy, France, Germany, Switzerland, Croatia, Turkey, West Asia |
| | | *Xyleborus dispar* (Fabricius) | Trunk and branches | Italy |
| | | *Xyleborinus attenuatus* (Blandford) | Trunk and branches | Europe, North America, Russia (Far East), Japan |
| | | *Curculio nucum* (Linnaeus) | Nuts | Europe |

(*Continued*)

**Table 4** (*Continued*)

| Pest | Order | Species | Damage | Country |
|---|---|---|---|---|
| | | *Agrilus viridis* (Linnaeus, 1758) | Trunk and branches | Europe |
| | | *Oberea linearis* (Linnaeus, 1761) | Twigs | Europe, Russia |
| | | *Megaplatypus mutatus* (Chapuis, 1865) | Trunk and branches | South America, Italy |
| | | *Oemona hirta* (Fabricius) | Sprouts and branches | New Zealand |
| | | *Altica brevicollis* (Foudras, 1860) | Leaves | Europe |
| | *Lepidoptera* | *Archips rosanus* (L.) | Leaves | Europe, North America, Asia |
| | | *Cydialatiferreana* (Walsingham) | Nuts | North America |
| | | *Choristoneura rosaceana* (Harris) | Leaves | United States |
| | | *Malacosoma americanum* (Fabricius) | Leaves | United States |
| | | *Malacosoma disstria* (Hübner) | Leaves | United States |
| | | *Lymantria dispar* (Linnaeus, 1758) | Leaves | North America, Europe, Asia |
| | | *Erannis defoliaria* (Clerck, 1759) | Leaves | Europe |
| | | *Operophtera brumata* (Linnaeus, 1758) | Leaves | Europe, West and East North America, Russia, Japan |
| | | *Zeuzera pyrina* (Linnaeus, 1761) | Branches | Europe, North, South and East European Russia, Ukraine |
| | | *Synanthedons puleri* (Fuchs, 1908) | Trunk and branches | Europe |
| | *Orthoptera* | *Oecanthus pellucens* (Scopoli, 1763) | Young branches | Europe, North Africa, Ukraine, Russia |
| | | *Leptophyes punctatissima* (Bosc, 1792) | Leaves | Europe |
| Mites | *Prostigmata* | *Phytoptus avellanae* (Nalepa, 1889) | Buds | Europe, Australian region |
| | *Acarida* | *Tetranycopsis horridus* (Canestrini and Fanzago, 1876) | Leaves | United States, Europe, Turkey, Argentina |

[1] Fauna Europaea, https://www.fauna-eu.org; EPPO Global Database, https://gd.eppo.int.

least sensitive is the 'Mortarella' cultivar. Generally, the plant is very tolerant of the presence of mites up to 10% of infested buds, beyond this value acaricide treatments are necessary during the leaf development period (stools 2-3 cm long with three to four leaves).

## 5.2 Gonocerus acuteangulatus

*Gonocerus acuteangulatus* species is reported in all Italian regions. It has six pre-imaginal stages including two stages of neanidae and three nymphae. It overwinters as an adult in natural shelters such as accumulations of wild herbs, bushes, shelters in hollow trunks and houses. They generally overwinter individually but in rare cases form an aggregation with more than four individuals. In spring, they come out of the wintering phase and go directly to receptive plants that in principle could be plants of *Prunus* spp. and then move on to hazelnut plants. They feed by pricking with their pungent and sucking mouthparts directly on the achenes from the beginning of seed development until they reach ripe fruit. During the first phases of hazelnut development, the infestations produce a *traumatic abortion*, which does not allow fruit growth. Late infestations, with already formed hazelnuts, form both white and dark spots on the fruit. These stains are often associated with unpleasant tastes commonly called '*bug flavour*'. This latter type of damage is the one that most affects product quality. After mating, which can occur several times, the females lay the eggs in small groups of usually two to three with a maximum of about 12 elements, on the bracts of the cupula. Laboratory tests have enabled verification that the female can live up to 400 days. Each stage of the pre-imaging phase requires a variable period depending on the climatic conditions, and we can assume an incubation of about 1 week in the summer months, with an average duration of the pre-imaging stages of 3 days. The complete cycle can last approximately 1.5–3 months. It lays its eggs from the end of May to September. Towards the end of July, the adults of the generation of the year emerge. Several authors hypothesize that the species can create two generations per year, but generally the insect is considered to be monovoltine. Given the potential plasticity of this type of feeding, it cannot be excluded that the second generation could occur on other hosts and only when the winter temperatures are milder than usual. The phytophage is parasitized by numerous species of *calcidoid Hymenoptera* and *tachinid Diptera*.

## 5.3 Nezara viridula

*Nezara viridula* is a light-green bug with three small light points on the back at the base of the scutellum. It overwinters as an adult in natural and artificial shelters. They lay from 40 to 100 eggs grouped on both herbaceous and arboreal plants and undergo two generations per year. The most susceptible host to its infestations is the tomato, where it forms yellowish areas corresponding to the sucking stings, transmitting an unpleasant taste to the berries. On the kernel, as with *G. acuteangulatus*, it pierces the hazelnuts to feed itself, producing the same damage.

## 5.4 Palomena prasina

*Palomena prasina* is an extremely polyphagous bug but very attached to the hazelnut. In southern Italy, it is the most important species for this plant. It overwinters in the adult stage in natural and artificial shelters. In April, adults mate preferably on plants used as winter shelter. In the same month, ovipositioning of the lower page of the leaves begins. Newborn neanids, after about 10 days of incubation, have a strong gregarious instinct while in the following stages they disperse on the plant. The next stages of development descend from the plant to prick the seeds in the formation of grasses and only in summer, when these plants dry up, they return to the hazelnut where they begin to prick the nucula causing the same damage as *G. acuteangulatus*. The first adults of the generation of the year emerge true in the second half of July. These individuals will wander between the hazelnut, bramble and chestnut tree and then position themselves in the winter shelters for wintering.

The hazelnut true bugs are monitored by shaking the branches in the early hours of the morning. The shaken insects fall to the ground in a state of 'thanatosis' and can be collected and counted to determine the adults. The intervention threshold, using proper pesticides, is 1.5 adults/plant/week, sampling ten representative trees per ha.

The pesticides approved in organic production such as kaolin, spinosad, neem and microbial formulations exhibited only partial control of the bug species present in hazelnut orchards.

## 5.5 Halyomorpha halys

*Halyomorpha halys* represents species represents a new emergency for the hazelnut. It is an invasive species native to Asia and is currently present in almost all of Europe (Fig. 5) and North America. It is one of the main species with economic importance because of its danger to the production of numerous crops. In its place of origin, it can undergo four to six generations per year, while in the United States it goes through two to three generations per year. They winter as adults in natural and artificial shelters. In spring, the wintering adults emerge from the shelters and after a short period begin ovipositioning on shrubs. Ovid laying takes place in small groups of about 28 eggs on the lower page of the leaves (Nielsen et al., 2008). It is extremely polyphagous and can feed on over 120 plant species including fruit trees, small fruits, vegetables, ornamental plants

**Figure 5** Adult *Halyomorpha halys* captured in the Cimini Mountains, in the province of Viterbo, Italy. Photo: Stefano Speranza 2018, DAFNE - UNITUS.

and plants of agricultural interest. It has strong food preferences for all plants with growing fruits or seeds. It is a species strongly attracted to wooded areas and therefore all adjacent agricultural fields will be used by this phytophagous insect as a place of feeding and ovipositioning (Rice et al., 2014). It disperses very quickly, and adults can fly up to 2 km in 24 h especially during the pre-wintering phase in late summer (Wiman et al., 2015). The pre-imaginal stages are very mobile from the second to the fifth stage by dispersing very quickly while walking (Lee et al., 2014). Males produce an aggregation pheromone. In winter they tend to cluster and hide in shelters such as rooms and houses near infested plots (Inkley, 2012) and in vegetation residues, dead plants, live trees with thick bark such as oak and the locust tree - Robinia (Lee et al., 2014). Their passive dispersion is facilitated by the gregarious behaviour that brings them together in shelters, such as cars or other vehicles, which will then transport the colonies to other areas. Typical damage is due to the insertion of the mouth stilettos into the fruits and seeds to suck food. The infested and affected parts are deformed, dried or with spongiform stains. The damage can vary:

- Reduction in fruit production due to early fall.
- Vector of bacterial stool fruit produced by the recently reported *Eremothecium coryli* bacterium in Bulgaria (Rice et al., 2014).
- The infested and spotted fruit or kernel may have an abnormal 'bug taste'.

## 5.6 Curculio nucum

The adult of the *Curculio nucum* beetle has a head that is characterized by a long rostrum with chewing mouthparts at the end. The length of the rostrum allows determination of females from males. The adults come out in spring, and feed on young cherry fruits. At the end of May, coupling takes place, which can last until the second week of July. At the beginning of summer, they move to the hazelnut where they drill holes before feeding, only with the rostrum on the growing fruits, and then begin ovipositioning towards the end of June. The female lays by piercing the base of the bract envelope up to 2–4 mm deep with the rostrum. In the hole, it lays an egg and pushes it deep using the rostrum. The emerged larvae feed on the seed and complete their cycle in a month. Infested hazelnuts tend to fall especially at the end of larval development. The mature larva pierces the shell of the hazelnut with its robust mandibles. It comes out in a few minutes and buries to a depth of 20–30 cm where it builds a wintering cell. The larvae overwinters and pupates the following spring. The insect is monitored by shaking the branches in the early hours of the morning. The shaken insect falls to the ground in a state of 'thanatosis' and can be collected and evaluated directly for sex determination. The intervention threshold is 1 female/plant/week. It is believed that sanitation (e.g. flailing the orchard floor after harvest and the surrounding oak habitats) would be helpful in reducing weevil populations.

Paparatti and Speranza (2005) have investigated *Beauveria bassiana* (Balsamo) with good results for the control of this pest and a commercial product (Naturalis) for *B. bassiana* is currently registered in many countries. Some entomopathogenic nematode species were also applied to nut weevil larvae and it was concluded that *Heterorhabditis bacteriophora* is the most promising organism for controlling the hazelnut weevil pest due to its good persistence and efficacy (Tuncer, 2009).

## 5.7 Xyleborus dispar

The *Xyleborus dispar* insect exhibits sexual dimorphism with the female being about 4–5 mm while the male is much smaller. In spring, when they reach a temperature of 20–22°C, the females begin to lay eggs by piercing the trunk in a perpendicular way and then digging a semi-circular gallery on a plane perpendicular to the branch itself. In the semi-circular gallery, the female digs small diverticules where it lays a maximum of 40 eggs. The newborn larvae feed on the mycelium of the fungus *Ambrosiella hartigii*, with the female transporting the spores inside its 'mycangia'. This insect causes damage both directly by weakening the plant and indirectly by predisposing it to attack by different phytopathogenic microorganisms. This species is monitored by funnel traps (multi-funnel traps) activated with ethyl alcohol. Phytophagous is associated with hazelnut death, a disease with complex aetiology (Speranza et al., 2009a,b).

The IPM practices used in hazelnut production help protect and maintain tree health and improve nut quality, and enable a substantial reduction in pesticide use. Maintaining and improving this successful IPM programme for insects, diseases and weeds is critical for the future of the hazelnut industry.

## 6 Exploitation of biomass

The exploitation of agricultural residual biomass as energy offers an interesting opportunity for diversification of production and farmers' income, as well as a solution to solve the long-standing problem of disposal, since the incorrect management of hazelnut by-products can have a strong environmental impact on air quality when they are burned on-field (Colantoni et al., 2016). Hazelnut prunings must be removed from the ground, avoiding the risks of pathological problems and facilitating mechanical harvest. Biomass production arising from the management of adult hazelnut orchards is equivalent to approximately 1.5 t/ha (Di Giacinto et al., 2014).

Agricultural hazelnut residues can be used as a raw material for the production of woodchips or pellets used in modern combustion devices taking into consideration the possible problems related to high ash content. Alternatively, a solution is represented by using pruning residues as a feedstock in a downdraft air gasifier, working with low values of equivalence ratio, in order to obtain syngas and biochar.

When to harvest the pruning residues depends on the quantity, annual yield, biomass produced per ha, and the chemical and physical properties of raw materials.

Hazelnut wood has an average calorific value of 15 MJ/kg, which is lower than coal and naphtha, with averages of 25 and 43 MJ/kg, respectively; this means that 1 L of diesel oil corresponds to 2.5–3 kg of hazelnut wood. Furthermore, hazelnut wood is characterized by lower volumetric mass density than other woods, such as beech or oak (650 kg/m$^3$ vs. 750 kg/m$^3$).

According to these data, hazelnut by-products could represent an interesting opportunity as the pruning material derived from a hectare produces the energy of about 500 kg of diesel oil.

The supply chain organization is fundamental: suitable machinery for harvest and chipping are necessary, and farmer cooperatives could represent a way to promote transport, storage and use at the territorial level.

A further way of exploiting pruning material and hazelnut by-products could be the hydrolysis of the cellulosic fraction and forming fermentable sugars that could be used in various processes for the production of ethanol, acetone and hydrogen using different microorganisms widely employed for cellulosic by-products, such as corn hub, but also applied to lignin-rich products such as wood. Another possible application is the use of wood for the production of edible mushrooms after enzymatic hydrolysis.

Recent studies have developed a method for agro-forestry waste management that is able to provide energy in terms of syngas to satisfy the energy demand of farms and at the same time replace commercial fertilizer with the use of biochar, obtained from waste biomasses, as a soil enhancer (Colantoni et al., 2016; Couto et al., 2013). Char from the pyrolysis of hazelnut biomass shows a high carbon content (about 75%) while the solid residue, that is the biochar resulting from the partial gasification of char via pyrolysis, is characterized by a carbon content ranging from 32% up to 47% for the higher and lower airflow values, respectively. The physico-chemical characterization of hazelnut biochar has been performed to assess the quality of biochar according to the standards of the European Biochar Certificate, demonstrating the high quality of biochar obtained from hazelnut-pruning residues (Zambon et al., 2016).

Hazelnut shells represent about half of the harvested fruit production, and are usually destined for thermal energy production, or mulching, since the particular shape promotes soil humidity conservation and its natural decomposition does not cause changes in soil pH.

Over the last few years, the shells, similar to the use of woody material, undergo a thermal hydrolysis process with acidified water, to produce hemicelluloses, FOS and XOS (fruit and xylo-oligosaccharides), or short-polymerized sugars with high nutritional interest, stimulating economic interest; however, these very complex processes are still under investigation (Surek and Buyukkileci, 2017).

Although these processes are currently still in the experimental phase, the large quantity of raw material available at no cost could lead to interesting economic and technological options.

The use of shells for heating residential and industrial spaces depends on a suitable calorific value (around 15–17 MJ/kg), its economy and ease of management and storage and commercial boilers that are available on the market, that is those that run on pellets. The chemical properties of the shell are very interesting (Table 5), as they also show a calorific value higher than the pellets. The application of shells and husks directly to the orchards has proved very useful for soil health with respect to saving moisture, preventing weeds and as a source of microminerals when they decay.

Furthermore, fruits not suitable for food processing could be used to extract the lipid and polyphenolic fractions for use in food or cosmetics; in addition, hazelnut oil is commonly commercialized in Turkey for its fatty-acid profiles classified into the oleic acid class as olive oil (Cristofori et al., 2008).

**Table 5** Chemical and physical characteristics of the pruning residues and hazelnut shells for energetic purposes

| Hazelnut-pruning wood | | Hazelnut shells | |
|---|---|---|---|
| Yield | 1.5–2.0 t/ha/year | Calorific value | 15.7 MJ/kg |
| Harvest humidity | 35–40% | Density | 360 kg/m$^3$ |
| Post-harvest humidity | 15–16% | Density (shredded shells) | 430 kg/m$^3$ |
| Carbon content | 46–47% | Humidity | 5% |
| Calorific value | 15–17 MJ/kg | Ash content | 2–4% |
| Ash content | 2–3% | | |

# 7 Summary and future trends

The European hazelnut is a species that contains a great amount of genetic diversity for climatic adaptation and resistance to pests and disease. However, the species is relatively poorly domesticated compared with other major fruit crops originating in temperate climates.

The breeding of *Corylus avellana* is still poor and only the OSU breeding programme is highly operational. Systematic exploration of the genus *Corylus*, collection in centres of diversity and germplasm exchanges between research centres is needed to establish 'new-generation breeding programmes' to obtain new cultivars with desirable nut and kernel traits, develop new non-sucker emission rootstocks, generate more adaptability to drought and cold temperatures including non-dormant genotypes suited for subtropical areas and new cultivars able to resist some high-pressure pathologies.

A new trend in developing a more friendly environment and profitable hazelnut cultivation, characterized by a high level of mechanization and high-density plantations, is interesting and some new, large orchards which are spreading around the world accompany a very fast diffusion of hazelnut cultivation both in traditional and new areas. This new trend is shepherded by the need to create new hazelnut cultivation techniques to develop an agricultural equivalent of an industrial Supervisory Control and Data Acquisition (SCADA) system to be used for the precision farming of large hazelnut orchards in a scenario of 'hazelnut cultivation 2.0'. Some experimental trials are being conducted with the aim of taking advantage of the technological advancements in the fields of robotics, remote sensing and big-data management, and the objective is to design integrated systems where a relatively limited number of heterogeneous unmanned robotic components (including terrestrial and aerial robots) move within the orchards to collect data and perform some of the most common farming operations. The information would be stored in central operative units that will integrate the data coming from the different robotic units to perform automatic feedback actions (e.g. to regulate the irrigation system) and to support the decisions of agronomists and farmers.

This system could be able to acquire information at the resolution of a single plant and could permit a drastic increase in the detection of possible limiting factors for each tree, such as lack of water or the presence of pests and diseases affecting plant health, and to react accordingly. Compared with the current state of the art in precision farming, the proposed SCADA infrastructures could represent a relevant step ahead in the context of hazelnut orchard management and the main advantages of this architecture would be:

- Increased hazelnut production;
- Decreased chemical input usage;
- Environmentally friendly water usage;
- Simplified orchard management.

## 8 Acknowledgements

This chapter has been supported by the European Commission under the Grant Agreement number 774571 (project PANTHEON –'Precision farming of hazelnut orchards') and by MIUR (Ministry for Education, University and Research), Law 28 232/2016, 'Department of excellence'.

## 9 Where to look for further information

A standard introduction to the subject is in book chapter 17 by S. A. Mehlenbacher, *Hazelnuts* (see Mehlenbacher, 1991a), and in book chapter 11 by S. D. Sharma, Fiberts (*Temperate Fruits*, edited by S. K. Mitra, T. K. Bose and D. S. Rathore, Horticulture and Allied Publishers).

The best sources of information on hazelnut research and cultivation are available on the website of the International Society for Horticultural Science (ISHS) (http://www.ishs.org/). The Publications section contains the peer-reviewed series of the *Proceedings of the International Congresses on Hazelnut* published in the series *Acta Horticulturae* (ISSN 0567-7572; https://www.actahort.org/). The book series is under the Editorial of the section 'Nuts and Mediterranean Climate Fruits' and of the related 'Hazelnut Working Group'.

The most important centres of expertise include:

- Oregon State University (OSU) 'Hazelnut Breeding Program and related Extension Services' (https://plantbreeding.oregonstate.edu/plantbreeding/research/hazelnut-breeding-program)

- National Clonal Germplasm Repository: Corvallis, OR (https://www.ars.usda.gov/pacific-west-area/corvallis-or/national-clonal-germplasm-repository/)
- International Nut and Dried Fruit Council: INC and related publications (https://www.nutfruit.org/)
- Hazelnut Research Institute, Giresun (Turkey) (https://arastirma.tarimorman.gov.tr/findik/Sayfalar/EN/Anasayfa.aspx)

The websites include information resources and key contacts.

There are current research projects designed to support innovation in the hazelnut chain including:

- 'Precision Agriculture for Hazelnut Orchards – PANTHEON' (H2020) (http://www.project-pantheon.eu/)
- 'Safeguard of hazelnut and almond genetic resources – SAFENUT' (AGRI GEN RES) (http://safenut.casaccia.enea.it/)
- 'Hazelnut Breeding Program', Oregon State University (see the website reported in the list of centres of expertise)

## 10 References

Akin, M., Eyduran, E. and Reed, B. M. 2017. Use of RSM and CHAID data mining algorithm for predicting mineral nutrition of hazelnut. *Plant Cell, Tissue and Organ Culture* 128(2), 303–16. doi:10.1007/s11240-016-1110-6.

Anderson, W. C. 1984. Micropropagation of filbert, *Corylus avellana*. *Combined Proceedings of International Plant Properties Society* 33, 132–7.

Anil, Ş., Kurt, H., Akar, A. and Köse, Ç. B. 2016. Hazelnut culture in Turkey. *Chronica Horticulturae* 56(4), 30–5.

Aramburu, J. and Rovira, M. 1998. The effects of apple mosaic ilarvirus (ApMV) on hazelnut (*Corylus avellana* L.). *The Journal of Horticultural Science and Biotechnology* 73(1), 97–101. doi:10.1080/14620316.1998.11510950.

Bacchetta, L., Aramini, M., Bernardini, C. and Rugini, E. 2008. *In vitro* propagation of traditional Italian hazelnut cultivars as a tool for the valorisation and conservation of local genetic resources. *HortScience* 43(2), 562–6. doi:10.21273/HORTSCI.43.2.562.

Bignami, C. and Natali, S. 1997. Influence of irrigation on the growth and production of young hazelnuts. *Acta Horticulturae* 445(445), 247–62. doi:10.17660/ActaHortic.1997.445.34.

Bignami, C., De Salvador, F. R. and Strabbioli, G. 1999. Aspetti agronomici e prospettive di valorizzazione della corilicoltura italiana. *Rivista di Frutticoltura* 11, 16–27.

Bignami, C., Bertazza, G., Bizzarri, S., Bruziches, A., Cammilli, C. and Cristofori, V. 2005a. Effect of high density and dynamic tree spacing on yield and quality of hazelnut. *Acta Horticulturae* 686, 263–70.

Bignami, C., Cammilli, C., Moretti, G. and Sallusti, L. 2005b. Growth analysis and nitrogen dynamics in hazelnut 'Tonda Gentile Romana'. *Acta Horticulturae* 686(686), 193–200. doi:10.17660/ActaHortic.2005.686.26.

Bignami, C., Cristofori, V., Ghini, P. and Rugini, E. 2009. Effects of irrigation on growth and yield components of hazelnut (*Corylus avellana* L.) in Central Italy. *Acta Horticulturae* 845(845), 309–14. doi:10.17660/ActaHortic.2009.845.47.

Boccacci, P., Akkak, A. and Botta, R. 2006. DNA-typing and genetic relationships among European hazelnut (*Corylus avellana* L.) cultivars using microsatellite markers. *Genome* 49(6), 598–611. doi:10.1139/g06-017.

Boccacci, P., Aramini, M., Valentini, N., Bacchetta, L., Rovira, M., Drogoudi, P., Silva, A. P., Solar, A., Calizzano, F., Erdoğan, V., et al. 2013. Molecular and morphological diversity of on-farm hazelnut (*Corylus avellana* L.) landraces from southern Europe and their role in the origin and diffusion of cultivated germplasm. *Tree Genetics and Genomes* 9(6), 1465–80. doi:10.1007/s11295-013-0651-7.

Cacka, J. and Sanguankeo, P. 2014. Calcium influence on hazelnut quality and yields in Oregon. *Acta Horticulturae* 1052(1052), 187–93. doi:10.17660/ActaHortic.2014.1052.24.

Cerovic, S., Ninic-Todorovc, J., Golosin, B., Ognjanov, V. and Bijelic, S. 2009. Grafting methods in nursery production of hazelnut grafted on *Corylus colurna* L. *Acta Horticulturae* 845(845), 279–82. doi:10.17660/ActaHortic.2009.845.40.

Colantoni, A., Longo, L., Gallucci, F. and Monarca, D. 2016. Pyro-gasification of hazelnut pruning using a downdraft gasifier for concurrent production of syngas and biochar. *Contemporary Engineering Sciences* 9(27), 1339–48. doi:10.12988/ces.2016.68142.

Contessa, C., Valentini, N. and Botta, R. 2011a. Decreasing the concentration of IBA or combination with ethylene inhibitors improve bud retention in semi-hardwood cuttings of hazelnut cv 'Tonda Gentile delle Langhe'. *Scientia Horticulturae* 131, 103–6. doi:10.1016/j.scienta.2011.09.029.

Contessa, C., Valentini, N., Caviglione, M. and Botta, R. 2011b. Propagation of *Corylus avellana* L. by means of semi-hardwood cutting: rooting and bud retention in four Italian cultivars. *European Journal of Horticultural Science* 76(5), 170–5.

Contessa, C., Valentini, N., Corte, M. and Botta, R. 2014. Investigation on effects of IBA treatments and ethylene inhibitors on the rooting and bud retention of semi-hardwood cuttings from 'Tonda Gentile delle Langhe'. *Acta Horticulturae* 1052, 151–6.

Couto, N., Rouboa, A., Silva, V., Monteiro, E. and Bouziane, K. 2013. Influence of the biomass gasification processes on the final composition of syngas. *Energy Procedia* 36, 596–606. doi:10.1016/j.egypro.2013.07.068.

Cristofori, V., Rouphael, Y., Mendoza-de Gyves, E. M. and Bignami, C. 2007. A simple model for estimating leaf area of hazelnut from linear measurements. *Scientia Horticulturae* 113(2), 221–5. doi:10.1016/j.scienta.2007.02.006.

Cristofori, V., Ferramondo, S., Bertazza, G. and Bignami, C. 2008. Nut and kernel traits and chemical composition of hazelnut (*Corylus avellana* L.) cultivars. *Journal of the Science of Food and Agriculture* 88(6), 1091–8. doi:10.1002/jsfa.3203.

Cristofori, V., Cammilli, C., Valentini, B. and Bignami, C. 2009. Effect of different pruning methods on growth, yield and quality of the hazelnut cultivar Tonda Gentile Romana. *Acta Horticulturae* 845, 315–22.

Cristofori, V., Rouphael, Y. and Rugini, E. 2010. Collection time, cutting age, IBA and putrescine effects on root formation in *Corylus avellana* L., cuttings. *Scientia Horticulturae* 124(2), 189–94. doi:10.1016/j.scienta.2009.12.034.

Cristofori, V., Bizzarri, S., Silvestri, C., Muleo, R., Rugini, E. and De Salvador, F. R. 2014a. First evaluations on vegetative and productive performance of many hazelnut cultivars in Latium region. *Acta Horticulturae* 1052(1052), 91–7. doi:10.17660/ActaHortic.2014.1052.11.

Cristofori, V., Muleo, R., Bignami, C. and Rugini, E. 2014b. Long term evaluation of hazelnut response, cv. Tonda Gentile Romana, to drip irrigation in Central Italy. *Acta Horticulturae* 1052, 179–85.

Cristofori, V., Bertazza, G. and Bignami, C. 2015. Changes in kernel chemical composition during nut development of three Italian hazelnut cultivars. *Fruits* 70(5), 311–22. doi:10.1051/fruits/2015025.

Cristofori, V., Blasi, E., Pancino, B., Stelliferi, R. and Lazzari, M. 2017. Recent innovations in the implementation and management of the hazelnut orchards in Italy. *Acta Horticulturae* 1160(1160), 165–72. doi:10.17660/ActaHortic.2017.1160.24.

Damiano, C., Catenaro, E., Giovinazzi, J., Frattarelli, A. and Caboni, E. 2005. Micropropagation of hazelnut (*Corylus avellana* L.). *Acta Horticulturae* 686(686), 221–6. doi:10.17660/ActaHortic.2005.686.29.

De Francesco, J. and Clarke, D. 2006. *Pest Management Strategic Plan for Hazelnuts in Oregon and Washington*, Aurora, Oregon, 29 November 2006. Summary of workshop, p. 61.

Di Giacinto, S., Leonardo, L., Giuseppina, M., Lavinia, D., Gianluca, E., De Benedictis, L., Riccioni, S. and Salvati, L. 2014. Model for estimating pruned biomass obtained from *Corylus avellana* L. *Applied Mathematical Sciences* 8, 6555–64. doi:10.12988/ams.2014.46441.

Dias, R., Conçalves, B., Moutinho-Pereira, J., Carvalho, J. L. and Silva, A. P. 2005. Effect of irrigation on physiological and biochemical traits of hazelnuts (*Corylus avellana* L.). *Acta Horticulturae* 686(686), 201–6. doi:10.17660/ActaHortic.2005.686.27.

Ellena, M., Sandoval, P., Montenegro, A., Gonzalez, A. and Azocar, G. 2014. Effect of foliar nutrient applications on fruit set in 'Chilean Barcelona' hazelnut, in southern Chile. *Acta Horticulturae* 1052(1052), 231–4. doi:10.17660/ActaHortic.2014.1052.31.

FAOSTAT. 2018. *FAO Statistical Yearbook*. Food and Agriculture Organization, Rome.

Garrison, W., Dale, A. and Saxena, P. K. 2013. Improved shoot multiplication and development in hybrid hazelnut nodal cultures by ethylenediamine di-2-hydroxy-phenylacetic acid (Fe-EDDHA). *Canadian Journal of Plant Science* 93(3), 511–21. doi:10.4141/cjps2012-218.

Germain, E. and Sarraquigne, J. P. 2004. *Le noisetier*. CTIFL, Paris, France, p. 296.

Girona, J., Cohen, M., Mata, M., Marsal, J. and Miravete, C. 1994. Physiological, growth and yield responses of hazelnut (*Corylus avellana* L.) to different irrigation regimes. *Acta Horticulturae* 351(351), 463–72. doi:10.17660/ActaHortic.1994.351.50.

Hand, C. and Reed, B. M. 2014. Minor nutrients are critical for the improved growth of *Corylus avellana* shoot cultures. *Plant Cell, Tissue and Organ Culture* 119(2), 427–39. doi:10.1007/s11240-014-0545-x.

Hand, C., Maki, S. and Reed, B. M. 2014. Modelling optimal mineral nutrition for hazelnut micropropagation. *Plant Cell, Tissue and Organ Culture* 119(2), 411–25. doi:10.1007/s11240-014-0544-y.

Hummer, K., Lagerstedt, H. B. and Kim, S. K. 1986. Filbert acclimation, maximum cold hardiness, and deacclimation. *Journal of the American Society of Horticultural Science* 111, 474–82.

Inkley, D. B. 2012. Characteristics of home invasion by the brown marmorated stink bug (Hemiptera: Pentatomidae). *Journal of Entomological Science* 47(2), 125–30. doi:10.18474/0749-8004-47.2.125.

Kantarci, M. and Ayer, M. 1994. Propagation of some important Turkish hazelnut varieties by cuttings. *Acta Horticulturae* 351(351), 353–60. doi:10.17660/ActaHortic.1994.351.39.

Karasawa, M. M. G., Chiancone, B., Gianguzzi, V., Abdelgalel, A. M., Botta, R., Sartor, C. and Germanà, M. A. 2016. Gametic embryogenesis through isolated microspore culture in *Corylus avellana* L. *Plant Cell, Tissue and Organ Culture* 124(3), 635–47. doi:10.1007/s11240-015-0921-1.

Köksal, A. İ. 2002. *Turkish Hazelnut Cultivars* (in Turkish). University of Ankara, Faculty of Agriculture, Ankara, Turkey, p. 136. ISBN: 975-92886-0-5.

Krpina, I., Curlje, M. and Vujevic, P. 1994. Influence of extremely low winter temperature on some hazelnut cultivars. *Acta Horticulturae* 351(351), 329–34. doi:10.17660/ActaHortic.1994.351.36.

Lagerstedt, H. 1993. Newberg and Dundee, two new filbert rootstocks. *Proceedings of the Nutrition Growers Society of Oregon* 78, 94–101.

Lee, D. H., Nielsen, A. L. and Leskey, T. C. 2014. Dispersal capacity and behaviour of nymphal stages of *Halyomorpha halys* (Hemiptera: Pentatomidae) evaluated under laboratory and field conditions. *Journal of Insect Behavior* 27(5), 639–51. doi:10.1007/s10905-014-9456-2.

Me, G., Radicati, C. and Salaris, C. 1994. Rejuvenation pruning of hazelnut cv. 'Tonda Gentile delle Langhe'. *Acta Horticulturae* 351, 439–46.

Mehlenbacher, S. A. 1991a. Chilling requirements of hazelnut cultivars. *Scientia Horticulturae* 47(3–4), 271–82. doi:10.1016/0304-4238(91)90010-V.

Mehlenbacher, S. A. 1991b. Hazelnuts. In: Moore, J. N. and Ballington, J. R. (Eds), *Genetic Resources in Temperate Fruit and Nut Crops. Acta Horticulturae* 290, 789–836.

Mehlenbacher, S. A. 2009. Genetic resources for hazelnut: state of the art and future perspectives. *Acta Horticulturae* 845(845), 33–8. doi:10.17660/ActaHortic.2009.845.1.

Mehlenbacher, S. A. 2014. Geographic distribution of incompatibility alleles in cultivars and selections of European hazelnut. *Journal of the American Society for Horticultural Science* 139(2), 191–212. doi:10.21273/JASHS.139.2.191.

Mehlenbacher, S. A., Brown, R. N., Davis, J. W., Chen, H., Bassil, N. V., Smith, D. C. and Kubisiak, T. L. 2004. RAPD markers linked to eastern filbert blight resistance in *Corylus avellana*. *Theoretical and Applied Genetics* 108(4), 651–6. doi:10.1007/s00122-003-1476-9.

Mehlenbacher, S. A., Smith, D. C. and McCluskey, R. L. 2011a. 'Jefferson' hazelnut. *HortScience* 46(4), 662–4. doi:10.21273/HORTSCI.46.4.662.

Mehlenbacher, S. A., Smith, D. C., McCluskey, R. L. and Thompson, M. M. 2011b. 'Tonda Pacifica' hazelnut. *HortScience* 46(3), 505–8. doi:10.21273/HORTSCI.46.3.505.

Mehlenbacher, S. A., Smith, D. C. and McCluskey, R. L. 2012. 'Eta' and 'Theta' hazelnut pollinizers. *HortScience* 47(8), 1180–1. doi:10.21273/HORTSCI.47.8.1180.

Mehlenbacher, S. A., Smith, D. C. and McCluskey, R. L. 2013. 'Dorris' hazelnut. *HortScience* 48(6), 796–9. doi:10.21273/HORTSCI.48.6.796.

Mehlenbacher, S. A., Smith, D. C. and McCluskey, R. L. 2014. 'Wepster' hazelnut. *HortScience* 49(3), 346–9. doi:10.21273/HORTSCI.49.3.346.

Mehlenbacher, S. A., Smith, D. C. and McCluskey, R. L. 2016. 'McDonald' hazelnut. *HortScience* 51(6), 757–60. doi:10.21273/HORTSCI.51.6.757.

Nas, M. N. and Read, P. E. 2004. An hypothesis for the development of a defined tissue culture medium for higher plants and micropropagation of hazelnuts. *Scientia Horticulturae* 101, 189–200.

Nielsen, A. L., Hamilton, G. C. and Matadha, D. 2008. Developmental rate estimation and life table analysis for *Halyomorpha halys* (Hemiptera: Pentatomidae). *Environmental Entomology* 37(2), 348–55. doi:10.1093/ee/37.2.348.

Olsen, J. 1997. Nitrogen management in Oregon hazelnuts. *Acta Horticulturae* 445(445), 263–8. doi:10.17660/ActaHortic.1997.445.35.

Olsen, J. L. 2002. Oregon State University's Integrated Pest Management Program for the Oregon Hazelnut Industry. *HortTechnology*, 12(4), 623–5.

Olsen, J. L. and Cacka, J. F. 2009. Foliar fertilizers on hazelnuts in Oregon, USA. *Acta Horticulturae* 845(845), 349–52. doi:10.17660/ActaHortic.2009.845.53.

Ozdemir, B. and Dumanoglu, H. 2018. Studies on propagation by stem cuttings of the Turkish hazelnut cultivar Tombul (*Corylus avellana* L.). *Acta Horticulturae* 1226, 219–24.

Pansecchi, A., Roversi, A. and Malvicini, G. L. 2014. Some years of trials on the effectiveness of hazelnut foliar fertilization. *Acta Horticulturae* 1052(1052), 195–9. doi:10.17660/ActaHortic.2014.1052.25.

Paparatti, B. and Speranza, S. 2005. Biological control of hazel weevil (*Curculio nucum* L., Coleoptera, Curculionidae) using the entomopathogenic fungus *Beauveria bassiana* (Balsamo) Vuill. (Deuteromycotina, Hyphomycetes). *Acta Horticulturae* 686, 407–12.

Rahemi, A. R., Taghavi, T. and Dale, A. 2016. A simple method to improve hazelnut grafting. *Journal of Nuts* 7(2), 83–7.

Rey, M., Diaz-Sala, C. and Rodriguez, R. 1994. Exogenous polyamines improve rooting of hazel microshoots. *Plant Cell, Tissue and Organ Culture* 36(3), 303–8. doi:10.1007/BF00046087.

Rice, K. B., Bergh, C. J., Bergmann, E. J., Biddinger, D. J., Dieckhoff, C., Dively, G., Fraser, H., Gariepy, T., Hamilton, G., Haye, T., et al. 2014. Biology, ecology, and management of brown marmorated stink bug (Hemiptera: Pentatomidae). *Journal of Integrated Pest Management* 5(3), 1–13. doi:10.1603/IPM14002.

Riggert, C. and McDonald, P. 1987. Managing filbert densities. *Nutrition Grower Society of Oregon and Washington Proceedings* 72, 68–73.

Romisondo, P., Me, G., Radicati, L. and Salaris, L. 1983. Importanza della potatura ai fini del ringiovanimento di piante di nocciolo in fase di senilità. In: *Atti del Convegno internazionale sul nocciolo*, Avellino, 22–24 Settembre, pp. 395–400.

Roversi, A. 2011. Concimazione del nocciolo. *L'Informatore Agrario* 49, 61–5.

Roversi, A. and Mozzone, G. 2005. Preliminary observations on the effects of renewal pruning in hazelnut orchards. *Acta Horticulturae* 686(686), 253–8. doi:10.17660/ActaHortic.2005.686.34.

Rovira, M., Tous, J. and Aramburu, J. 2006. El clon de avellano 'Negret IRTA-N·9'. *Vida Rural* 232, 18–20.

Rovira, M., Cristofori, V., Silvestri, C., Celli, T., Hermoso, J. F., Tous, J. and Romero, A. 2014. Evaluation of 'Negret' hazelnut cultivar grafted on non-suckering rootstocks in Spain: kernel quality. *Acta Horticulturae* 1052, 145–50.

Sarraquigne, J. P. 1989. Effet de quelques pratiques d'élagage sur le comportement de noisetiers entrés en plein production. In: *2ième Colloque AGRIMED 'Noyer-noysetier'*, Bordeaux, Septembre 1988. Ed. Commission des Communautés Européennes, Luxembourg. Rapport EUR 12005 FR-EN, pp. 176–84.

Serdar, U. and Akyuz, B. 2018. Sucker management methods in hazelnut cultivation. *Acta Horticulturae* 1226(1226), 309–14. doi:10.17660/ActaHortic.2018.1226.46.

Silvestri, C., Cristofori, V., Ceccarelli, M., Caceres, M. E., Escribà-Lacuesta, J. and Rugini, E. 2016. Adventitious shoot organogenesis from leaf and petiole explants of European hazelnut. *Plant Cell, Tissue and Organ Culture* 126(1), 59–65. doi:10.1007/s11240-016-0976-7.

Silvestri, C., Rugini, E. and Cristofori, V. 2018a. Further progress with *in vitro* anther culture of European hazelnut. *Acta Horticulturae* 1226(1226), 237–42. doi:10.17660/ActaHortic.2018.1226.35.

Silvestri, C., Rugini, E. and Cristofori, V. 2018b. The effect of $CuSO_4$ for establishing *in vitro* culture, and the role of nitrogen and iron sources in *in vitro* multiplication of *Corylus avellana* L. cv Tonda Gentile Romana. *Plant Biosystem*. doi:10.1080/11263504.2018.1549610.

Solar, A., Smole, J. and Stampar, F. 1994. Investigations of different methods of propagation of hazelnut (*Corylus avellana* L.). *Acta Horticulturae* 351, 381–6.

Soylu, A. and Erturk, Ü 1997. Some factors affecting the rooting of filbert hardwood cuttings. *Acta Horticulturae* 445(445), 459–68. doi:10.17660/ActaHortic.1997.445.59.

Speranza, S., Bucini, D. and Paparatti, B. 2009a. European shot-hole borer [*Xyleborus dispar* (f.)]: comparison between capture with Chemio-Chromotropic Rebell® Rosso traps and modified Mastrap®L traps. *Acta Horticulturae* 845(845), 535–8. doi:10.17660/ActaHortic.2009.845.83.

Speranza, S., Bucini, D. and Paparatti, B. 2009b. New observation on biology of European shot-hole borer [*Xyleborus dispar* (f.)] on hazel in northern Latium (Central Italy). *Acta Horticulturae* 845(845), 539–42. doi:10.17660/ActaHortic.2009.845.84.

Surek, E. and Buyukkileci, A. O. 2017. Production of xylooligosaccharides by autohydrolysis of hazelnut (*Corylus avellana* L.) shell. *Carbohydrate Polymers* 174, 565–71. doi:10.1016/j.carbpol.2017.06.109.

Thompson, M. M., Lagerstedt, H. B. and Mehlenbacher, S. A. 1996. Hazelnuts. In: Janick, J. and Moore, J. N. (Eds), *Fruit Breeding: Nuts* (vol. 3). Wiley, New York, pp. 125–84.

Tombesi, A. 1994. Influence of soil water levels on assimilation and water use efficiency in hazelnut. *Acta Horticulturae* 351(351), 247–56. doi:10.17660/ActaHortic.1994.351.25.

Tombesi, A. 2005. World hazelnut situation and perspectives: Italy. *Acta Horticulturae* 686(686), 649–58. doi:10.17660/ActaHortic.2005.686.86.

Tombesi, A. and Rosati, A. 1997. Hazelnut response to water levels in relation to productive cycle. *Acta Horticulturae* 445(445), 269–78. doi:10.17660/ActaHortic.1997.445.36.

Tombesi, S., Palliotti, A., Poni, S. and Farinelli, D. 2015. Influence of light and shoot development stage on leaf photosynthesis and carbohydrate status during the adventitious root formation in cuttings of *Corylus avellana* L. *Frontiers in Plant Science* 6, 973. doi:10.3389/fpls.2015.00973.

Tous, J., Romero, A., Rovira, M. and Clavé, J. 1994a. Comparison of different training systems on hazelnut. *Acta Horticulturae* 351(351), 455–62. doi:10.17660/ActaHortic.1994.351.49.

Tous, J., Girona, J. and Tasias, J. 1994b. Cultural practices and costs in hazelnut production. *Acta Horticulturae* 351(351), 395–418. doi:10.17660/ActaHortic.1994.351.44.

Tous, J., Romero, A., Rovira, M. and Hermoso, J. F. 2009. Performance of 'Negret' hazelnut cultivar grafted on 4 rootstocks in Catalonia (Spain). *Acta Horticulturae* 845(845), 89–94. doi:10.17660/ActaHortic.2009.845.8.

Tuncer, C. 2009. Arthropod pest management in organic hazelnut growing. *Acta Horticulturae* 845(845), 571–8. doi:10.17660/ActaHortic.2009.845.90.

Turan, A. and İslam, A. 2018. Effect of drying methods on some chemical characteristics of hazelnuts (*Corylus avellana* L.) during storage. *Journal of the Institute of Science and Technology* 8(3), 11–9. doi:10.21597/jist.458541.

Ughini, V., Roversi, A., Malvicini, G. L. and Sonnati, C. 2009. Effects of hazelnut summer pruning made in different months. *Acta Horticulturae* 845, 363–6.

Wiman, N. G., Walton, V. M., Shearer, P. W., Rondon, S. I. and Lee, J. C. 2015. Factors affecting flight capacity of brown marmorated stink bug, *Halyomorpha halys* (Hemiptera: Pentatomidae). *Journal of Pest Science* 88(1), 37–47. doi:10.1007/s10340-014-0582-6.

Yu, X. and Reed, B. M. 1993. Improved shoot multiplication of mature hazelnut (*Corylus avellana* L.) *in vitro* using glucose as a carbon source. *Plant Cell Reports* 12(5), 256–9. doi:10.1007/BF00237130.

Zambon, I., Colosimo, F., Monarca, D., Cecchini, M., Gallucci, F., Proto, A. R., Lord, R. and Colantoni, A. 2016. An innovative agro-forestry supply chain for residual biomass: physicochemical characterisation of biochar from olive and hazelnut pellets. *Energies* 9(7): doi:10.3390/en9070526.

# Chapter 17

# Advances in breeding of hazelnuts

*Ali Islam, Ordu University, Turkey*



## 1 Introduction

The hazelnut is widely distributed in temperate regions of the northern hemisphere. It is typically grown between 42 and 45 degrees of latitude. The crop is also starting to be grown in southern hemisphere countries such as Chile. Hazelnut plants usually grow well in coastal regions and temperate climates. Hazelnut is grown, for example, in the Black Sea costal region of Turkey, as well as in Italy and Spain. Other Eurasian areas of cultivation include the Caucasus and Balkans (Georgia and Azerbaijan). Cultivation of European hazelnut dates back to the classical period in the Mediterranean region but occurred much earlier in Turkey and the Caucasus. In the United States, hazelnut production is concentrated in the Willamette Valley of Oregon State, where the climate is affected by the Pacific Ocean. Hazelnut can also be grown in different climatic conditions such as China and India. World hazelnut production is about 900 thousand tons (FAO, 2016). Turkey is the leading producer and exporter of hazelnuts, accounting for approximately 70% of total global production, with Italy ranked second with about 22% of world production. Most production is for industrial use in the food and beverage industries as a flavoring, as well as in the cosmetics and pharmaceutical industries.

Hazelnut (*Corylus* spp.) is a member of the Betulaceae family. The number of species varies depending on the taxonomic authority. Most of the *Corylus* spp. are in the form of a bush though some species form trees. The most commonly known shrub species are: *Corylus avellana* L., *C. americana* Marshall, *C. comuta* Marshall, *C. heterophylla* Fischer and *C. sieboldiana* Blume. Common examples of the tree form of hazelnut are: *C. colurna* L, *C. jacquemontii* Decaisne,

http://dx.doi.org/10.19103/AS.2018.0042.15

*C. chinensis* Franchet and *C. ferox* Wallich. *Corylus* contains 11 commonly recognized species distributed in the northern hemisphere. Of these 11 species, two occur in Europe and Asia Minor (*C. avellana* and *C. colurna*), three ocur in North America (*C. americana*, *C. cornuta* and *C. californica*) and one in the Himalayas (*C. jacquemontii*). The remaining species occur in eastern Asia. They include the tree hazel species: *C. chinensis*, *C. fargesii* Schneid. and *C. ferox*, and the shrub hazel species: *C. heterophylla* and *C. sieboldiana* (Whitcher and Wen, 2001).

Taxonomists consider many transitions to be part of a large polymorphic species (Mehlenbacher, 1991; Thompson et al., 1996; Gökirmak et al., 2009; Bassil et al., 2013). Studies of diversity using simple sequence repeat (SSR) markers place many accessions into one of four groups corresponding to their geographic origin: Black Sea, Spanish-Italian, Central European and English (Boccacci et al., 2006; Gökirmak et al., 2009). The number of chromosomes is 2n = 22 diploid in *Corylus* species, while x = 12 or triploid species in hybrids is also known. Tetraploid species have been obtained in *Corylus* species (Danielsson-Santesson, 1951; Kasapligil, 1968; Salesses, 1973; Botta et al., 1997). All of the important commercial cultivars in Europe and Turkey have been selected from local populations of *C. avellana*. These high-quality varieties are the basis of all modern breeding programs. These varieties been used alone in such breeding programs or in combination with other species.

Cultivated varieties differ from each other in such characteristics as size, shape, kernel ratio, shell thickness, etc. Key target characteristics for breeding include high yield as well as good kernel characteristics which include: a high kernel percentage and pellicle removal; thin shell; uniform kernel size and shape; a low number of kernel defects (such as shriveled nuts, necrotic spots, double kernel, poorly filled nuts or mouldy kernels) (Thompson et al., 1996; Mehlenbacher, 2003; Islam, 2000, 2003). Vegetative (morphological and phenological) characteristics include: non-suckering or with few suckers, good grafting affinity, erect growing habit to facilitate machine harvesting and late bud burst to avoid spring frosts. Important features for rootstock breeding include non-suckering and erect habit.

Other desirable characteristics include resistance to biotic threats. This includes resistance to the following pests: Big bud mite, *Phytoptus avellanae* Nal. (in all regions where hazelnuts are grown), Filbert weevil, *Curculio nucum* L. (in Europe and Asia Minor), bark beetles, ambrosia beetles (*Xyleborus* spp) (in Turkey) and *Haliomorpha halys* (in Europe and other countries). Desirable varieties also need resistance to the following major diseases: Eastern filbert blight (EFL) (*Anisogramma anomala* (Peck) E. Muller, in North America), Bacterial blight (*Xanthomonas campestris* pv. Corylina, in the United States and western Europe), *Gloeosporium coryli* (Desm.) Sacc. (syn. *Monostichella coryli* (Desm.) Hohnel) (in Europe and Asia) and powdery mildew (*Erysiphe corylacearum,* in Turkey, Iran).

## 2 Floral biology, pollination and compatibility

Hazelnuts are monocious, dichogamous, anemophilous and self-incompatible. The male inflorescences are catkins, each composed of 130–290 flowers. Each flower includes one bract and two bracteoles protecting four bifurcated stamens, each terminated in two anthers. A single catkin releases four to five million pollen grains, each of which is 25–40 microns in size (Janick and Paull, 2008). About 30–50% are not viable in some varieties. This abnormality is closely related to heterozygous translocation. However, the live pollens produced are sufficient for good pollination, since male flowers produce 4–40 million pollen grains. Stigma can remain receptive for up to 2 months though the stigma surface remains most receptive for 15 days from the beginning of flowering. After pollination, the male flowers (pollen) germinate. Germinated pollens enter from the stigma surface. It completes movement into the style in 4–10 days. When it reaches the style base, the pollen tube expands and is covered with a thick callus layer where it rests until the ovarian tissue develops. Pollination promotes ovule development which does not occur before this point (Germain, 1994; Beyhan, 2000; Liu et al., 2014). After pollination is completed, the stigma gets darker by losing turgor. The non-pollinated stigma extends a little more, creating more receptive tissue and may remain receptive for up to 3 months (Beyhan, 1993). About 90% of cultivars are protandrous (Olsen et al., 2000). Dichogamy is identical in key hazelnut cultivars and protandry is common (Fattahi et al., 2014). These characteristics may vary, depending on ecological conditions (Crepinsek et al., 2012). In the Oltenia region in Turkey, for example, most hazelnut cultivars exhibit protogynous behavior. A change in the type of dichogamy has been observed for 'Hall's Giant', 'Romavel' and 'Du Chilly' (Botu et al., 2017).

The number of female flowers in the buds varies depending on the variety; for example, there are up to 34 in the Kalınkara variety. In general, varieties with a higher number of female flowers in the buds have been observed to be more efficient (Beyhan, 1998). Hazelnut has a monoic flower structure and is pollinated in windy conditions during the winter months. The opening time and duration of male and female flowers may vary depending on factors such as variety, age and ecological conditions (Beyhan, 2000). About 50–67% of male flowers are aborted in *Corylus heterophylla* L. species (Liu et al., 2014).

Nearly all hazelnut cultivars are self-incompatible and many cultivars are cross-incompatible. Incompatibility in hazelnuts is a sporophytic type controlled by a single locus with S-alleles. The allele genes have been determined to be co-dominates in the male flowers and dominate or co-dominant in the pollen (Mehlenbacher and Thompson, 1988). Studies of crossing by Erdoğan and Mehlenbacher (2000) showed that the average percent cluster set varied among genotypes and that reciprocal differences were common. The European hazelnut *C. avellana* showed fairly good set with *C. colurna*, and the North American species *C. cornuta* and *C. californica* were highly cross-compatible. In 2001, Erdoğan and Mehlenbacher (2001) reported their study of incompatibility in wild *Corylus* species, indicating that sporophytic incompatibility exists in wild *Corylus* species with a large number of S-alleles involved.

Wild *Corylus* species play an important role in breeding research with the need to introduce specific traits from wild *Corylus* species for improvement of existing hazelnut cultivars (Thompson et al., 1996). Self-pollination of *Corylus avellana* L. cultivars in 1988 and 1990 by Mehlenbacher and Smith (1991) revealed the existence of partial self-compatibility in 'Tombul' and 'Montebello'. Cluster set in these cultivars averaged between 44% and 20%, respectively. Mehlenbacher and Smith found a new S-allele (S28) that appeared to be a style-part mutation of S30. The results showed that *C. heterophylla* and *C. mandshurica* were self-compatible, due to the adaptability created by a wild environment (Zhai et al., 2009). This suggests that self-compatibility of hazelnut species deserves further investigation.

Erdoğan and Mehlenbacher (2001) identified pollen tubes that were short, usually 'comma-shaped', often terminating in pronounced bulbs, and failing to penetrate the stigma. Both compatible and incompatible reactions were observed in intra- and inter-specific pollinations with some reciprocal differences. Controlled self-pollinations gave low set in representatives of *C. americana* and *C. sieboldiana*, but one of four *C. californica* and two of three *C. colurna* selections set well. Similar studies were undertaken by Erdoğan et al. (2005); Ünal (2009); Ma et al. (2013); Xie and Liu (2014); Korkmaz et al. (2015); and Hosseinpour et al. (2015).

Balık (2018) has found that pollinizer cultivars cause changes in some nut and kernel characteristics due to xenia or metaxenia effects. It is thought that changes in nut characteristics may be related to the cluster set as well as pollen sources. These characteristics include improved biochemical properties and color as well as nut and kernel shape. When the effects of pollinizers on cluster set and nut number per clusters and flowering times were evaluated together, the most suitable pollinators were: 'Allahverdi' for 'Tombul', 'Foşa' for 'Palaz', 'Tombul' for 'Çakıldak', 'Çakıldak' and 'Allahverdi' for 'Foşa', and 'Foşa' for 'Allahverdi'. Fattahi et al. (2014) observed self-pollination reduced values of nut and kernel traits and increased percentage of blank nuts. Another study by Rahemi and Mojadad (2001) evaluated xenia and metaxenia in hazelnut. 'Fertile de Coutard', 'Segorbe' and 'Ronde du Piedmont' flowers were crossed with 'Daviana', 'Negret' and 'Cosford' pollen. The results of this study showed that nut and kernel weight and blank percentage could be affected by pollen sources. Ava et al. (2010) evaluated the viability of pollen grains of five hazelnut cultivars ('Gercheh', 'Negret', 'Pashmineh', 'Gerdoii' and 'Gerd-e-Eshkevarat') and the selection of the best pollinator. Cross pollination between 'Gercheh' with 'Dubase' and 'Secorp' achieved the highest mean of fruit set.

# 3 Advances in hazelnut breeding

## 3.1 Variety development

Common traditional breeding methods for hazelnut are selection and hybridization. A major hybrid breeding programme in Turkey has been undertaken by Okay (1999) crossing high-yielding and thin-shelled 'Tombul' cultivars with large-fruiting 'Kargalak' cultivars. The hybrids were subjected to two-stage selection with 44 individuals obtained from the first selection. In 2012 two of the selected types were registered as 'Giresun Melezi' and 'Okay 28' (Balık et al., 2015; İslam, 2018). *Corylus avellana* (European hazel) is sensitive to cold. Young shoots (early bud burst) are subject to injury by spring frost. In contrast *C. heterophylla* (Siberian hazel) is quite hardy. Interspecific hybridization has been conducted with *C. heterophylla* as the female parent and *C. avellana* as the male parent. The resulting hybrid types are 'Ping'ou' ('Pingzhen' × 'Ouzhen'). The selected hybrids are cold hardy with large nut size and high yield. Fourteen cultivars of 'Ping'ou' hybrids have been released (Wang et al., 2018).

Selection breeding is the most widely used breeding method in Turkey. Clonal selection has been used for a range of cultivars: 'Tombul' (İslam, 2000; Demir and Beyhan, 2000; Turan, 2007; Balık et al., 2014); 'Uzunmusa' (Islam, 2003); 'Palaz' and 'Kalınkara' (Islam and Ozguven, 2001; Demir and Beyhan, 2000); and 'Çakıldak' (Çayan and İslam, 2019). Clonal selection programmes in Italy include: 'Tonda di Giffoni' (Petriccione et al., 2010); 'Tonda Gentile Delle Langhe' (Marinoni et al., 2000; Valentini et al., 2014); and 'Tonda Gentile Romana' (Monastra et al., 1996). Clonal selection has also been applied to the 'Gironel' and 'Negret' cultivars (Rovira et al., 1996) in Spain. Some new Italian cultivars ('Madonella', 'Romanella', 'Tonda Etrucsa' and 'Tonda Francescana') have recently been released (Mehlenbacher, 2018a). Clonal selection can be considered a very conservative approach to genetic improvement, as the amount of genetic variation among clones of a cultivar is much less than that available in *C. avellana* as a whole (Mehlenbacher, 2018a). Farinelli et al. (2009) have also undertaken selection by crossing 'Tonda Gentile Romana' and 'Tonda di Giffoni'. The crosses F25P29 and F13P9 were found to have a very good kernel percentage (around 51%). The other crosses (F4P32, F25P29 and F13P9) demonstrated good kernel quality and kernel percentage, but their yield needs to be improved through cultural techniques or by further breeding. As has been noted, wild *Corylus* species play an important role in breeding research with the need to introduce specific traits from wild *Corylus* species to improve existing hazelnut cultivars. This importance can be seen in the breeding programme started in Vâlcea in Romania in the 1970s which, it has been argued, would benefit particularly from identification and evaluation of wild hazelnut genetic resources (Botu et al., 2015). In recent years nine new hazelnut cultivars have been released (Botu et al., 2017).

The United States has been a major recent centre for hazelnut breeding with a particular focus on resistance to Eastern filbert blight (EFB) which has caused serious damage in the United States in the last thirty years. Techniques have included hybridization and marker-based selection (Lunde et al., 2000). Research has included identification of sources of genetic resistance to EFB in hazelnut germplasm (Coyne et al., 1998; Chen et al., 2007). Resistance in 'Gasaway' is conferred by a dominant allele at a single locus on linkage group 6 (LG6), and five candidate genes have been identified. 'Jefferson', 'Yamhill', 'Dorris', 'Wepster' and 'McDonald' and several pollinizers carrying 'Gasaway' resistance have been released by Oregon State University (Mehlenbacher, 2018b). Resistant varieties have also been developed in Russia, Ukraine and Poland (Capik et al., 2013). There are also ongoing studies to find sources of resistance to apple mosaic virus (ApMV) which causes significant damage in Spain, Italy and Turkey (Akbas and Degirmenci, 2010; Grimová et al., 2016).

Microsatellite markers are useful for marker-assisted selection (MAS) in plant breeding. Many microsatellite markers have been developed for hazelnut (Bassil et al., 2005; Boccacci et al., 2005; Gürcan et al., 2010). A search of the genome sequence of 'Jefferson' hazelnut identified 9094 fragments with long repeat motifs of 4, 5 or 6 base pairs (bp), from which polymorphic SSR markers were developed (Bhattarai and Mehlenbacher, 2018). Beltramo et al. (2016) have investigated quantitative trait loci (QTL) associated with vegetative traits to allow MAS. An F1 progeny (275 plants) of 'Tonda Gentile delle Langhe' × 'Merveille de Bollwiller' was used to develop a QTL linkage map for vigor, sucker habit and time of bud burst. A set of 163 plants was analysed using 152 microsatellite markers. Of these, 10 were 'major' QTL for leaf bud burst. Suckers/trunk circumference ratios with high inter-trait correlations were observed on LG_05, suggesting a single pleiotropic locus. Colburn et al. (2017) have analysed the transcriptome sequence of the 'Jefferson' variety to develop microsatellite markers for hazelnut. 109 marker loci mostly showed clustering of accessions from similar geographic origins and confirmed the tremendous genetic diversity present within *C. avellana*. Identification of chromosomal regions carrying genes will be important for MAS breeding programs in the future.

### 3.2 Rootstock breeding

The use of rootstocks is important for nut and fruit trees. Cultivars should be grafted onto non-suckering rootstocks in order to increase tree vigor, early bearing precocity and field as well as reduce cultivation costs, for example, by facilitating mechanization. The first studies on rootstock breeding were initiated in the United States. *C. colurna* × *C. avellana* hybrids were obtained by free pollination. Among these, 'Newberg' and 'Dundee' were subsequently selected and registered (Lagerstedt, 1993). These rootstocks showed sensitivity for EFB which has prompted a search for new EFB-resistant varieties of rootstock. As noted earlier, this led to the identification of 'Gazaway' as heterozygous for the single dominant resistance gene (Mehlenbacher et al., 1991). Pollenizers resistant to EFB were then released (Mehlenbacher et al., 1991), and other sources of resistance in other species have been identified (Coyne et al., 1998). Patterns of interspecific hybridization have been clarified (Erdogan and Mehlenbacher, 2000) which allow identification of sources of resistance in crosses.

It is also important to develop improved non-suckering rootstock materials. Most research has focused on *Corylus colurna* as a non-suckering rootstock (Ninić-Todorović et al., 2012; Miljković, 2017; İslam, 2018). Clonal selection using the 'Negret' cultivar has led to a rootstock that non-suckering (Rovira et al., 2012). Rootstock studies for hazelnuts have just started in Turkey based on the huge genetic diversity in northern Anatolia (İslam, 2018). Salimia and Hoseinovab (2012) have used selection to develop a non-suckering rootstock for climatic conditions in Iran. Other rootstock studies have been undertaken in Italy (Salvador et al., 2009; Farinelli et al., 2009).

## 4 New cultivars of hazelnuts

A number of new key cultivars are:

- 'Okay 28': developed by the Hazelnut Research Institute during 1981–1999 registered in 2012. Nut weight is 2, 85 g, shell thickness is 0, 74 mm, kernel percentage 54%, blanching is 92%.
- 'Giresun Melezi': developed by the Hazelnut Research Institute during 1981–1999 and registered in 2012. Nut weight 2, 39 g, kernel percentage 51%, blanching is 89%.
- 'Allahverdi': bud break is 15 days late, higher yield, erect bearing in the 'Tombul' hazelnut variety (Anon., 2019).
- 'Yamhill': a new cultivar for the kernel market released by the Oregon Agricultural Experiment Station in 2008. It combines complete resistance to EFB caused by the fungus *Anisogramma anomala* (Peck) E. Müller with high nut yield, early nut maturity, high kernel percentage, and good kernel quality (Mehlenbacher et al., 2009)
- 'Jeferson': a new cultivar for the in-shell market released by the Oregon Agricultural Experiment Station in 2011. It combines complete resistance to EFB caused by the fungus Anisogramma anomala (Peck) with high nut yield, large nut size, and good kernel quality (Mehlenbacher et al., 2011).

- 'Wepster': a new cultivar for the blanched kernel market by the Oregon Agricultural Experiment Station in Jan. 2013 as an alternative to 'Yamhill'. It combines complete resistance to EFB with high nut yield, small nut size, early nut maturity, very good kernel quality, and high tree vigor. 'Wepster' is recommended for Oregon's Willamette Valley and other similar climate areas (Mehlenbacher et al., 2014).
- 'York' and 'Felix': these are two new hazelnut pollenizers for *Corylus avellana* (Mehlenbacher, 2018a). 'York' sheds pollen in mis season, 'Felix' sheds pollen in late season.
- 'Ping'ou': these are selected hybrids from crossbreeding. 'Ping'ou' is high yield and with large nut size and is cold resistant (Wang et al., 2018).

These varieties indicate the potential for future breeding programmes.

## 5 Future trends

As has been noted, current world production is about 900 thousand tonnes (FAO, 2016). This needs to increase to 1.5 million tonnes to meet future demand. There is a need to improve yield in existing areas of cultivation (Islam et al., 2005) rather than extend the area of cultivation (which threatens damaging the environment). This can best be achieved through cultivar breeding. Breeding targets need to include not just improved yields but pest and disease resistance, non-suckering cultivars or rootstocks as well as nut quality. Improving tree cultivars will be important because these are best suited to more intensive and large-scale orchard systems, with potentially easier pruning, sucker control, plant protection and harvesting. However, it will also be important to improve shrub varieties suited to the more traditional cultivation still carried out in many parts of the world. Another requirement is better linkage between public breeding programmes and commercial breeders to ensure better transfer of new varieties into cultivation. In addition, there is a need for improved varieties suited to different consumption habits such as table nuts.

## 6 Where to look for further information

*International Societies*

- International Society for Horticultural Science (ISHS) (http://www.ishs.org/)
- International Nut and Dried Fruit Council: www.nutfruit.org
- ISHS Acta Horticulturae https://actahort.org

*Institution*

- Ordu University, https://odu.edu.tr
- Oregon State University, https://oregonstate.edu

- Tuscia University http://www.unitus.it/
- Hazelnut Research Institute, Giresun. https://arastirma.tarimorman.gov.tr/findik/
- National Hazelnut Council www.ufk.gov.tr
- IRTA Institut de Recerca i Tecnologia Agroalimentàries: www.irta.cat
- Nocciolare portal, https://www.nocciolare.it/

*Project*

- Examination of traditional and modern hazelnut cultivation techniques, Erasmus+ Project. 2016-1-Tr01-KA202-034979. Hazelnut cultivation in Europe. 2018. Ed. by İslam, A. Cristofori, V. Rovira, M. MKB Publishing, Ordu, Turkey. 196 p ISBN:9786058099005
- Precision farming of hazelnut orchard - PANTHEON" (H2020 - EU Project) (http://www.project-pantheon.eu/).
- Hazelnut Breeding Programme in Oregon State university https://plantbreeding.oregonstate.edu/

## 7 References

Akbas, B. and Degirmenci, K. 2010. Simultaneous detection of *Apple mosaic virus* in cultivated hazelnuts by one-tube RT-PCR. *African Journal of Biotechnology* 9(12), 1753–7. doi:10.5897/AJB10.1749.

Anon. 2019. 'Allahverdi' hazelnut cultivar. Available at: https://arastirma.tarimorman.gov.tr/findik.

Ava, H., Javadi, S. and Bahman, D. T. 2010. *Selection of Suitable Polinazer for Selected Hazelnut Cultivars*. Seed and Plant Improvement Institute, Iran, 30p.

Balık, H. İ. 2018. Investigations of xenia and metaxenia in hazelnut. PhD thesis. Ondokuz Mayis University Department of Horticulture, Samsun, 274p.

Balık, H. İ., Balık, S. K., Köse, Ç., Duyar, Ö., Sıray, E., Sezer, A., Turan, A., Beyhan, N., Erdoğan, V., İslam, A., Kurt, H., Ak, K. and Kalkışım, Ö. 2014. Development of the new cultivars of hazelnut by selection from Tombul hazelnut populations in Giresun and Trabzon Provinces. In: *International Mesopotamia Agriculture Congress, Proceeding Books*. GAP International Research and Training Center, Diyarbakır, Turkey, pp. 172–9.

Balık, H. İ., Kayalak Balık, S. and Okay, A. N. 2015. New hazelnut cultivars (Okay 28 and Giresun Melezi). *The Journal of Harran. Tarım ve Gıda Bilimleri* 19(2), 104–9.

Bassil, N. V., Botta, R. and Mehlenbacher, S. A. 2005. Microsatellite markers in hazelnut: isolation, characterization, and cross-species amplification. *Journal of the American Society for Horticultural Science* 130(4), 543–9. doi:10.21273/JASHS.130.4.543.

Bassil, N., Boccacci, P., Botta, R., Postman, J. and Mehlenbacher, S. 2013. Nuclear and chloroplast microsatellite markers to assess genetic diversity and evolution in hazelnut species, hybrids and cultivars. *Genetic Resources and Crop Evolution* 60(2), 543–68. doi:10.1007/s10722-012-9857-z.

Beltramo, C., Valentini, N., Portis, E., Torello Marinoni, D., Boccacci, P., Sandoval Prando, M. A. and Botta, R. 2016. Genetic mapping and QTL analysis in European hazelnut (*Corylus avellana* L.). *Molecular Breeding* 36(3), 27. doi:10.1007/s11032-016-0450-6.

Beyhan, N. 1993. Flower biology of some important nut varieties. PhD thesis. Ondokuzmayıs University, Institute of Science, Department of Horticulture, Samsun, 175p (in Turkish).

Beyhan, N. 1998. Some biological observations and investigations in hazelnut. Project results. Ondokuz Mayıs University Faculty of Agriculture Department of Horticulture, Samsun (in Turkish).

Beyhan, N. 2000. Fertilization biology of hazelnut. *Ondokuzmayıs University Journal of Agricultural Faculty* 15(2), 116–22 (in Turkish).

Bhattarai, G. and Mehlenbacher, S. A. 2018. Discovery, characterization, and linkage mapping of simple sequence repeat markers in hazelnut. *Journal of the American Society for Horticultural Science* 143(5), 347–62. doi:10.21273/JASHS04461-18.

Boccacci, P., Akkak, A., Bassil, N. V., Mehlenbacher, S. A. and Botta, R. 2005. Characterization and evaluation of microsatellite loci in European hazelnut (*Corylus avellana* L.) and their transferability to other *Corylus* species. *Molecular Ecology Notes* 5(4), 934–7. doi:10.1111/j.1471-8286.2005.01121.x.

Boccacci, P., Akkak, A. and Botta, R. 2006. DNA typing and genetic relations among European hazelnut (*Corylus avellana* L.) cultivars using microsatellite markers. *Genome* 49(6), 598–611. doi:10.1139/g06-017.

Botta, R., Gianotti, C. and Me, G. 1997. Kernel quality in hazelnut cultivars and selections analysed for sugars, lipids and fatty acid composition. In: *IV International Symposium on Hazelnut* (vol. 445), pp. 319–26.

Botu, M., Achim, G. and Cosmulescu, S. 2015. Germplasm evaluation, conservation and breeding of nut crops in Romania. In: *1st European Fruit Research Institutes Network Shell Fruit Species Meeting*, Abstract Book, 23.

Botu, M., Botu, I., Achim, G., Preda, S., Scutelnicu, A. and Giura, S. 2017. Conservation of fruit tree genetic resources and their use in the breeding process. *Annals "Valahia" University of Targoviste – Agriculture* 11(1), 66–9. doi:10.1515/agr-2017-0013.

Capik, J. M., Muehlbauer, M., Novy, A., Honig, J. A. and Molnar, T. J. 2013. Eastern filbert blight-resistant hazelnuts from Russia, Ukraine and Poland. *HortScience* 48(4), 466–73. doi:10.21273/HORTSCI.48.4.466.

Çayan, M. and İslam, A. 2019. Clonal selection in 'Çakıldak' cultivar grown in Gürgentepe district. *Akademik Ziraat Dergisi* 8(1), 25–32 (in Turkish).

Chen, H., Mehlenbacher, S. A. and Smith, D. C. 2007. Hazelnut accessions provide new sources of resistance to eastern filbert blight. *HortScience* 42(3), 466–9. doi:10.21273/HORTSCI.42.3.466.

Colburn, B. C., Mehlenbacher, S. A. and Sathuvalli, V. R. 2017. Development and mapping of microsatellite markers from transcriptome sequences of European hazelnut (*Corylus avellana* L.) and use for germplasm characterization. *Molecular Breeding* 37(2), 16. doi:10.1007/s11032-016-0616-2.

Coyne, C. J., Mehlenbacher, S. A. and Smith, D. C. 1998. Sources of resistance to eastern filbert blight in hazelnut. *Journal of the American Society for Horticultural Science* 123(2), 253–7. doi:10.21273/JASHS.123.2.253.

Crepinsek, Z., Stampar, F., Bogataj, L. and Solar, A. 2012. The response of *Corylus avellana* L. phenology to rising temperature in north-eastern Slovenia. *International Journal of Biometeorology* 56(4), 681–94. doi:10.1007/s00484-011-0469-7.

Danielsson-Santesson, B. 1951. Fortsatta Undersökningar av polyploid hassel. (Investigations of polyploid hazels) *Sveriges Pom. Fören. Årsskrift* 52, 84–94. (In Germain)

Demir, T. and Beyhan, N. 2000. Research on the selection of hazelnuts grown in Samsun. *Turkish Journal of Agriculture and Forestry* 24, 173–83.

Erdoğan, V. and Mehlenbacher, S. A. 2000. Interspecific hybridization in hazelnut (*Corylus*). *Journal of the American Society for Horticultural Science* 125(4), 489–97. doi:10.21273/JASHS.125.4.489.

Erdoğan, V. and Mehlenbacher, S. A. 2001. Incompatibility in wild *Corylus* species. In: *Proceedings of the Vth International Congress on Hazelnut*, 30 July 2001, ISHS. Acta Horticulturae (vol. 556), pp. 163–9. doi:10.17660/ActaHortic.2001.556.22.

Erdoğan, V., Mehlenbacher, S. A., Köksal, A. İ. and Kurt, H. 2005. Incompatibility alleles expressed in pollen of Turkish hazelnut cultivars. *Turkish Journal of Biology* 29, 111–6.

FAO. 2016. Food and Agricultural Organization. Available at: www.fao.org.

Farinelli, D., Boco, M. and Tombesi, A. 2009. Productive and organoleptic evaluation of new hazelnut crosses. *Acta Horticulturae* 845(845), 651–6. doi:10.17660/ActaHortic.2009.845.102.

Fattahi, R., Mohammadzedeh, M. and Khadivi-Khubb, A. 2014. Influence of different pollen sources on nut and kernel characteristics of hazelnut. *Scientia Horticulturae* 173, 15–9. doi:10.1016/j.scienta.2014.04.031.

Germain, E. 1994. The reproduction of hazelnut (*Corylus avellana*): a review. In: *3rd International Congress on Hazelnut*, 14–18 September, Alba, Italy.

Grimová, L., Winkowska, L., Konrady, M. and Ryšán, P. 2016. Apple mosaic virus. *Phytopathologia Mediterranea* 55(1), 1–19.

Gökirmak, T., Mehlenbacher, S. A. and Bassil, N. V. 2009. Characterization of European hazelnut (*Corylus avellana*) cultivars using SSR markers. *Genetic Resources and Crop Evolution* 56(2), 147–72. doi:10.1007/s10722-008-9352-8.

Gürcan, K., Mehlenbacher, S. A., Botta, R. and Boccacci, P. 2010. Development, characterization, segregation, and mapping of microsatellite markers for European hazelnut (*Corylus avellana* L.) from enriched genomic libraries and usefulness in genetic diversity studies. *Tree Genetics and Genomes* 6(4), 513–31. doi:10.1007/s11295-010-0269-y.

Hosseinpour, A., Seifi, E., Javadi, D. and Ramezanpour, S. S. 2015. A preliminary study on pollen compatibility of some hazelnut cultivars in Iran. *Advances in Horticultural Science* 29(1), 13–6.

İslam, A. 2000. Clonal selection in hazelnut cultivars grown in Ordu Province. PhD thesis. University of Çukurova, Turkey (in Turkish).

Islam, A. 2003. Clonal selection in 'Uzunmusa' hazelnut. *Plant Breeding* 122(4), 368–71. doi:10.1046/j.1439-0523.2003.00853.x.

İslam, A. 2018. Hazelnut culture in Turkey. *Akademik Ziraat Dergisi* 7(2), 259–66. doi:10.29278/azd.476665.

Islam, A. and Ozgüven, A. I. 2001. Clonal selection in the Turkish hazelnut cultivars grown in Ordu Province. *Acta Horticulturae* 556(556), 203–8. doi:10.17660/ActaHortic.2001.556.29.

Islam, A., Turan, A. and Kurt, H. 2005. Effect of Ocak and single trunk training systems on yield and nut quality. *Acta Horticulturae* 686, 259–62.

Janick, J. and Paull, E. R. 2008. *The Encyclopedia of Fruit & Nuts*. Cambridge University Press, UK.

Kasapligil, B. 1968. Chromosome studies in the genus *Corylus*. Sci. Rept. Faculty Sci. Ege University. İzmir. No. 59. (in Turkish)

Korkmaz, Ş., Ak, B. E., Sakar, E., Turanoğlu, İ. and Söylemez, S. 2015. Incompatibility mechanism in fruit trees. *Journal of Harran Agriculture and Food Sciences* 19(3), 180–6 (In Turkish).

Lagerstedt, H. 1993. The evolution of a clonal filbert rootstock. *Proceedings of the Nut Growers Association of Oregon, Washington and British Columbia* 78, 89–101.

Liu, J., Zhang, H., Cheng, Y., Wang, J., Zhao, Y. and Geng, W. 2014. Comparison of ultrastructure, pollen tube growth pattern and starch contain in developing abortive ovaries during the progamic phase in hazel. *Frontiers in Plant Science* 5, 528.

Lunde, C. F., Mehlenbacher, S. A. and Smith, D. C. 2000. Survey of hazelnut cultivars for response to eastern filbert blight inoculation. *HortScience* 35, 729–31.

Ma, Q., Wang, G., Liang, W., Chen, X., Liang, L. and Zhao, T. 2013. Progress on pollen-stigma compatibility in *Corylus* (hazelnuts): a review. *Journal of Forestry Research* 24(3), 397–402. doi:10.1007/s11676-013-0372-7.

Marinoni, D., Me, G. and Valentini, N. 2000. Evaluation of 'Tonda Gentile Delle Langhe' clones. In: *Proceedings of the Vth International Congress on Hazelnut*. Acta Horticulturae (vol. 556), pp. 209–18.

Mehlenbacher, S. A. 2003. Progress and prospects in nut breeding. *Acta Horticulturae* 622(622), 57–79. doi:10.17660/ActaHortic.2003.622.4.

Mehlenbacher, S. A. 2018a. Advances in genetic improvement of hazelnut. *Acta Horticulturae* 1226(1226), 1–12. doi:10.17660/ActaHortic.2018.1226.1.

Mehlenbacher, S. A. 2018b. Breeding hazelnuts resistant to eastern filbert blight. *Acta Horticulturae* 1226, 73–8. doi:10.17660/ActaHortic.2018.1226.10.

Mehlenbacher, S. A. and Smith, D. C. 1991. Partial self-compatibility in 'Tombul' and 'Montebello' hazelnuts. *Euphytica* 56(3), 231–6. doi:10.1007/BF00042369.

Mehlenbacher, S. A. and Thompson, M. M. 1988. Dominance relationships among S alleles in *Corylus avellana* L. *Theoretical and Applied Genetics* 76, 669–72.

Mehlenbacher, S. A., Thompson, M. M. and Cameron, H. R. 1991. Occurrence and inheritance of resistance to eastern filbert blight in 'Gasaway' hazelnut. *HortScience* 26(4), 410–1. doi:10.21273/HORTSCI.26.4.410.

Mehlenbacher, S. A., Smith, D. C. and McCluskey, R. L. 2009. 'Yamhill' hazelnut. *HortScience* 44(3), 845–7. doi:10.21273/HORTSCI.44.3.845.

Mehlenbacher, S. A., Smith, D. C. and McCluskey, R. 2011. 'Jefferson' hazelnut. *HortScience* 46(4), 662–4.

Mehlenbacher, S. A., Smith, D. C. and McCluskey, R. 2014. 'Wepster' hazelnut. *HortScience* 49(3), 346–9.

Miljković, I. 2017. Growth and yield of hazelnut varieties on the rootstock *Corylus colurna* and their own root in ecological conditions of Istria. *Pomologia Croatica* 21(1–2), 3–12. doi:10.33128/pc.21.1-2.1.

Monastra, F., Raparelli, E. and Fanigliulo, R. 1996. Clonal selection of 'Tonda Gentile Romana'. In: *Symposium on Hazelnut* (vol. 445), pp. 52–6.

Ninić-Todorović, J., Čukanović, J., Kurjakov, A., Mladenović, E., Lazović, R. and Todorović, I. 2012. Turkish hazel (*Corylus colurna* L.) seedling characteristics as rootstock for hazelnut cultivar grafting. *Contemporary Agriculture* 61(3–4), 240–6.

Okay, A. N. 1999. Hazelnut crossbreeding project report. Ministry of Agriculture and Rural Affairs, Hazelnut Research Directorate, Giresun (in Turkish).

Olsen, J. L., Mehlenbacher, S. A. and Azarenko, A. N. 2000. Hazelnut pollination. *HortTechnology* 10(1), 113–5. doi:10.21273/HORTTECH.10.1.113.

Petriccione, M., Ciarmiello, L. F., Boccacci, P., De Luca, A. and Piccirillo, P. 2010. Evaluation of 'Tonda di Giffoni' hazelnut (*Corylus avellana* L.) clones. *Scientia Horticulturae* 124(2), 153–8. doi:10.1016/j.scienta.2009.12.019.

Rahemi, M. and Mojadad, D. 2001. Effect of pollen source on nut and kernel characteristics of hazelnut. *Acta Horticulturae* 556, 371–6. doi:10.17660/Acta Hortic.2001.556.55.

Rovira, M., Romero, M. and Clave, J. 1996. Clonal selection of 'Gironell' and 'Negret' hazelnut cultivars. In: *IV International Symposium on Hazelnut* (vol. 446), pp. 35–9.

Rovira, M., Cristofori, V., Silvestri, C., Celli, T., Hermoso, J. F., Tous, J. and Romero, A. 2012. Last results in the evaluation of 'Negret' hazelnut cultivar grafted on non-suckering rootstocks in Spain. *Acta Horticulturae* 1052, 145–50. doi:10.17660/ActaHortic.2014.1052.18

Salesses, G. 1973. Cytological study of genus *Corylus*. A heterozygotic translocation in some low male fertile varieties of hazelnut (*C. avellana*). *Annales de l'amélioration des plantes* 23, 59–66.

Salimia, S. and Hoseinovab, S. 2012. Selecting hazelnut (*Corylus avellana* L.) rootstocks for different climatic conditions of Iran. *Crop Breeding Journal* 2(2), 139–44.

Salvador, F. R., Lolletti, D. and Sabelli, A. 2009. Current progress in the hazelnut breeding program at the fruit tree research Centre-Rome. *Acta Horticulturae* 845(845), 133–8. doi:10.17660/ActaHortic.2009.845.15.

Thompson, M. M., Lagerstedt, H. B. and Mehlenbacher, S. A. 1996. *Hazelnuts Fruit Breeding, Volume III: Nuts*. Janick, J. and Moore, J. N. (Eds). John Wiley & Sons, Inc., New York. ISBN: 0-471-12669-1.

Turan, A. 2007. Clonal selection in Tombul cv grown in Giresun. BSc thesis. OMÜ Fen Bilimleri Enst, Samsun (in Turkish).

Ünal, M. 2009. Bitki Angiosperm Embriyolojisi. Nobel yayın no: 1024:4, 141–58 (in Turkish).

Valentini, N., Me, G., Vallania, R. and Zeppa, G. 2001. New hazelnuts selections for direct consumption. *Acta Horticulturae* 556(556), 103–8. doi:10.17660/ActaHortic.2001.556.13.

Valentini, N., Calizzano, F., Boccacci, P. and Botta, R. 2014. Investigation on clonal variants within the hazelnut (*Corylus avellana* L.) cultivar 'Tonda Gentile delle Langhe'. *Scientia Horticulturae* 165, 303–10. doi:10.1016/j.scienta.2013.11.027.

Wang, G. X., Ma, Q. H., Zhao, T. T. and Liang, L. S. 2018. Resources and production of hazelnut in China. *Acta Horticulturae* 1226(1226), 59–64. doi:10.17660/ActaHortic.2018.1226.8.

Whitcher, I. N. and Wen, J. 2001. Phylogeny and biogeography of *Corylus* (Betulaceae): inference from ITS sequences. *Systematic Botany* 26, 283–98.

Xie, M. and Liu, Z. P. 2014. Studies on pollen viability and pollen-stigma compatibility of hybrid hazelnut. In: *Proceedings of the VIIIth International Congress on Hazelnut*. Acta Horticulturae (vol. 1052), pp. 117–9. doi:10.17660/ActaHortic.2014.1052.14.

Zhai, X. J., Dong, F. X., Zhang, R. Q., Wang, G. X., Yi, M. P. and Liang, L. S. 2009. Research on the compatibility of five *Corylus* specie. *Journal of Central South University of Forestry and Technology* 29(4), 26–30.

# Index